I0054699

PERIODIC CONTROL SYSTEMS 2001
(PSYCO 2001)

A Proceedings volume from the IFAC Workshop,
Cernobbio-Como, Italy, 27 - 28 August 2001

Edited by

S. BITTANTI and P. COLANERI
Dipartimento di Elettronica e Informazione,
Politecnico di Milano, Milano, Italy

Published for the

INTERNATIONAL FEDERATION OF AUTOMATIC CONTROL

by

PERGAMON
An Imprint of Elsevier Science

ELSEVIER SCIENCE Ltd
The Boulevard, Langford Lane
Kidlington, Oxford OX5 1GB, UK

Elsevier Science Internet Homepage
http://www.elsevier.com

Consult the Elsevier Homepage for full catalogue information on all books, journals and electronic products and services.

IFAC Publications Internet Homepage
http://www.elsevier.com/locate/ifac

Consult the IFAC Publications Homepage for full details on the preparation of IFAC meeting papers, published/forthcoming IFAC books, and information about the IFAC Journals and affiliated journals.

Copyright © 2002 IFAC

All Rights Reserved. No part of this publication may be reproduced, stored in a retrieval system or transmitted in any form or by any means: electronic, electrostatic, magnetic tape, mechanical, photocopying, recording or otherwise, without permission in writing from the copyright holders.

First edition 2002

Library of Congress Cataloging in Publication Data

A catalogue record for this book is available from the Library of Congress

British Library Cataloguing in Publication Data

A catalogue record for this book is available from the British Library

ISBN 0-08-043682 X
ISSN 1474-6670

These proceedings were reproduced from manuscripts supplied by the authors, therefore the reproduction is not completely uniform but neither the format nor the language have been changed in the interests of rapid publication. Whilst every effort is made by the publishers to see that no inaccurate or misleading data, opinion or statement appears in this publication, they wish to make it clear that the data and opinions appearing in the articles herein are the sole responsibility of the contributor concerned. Accordingly, the publisher, editors and their respective employers, officers and agents accept no responsibility or liability whatsoever for the onsequences of any such inaccurate or misleading data, opinion or statement.

Transferred to digital printing 2005

To Contact the Publisher

Elsevier Science welcomes enquiries concerning publishing proposals: books, journal special issues, conference proceedings, etc. All formats and media can be considered. Should you have a publishing proposal you wish to discuss, please contact, without obligation, the publisher responsible for Elsevier's industrial and control engineering publishing programme:

Dr Martin Ruck
Publishing Editor
Elsevier Science Ltd
The Boulevard, Langford Lane
Kidlington, Oxford
OX5 1GB, UK

Phone: +44 1865 843230
Fax: +44 1865 843920
E.mail: m.ruck@elsevier.co.uk

General enquiries, including placing orders, should be directed to Elsevier's Regional Sales Offices – please access the Elsevier homepage for full contact details (homepage details at the top of this page).

IFAC WORKSHOP ON PERIODIC CONTROL SYSTEMS 2001

Sponsored by
International Federation of Automatic Control (IFAC)
Technical Committees on:
- Linear Systems
- Control Design

Organized by
Italian National Member Organization of IFAC: CNR (Consiglio Nazionale delle Ricerche)

Supported by
Dipartimento di Elettronica e Informazione, Politecnico di Milano, Italy
CNR-CESTIA, Milano, Italy

International Programme Committee (IPC)
Bittanti, S. (Italy) (Chairman)

Albertos, P. (Spain)
Araki, M. (Japan)
Callier, F. (Belgium)
Colaneri, P. (Italy)
Dasgupta, S. (USA)
de Souza, C. (Brazil)
Grasselli, O.M. (Italy)
Guardabassi, G.O. (Italy)
Hernandez, V. (Spain)
Khargonekar, P.P. (USA)
Kucera, V. (Czech Republic)
Longhi, S. (Italy)
Qiu, L. (PRC)
Schiavoni, N. (Italy)
Speyer, J. (USA)
Tomizuka, M. (USA)
Van Dooren, P. (Belgium)
Varga, A. (Germany)
Verriest, E. (USA)
Yakubovich, V.A. (Russia)
Zhang, C. (Australia)
Zhang, J. (Australia)

National Organizing Committee (NOC)
Colaneri, P. (Chairman)

Astolfi, A.
Lovera, M.

FOREWORD

Periodic control has reached a notable degree of maturity thanks to developments over the last few decades. We have seen not only major theoretical achievements but also new significant applications.

The IFAC workshop on *Periodic Control Systems* (PSYCO 2001), held at the Villa Erba Congress Centre in Cernobbio-Como (Italy), August 27-28 2001, aimed at presenting the full picture of the area by gathering experts in the field and all interested researchers, coming from universities, research institutions and industries.

The program consisted of technical sessions, organized in two parallel streams and two plenary lectures, given by Jason L. Speyer (University of California at Los Angeles, USA) and Yutaka Yamamoto (Kyoto University, Japan). The technical sessions included 42 papers covering the following subjects:

- Periodic Systems Analysis
- Application
- Time-Series
- Hybrid and Sampled-Data
- Aerospace Applications
- Periodic Systems Control
- Numerical Methods
- Multirate and Batch Processes
- Repetitive and Nonlinear Control

A dozen of the papers were devoted to a number of applications, including aerospace, jet and diesel engines, gas turbines, nuclear reactors, power systems, satellites, environmental sciences and finance.

There were 62 participants, coming from Australia (1), Belgium (1), China (2), Croatia (1), Denmark (2), France (1), Germany (6), India (1), Israel (2), Italy (11), Japan (13), Korea (1), Poland (1), Portugal (1), Romania (1), Russia (1), Spain (1), Sweden (2), Switzerland (3), UK (2), USA (8).

The meeting was the first conference devoted entirely to *periodic control systems*. The participants had the opportunity to be exposed to new ideas, meet friends and enjoy the lake atmosphere.

The interest raised by the workshop is also witnessed by the fact that there will be a follow up in Japan. Indeed, the next meeting is planned for Yokohama, Japan, between 30 August and 1 September 2004. Hopefully, this will give rise to a series, which will prosecute in the years to come.

S. Bittanti and P. Colaneri (Editors)

CONTENTS

PERIODIC SYSTEMS ANALYSIS

APPLICATION I

TIME-SERIES

APPLICATION II

HYBRID AND SAMPLED-DATA

AEROSPACE APPLICATIONS

PERIODIC SYSTEMS CONTROL

NUMERICAL METHODS

MULTIRATE AND BATCH PROCESSES

REPETITIVE AND NONLINEAR CONTROL

Copyright © IFAC Periodic Control Systems,
Cernobbio-Como, Italy, 2001

TRACE FORMULAS FOR THE H_2 NORM OF LINEAR CONTINUOUS-TIME PERIODIC SYSTEMS

Jun Zhou,* Tomomichi Hagiwara and Mituhiko Araki

Department of Electrical Engineering, Kyoto University

Abstract: A trace formula is established for the H_2 norms of a class of finite-dimensional linear continuous-time periodic (FDLCP) systems based on the solution of the so-called harmonic Lyapunov equations. The trace formula is similar to what we have for the H_2 norm of an LTI continuous-time system apart from the fact that infinite-dimensional matrices are involved in the FDLCP setting. Based on this formula, trace formulas are developed via approximate modeling approach, which are numerically implementable in most practical systems.
Copyright © 2001 IFAC

Key Words— H_2 norm; continuous-time periodic systems; frequency response operator; trace formula; computations.

1 INTRODUCTION

The H_2 norm is one of the performance measures of control systems, which is frequently used as an objective for control system synthesis (Bamieh and Pearson, 1992; Green and Limebeer, 1995; Zhang, C. and Zhang, J., 1997; Zhou, K., 1998). The computation of the H_2 norm of an LTI continuous-time system can be carried out by the trace formula via the solution of an algebraic Lyapunov equation. However, in finite-dimensional linear continuous-time periodic (FDLCP) systems, it is much harder to deal with though the definitions of the H_2 norm of FDLCP systems have been given both in the time-domain and the frequency-domain for a long time (Wereley, 1990; Zhang *et al.*, 1997). The lifting technique (Bamieh *et al.*, 1992; Yamamoto, 1996) is a powerful tool to tackle the H_2 norms in FDLCP systems, and with this technique, it is shown (Bamieh *et al.*, 1992; Colaneri, 2000) in FDLCP systems that the time-domain definition of the H_2 norm and the frequency-domain counterpart are equivalent; this fact is also verified via the frequency response operator (defined via steady-state analysis) recently by Zhou and Hagiwara (2000b). As for the lifting technique, most works are devoted to sampled-data systems, which are periodic (Chen and Francis, 1995; Dullerud, 1996). Unfortunately, however, the H_2 norm formulas developed there are difficult to apply directly to FDLCP systems. The obstacles in the H_2 norm computation are that if we deal with the time-domain definition, the solution of a periodic Lyapunov differential equation is needed (Green *et al.*, 1995; Colaneri, 2000); on the other hand, if the frequency-domain approach were adopted, one would encounter the difficulties caused by the operator composition if the frequency response relation is defined by continuous-time lifting (Bamieh *et al.*, 1992) or the infinite-dimensional structure of the frequency response operator if defined via steady-state analysis (Hall and Wereley, 1990; Wereley and Hall, 1990; Wereley, 1990).

In the frequency response operator approach, the 'square' truncation is proposed in (Wereley, 1990) to overcome the infinite-dimensional structure of the frequency response operator as we just mentioned. However, there has been no proof to show the convergence for this truncation; besides, a numerical integration problem involved remains untouched. As an alternative truncation approach, the so-called skew truncation is introduced in (Zhou *et al.*, 2000a) for the H_2 norm computation via the frequency response operator of the FDLCP system with rigorous proofs for the convergence involved, which leads to an 'asymptotic trace formula' based on the solution of a finite-dimensional algebraic Lyapunov equation. In contrast, this paper establishes an 'exact trace formula' via the harmonic Lyapunov equation, which is an infinite-dimensional Lyapunov equation and was introduced in (Zhou, J. and Hagiwara, T., 2001a) in the stability analysis of FDLCP systems. The well-known trace formula (Green *et al.*, 1995; Zhou, K., 1998) of the H_2 norms of LTI continuous-time systems is just a special case here. The trace formula can also be regarded as the limit of the above-mentioned approximate trace formula developed via the skew truncation on the frequency response operator (Zhou *et al.*, 2000a).

Now we outline this paper. In Section 2 we quickly

*Corresponding Author, Address: Yoshida, Sakyo-ku, Kyoto 606-8501, JAPAN. Email: zhouj@jaguar.kuee.kyoto-u.ac.jp, Fax: +81-75-7533338 review the Floquet theorem so that the reader can get a better understanding to our arguments. The similarity transformation formulas and the harmonic Lyapunov equation of FDLCP systems are also summarized. Section 3 is devoted to establishing the (infinite-dimensional and thus exact) trace formula for the H_2 norm of an FDLCP system via the solution of a harmonic Lyapunov equation. To implement this formula numerically, several algorithms are considered based on the approximate modeling and truncations in Section 4. There are numerical examples in Section 5.

We say $F(t) \in L_2[0,h]$ to mean that F is a matrix function, each element of which is h-periodic and belongs to $L_2[0,h]$ when its domain is restricted to the interval $[0,h]$. Similarly for other function sets defined over $[0,h]$. $\mathcal{S}^+$ denotes the set of all strictly positive definite self-adjoint bounded operators on l_2. $\mathcal{Z}$ is the set of all integers.

2 PRELIMINARIES

Consider the strictly proper FDLCP system

$$G : \begin{cases} \dot{x} = A(t)x + B(t)u \\ y = C(t)x \end{cases} \tag{1}$$

where $A(t) \in L_{\mathrm{PCD}}[0,h]$, and $B(t), C(t) \in L_{\mathrm{CAC}}[0,h]$ with $L_{\mathrm{PCD}}[0,h]$ and $L_{\mathrm{CAC}}[0,h]$ being the sets of h-periodic functions given by

$$L_{\mathrm{PCD}}[0,h] := \left\{ f(t) : \begin{array}{l} f \text{ is piecewise continuous and} \\ \text{differentiable at a.e. } t \in [0,h]. \end{array} \right\}$$

$$L_{\mathrm{CAC}}[0,h] := \left\{ f(t) : \begin{array}{l} f \text{ is continuous and its Fourier} \\ \text{series is absolutely convergent} \end{array} \right\}$$

where PCD stands for piecewise continuous and differentiable and CAC stands for continuous and absolutely convergent. The following Floquet theorem plays a key role in understanding the transition matrix $\Phi(t,t_0)$ (t_0 is the initial time) and asymptotic stability.

Proposition 1 (Lukes, 1982) Assume $A(t) \in L_2[0,h]$. Then $\Phi(t,t_0)$ is continuous with respect to t and can be expressed as $\Phi(t,t_0) = P(t,t_0)e^{Q(t-t_0)}$ where $P(t,t_0)$ is a nonsingular h-periodic matrix and Q is a constant matrix. Moreover, the system is asymptotically stable if and only if the eigenvalues of the monodromy matrix, $\Phi(h+t_0,t_0)$, are in the open unit disk.

Now let us expand $A(t)$ to its Fourier series $A(t) = \sum_{m=-\infty}^{+\infty} A_m e^{jm\omega_h t}$ with $\omega_h = \frac{2\pi}{h}$. The Toeplitz transformation on $A(t)$ (Wereley, 1990), denoted by $\mathcal{T}\{A(t)\}$, maps $A(t)$ into a doubly infinite-dimensional block Toeplitz operator (Wereley, 1990) (or to be more precise, block Laurent operator (Gohberg, Goldberg and Kaashoek, 1993)):

$$\mathcal{T}\{A(t)\} := \begin{bmatrix} \ddots & \vdots & \vdots & \vdots & \cdot^{\cdot^{\cdot}} \\ \cdots & A_0 & A_{-1} & A_{-2} & \cdots \\ \cdots & A_1 & A_0 & A_{-1} & \cdots \\ \cdots & A_2 & A_1 & A_0 & \cdots \\ \cdot^{\cdot^{\cdot}} & \vdots & \vdots & \vdots & \ddots \end{bmatrix}$$

Based on the Toeplitz transformation and the Floquet theorem, the so-called similarity transformation formulas in the FDLCP system (1) can be established (Wereley, 1990; Zhou *et al.*, 2000b) as in Proposition 2. Here, we define $\underline{E}(j\varphi) = \mathrm{diag}[\cdots, j\varphi_{-1}I, j\varphi_0 I, j\varphi_1 I, \cdots]$ with $\varphi_m := \varphi + m\omega_h, \varphi \in \mathcal{I}_0 := [-\frac{\omega_h}{2}, +\frac{\omega_h}{2})$, $m \in \mathcal{Z}$, and $l_E := \{\underline{x} \in l_2 : \underline{E}(j0)\underline{x} \in l_2\} \subset l_2$.

Proposition 2 Assume in the system (1) that $A(t) \in L_{\mathrm{PCD}}[0,h]$ and $B(t), C(t) \in L_{\mathrm{CAC}}[0,h]$. Let $\underline{P} := \mathcal{T}\{P(t,0)\}$ and $\underline{Q} := \mathcal{T}\{Q\}$. Then, it holds on $l_E \subset l_2$ that

$$\underline{P}(\underline{E}(j0) - \underline{Q})\underline{P}^{-1} = \underline{E}(j0) - \underline{A} \tag{2}$$

Moreover, it holds on l_2 that $\hat{\underline{B}} = \underline{P}^{-1}\underline{B}$ and $\hat{\underline{C}} = \underline{C}\,\underline{P}$ where $\hat{\underline{B}} := \mathcal{T}\{P^{-1}(t,0)B(t)\}$, $\hat{\underline{C}} := \mathcal{T}\{C(t)P(t,0)\}$, $\underline{B} := \mathcal{T}\{B(t)\}$, and $\underline{C} := \mathcal{T}\{C(t)\}$. Furthermore, l_E is $\underline{P}$-invariant and $\underline{P}^{-1}$-invariant, and $\underline{P}$ is invertible on l_E and the unique inverse of $\underline{P}$ on l_E is $\underline{P}^{-1}$ restricted to l_E. Finally, if (1) is asymptotically stable, then $\underline{E}(j\varphi) - \underline{A}$ is invertible for all $\varphi \in \mathcal{I}_0$ and

$$\underline{P}(\underline{E}(j\varphi) - \underline{Q})^{-1}\underline{P}^{-1} = (\underline{E}(j\varphi) - \underline{A})^{-1}$$

Furthermore, $(\underline{E}(j\varphi) - \underline{A})^{-1}$ is compact and bounded uniformly over $\varphi \in \mathcal{I}_0$.

Since l_E is dense in l_2 (Zhou *et al.*, 2000b), (2) should be interpreted as an operator-valued relation densely defined on l_2 (Naylor and Sell, 1982, p. 486). Proposition 2 is important in understanding the frequency response operator of the FDLCP system and the harmonic Lyapunov equation of Proposition 3 (Zhou *et al.*, 2001a).

Proposition 3 Suppose in the system (1) that $A(t) \in L_{\mathrm{PCD}}[0,h]$. Then the system is asymptotically stable if and only if for any $\underline{W} \in \mathcal{S}^+$, there exists a unique $\underline{V} \in \mathcal{S}^+$ satisfying

$$(\underline{A} - \underline{E}(j0))^*\underline{V} + \underline{V}(\underline{A} - \underline{E}(j0)) = -\underline{W} \tag{3}$$

on $l_E \subset l_2$, which is called the harmonic Lyapunov equation. Moreover, if the system is asymptotically stable, the unique solution of (3) is

$$\underline{V} = \underline{P}^{-*} \int_0^{\infty} \underline{e}(Q,\tau)^* \underline{P}^* \underline{W}\,\underline{P}\,\underline{e}(Q,\tau)\, d\tau\, \underline{P}^{-1}$$

with $\underline{e}(Q,t)$ being an infinite-dimensional matrix given by $\mathrm{diag}[\cdots, e^{(Q+j\omega_h I)t}, e^{Qt}, e^{(Q-j\omega_h I)t}, \cdots]$. The e^{Qt}-block is at the center of $\underline{e}(Q,t)$.

Next let us quickly review the definition of the frequency response operator of the FDLCP system (1) defined via the steady-state input-output analysis, which is first introduced in (Hall *et al.*, 1990; Wereley *et al.*, 1990; Wereley, 1990) and whose existence conditions and basic properties are analyzed in (Zhou *et al*, 2000b). Assume in the system (1) that $A(t)$ belongs to $L_{\mathrm{PCD}}[0,h]$, and $B(t)$ and $C(t)$ belong to $L_{\mathrm{CAC}}[0,h]$. Assume that the system is asymptotically stable. Then the frequency response operator of (1) can be expressed by

$$\underline{G}(j\varphi) := \underline{C}(\underline{E}(j\varphi) - \underline{A})^{-1}\underline{B} = \hat{\underline{C}}(\underline{E}(j\varphi) - \underline{Q})^{-1}\hat{\underline{B}}$$

which is compact and bounded on l_2 uniformly over $\varphi \in \mathcal{I}_0$. In deriving the last equality, Proposition 2 is used. This equality gives some mathematical convenience when computing the inverse of $\underline{E}(j\varphi) - \underline{A}$ in our discussions.

3 TRACE FORMULA FOR THE H_2 NORM

In this section, the relation is discussed between the H_2 norm of the frequency response operator of the FDLCP system and the harmonic Lyapunov equation. The purpose is to express the H_2 norm by a trace formula via the solution of the (harmonic) Lyapunov equation so that the well-know trace formula is established in the LTI continuous-time fashion but with an infinite-dimensional matrix expression. In some less rigorous sense, the H_2 norm of an FDLCP system can be 'computed' just as we do in LTI continuous-time systems.

First, let us introduce the H_2 norm of the system (1) via the frequency response operator (Green *et al.*, 1995; Hagiwara and Araki,1995; Wereley, 1990; Zhang *et al.*, 1997). The H_2 norm of the FDLCP system (1) is the quantity

$$||\mathcal{G}||_2 = \left\{\frac{1}{2\pi}\int_{-\frac{\omega_h}{2}}^{+\frac{\omega_h}{2}} \mathrm{trace}\Big(\underline{G}(j\varphi)^*\underline{G}(j\varphi)\Big)d\varphi\right\}^{\frac{1}{2}}$$

Remark 1 It is not hard to show that $\underline{G}(j\varphi)$ is a Hilbert-Schmidt operator (Naylor *et al.*, 1982, p. 387) for each $\varphi \in \mathcal{I}_0$. Therefore, the H_2 norm is well-defined for FDLCP systems. Indeed, a proof is given in (Zhou *et al.*, 2000a) to verify that $\mathrm{trace}\Big(\underline{G}(j\varphi)^*\underline{G}(j\varphi)\Big) \le M < \infty$ for all $\varphi \in \mathcal{I}_0$. In (Zhou *et al.*, 2000b), the equivalence between the frequency-domain H_2 norm and its time-domain counterpart is verified via the frequency response operator $\underline{G}(j\varphi)$ of FDLCP systems. It is worth mentioning that this equivalence relation in FDLCP systems is also shown to be true via the well-known continuous-time lifting technique (Bamieh *et al.*, 1992; Colaneri, 2000).

Now we state and prove the trace formula, which is the main contribution of this paper.

Theorem 1 Suppose in the system (1) that $A(t) \in L_{\mathrm{PCD}}[0,h]$, $B(t), C(t) \in L_{\mathrm{CAC}}[0,h]$ and that the system is stable. Then

$$||\mathcal{G}||_2^2 = \mathrm{trace}(\underline{b}^*\underline{V}\,\underline{b}) = \mathrm{trace}(\underline{c}\,\underline{W}\,\underline{c}^*) \qquad (4)$$

where $\underline{b} := [\cdots, B_{-1}^T, B_0^T, B_1^T, \cdots]^T$, while $\underline{c} := [\cdots, C_1, C_0, C_{-1}, \cdots]$ with $\{B_m\}_{m=-\infty}^{+\infty}$ and $\{C_m\}_{m=-\infty}^{+\infty}$ being the Fourier coefficients of $B(t)$ and $C(t)$, respectively. The (infinite-dimensional) matrices $\underline{V}$ and $\underline{W}$ are, respectively, the unique solutions of the harmonic Lyapunov equations:

$$(\underline{A} - \underline{E}(j0))^*\underline{V} + \underline{V}(\underline{A} - \underline{E}(j0)) = -\underline{C}^*\underline{C} \qquad (5)$$

$$(\underline{A} - \underline{E}(j0))\underline{W} + \underline{W}(\underline{A} - \underline{E}(j0))^* = -\underline{B}\,\underline{B}^* \qquad (6)$$

Because of the space limitation, we give only a few remarks to indicate the difficulties one would encounter in the arguments. A complete proof can be found in (Zhou, Hagiwara and Araki, 2001b). Firstly, it is shown (Zhou, Hagiwara and Araki, 2001a) that the harmonic Lyapunov equations is an operator-valued equation densely defined on l_2, or more precisely on the whole l_E. It is also clarified that the adjoint operator of the unbounded operator $\underline{A}-\underline{E}(j0)$ which is defined on l_E, denoted by $(\underline{A} - \underline{E}(j0))^*$, is also defined on the whole l_E and that the matrix expression of $(\underline{A} - \underline{E}(j0))^*$ is just the complex conjugate transpose of the matrix expression of $\underline{A} - \underline{E}(j0)$. Therefore, the relation between the second equality of (4) and the harmonic Lyapunov equation (6) can be proved exactly in the same way as in showing the relation of the first equality of (4) and (5) by introducing a complex conjugate transpose dual system. Secondly, the proof will follow some similar idea to what we use in LTI continuous-time systems by using the similarity transformation relations of Proposition 1 and partitioning the infinite-dimensional matrix $\underline{\hat{B}}$. Because of the infinite-dimensional structure of the frequency response operator $\underline{G}(j\varphi)$, however, there are frequent order interchanges between infinite integrals and infinite summations so that one must pay attention to the validity of such order interchanges.

Remark 2 If the system (1) is LTI continuous-time, the harmonic Lyapunov equation can be seen as the 'lifted' version of the usual algebraic Lyapunov equation. Hence the trace formula of Theorem 1 reduces to that in the LTI continuous-time case (Green *et al.*, 1995; Zhou, K., 1998). Unfortunately, however, it is hard to find the solutions of (5) and (6) and the trace formulas for general FDLCP systems involves the infinite-dimensional matrices $\underline{b}$ and $\underline{V}$. These difficulties more or less confine the value of Theorem 1 to the theoretical analysis. In the next section, we derive some modified trace formulas for the H_2 norm estimation via the approximate modeling approach, in which Theorem 1 plays a central role.

4 TRACE FORMULAS BASED ON APPROXIMATE MODELS

By Theorem 1, there exist two obstacles for one to apply the trace formula to compute the H_2 norm of an FDLCP system. The first is that one has to determine the solutions of the harmonic Lyapunov equations (5) or (6), which are infinite-dimensional matrices. Strictly speaking, this task can not be completed even though Proposition 3 really gives the closed form of the solution, since it unfortunately relies on the knowledge of the transition matrix of the FDLCP system that is equally hard to find. The second obstacle is that, even if we know the solution matrix from the related harmonic Lyapunov equation, we still face the multiplication of infinite-dimensional matrices in the trace formula itself. To overcome these two problems, in this section we discuss trace formulas based on approximate modeling, which can be used for asymptotic estimation of the H_2 norm of the original FDLCP system. The relation between the trace formula of Theorem 1 and a finite-dimensional trace formula proposed in (Zhou *et al.*, 2000a) is also considered.

To compute the H_2 norm of an FDLCP system, one can resort to the formulas established via time-varying Lyapunov equations (Green *et al.*, 1995, pp. 94). Although the existence of such solutions can be guaranteed under proper assumptions (Bolzern and Colaneri, 1988), one needs to find solutions of time-varying Lyapunov equations and compute integrals. In contrast to this, the trace formulas we will introduce below are based on (only finitely many) Fourier series coefficients of FDLCP system matrices $(A(t), B(t), C(t))$ and solutions of finite-dimensional algebraic Lyapunov equations with desired convergence under standard assumptions.

4.1 *Approximate Modeling and Convergence*

To express the approximate modeling idea, we truncate the vector $\underline{b}$ into $\underline{b}_N$ which is given by

$$\underline{b}_N := [\cdots, 0, B_{-N}^T, \cdots, B_0^T, \cdots, B_N^T, 0, \cdots]^T$$

Now we define

$$||\mathcal{G}_N||_2^2 := \text{trace}\left(\underline{b}_N^* \underline{V}\, \underline{b}_N\right) \qquad (7)$$

Note that by the structure of $\underline{b}_N$, only finitely many block matrix entries of $\underline{V}$ are actually involved in the computation of $||\mathcal{G}_N||_2$. Therefore, by truncating $\underline{b}$, we get two benefits: on one hand, we do not need to know all the components of $\underline{V}$ (thereupon, only finitely many variables of the harmonic Lyapunov equation need to be determined); on the other hand, the trace formula computation itself is reduced to some finite operations. However, before we take any advantage of this truncation merit, the convergence problem should be scrutinized first: does $||\mathcal{G}_N||_2$ converge to $||\mathcal{G}||_2$ as $N \to \infty$? Proposition 4 gives the answer.

Proposition 4 Suppose in the system (1) that $A(t)$ belongs to $L_{\text{PCD}}[0, h]$, $B(t)$ and $C(t)$ belong to $L_{\text{CAC}}[0, h]$, and that the system is asymptotically stable. Then $||\mathcal{G}||_2 = \lim_{N\to\infty} ||\mathcal{G}_N||_2$.

Equipped with Proposition 4, one might optimistically feel that all the difficulties to implement the trace formula numerically have been removed by truncating $\underline{b}$. However, careful observations indicate that it is hopeless, strictly speaking, to try to find some finitely many matrix entries of $\underline{V}$ by working only on some finite-dimensional portion of (5). This is because there are multiplications of infinite-dimensional matrices in (5).

To reduce the problem to the solution of a finite-dimensional algebraic Lyapunov equation, we need to rewrite the harmonic Lyapunov equation (5) in terms of the block-diagonal matrix $\underline{Q} - \underline{E}(j0)$ instead of $\underline{A} - \underline{E}(j0)$. However, this again requires us to have the knowledge on the transition matrix of the original FDLCP system. To avoid this, we resort to the approximate modeling approach. To be more precise, we construct the following FDLCP approximate model for the original FDLCP system (1).

$$G_a : \begin{cases} \dot{x} = A_a(t)x + B(t)u \\ y = C(t)x \end{cases} \qquad (8)$$

Here $A_a(t)$ is taken such that $A_a(t) \in L_{\text{PCD}}[0, h]$, and the error matrix $A_\Delta(t)$ is defined as $A_\Delta(t) := A(t) - A_a(t)$, $(\forall t \in [0, h])$. It should be pointed out that no further approximate treatments are imposed on $B(t)$ and $C(t)$. We also suppose that the system (8) has the (explicit) transition matrix $\Phi_a(t, t_0) = P_a(t, t_0)e^{Q_a(t-t_0)}$. It is well-known (Farkas, 1994; Willems, 1970) that if the approximate state matrix $A_a(t)$ is given by piecewise constant functions, the transition matrix $\Phi_a(t, 0)$ can be explicitly determined.

Since $A_a(t) \in L_{\text{PCD}}[0, h]$, $B(t), C(t) \in L_{\text{CAC}}[0, h]$, the frequency response operator $\underline{G}_a(j\varphi)$ of (8) is well-defined if the system (8) is asymptotically stable (this can be checked by the eigenvalues of Q_a), and given by $\underline{G}_a(j\varphi) := \underline{C}(\underline{E}(j\varphi) - \underline{A}_a)^{-1}\underline{B}$ with $\underline{A}_a := \mathcal{T}\{A_a(t)\}$. It is also clear that $\underline{A} = \underline{A}_a + \underline{A}_\Delta$ with $\underline{A}_\Delta := \mathcal{T}\{A_\Delta(t)\}$. Therefore, by the Fourier expansion operator from $L_2[0, h]$ to l_2, which is an isometric isomorphism, it can be shown (Zhou *et al.*, 2000*b*) readily that

$$\begin{aligned} ||\underline{A}_\Delta||_{l_2/l_2} &= ||A_\Delta(\cdot)||_{L_2[0,h]/L_2[0,h]} \\ &= \sup_{t\in[0,h]} ||A_\Delta(t)|| \\ &=: ||A_\Delta(\cdot)|| \end{aligned} \qquad (9)$$

Since $A_\Delta \in L_{\text{PCD}}[0, h]$, $||A_\Delta(\cdot)||$ is well-defined. Proposition 5 below shows that the H_2 norm of the approximate FDLCP system (8) can approach that of the original FDLCP system as close as desired by making $||A_\Delta(\cdot)||$ small enough.

Proposition 5 Assume in the system (1) that $A(t) \in L_{\text{PCD}}[0, h]$, $B(t), C(t) \in L_{\text{CAC}}[0, h]$, and that the system (1) is asymptotically stable. Then if $||A_\Delta(\cdot)||$ is small enough, the approximate model (8) is also asymptotically stable and $||\mathcal{G}||_2 = \lim_{||A_\Delta(\cdot)||\to 0} ||\mathcal{G}_a||_2$.

4.2 *Trace Formulas via Approximate Models*

Now based on Propositions 4 and 5, we develop another trace formula for the H_2 norm estimation of (1) via that of the approximate model (8). It is clear from Theorem 2 that the H_2 norm of the approximate model (8) can be expressed as

$$||\mathcal{G}_a||_2^2 = \text{trace}\left(\hat{\underline{b}}_a^* \hat{\underline{V}}_a \hat{\underline{b}}_a\right) \qquad (10)$$

where $\hat{\underline{b}}_a := [\cdots, \hat{B}_{a-1}^T, \hat{B}_{a0}^T, \hat{B}_{a1}^T, \cdots]^T$ with $\{\hat{B}_{an}\}_{n=-\infty}^{+\infty}$ being the Fourier coefficients sequence of $P_a^{-1}(t, 0)B(t) =: \hat{B}_a(t)$, and $\underline{V}_a$ is the solution of the harmonic Lyapunov equation

$$(\underline{Q}_a - \underline{E}(j0))^* \hat{\underline{V}}_a + \hat{\underline{V}}_a(\underline{Q}_a - \underline{E}(j0)) = -\hat{\underline{C}}_a^* \hat{\underline{C}}_a \qquad (11)$$

with $\hat{\underline{C}}_a := \mathcal{T}\{P_a(t, 0)C(t)\} =: \mathcal{T}\{\hat{C}_a(t)\}$. From (10), let us further truncate $\hat{\underline{b}}_a$ to $\hat{\underline{b}}_{aN}$. The discussions in Proposition 4 have already revealed that it makes sense to define the H_2 norm of the truncated FDLCP system $(Q_a, \hat{B}_{aN}(t), \hat{C}_a(t))$, which can be written as $||\mathcal{G}_{aN}||_2^2 := \text{trace}(\hat{\underline{b}}_{aN}^* \hat{\underline{V}}_a \hat{\underline{b}}_{aN})$.

that the system is asymptotically stable in the Floquet theorem sense. Let $||\mathcal{G}_{aN}||_2$ denote the H_2 norm of the stable truncated approximate FDLCP system $(Q_a, \hat{B}_{aN}(t), \hat{C}_a(t))$. Then

$$\lim_{||A_\Delta(\cdot)||\to 0} \lim_{N\to\infty} ||\mathcal{G}_{aN}||_2 = ||\mathcal{G}||_2 \qquad (12)$$

Furthermore, $||\mathcal{G}_{aN}||_2$ can be computed as the square root of $\text{trace}(\hat{\underline{b}}^*_{a[N]}\hat{V}_{aN}\hat{\underline{b}}_{a[N]})$ with $\hat{V}_{aN}$ being the unique solution of the following finite-dimensional algebraic Lyapunov equation.

$$\begin{aligned}&((\underline{Q}_a - \underline{E}(j0))^*)_{[NN]}\hat{V}_{aN}\\&+\hat{V}_{aN}(\underline{Q}_a - \underline{E}(j0))_{[NN]} = -(\hat{\underline{C}}^*_a\hat{\underline{C}}_a)_{[NN]}\end{aligned} \qquad (13)$$

In the above, $(\cdot)_{[N]}$ denotes the sub-vector consisting of the $(2N+1)$ block entries of the infinite-dimensional vector $(\cdot)$ at the center while $(\cdot)_{[NN]}$ denotes the $(2N+1)\times(2N+1)$ sub-matrix of the infinite-dimensional matrix $(\cdot)$ at the center.

Proof By $||\mathcal{G}_{aN}||_2^2 = \text{trace}(\hat{\underline{b}}^*_{aN}\hat{\underline{V}}_a\hat{\underline{b}}_{aN})$, (12) follows readily from Propositions 4 and 5. On the other hand, since for each fixed N, only the $(2N+1)\times(2N+1)$ sub-matrix of the infinite-dimensional matrix $\hat{\underline{V}}_a$ at the center is actually used in the computation of $||\mathcal{G}_{aN}||_2$, it is enough to determine only that sub-matrix from (11). Noting that $\underline{Q}_a - \underline{E}(j0)$ is block-diagonal, the reduced-order Lyapunov equation (13) follows, where $\hat{V}_{aN}$ is nothing but $(\hat{\underline{V}}_a)_{[NN]}$. □

4.3 *Comparison with the Trace Formula via the Skew Truncation*

In (Zhou *et al.*, 2000a), a trace formula is suggested for the H_2 norm estimation in the FDLCP system setting via the skew truncation on the frequency response operator $\underline{G}(j\varphi)$. A convergence property regarding this truncation has been proved under the Lipschitz condition assumptions (Naylor *et al.*, 1982, p. 595) on $B(t)$ and $C(t)$. However, our current study reveals that the Lipschitz condition is an unnecessarily too strong assumption for the convergence of the skew truncation. Indeed the Bernstein theorem (Igari, 1975, p.118) indicates that the Lipschitz condition on a periodic function implies the absolute convergence of the Fourier series expansion of the function. Thus it would not be surprising that the conclusion in (Zhou *et al.*, 2000a) can be re-stated under the weaker assumptions that $B(t)$ and $C(t)$ belong to $L_{\text{CAC}}[0,h]$. Now we state the improved results (Theorem 1 of (Zhou *et al.*, 2000a)) and do some comparisons with the trace formulas here.

Theorem 3 Suppose in the system (1) that $A(t)$ belongs to $L_{\text{PCD}}[0,h]$, $B(t)$ and $C(t)$ belong to $L_{\text{CAC}}[0,h]$, and that the system is asymptotically stable. Then $\lim_{N\to\infty}\text{trace}(\mathcal{B}^*_N\mathcal{V}_N\mathcal{B}_N) = ||\mathcal{G}||_2^2$. Here $\mathcal{V}_N$ is the unique solution of the finite-dimensional Lyapunov equation

$$\mathcal{Q}^*_N\mathcal{V}_N + \mathcal{V}_N\mathcal{Q}_N = -\mathcal{C}^*_N\mathcal{C}_N \qquad (14)$$

where $\mathcal{Q}_N := \text{diag}[Q + jN\omega_h I, \cdots, Q + j\omega_h I, Q, Q - j\omega_h I, \cdots, Q - jN\omega_h I]$, $\mathcal{B}_N := [\hat{B}^T_{-N}, \cdots, \hat{B}^T_{-1}, \hat{B}^T_0, \hat{B}^T_1, \cdots, \hat{B}^T_N]^T$ and $\mathcal{C}_N := (\hat{\underline{C}}_N)_{[(2N)N]}$ with $\hat{\underline{C}}_N := \mathcal{T}\{\hat{C}_N(t)\} = \mathcal{T}\{\sum_{|k|\le N}\hat{C}_k e^{jk\omega_h t}\}$.

An underlying assumption of Theorem 3 is that the transition matrix of the system (1) is known so that no approximation treatments are needed. Hence it follows from Theorem 2 that $||\mathcal{G}||_2^2 = \lim_{N\to\infty}\text{trace}(\hat{\underline{b}}^*_{[N]}\hat{V}_N\hat{\underline{b}}_{[N]})$ with $\hat{V}_N$ satisfying

$$\begin{aligned}&((\underline{Q} - \underline{E}(j0))^*)_{[NN]}\hat{V}_N\\&+\hat{V}_N(\underline{Q} - \underline{E}(j0))_{[NN]} = -(\hat{\underline{C}}^*\hat{\underline{C}})_{[NN]}\end{aligned}$$

which is derived from (13) by replacing Q_a and $(\hat{\underline{C}}^*_a\hat{\underline{C}}_a)_{[NN]}$ with Q and $(\hat{\underline{C}}^*\hat{\underline{C}})_{[NN]}$, respectively. On the other hand, by the dual arguments to Proposition 4, the convergence regarding the 'skew truncation' on $\hat{\underline{C}}$ (that corresponds to replacing $\hat{C}(t)$ with $\hat{C}_N(t)$) follows readily. Namely, $||\mathcal{G}||_2^2 = \lim_{N\to\infty}\text{trace}(\hat{\underline{b}}^*_{[N]}\check{V}_N\hat{\underline{b}}_{[N]})$ with $\check{V}_N$ satisfying

$$\begin{aligned}&((\underline{Q} - \underline{E}(j0))^*)_{[NN]}\check{V}_N\\&+\check{V}_N(\underline{Q} - \underline{E}(j0))_{[NN]} = -(\hat{\underline{C}}^*_N\hat{\underline{C}}_N)_{[NN]}\end{aligned}$$

Finally, noting that $\hat{\underline{b}}_{[N]} = \mathcal{B}_N$, $(\hat{\underline{C}}^*_N\hat{\underline{C}}_N)_{[NN]} = \mathcal{C}^*_N\mathcal{C}_N$ and letting $\mathcal{V}_N = \check{V}_N$, Theorem 3 follows immediately. We can also say that the trace formula of Theorem 1 is the limit of the trace formula of Theorem 3 as $N\to\infty$. Theorem 1 implies that this limit does exist and is related to the unique solution of the harmonic Lyapunov equation.

5 NUMERICAL EXAMPLES

In this section, we consider the H_2 norm computation and estimation for the following π-periodic FDLCP system when the input weighting parameter β varies from 0 to 0.4.

$$\begin{cases}\dot{x} = \begin{bmatrix} -1-\sin^2(2t) & 2-\frac{1}{2}\sin(4t)\\ -2-\frac{1}{2}\sin(4t) & -1-\cos^2(2t)\end{bmatrix}x\\ \quad + \begin{bmatrix} 0\\ 1-2\beta\rho(t)\end{bmatrix}u\\ y = [\,1 \quad 1\,]x\end{cases}$$

Here the function ρ is given by

$$\rho(t) = \begin{cases}\sin(2t) & (0\le t\le\frac{\pi}{2})\\ 0 & (\frac{\pi}{2} < t \le \pi)\end{cases}$$

The transition matrix of the above system has a Floquet factorization of the form

$$P(t,0) = \begin{bmatrix}\cos(2t) & \sin(2t)\\ -\sin(2t) & \cos(2t)\end{bmatrix}$$

$$Q = \begin{bmatrix}-1 & 0\\ 0 & -2\end{bmatrix}$$

Since the transition matrix is available and $C(t)$ is constant, Theorem 1 can be applied to get the exact value of the H_2 norm of the above system. Corresponding to the parameter β of 0, 0.1, 0.2, 0.3 and 0.4, the H_2 norms are 0.7323, 0.6836, 0.6408, 0.6053 and 0.5783, respectively.

Table 1: H_2 Norm Estimation: Truncation-Only Approach

H_2	$N=1$	6	10	14
$\beta=0$	0.7323	0.7323	0.7323	0.7323
0.1	1.0095	0.8739	0.6836	0.6836
0.2	0.9423	0.8186	0.6408	0.6408
0.3	0.8885	0.7727	0.6053	0.6053
0.4	0.8448	0.7379	0.5783	0.5783

Table 2: H_2 Norm Estimation: Approximate Modeling

H_2	$N=1$	6	10	14
$\beta=0$	0.7323	0.7323	0.7323	0.7323
0.1	1.0010	1.0017	0.6835	0.6835
0.2	0.9331	0.9365	0.6408	0.6408
0.3	0.8715	0.8794	0.6052	0.6052
0.4	0.8178	0.8318	0.5783	0.5783

We can apply Proposition 4 to give an asymptotic estimation of the H_2 norm by only truncating the input matrix $B(t)$. The estimation results are listed in Table 1. For the given FDLCP system, we can also use the approximate modeling approach to get estimations for the H_2 norm. The approximation FDLCP model is constructed through a piecewise constant approximation of $A(t)$. To be more precise, the period π is divided into $M(=100)$ segments with the same length, during each of which $A(t)$ is treated as a constant matrix. The transition matrix can be computed explicitly (Willems, 1970) and Theorem 2 applies. The estimation results are listed in Table 2.

6 CONCLUSION

In this paper, the H_2 norm of a class of general FDLCP systems was considered. Based on what we called the harmonic Lyapunov equation in (Zhou *et al.*, 2001a), we established for the first time the (infinite-dimensional) trace formula in the FDLCP setting, which includes the trace formula for the H_2 norm of LTI continuous-time systems as a special case. The interpretation of this formula is that FDLCP systems are essentially LTI when the H_2 problem is concerned. As another contribution of this study, we developed some trace formulas for the H_2 norm asymptotic estimation in FDLCP systems via the approximate modeling approach, which is highly applicable to most practical FDLCP systems.

It is expected that a trace formula can be developed for the H_2 norm estimation by directly truncating the harmonic Lyapunov equation (5) or (6) so that even the transition matrix of the original and/or approximate models is circumvented. This will be pursued in our subsequent study.

References

Bamieh, B. and J. B. Pearson (1992). The H^2 problem for sampled-data systems. *Systems and Control Lett.*, **19**, 1–12.

Bolzern P. and Colaneri, P (1988). The periodic Lyapunov equation,", *SIAM J. Anal. Appl.*, **9**, 499–512.

Chen, T. and Francis, B. A (1995). *Optimal Sampled-Data Control Systems.*

Colaneri, P (2000). Continuous-time periodic systems in H_2 and H_∞. *Kybernetika*, **36**, Part I: 211–242, Part II: 329–350.

Dullerud, G. E. (1996). *Control of Uncertain Sampled-Data Systems.* Birkhäuser, Boston, 19–23.

Farkas, M (1994). *Periodic Motions,* Springer-Verlag.

Gohberg, I., Goldberg, S. and Kaashoek, M. A (1993). *Classes of Linear Operators,* Birkhäuser, Vol. II.

Green, M. and D. J. N. Limebeer (1995). *Linear Robust Control.* Prentice-Hall, 93–96.

Hagiwara, T. and M. Araki (1995). FR-operator approach to the H_2 analysis and synthesis of sampled-data systems. *IEEE Trans. Automat. Contr.*, **AC-40**, 1411–1421.

Hall, S. R. and N. M. Wereley (1990). Generalized Nyquist criterion for linear time systems.*Proc. ACC,* 673–679.

Igari, S (1975). *Fourier Series,* Iwanami(in Japanese).

Lukes, D. L. (1982). *Differential Equations: Classical to Controlled.* Academic Press, 106-179.

Naylor, A. W. and Sell, G. R (1982). *Linear Operator Theory in Engineering and Science.* Springer-Verlag.

Wereley, N. M. and S. R. Hall (1990). Frequency response of linear time periodic systems. *Proc. CDC,* 3650–3655.

Wereley, N. M. (1990). *Analysis and Control of Linear Periodically Time Varying Systems,* Ph. D. Thesis. Dept. of Aeronautics and Astronautics, M.I.T.

Yamamoto, Y. (1996). Frequency response of sampled-data systems. *IEEE Trans. Automat. Contr.*, **AC-41**, 166–176.

Willems, J. L (1970). *Stability Theory of Dynamical Systems,* Nelson.

Zhang, C. and J. Zhang (1997). H_2 performance of continuous time periodically time varying controllers. *Systems and Control Lett.*, **32**, 209–221.

Zhou, J. and Hagiwara, T (2000a). H_2 and H_∞ norm computations of linear continuous-time periodic systems via the skew analysis of frequency response operators. Automatic Control Engineering Group, Dept. Electrical Engineering, Kyoto Univ., Tech. Rep., No. 00-04.

Zhou, J. and T. Hagiwara (2000b). Existence conditions and properties of frequency response operators of continuous-time periodic systems. Automatic Control Engineering Group, Dept. Electrical Engineering, Kyoto Univ., Tech. Rep., No. 00-05.

Zhou, J. and Hagiwara, T (2001a). Stability analysis of continuous-time periodic systems via the harmonic analysis. Proceedings of the American Control Conference (to appear).

Zhou, J., T. Hagiwara and M. Araki (2001b). Trace formulas for the H_2 norm of linear continuous-time periodic systems. Automatic Control Engineering Group, Dept. Electrical Engineering, Kyoto Univ., Tech. Rep., No. 01-04.

Zhou, K. (1998). *Essentials of Robust Control,* Prentice Hall, 45–62.

Copyright © IFAC Periodic Control Systems,
Cernobbio-Como, Italy, 2001

STATISTICAL ANALYSIS AND H_2–NORM OF FINITE DIMENSIONAL LINEAR TIME-PERIODIC SYSTEMS

B.P. Lampe *, **E.N. Rosenwasser** **

* *University of Rostock, D-18051 Rostock, Germany*
fax : +49 381/498-3563
E-mail : bernhard.lampe@etechnik.uni-rostock.de
** *St. Petersburg State University of Ocean Technology*
Lotsmanskaya ul. 3, St. Petersburg, 190008, Russia
E-mail : k10@smtu.ru

Abstract: The paper presents the theoretical foundations for the investigation of linear time-periodic (LTP) systems in the frequency domain. This is done on the basis of the parametric transfer function concept. Closed formulae for the $\mathcal{H}_2$–norm and associated $\mathcal{H}_\infty$–norm of an LTP system are given as norms of finite-dimensional matrices. *Copyright © 2001 IFAC*

Keywords: Linear systems, Time-varying systems, Periodic structures, Stochastic variables, Transfer functions, Norms

1. INTRODUCTION

A multitude of physical and technological tasks consequently leads to control problems for linear systems with periodically time-varying coefficients (LTP systems), see for instance (Yakubovich and Starzhinskii 1975, Schilkin 1978, Richards 1983, Chen and Francis 1995, Möllerstedt and Bernhardsson 2000*b*). Especially, we find among them, the stability problem for elastic dynamic systems, the parametric resonance in main lines of transmitters and cyclotrons, a number of tasks in celestial mechanics, stability of servo systems for alternating current, investigations of vibrations in milling machines, stability considerations for self-tuning systems with tuneable amplifiers, investigations of power converters of locomotives, research on rotor oscillations of helicopters, and many others. A general overview and further literature to the above questions can be found in (Yakubovich and Starzhinskii 1975, Schilkin 1978, Richards 1983, Chen and Francis 1995, Möllerstedt and Bernhardsson 2000*b*).

For the solution of such kinds of problems special mathematical instruments are necessary, because the standard methods are developed for the class of linear time-invariant (LTI) systems, but we want to investigate LTP systems now.

Best known methods for the investigation of LTP systems are based on the theory of linear differential equations with periodic coefficients, introduced by Floquet (Floquet 1883, Yakubovich and Starzhinskii 1975), and the method of Hill (Hill 1886, Whittaker and Watson 1927). These theories are connected with the application of infinite-dimensional matrices and infinite determinants. An important point of Hill's method is the fact that it allows for the solution of LTP control problems to use frequency representations that are costumary for engineers. Nevertheless, there it is also necessary to operate with infinite-dimensional matrices and determinants, what means that the employer has to overcome essential technical obstacles.

As alternative to Hill's method, the methods based on the parametric transfer function (PTF)

concept offer new possibilities. These methods, like Hill's method, are also operating in the frequency domain. But in contrast to Hill's method, they use only matrices of finite dimensions what essentially simplifies the solution of analysis and design problems. Thereby, as in case of standard frequency domain methods for investigation of LTI systems, the PTF based methods permit the possibility of using experimental data (Sommer *et al.* 1994, Volovodov *et al.* 2000).

L. Zadeh introduced in the papers (Zadeh 1950, Zadeh 1951) the name parametric transfer function into control theory, but without a strictly mathematical foundation.

The monographs (Rosenwasser 1970, Rosenwasser 1973, Rosenwasser 1977) deal with issues about theory and application of the PTF concept for LTP systems of various types. The general PTF theory for arbitrary linear non-stationary systems is presented in (Rosenwasser and Lampe 2000). A number of questions about analysis of LTP systems under deterministic disturbances are investigated by means of the PTF in (Möllerstedt and Bernhardsson 2000*b*, Wereley and Hall 1991, Möllerstedt and Bernhardsson 2000*a*).

The present paper considers applications of the PTF method for statistical analysis of finite-dimensional LTP systems, given by differential equations in state space. The obtained relations generalize the results of (Rosenwasser 1977, Rosenwasser and Lampe 2000). The fundamental results will be formulated in a number of theorems, that are given without proofs, because the proofs could be constructed in the same way as in (Rosenwasser 1977, Rosenwasser and Lampe 2000, Rosenwasser 1997).

2. BASIC RELATIONS

2.1 *Process description*

The paper considers LTP systems described by the state equations

$$\frac{\mathrm{d}x(t)}{\mathrm{d}t} = A(t)x(t) + B(t)u(t)$$
$$y(t) = C(t)x(t) \tag{1}$$

where $y(t)$, $x(t)$, $u(t)$ are vectors of dimensions $n \times 1$, $p \times 1$, $m \times 1$, respectively, and $C(t)$, $A(t)$, $B(t)$ are T–periodic matrices of appropriate size. The matrices $A(t)$ and $B(t)$ are supposed to be continuous, and to have bounded variation. The matrix $C(t)$ should be piece-wise smooth. Let us denote by $H(t)$ the $p \times p$ transition matrix, satisfying

$$\frac{\mathrm{d}H(t)}{\mathrm{d}t} = A(t)H(t), \quad H(0) = I_p \tag{2}$$

where I_p stands for the identity $p \times p$ matrix.

The matrix

$$M = H(T) \tag{3}$$

is called the *matrix of monodromy* for the system (1).

The eigenvalues z_i $(i = 1, \ldots, q)$ of the matrix M are called the *multipliers* of the system (1), and the numbers

$$s_{ik} = \frac{1}{T}\ln z_i + \frac{2k\pi \mathrm{j}}{T}$$
$$(k = 0, \pm 1, \ldots\ ;\ i = 1, \ldots, q) \tag{4}$$

with $\mathrm{j} = \sqrt{-1}$, are called its characteristic indices. The characteristic indices of the system (1) coincide with the roots of the equation

$$\det(I_p - \mathrm{e}^{-sT} M) = 0\,. \tag{5}$$

Definition 1 If all characteristic indices of the system (1) have negative real parts then this system is called *stable.*

2.2 *Bilateral Laplace transforms*

Suppose α, β to be real numbers with $\alpha < \beta$. Denote by $\Lambda(\alpha, \beta)$ the set of functions $f(t)$ of bounded variation satisfying for $\alpha < \operatorname{Re} s < \beta$ the estimation

$$\left|f(t)\,\mathrm{e}^{-st}\right| < N\,\mathrm{e}^{-\gamma|t|}, \qquad -\infty < t < \infty \tag{6}$$

where N, γ are positive constants. The functions $f(t) \in \Lambda(\alpha, \beta)$ possess absolute convergent (bilateral) Laplace transforms

$$F(s) = \int_{-\infty}^{\infty} f(t)\,\mathrm{e}^{-st}\,\mathrm{d}t, \quad \alpha < \operatorname{Re} s < \beta \tag{7}$$

and the inversion formula

$$f(t) = \frac{1}{2\pi \mathrm{j}} \lim_{\lambda\to\infty} \int_{c-\mathrm{j}\lambda}^{c+\mathrm{j}\lambda} F(s)\,\mathrm{e}^{st}\,\mathrm{d}s$$
$$\alpha < c < \beta \tag{8}$$

takes place. Henceforth, for a matrix $Q(t)$ we will write $Q(t) \in \Lambda(\alpha, \beta)$ if each of its entries $q_{ik}(t)$ fulfills the condition $q_{ik}(t) \in \Lambda(\alpha, \beta)$.

3. CHARACTERISTIC FUNCTIONS

3.1 *Parametric transfer function (PTF)*

Theorem 1. The matrix equation

$$\frac{\mathrm{d}w_0(s,t)}{\mathrm{d}t} = [A(t) - sI_p]\, w_0(s,t) + B(t) \quad (9)$$

where s is a complex parameter, possesses for each s, that is different from all characteristic indeces of the system (1), a unique periodic solution with

$$w_0(s,t) = w_0(s,t+T) \quad (10)$$

that takes for $0 \le t \le T$ the form

$$w_0(s,t) = \int_0^T H(t)\left(I_p - \mathrm{e}^{-sT} M\right)^{-1} \mathrm{e}^{-sT} M \, *$$
$$*H^{-1}(\tau)B(\tau)\,\mathrm{e}^{-s(t-\tau)}\,\mathrm{d}\tau + \quad (11)$$
$$+ \int_0^t H(t)H^{-1}(\tau)B(\tau)\,\mathrm{e}^{-s(t-\tau)}\,\mathrm{d}\tau .$$

■

Definition 2 The matrix

$$w(s,t) = C(t)w_0(s,t) \quad (12)$$

is called the *parametric transfer function* (PTF) of the system (1).

Theorem 2. The PTF $w(s,t)$ is, for each value of the real parameter t, a meromorphic function of the argument s. The set of poles for $w(s,t)$ coincides with the set of characteristic indices of the system (1). ■

Corrolary For the PTF $w(s,t)$ there exists a real constant μ such that for all poles s_i the inequality holds

$$\mathrm{Re}\, s_i < \mu . \quad (13)$$

3.2 *Weighting function and impulse response*

Definition 3 The matrix

$$G(t,\tau) = \begin{cases} C(t)H(t)H^{-1}(\tau)B(\tau) & t > \tau \\ O_{nm} & t < \tau \end{cases} \quad (14)$$

where O_{nm} is the zero matrix of size $n \times m$, is called the *weighting function* of the system (1). Furthermore, the matrix

$$R(t,\nu) = G(t,\tau)\,|_{\tau = t-\nu} \quad (15)$$
$$= \begin{cases} C(t)H(t)H^{-1}(t-\nu)B(t-\nu) & \nu > 0 \\ O_{nm} & \nu < 0 \end{cases}$$

is called the *impulse response* of the system (1).

It can easily be seen that

$$G(t,\tau) = G(t+T, \tau+T)$$
$$R(t,\nu) = R(t+T,\nu) . \quad (16)$$

Theorem 3. Suppose

$$\max_i \mathrm{Re}\, s_{ik} = \mu . \quad (17)$$

Then, for each t we have

$$R(t,\nu) \in \Lambda(\mu, \infty) . \quad (18)$$

For $\mathrm{Re}\, s > \mu$ the following equation takes place:

$$\int_{-\infty}^t G(t,\tau)\,\mathrm{e}^{-s(t-\tau)}\,\mathrm{d}\tau =$$
$$= \int_0^\infty R(t,\nu)\,\mathrm{e}^{-s\nu}\,\mathrm{d}\nu \quad (19)$$
$$= w(s,t) .$$

Moreover, for $c > \mu$ the following relations are valid:

$$\frac{1}{2\pi \mathrm{j}} \lim_{\lambda\to\infty} \int_{c-\mathrm{j}\lambda}^{c+\mathrm{j}\lambda} w(s,t)\,\mathrm{e}^{s\nu}\,\mathrm{d}s = R(t,\nu) \quad (20)$$

$$\frac{1}{2\pi \mathrm{j}} \lim_{\lambda\to\infty} \int_{c-\mathrm{j}\lambda}^{c+\mathrm{j}\lambda} w(s,t)\,\mathrm{e}^{s(t-\tau)}\,\mathrm{d}s = G(t,\tau) . \quad (21)$$

■

4. QUASI-STATIONARY MOTIONS

4.1 *Definition*

Consider the integral operator

$$y_0(t) = \int_{-\infty}^t G(t,\tau)u(\tau)\,\mathrm{d}\tau$$
$$= \int_0^\infty R(t,\nu)u(t-\nu)\,\mathrm{d}\nu \triangleq \mathsf{L}[u(t)] \quad (22)$$

where $u(t)$ is an $m \times 1$ input vector, or an $m \times \ell$ matrix of collected input vectors.

Theorem 4. Suppose the system (1) to be stable, that means (17) is valid with $\mu = -\mu_0$, $\mu_0 > 0$. Suppose, furthermore,

$$\|u(t)\| < N_1 \, e^{\mu_1 |t|}, \quad -\infty < t < \infty \tag{23}$$

where $\| \cdot \|$ is a certain norm for matrices of numbers, and N_1, μ_1 are constants with $N_1 > 0$ and

$$\mu_1 < \mu_0 \, . \tag{24}$$

Then, the integral (22) converges, and the vector (resp. matrix) $y_0(t)$ is a solution of the equation (1). Thereby, the estimation

$$\|y_0(t)\| < N_2 \, e^{\mu_0 |t|}, \quad -\infty < t < \infty \qquad N_2 = \text{const.} \tag{25}$$

is valid. ∎

Definition 4 In case of convergence of the integrals in (22) the vector (resp. matrix) $y_0(t)$ is named a *quasi-stationary* motion of the system (1).

4.2 *Existence*

Theorem 5. Suppose the system (1) to be stable, and

$$u(t) = I_m t^r \tag{26}$$

with a certain non-negative integer r. Then, there exists the quasi-stationary output, and it can be determined by the relation

$$y_0(t) = \left\{ \frac{\partial^{r-1}}{\partial s^{r-1}} \left[w(s,t) \, e^{st} \right] \right\}_{|s=0} \tag{27}$$

∎

Corollary For $r = 0$, i.e. $u(t) = I_m$, the quasi-stationary output exists, and it can be determined by the relation

$$y_0(t) = w(0,t) = w(0,t+T) \tag{28}$$

that means, it is a $T-$periodic function.

5. ANALYSIS OF LTP SYSTEMS UNDER STOCHASTIC DISTURBANCES

5.1 *Correlation in LTP systems*

Theorem 6. Suppose the system (1) to be stable, and $u(t)$ to be a centered stationary stochastic vector process, such that for each of its realizations there exists the quasi-stationary output (22). Then, $y_0(t)$ is a periodic non-stationary centered stochstic vector process with an auto-correlation matrix $K_0(t_1, t_2)$ that satisfies the relation

$$K_0(t_1, t_2) = K_0(t_1 + T, t_2 + T) \, . \tag{29}$$

Moreover, if $K_u(\tau)$ denotes the auto-correlation matrix of the stationary stochstic input $u(t)$, and

$$\Phi_u(s) = \int_{-\infty}^{\infty} K_u(\tau) \, e^{-s\tau} \, d\tau \tag{30}$$

denotes its power spectrum, then the matrix $K_0(t_1, t_2)$ can be determined by the formula

$$K_0(t_1, t_2) = \frac{1}{2\pi j} \int_{-j\infty}^{j\infty} w(-s, t_1) \Phi_u(s) w'(s, t_2) \, e^{s(t_2 - t_1)} \, ds \tag{31}$$

where the dash stands for the transposition operation. ∎

5.2 *Output evaluation*

Definition 5 The scalar function

$$d_0(t) = \text{trace}\, K_0(t,t) \tag{32}$$

is called the *variance* of the output $y_0(t)$. It follows from (31) that

$$d_0(t) = d_0(t+T) \, . \tag{33}$$

Definition 6 The number

$$\overline{d_0} = \frac{1}{T} \int_0^T d_0(t) \, dt \tag{34}$$

is called the *mean variance* of the output $y_0(t)$.

Definition 7 For $\Phi_u(s) = I_m$ the number

$$\|\mathsf{L}\|_2 = \sqrt{\overline{d_0}} \tag{35}$$

is called the $\mathcal{H}_2-$*norm* of the system (1).

5.3 *Determination of the* $\mathcal{H}_2-$*norm*

Theorem 7. The $\mathcal{H}_2-$norm of the system (1) can be determined by the formula

$$\|\mathsf{L}\|_2^2 = \frac{1}{2\pi j} \oint \text{trace}\, R(\zeta) \frac{d\zeta}{\zeta} \tag{36}$$

where the integral has to be taken over the unit circle in positive direction, and

$$R(\zeta) = \mathcal{D}_{B_0}(T, s, 0)\,\big|_{e^{-sT}=\zeta} \triangleq \frac{1}{T}\sum_{k=-\infty}^{\infty} B_0(s + kj\omega_0)\,\big|_{e^{-sT}=\zeta} \tag{37}$$

with

$$B_0(s) = \frac{1}{T}\int_0^T w(-s,t)w'(s,t)\,dt \qquad \omega_0 = \frac{2\pi}{T}\,. \tag{38}$$

■

Remark The convergence properties of the series (37) are investigated in (Rosenwasser 1997).

Definition 8 The quantity

$$\|\mathsf{L}\|_{a\infty} \triangleq \sup_{|\zeta|=1} \sigma_{\max}[R(\zeta)] \tag{39}$$

where $\sigma_{\max}[R(\zeta)]$ is the greatest singular value of the rational matrix $R(\zeta)$, will be called the *$\mathcal{H}_\infty$−norm* of the system (1), *associated* with the functional (36).

The idea for that definition is derived in (Rosenwasser 1997), where it is also shown that the quantity $\|\mathsf{L}\|_{a\infty}$ can be used as a measure for the robustness of the system (1).

6. SUPPLEMENTARY RELATIONS

Theorem 8. Let us have the Fourier series

$$R(t,\nu) = \sum_{k=-\infty}^{\infty} r_k(\nu)\,e^{kj\omega_0 t} \qquad r_k(\nu) = \frac{1}{T}\int_0^T R(t,\nu)\,e^{-kj\omega_0 t}\,dt\,. \tag{40}$$

Then, for a stable system (1), the formula

$$\|\mathsf{L}\|_2^2 = \sum_{k=-\infty}^{\infty}\int_0^\infty \operatorname{trace}[r_k(\nu)r'_{-k}(\nu)]\,d\nu \tag{41}$$

is valid. ■

Theorem 9. Let us have the Fourier series

$$w(s,t) = \sum_{k=-\infty}^{\infty} w_k(s)\,e^{kj\omega_0 t} \qquad w_k(s) = \frac{1}{T}\int_0^T w(s,t)\,e^{-kj\omega_0 t}\,dt\,. \tag{42}$$

Then, for a stable system (1), the formula

$$\|\mathsf{L}\|_2^2 = \frac{1}{2\pi j}\sum_{k=-\infty}^{\infty}\int_{-j\infty}^{j\infty} \operatorname{trace}[w_k(-s)w'_k(s)]\,ds \tag{43}$$

is valid. ■

Theorem 10. Denote

$$Q(s,t,\tau) = \begin{cases} C(t)H(t)\left(I_p - e^{-sT}M\right)^{-1}\cdot \\ \quad\cdot e^{-sT}MH^{-1}(\tau)B(\tau) + \\ \quad + C(t)H(t)H^{-1}(\tau)B(\tau) & t > \tau \\ C(t)H(t)\left(I_p - e^{-sT}M\right)^{-1}\cdot \\ \quad\cdot e^{-sT}MH^{-1}(\tau)B(\tau) & t < \tau\,. \end{cases} \tag{44}$$

Then, for a stable system (1), we have

$$\mathcal{D}_{B_0}(T,s,0) = \int_0^T\int_0^T Q(-s,t,\tau)Q'(s,t,\tau)\,dt\,d\tau\,. \tag{45}$$

Therefore, the associated $\mathcal{H}_\infty$−norm is equal to

$$\|\mathsf{L}\|_{a\infty} = \sup_{-\frac{\omega_0}{2}\le\omega\le\frac{\omega_0}{2}} \sigma_{\max}\left[\mathcal{D}_{B_0}(T,j\omega,0)\right]\,. \tag{46}$$

■

ACKNOWLEDGMENTS

The work has been supported by the German Science Foundation (DFG). This support is very gratefully acknowledged.

7. REFERENCES

Chen, T. and B.A. Francis (1995). *Optimal sampled-data control systems.* Springer-Verlag. Berlin, Heidelberg, New York.

Floquet, G. (1883). Sur les équations différentielles linéaires a coefficients périodiques. *Annales de L'Ecole Normale Supérieure* **12**, 47–89.

Grimble, M.J. (1995). *Robust Industrial Control.* Prentice-Hall. UK.

Hill, G.W. (1886). On the part of the lunar perigee which is a function of the mean motions of the sun and the moon. *Acta Mathematica* **8**, 1–36.

Möllerstedt, E. and B. Bernhardsson (2000*a*). A harmonic transfer function model for a diode converter train. In: *IEEE PES Winter Meeting*. Singapore.
Möllerstedt, E. and B. Bernhardsson (2000*b*). Out of control because of harmonics. *IEEE Control Systems Magazine* **20**(4), 70–81.
Richards, J. (1983). *Analysis of periodically time varying systems*. Springer-Verlag. Berlin.
Rosenwasser, E.N. (1970). *Vibrations of nonlinear systems, the method of integral equations*. Techn. Information Service. Springfield, VA.
Rosenwasser, E.N. (1973). *Periodically nonstationary control systems*. Nauka. Moscow. (in Russian).
Rosenwasser, E.N. (1977). *Lyapunov indices in linear control theory*. Nauka. Moscow. (in Russian).
Rosenwasser, E.N. (1997). Frequency analysis and the ℓ_2–norm of linear periodic operators. *Automation and Remote Control* **58**(9), 1437–1458.
Rosenwasser, E.N. and B.P. Lampe (1997). *Digitale Regelung in kontinuierlicher Zeit - Analyse und Entwurf im Frequenzbereich*. B.G. Teubner. Stuttgart.
Rosenwasser, E.N. and B.P. Lampe (2000). *Computer Controlled Systems - Analysis and Design with Process-orientated models*. Springer-Verlag. London, Berlin, Heidelberg.
Schilkin, S.V. (1978). *Generating function method in theory of dynamic systems*. Nauka. Moscow. (in Russian).
Sommer, V.B., B.P. Lampe and E.N. Rosenwasser (1994). Experimental investigations of analog-digital control systems by frequency methods. *Automation and Remote Control* **55**(Part 2), 912–920.
Volovodov, S.K., B.P. Lampe, E.N. Rosenwasser and A.V. Smolnikov (2000). Frequency analysis of linear periodical systems - Theory and experiment. In: *Proc. 2nd IEEE Int. Conf. Control Oscill. Chaos*. Vol. 3. Saint Petersburg, Russia. pp. 408–413.
Wereley, N. and S. Hall (1991). Linear time periodic systems: transfer function, poles, transmission zeros and directional properties. In: *Proc. Amer. Control Conf.*. Boston, MA.
Whittaker, E.T. and G.N. Watson (1927). *A course of modern analysis*. 4 ed.. University Press. Cambridge.
Yakubovich, V.A. and V.M. Starzhinskii (1975). *Linear differential equations with periodic coefficients*. Vol. 1. John Wiley & Sons. New York.
Zadeh, L.A. (1950). Frequency analysis of variable networks. *Proc. IRE* **39**(March), 291–299.
Zadeh, L.A. (1951). Stability of linear varying-parameter systems. *J. Appl. Phys.* **22**(4), 202–204.

Copyright © IFAC Periodic Control Systems,
Cernobbio-Como, Italy, 2001

PARAMETRIC FREQUENCY RESPONSE OF LINEAR PERIODIC SYSTEMS – THEORY AND EXPERIMENT

B.P. Lampe *,
S.K. Volovodov, E.N. Rosenwasser, A.V. Smolnikov **

* *University of Rostock, D-18051 Rostock, Germany*
fax : +49 381/498-3563
E-mail : bernhard.lampe@etechnik.uni-rostock.de
** *St. Petersburg State University of Ocean Technology*
Lotsmanskaya ul. 3, St. Petersburg, 190008, Russia
E-mail : k10@smtu.ru

Abstract: The paper presents the theoretical and experimental foundations for the investigation of linear time-periodical (LTP) systems in the frequency domain. This is done on the basis of the parametric frequency response (PFR). Two basic method for an experimental determination of the PFR are derived, and tested for examples.
Copyright © 2001 IFAC

Keywords: Linear systems, Time-varying systems, Periodic structures, Frequency responses, Measured values

1. PROBLEM

The wide field of application of the frequency response is well known for linear time-invariant (LTI) systems. In addition to its great theoretical importance these methods create the possibility for the analysis and design of LTI systems on the basis of experimental data. This gives a strong motivation for extending the concept of the frequency response to LTP systems.

Very useful results in this direction can be held by the parametric frequency response (PFR). Practically, the PFR concept (with different terminology) was introduced into control theory by the papers of L. Zadeh (Zadeh 1950, Zadeh 1951).

Different aspects of the PFR theory for various types of LTP systems are given in the papers (Rosenwasser 1970, Rosenwasser 1973, Rosenwasser and Lampe 1997, Rosenwasser and Lampe 2000). A general theory for arbitrary finite dimensional linear non-stationary systems is developped in (Rosenwasser 1977). The papers (Sommer *et al.* 1994, Lampe and Richter 1997, Lampe *et al.* 1999, Volovodov *et al.* 2000, Volovodov *et al.* 1991) present methods for the experimental determination of the PFR.

The present paper has a look from a general position to the theoretical and experimental determination of the PFR for finite-dimensional LTP systems given in an input-output matrix description. The received results are more general than those given in (Sommer *et al.* 1994, Lampe and Richter 1997, Volovodov *et al.* 2000, Volovodov *et al.* 1991).

2. THEORETICAL FUNDAMENTALS

2.1 *System description*

The present paper considers finite-dimensional systems that can be described by the input-output relations

$$D(s,t)y = M(s,t)u \tag{1}$$

with

$$D(s,t) = I_n s^q + D_1(t)s^{q-1} + \ldots + D_q(t)$$
$$M(s,t) = M_1(t)s^{q-1} + \ldots + M_q(t) \tag{2}$$

In (2) I_n is the identity matrix of dimension $n \times n$, $D_i(t)$ and $M_i(t)$, $(i = 1, \ldots, q)$ are $T-$periodic matrices of type $n \times n$ and $n \times m$, respectively. Moreover, $s = \frac{\mathrm{d}}{\mathrm{d}t}$ is the differential operator, and $y(t)$, $u(t)$ are vectors of dimensions $n \times 1$ or $m \times 1$, respectively. The matrices $D_i(t)$ and $M_i(t)$ are assumed to be continuous functions of t.

Furthermore, suppose that all multipliers $\mu_1, \mu_2, \ldots, \mu_\ell$, (Yakubovich and Starzhinskii 1975), of the homogenious equation

$$D(s,t)y = 0 \tag{3}$$

are located inside the unity circle, i.e.

$$|\mu_i| < 1, \quad (i = 1, \ldots, \ell). \tag{4}$$

If (4) is valid the system (1) will be called stable.

2.2 *Parametric Frequency Response*

Theorem 1. Suppose the system (1) to be stable. Than for all real $-\infty < \omega < \infty$ the matrix equation

$$D(s + \mathrm{j}\omega, t)W(\mathrm{j}\omega, t) = M(\mathrm{j}\omega, t) \tag{5}$$

has a unique periodic solution $W(\mathrm{j}\omega, t) = W(\mathrm{j}\omega, t+T)$.

Definition 1 The matrix $W(\mathrm{j}\omega, t)$ is called the *parametric frequency response* (PFR) of the system (1). The matrices $P(\omega, t) = \operatorname{Re} W(\mathrm{j}\omega, t)$, $Q(\omega, t) = \operatorname{Im} W(\mathrm{j}\omega, t)$ are called the real and imaginary parametric frequency response, and the matrix $R(\omega, t) = \sqrt{P^2(\omega, t) + Q^2(\omega, t)}$ is called the parametric frequency response magnitude.

Theorem 2. Under the conditions of Theorem 1 the relation

$$\lim_{\lambda \to \infty} \frac{1}{2\pi \mathrm{j}} \int_{-\lambda}^{\lambda} W(\mathrm{j}\omega, t)\, \mathrm{e}^{\mathrm{j}\omega(t-\tau)}\, \mathrm{d}\omega = G(t,\tau)$$
$$= \begin{cases} h(t,\tau) & t > \tau \\ 0 & t < \tau \end{cases} \tag{6}$$

holds with an $n \times m$ matrix

$$h(t,\tau) = h(t+T, \tau+T). \tag{7}$$

Methods for the determination of the matrix $h(t,\tau)$ are considered in (Rosenwasser 1977).

Definition 2 The matrix $G(t,\tau)$ is called the *weighting function* of the system (1).

Consider the linear operator

$$y_\infty(t) = \int_{-\infty}^{\infty} G(t,\tau)u(\tau)\,\mathrm{d}\tau$$
$$= \int_{-\infty}^{t} h(t,\tau)u(\tau)\,\mathrm{d}\tau = \mathsf{L}[u(t)]. \tag{8}$$

Definition 3 For all $u(t)$ for which the integral (8) converges, it defines a special solution of the equation (1) that is called the *quasi-stationary output*.

Theorem 3. Suppose the system (1) to be stable and

$$u(t) = I_m\, \mathrm{e}^{\mathrm{j}\omega t}. \tag{9}$$

Then, the quasi-stationary output exists for all $-\infty < \omega < \infty$, and it has the form

$$y_\infty(t) = \mathrm{e}^{\mathrm{j}\omega t}\, W(\mathrm{j}\omega, t) \tag{10}$$

where $W(\mathrm{j}\omega, t)$ is the PFR of the system.

2.3 *Experimental determination of the PFR*

From Theorem 3 follows on principle the possibility to determine the PFR of stable systems (1). Indeed, for $u(t) = I_m\, \mathrm{e}^{\mathrm{j}\omega t}$ the general solution of the equation (1) has the form

$$y(t) = \mathrm{e}^{\mathrm{j}\omega t}\, W(\mathrm{j}\omega, t) + y_0(t) \tag{11}$$

where $y_0(t)$ is a certain function that depends on the initial conditions. Due to the stability of the system we have

$$\lim_{t \to \infty} \|y_0(t)\| = 0 \tag{12}$$

with any norm $\|\cdot\|$. Thus, after a sufficiently long waiting period we will have

$$y(t) \approx \mathrm{e}^{\mathrm{j}\omega t}\, W(\mathrm{j}\omega, t) = y_\infty(t) \tag{13}$$

at the output of the system, and therefore

$$W(\mathrm{j}\omega, t) \approx y(t)\, \mathrm{e}^{-\mathrm{j}\omega t}. \tag{14}$$

2.4 *Fourier series of the PFR*

Designate

$$g(t,\nu) = h(t,\tau)\,|_{\tau = t-\nu}. \tag{15}$$

Then, by construction we have

$$g(t,\nu) = g(t+T,\nu)\,. \tag{16}$$

Consider the representation

$$g(t,\nu) = \sum_{k=-\infty}^{\infty} g_k(\nu)\,e^{kj\omega_0 t} \tag{17}$$

with

$$g_k(\nu) = \frac{1}{T}\int_0^T g(t,\nu)\,e^{-kj\omega_0 t}\,dt$$

$$\omega_0 = \frac{2\pi}{T}$$

and designate

$$H_k(j\omega) = \int_{-\infty}^{\infty} g_k(\nu)\,e^{-j\omega\nu}\,d\nu\,. \tag{18}$$

For the design of the LTP system the paper (Möllerstedt and Bernhardsson 2000) used the harmonic transfer function $H(j\omega)$ represented by the infinite dimensional matrix

$$H(j\omega) = \tag{19}$$

$$\begin{bmatrix} \ddots & \ddots & \ddots & \ddots & \ddots \\ \ddots & H_0(j\omega - j\omega_0) & H_{-1}(j\omega) & H_{-2}(j\omega + j\omega_0) & \ddots \\ \ddots & H_1(j\omega - j\omega_0) & H_0(j\omega) & H_{-1}(j\omega + j\omega_0) & \ddots \\ \ddots & H_2(j\omega - j\omega_0) & H_1(j\omega) & H_0(j\omega + j\omega_0) & \ddots \\ \ddots & \ddots & \ddots & \ddots & \ddots \end{bmatrix}$$

The connection between the harmonic frequency response and the PFR is given by the relation

$$H_k(j\omega) = W_k(j\omega), \qquad k = 0, \pm 1, \ldots \tag{20}$$

where

$$W_k(j\omega) = \frac{1}{T}\int_0^T W(j\omega,t)\,e^{-kj\omega_0 t}\,dt\,. \tag{21}$$

It follows from (21) that the matrices $H_k(j\omega)$ are the Fourier coefficients of the PFR $W(j\omega,t)$.

2.5 *Application of the PFR*

The application field of the PFR for LTP systems practically coincides with the application field of the ordinary frequency response for LTI systems. Thus, for instance:

a) For calculating the transient process $y(t)$ at the output of a stable system for the input signal $u(t) = \mathbb{1}(t)$ the well-known relation

$$y(t) = \frac{2}{\pi}\int_0^{\infty} \frac{P(\omega,t)}{\omega}\sin(\omega t)\,d\omega \tag{22}$$

can be used, where $P(\omega,t) = \mathrm{Re}[W(j\omega,t)]$.

b) In case of stable systems the relation

$$d_y(t) = \tag{23}$$

$$\frac{1}{2\pi}\int_{-\infty}^{\infty} \mathrm{trace}[W(-j\omega,t)S_u(\omega)W'(j\omega,t)]\,d\omega$$

takes place, where $d_y(t)$ is the variance of the quasi-stationary vector output, and $S_u(\omega)$ is the power spectral matrix of the centered stationary input signal $u(t)$.

It follows from (23) that

$$d_y(t) = d_y(t+T)\,. \tag{24}$$

Definition 4 The value

$$\overline{d}_y = \frac{1}{T}\int_0^T d_y(t)\,dt \tag{25}$$

is called the *mean variance* of the output.

Definition 5 Assume $S_u(\omega) = I_m$. In that case the value

$$\|\mathsf{L}\|_2 = \sqrt{\overline{d}_y} \tag{26}$$

will be called the $\mathcal{H}_2$*−norm* of the system (1).

Theorem 4. The $\mathcal{H}_2$−norm of the stable system (1) can be determined by the formula

$$\|\mathsf{L}\|_2^2 = \frac{1}{2\pi}\int_{-\infty}^{\infty} \mathrm{trace}\,B_0(j\omega)\,d\omega \tag{27}$$

where

$$B_0(j\omega) = \frac{1}{T}\int_0^T W(-j\omega,t)W'(j\omega,t)\,dt\,.$$

3. METHODS FOR NUMERICAL AND EXPERIMENTAL PFR DETERMINATION OF STABLE SYSTEMS

Different approaches are possible for the experimental (including also the numerical) determination of the PFR of LTP systems. One method applies the stationary response to Sinus or Cosinus input signals. It is presented in (Sommer *et al.* 1994, Lampe and Richter 1997, Volovodov *et al.* 2000, Volovodov *et al.* 1991), and was refered to as *two-signal method.* The accuracy of this method only depends on the accuracy of the measurements during the experiment.

A second way consists in the harmonic analysis on the basis of (20), (21) using the response to only a single sin-signal for every frequency ω. This method is called *One-signal method,* (Lampe *et al.* 1999). Its accuracy does not only depend on the accuracy of the measurements during the experiments, but also on the number of elements in the Fourier series used for the PFR.

In the following, the fundamental relations are given that are necessary for understanding the basic ideas of these methods.

3.1 *Two-signal method for PFR determination*

Apply a sin-signal $x_s(t)$ to the input of the LTP system under investigation

$$x_s(t) = \sin \omega t = \frac{1}{2j}(e^{j\omega t} - e^{-j\omega t}). \quad (28)$$

The quasi-stationary reaction to this input signal has the form

$$y_s(t) = P(\omega, t) \sin \omega t + Q(\omega, t) \cos \omega t. \quad (29)$$

Here, $P(\omega, t)$ is the real, and $Q(\omega, t)$ is the imaginary part of the PFR $W(\omega, t)$. After that, a cos-signal

$$x_c(t) = \cos \omega t = \frac{1}{2}(e^{j\omega t} + e^{-j\omega t}) \quad (30)$$

is given on the input, and the system answers quasi-stationary with

$$y_c(t) = P(\omega, t) \cos \omega t - Q(\omega, t) \sin \omega t. \quad (31)$$

This equations are solved with respect to $P(\omega, t)$ and $Q(\omega, t)$. As a result the next expressions are received

$$\begin{aligned} P(\omega, t) &= y_s(t) \sin \omega t + y_c(t) \cos \omega t \\ Q(\omega, t) &= y_s(t) \cos \omega t - y_c(t) \sin \omega t. \end{aligned} \quad (32)$$

It is easy to show that

$$\begin{aligned} R^2(\omega, t) &= |W(j\omega, t)|^2 \\ &= P^2(\omega, t) + Q^2(\omega, t) \\ &= y_s^2(t) + y_c^2(t). \end{aligned} \quad (33)$$

The simplest scheme for the experimental determination of the PFR by the two-signal method is shown on Figure 1.

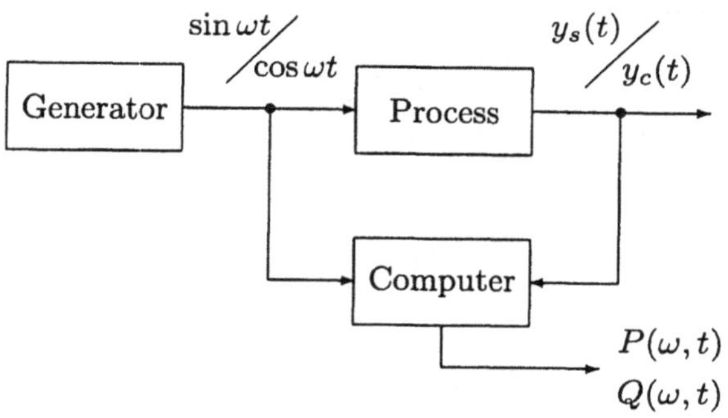

Figure 1. Experimental determination of PFR, two-signal method

3.2 *One-signal method for PFR determination*

The main difficulty of the two-signal method consists in synchronizing the two system reactions at different times. (This problem does not occur for numerical simulation, because in that case the two responses are calculated with the same model.)

For physical experiments, however, the one-signal method is to prefer. Here the quasi-stationary reaction $y_s(t)$ of the system to only a single signal $x_s(t) = \sin \omega t$ is needed for every ω of an appropriate set of frequencies. In this case the periodicity of the PFR $W(j\omega, t) = W(j\omega, t + T)$ is used.

As a result, the elements (21) of the PFR can be found for parameters $\omega \neq \lambda\omega_0/2$, where λ is an integer, by the following relations:

$$\begin{aligned} P_k(\omega) &= \operatorname{Re} W_k(j\omega) \\ &= \lim_{T\to\infty} \frac{1}{T} \int_0^T y_s(\sigma) \cos(k\omega_0 - \omega)\sigma \, d\sigma \end{aligned} \quad (34)$$

$$\begin{aligned} Q_k(\omega) &= \operatorname{Im} W_k(j\omega) \\ &= \lim_{T\to\infty} \frac{1}{T} \int_0^T y_s(\sigma) \sin(k\omega_0 - \omega)\sigma \, d\sigma. \end{aligned} \quad (35)$$

These expressions are permit to determine the amplitudes for a continuous set of the time-parameter t with adaptable accuracy. Moreover, no synchronisation conditions concerning the rational relation between the frequency ω of the input signal and the sampling frequency ω_0 must be satisfied.

The scheme in Figure 2 shows in principle the one-signal experiment.

Usually, for technical systems the first harmonics up to the third of the frequency response are most important. These components are determined by (34) for $k = 0, 1, 2, 3$. The following example should illustrate the application of the above derived methods.

3.3 *Example: Power converter control of a locomotive*

In (Möllerstedt and Bernhardsson 2000) an example of a linearized LTP control system for power converters of locomotives is given. There was

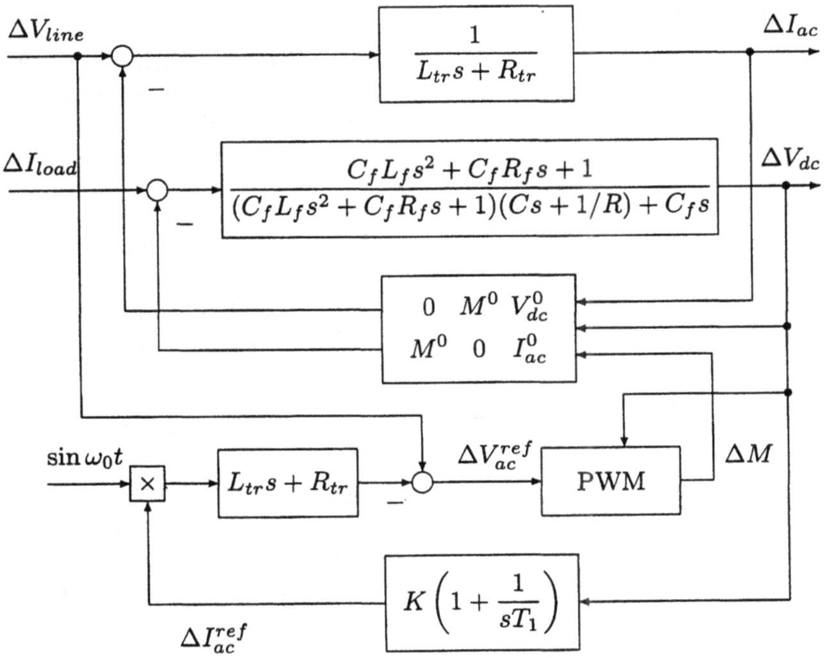

Figure 3. LTP Model for the power converter system

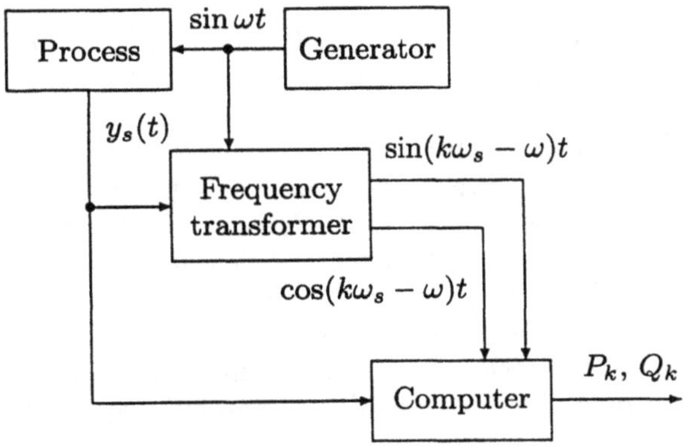

Figure 2. Experimental determination of PFR, one-signal method

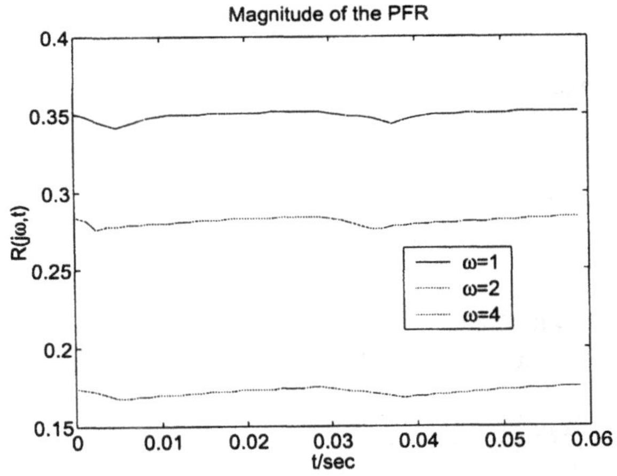

Figure 4. Magnitude of the PFR in dependence of the parameter t

shown that such a system can be described by a system of linear differential equations with periodically varying parameters, where the period is that of the net. Furthermore, it was demonstrated that only the LTP system has the ability to describe the complex interaction between the power net and the converter control system. As investigation method in (Möllerstedt and Bernhardsson 2000) the harmonic transfer function was used, that appears as a series like (20),(21) for some harmonics.

With respect to (Möllerstedt and Bernhardsson 2000) the considerred system can be represented by scheme of Figure 3.

The PFR with input ΔV_{line} and output ΔI_{ac} was considered. The investigations show that the system loses stability for $K = 2.1$. In the paper (Möllerstedt and Bernhardsson 2000) the value $K = 2.23$ was calculated as limit value for stability. The discrepancy is surely originated in deviations of the various system parameters.

The Figures 4 and 5 show the dependence of the magnitude of the PFR $R(j\omega, t)$ on the parameter t for some values of ω. Figures 6 and 7 represent the real and imaginary parts of the first harmonics, namely $P_k(\omega)$ and $Q_k(\omega)$ as functions of ω for $k = 0, 1, 2$.

The figures point out that for low frequencies the instationary behaviour is weak.

The dependence on the frequency ω is presented in the Figures 6 and 7 for the steady-state PFR, calculated by the formulae (34) with k up to 2. The spectator can see that in the low frequency range the zeroth harmonic dominates. However, with increasing frequency of the net excitation its role decreases in such a degree as the role of the higher frequent parts of the PFR grow up.

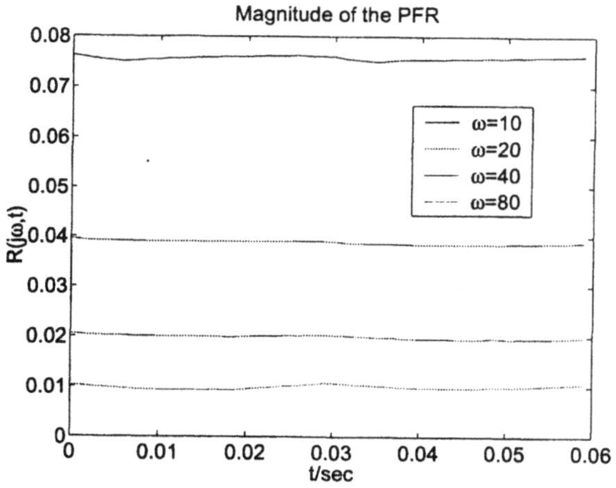

Figure 5. Magnitude of the PFR in dependence of the parameter t

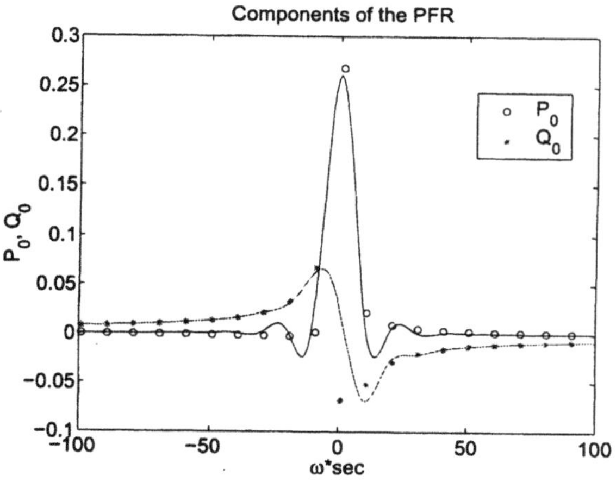

Figure 6. Real and imaginary part of the constant contribution to the PFR

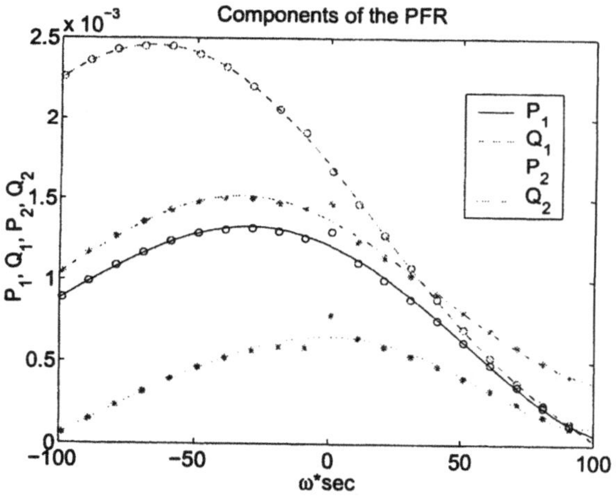

Figure 7. Real and imaginary part of the 1. and 2. harmonic components of the PFR

ACKNOWLEDGMENTS

The work has been supported by the German Science Foundation (DFG). This support is very gratefully acknowledged.

4. REFERENCES

Lampe, B.P. and U. Richter (1997). Experimental investigation of parametric frequency response. In: *Proc. 4. Int. Symp. Methods Models Autom. Robotics.* Miedzyzdroje, Poland. pp. 341–344.

Lampe, B.P., E.N. Rosenwasser, S.K. Volovodov and A.V. Smolnikov (1999). Parametric frequency response of sampled-data systems – theory and experiment. In: *Proc. 5. European Control Conf.* Karlsruhe, Germany. pp. BP6-5, F1040-5.

Möllerstedt, E. and B. Bernhardsson (2000). Out of control because of harmonics. *IEEE Control Systems Magazine* **20**(4), 70–81.

Rosenwasser, E.N. (1970). *Vibrations of nonlinear systems, the method of integral equations.* Techn. Information Service. Springfield, VA.

Rosenwasser, E.N. (1973). *Periodically nonstationary control systems.* Nauka. Moscow. (in Russian).

Rosenwasser, E.N. (1977). *Lyapunov indices in linear control theory.* Nauka. Moscow. (in Russian).

Rosenwasser, E.N. and B.P. Lampe (1997). *Digitale Regelung in kontinuierlicher Zeit - Analyse und Entwurf im Frequenzbereich.* B.G. Teubner. Stuttgart.

Rosenwasser, E.N. and B.P. Lampe (2000). *Computer Controlled Systems - Analysis and Design with Process-orientated models.* Springer-Verlag. London, Berlin, Heidelberg.

Sommer, V.B., B.P. Lampe and E.N. Rosenwasser (1994). Experimental investigations of analog-digital control systems by frequency methods. *Automation and Remote Control* **55**(Part 2), 912–920.

Volovodov, S.K., B.P. Lampe, E.N. Rosenwasser and A.V. Smolnikov (2000). Frequency analysis of linear periodical systems - Theory and experiment. In: *Proc. 2nd IEEE Int. Conf. Control Oscill. Chaos.* Vol. 3. Saint Petersburg, Russia. pp. 408–413.

Volovodov, S.K., E.N. Rosenwasser and Smolnikov A.V. (1991). Device for determination of frequency response of sampled-data control systems. *Russian patent.* Disclosure and registration.

Yakubovich, V.A. and V.M. Starzhinskii (1975). *Linear differential equations with periodic coefficients.* Vol. 1. John Wiley & Sons. New York.

Zadeh, L.A. (1950). Frequency analysis of variable networks. *Proc. IRE* **39**(March), 291–299.

Zadeh, L.A. (1951). Stability of linear varying-parameter systems. *J. Appl. Phys.* **22**(4), 202–204.

Copyright © IFAC Periodic Control Systems,
Cernobbio-Como, Italy, 2001

PERIODIC INVARIANT SUBSPACES IN CONTROL

Wen-Wei Lin *, Paul Van Dooren **, Quan-Fu Xu ***

* *Department of Mathematics*
Tsing Hua University, Hsin-Chu
Taiwan, R.O.C.
wwlin@am.nthu.edu.tw
** *Centre for Engineering Systems and Applied Mechanics*
Université catholique de Louvain
B-1348 Louvain-la-Neuve, Belgium
VDooren@csam.ucl.ac.be
*** *Department of Mathematics*
Tsing Hua University, Hsin-Chu
Taiwan, R.O.C.

Abstract: In this paper we present several different characterizations of invariant subspaces of periodic eigenvalue problems. We analyze their equivalence and discuss their use in control theory. *Copyright ©2001 IFAC*

Keywords: Periodic systems, discrete-time systems, canonical forms, eigenvectors, eigenvalues

1. MAIN RESULTS

Consider the (homogeneous) linear time varying system :

$$E_k x_{k+1} = A_k x_k + B_k u_k, \qquad k \in \mathbf{N} \qquad (1)$$

where $\mathbf{N}$ is the set of natural numbers, x_k is an n-dimensional vector of descriptor variables, u_k is an m-dimensional vector of input variables, the matrices E_k and A_k are $n \times n$, and B_k is $n \times m$. This system is said to be periodic with period K if $E_k = E_{k+K}$, $A_k = A_{k+K}$ and $B_k = B_{k+K}$, for all $k \in \mathbf{N}$, and K is the smallest positive integer for which this holds. If we allow both the E_k and A_k matrices to be singular, then x_k may still uniquely be defined in the context of a two point boundary value problem, as e.g. in optimal control of periodic systems. It is shown in (Sreedhar and Van Dooren, 1999) that a necessary and sufficient condition for this is that the pencil

$$\lambda\mathcal{E} - \mathcal{A} := \begin{bmatrix} \lambda E_1 & & & -A_1 \\ -A_2 & \lambda E_2 & & \\ & \ddots & \ddots & \\ & & -A_K & \lambda E_K \end{bmatrix} \qquad (2)$$

is regular (i.e. $\det(\lambda\mathcal{E} - \mathcal{A}) \not\equiv 0$). We call such periodic systems *regular*.

We can always define a periodic similarity transformation of a regular periodic system (1.1) by multiplying it from the left by the invertible transformation S_k and substituting x_k by $x_k \doteq T_{k-1}\hat{x}_k$, where S_k and T_k are again K-periodic. Let us denote diagonal block transformations $\mathcal{S}$ and $\mathcal{T}$ as follows :

$$\mathcal{S} \doteq \begin{bmatrix} S_1 & & & \\ & S_2 & & \\ & & \ddots & \\ & & & S_K \end{bmatrix}, \ \mathcal{T} \doteq \begin{bmatrix} T_1 & & & \\ & T_2 & & \\ & & \ddots & \\ & & & T_K \end{bmatrix}.$$

Defining the *periodic similarity transformation*

$$\mathcal{S}\begin{bmatrix} \lambda E_1 & & & -A_1 \\ -A_2 & \lambda E_2 & & \\ & \ddots & \ddots & \\ & & -A_K & \lambda E_K \end{bmatrix}\mathcal{T}$$

$$\doteq \begin{bmatrix} \lambda \hat{E}_1 & & & -\hat{A}_1 \\ -\hat{A}_2 & \lambda \hat{E}_2 & & \\ & \ddots & \ddots & \\ & & -\hat{A}_K & \lambda \hat{E}_K \end{bmatrix} \quad (3)$$

one then obtains a new system

$$\hat{E}_k \hat{x}_{k+1} = \hat{A}_k \hat{x}_k + \hat{B}_k u_k, \qquad k \in \mathbf{N}, \quad (4)$$

where $\hat{B}_k \doteq S_k B_k$.

One important case of such a periodic similarity transformation is the so-called "Floquet transform" of a periodic system :

$$\begin{bmatrix} \lambda \hat{E}_1 & & & -\hat{A}_1 \\ -\hat{A}_2 & \lambda \hat{E}_2 & & \\ & \ddots & \ddots & \\ & & -\hat{A}_K & \lambda \hat{E}_K \end{bmatrix} = \begin{bmatrix} \lambda I & & & -M \\ -M & \lambda I & & \\ & \ddots & \ddots & \\ & & -M & \lambda I \end{bmatrix} (5)$$

This reduces the homogeneous problem to a time-invariant one, but the transformation may not always exist (Van Dooren and Sreedhar, 1994; Sreedhar and Van Dooren, 1997). Another important case is the Periodic Schur Form (Bojanczyk et al., 1992), where the transformation matrices S_k and T_k are constrained to be unitary, and the resulting matrices $\hat{E}_k$ and $\hat{A}_k$ can all be chosen upper triangular. This second form always exists and can be computed in a numerically stable manner (Bojanczyk et al., 1992).

Both forms are closely linked to the concept of ***periodic invariant subspaces***, which play a fundamental role in the solution of important control problem of periodic systems (Sreedhar and Van Dooren, 1994), (Sreedhar and Van Dooren, 1997), (Ferng et al., 1998) :

- solution of periodic Riccati equations and their use in optimal and robust control,
- solution of periodic Lyapunov equations and their use in stability analysis,
- solution of periodic Sylvester equations and their use in decoupling,
- inverse eigenvalue problems and their use in pole placement.

We now describe several characterizations of invariant subspaces of a regular periodic system. A first characterization uses the classical concept of deflating subspace, directly applied to the pencil (2) (we assume $K = 3$ for the sake of simplicity) :

$$\begin{bmatrix} \lambda E_1 & & -A_1 \\ -A_2 & \lambda E_2 & \\ & -A_3 & \lambda E_3 \end{bmatrix}\begin{bmatrix} X_1 \\ X_2 \\ X_3 \end{bmatrix} = \begin{bmatrix} Y_1 \\ Y_2 \\ Y_3 \end{bmatrix}(\lambda I - M)(6)$$

where $X \doteq \begin{bmatrix} X_1^T & X_2^T & X_3^T \end{bmatrix}^T, Y \doteq \begin{bmatrix} Y_1^T & Y_2^T & Y_3 \end{bmatrix}^T$ are assumed to have full rank d. With

$$\mathcal{X} \doteq \begin{bmatrix} X_1 & & \\ & X_2 & \\ & & X_3 \end{bmatrix}, \mathcal{Y} \doteq \begin{bmatrix} Y_1 & & \\ & Y_2 & \\ & & Y_3 \end{bmatrix}$$

we can also rewrite this as follows :

$$\begin{bmatrix} \lambda E_1 & & -A_1 \\ -A_2 & \lambda E_2 & \\ & -A_3 & \lambda E_3 \end{bmatrix}\mathcal{X} = \mathcal{Y}\begin{bmatrix} \lambda I & & -M \\ -M & \lambda I & \\ & -M & \lambda I \end{bmatrix}(7)$$

so that every d-dimensional eigenspace of the type (6) induces a $d{\cdot}K$-dimensional eigenspace of the type (7), *provided* all individual matrices X_k and Y_k have full column rank. The following example shows that one must impose certain conditions on M, since this does not hold in general.

Example 1 Consider $K = 3$ and $A_k = E_k = I_2$. And let $\mathbf{1} = (1,1)^T, M = \begin{bmatrix} 1 & 0 \\ 0 & \omega \end{bmatrix}$ with $\omega \neq 1$ and $\omega^3 = 1$. Then we have

$$\begin{bmatrix} \lambda I_2 & 0 & -I_2 \\ -I_2 & \lambda I_2 & 0 \\ 0 & -I_2 & \lambda I_2 \end{bmatrix}\begin{bmatrix} \mathbf{1} & \omega^2\mathbf{1} \\ \mathbf{1} & \omega^1\mathbf{1} \\ \mathbf{1} & \omega^0\mathbf{1} \end{bmatrix}$$

$$= \begin{bmatrix} \mathbf{1} & \omega^2\mathbf{1} \\ \mathbf{1} & \omega^1\mathbf{1} \\ \mathbf{1} & \omega^0\mathbf{1} \end{bmatrix}\left(\lambda I_2 - \begin{bmatrix} 1 & 0 \\ 0 & \omega \end{bmatrix}\right).$$

So (6) and (7) hold with

$$\begin{cases} X_1 = Y_1 = [\mathbf{1}, \omega^2\mathbf{1}] \\ X_2 = Y_2 = [\mathbf{1}, \omega^1\mathbf{1}] \\ X_3 = Y_3 = [\mathbf{1}, \omega^0\mathbf{1}]. \end{cases}$$

Although both the matrices

$$\begin{bmatrix} X_1 \\ X_2 \\ X_3 \end{bmatrix} \quad \text{and} \quad \begin{bmatrix} Y_1 \\ Y_2 \\ Y_3 \end{bmatrix}$$

are of full column rank 2, all X_k and Y_k matrices are only of rank 1. □

This example illustrates well that M cannot be arbitrary. The following theorem characterizes which matrix M will yield full column rank matrices X_k and Y_k.

Theorem 1 Let A_k and E_k be nonsingular $n \times n$ matrices, for $k = 1, 2, \cdots, K$. Assume that $X = [X_1^T, X_2^T, \cdots, X_K^T]^T$ is a full column rank matrix that generates a d-dimensional deflating subspace of $\lambda\mathcal{E} - \mathcal{A}$:

$$\lambda\mathcal{E} - \mathcal{A} \begin{bmatrix} X_1 \\ X_2 \\ \vdots \\ X_K \end{bmatrix} = \begin{bmatrix} Y_1 \\ Y_2 \\ \vdots \\ Y_K \end{bmatrix} (\lambda I - M). \quad (8)$$

If M has the property that $\lambda^K \neq \mu^K$ whenever λ and μ are two distinct eigenvalues of M, i.e.,

$$Ker(M - \lambda I) = Ker\left(M^K - \lambda^K I\right), \quad (9)$$

$\forall \lambda \in \sigma(M)$ (the spectrum of M), then the X_k and Y_k matrices also have full column rank d, for $k = 1, 2, \cdots, K$.

Proof. Expanding (8), yields

$$E_k X_k = Y_k \quad \text{and} \quad A_k X_{k-1} = Y_k M, \quad (10)$$

for $k = 1, 2, \cdots, K$. Thus we have

$$A_k X_{k-1} = E_k X_k M, \quad \forall k. \quad (11)$$

Since the E_k matrices are nonsingular, we can define $S_k = E_k^{-1} A_k$. Then (11) is equivalent to

$$\begin{bmatrix} 0 & \cdots & 0 & S_1 \\ S_2 & \ddots & & 0 \\ & \ddots & \ddots & \vdots \\ 0 & & S_k & 0 \end{bmatrix} \begin{bmatrix} X_1 \\ X_2 \\ \vdots \\ X_k \end{bmatrix} = \begin{bmatrix} X_1 \\ X_2 \\ \vdots \\ X_K \end{bmatrix} M$$

or also

$$\begin{bmatrix} 0 & \cdots & 0 & S_1 \\ S_2 & \ddots & & 0 \\ & \ddots & \ddots & \vdots \\ 0 & & S_k & 0 \end{bmatrix} \mathcal{X} = \mathcal{X} \begin{bmatrix} 0 & \cdots & 0 & M \\ M & \ddots & & 0 \\ & \ddots & \ddots & \vdots \\ 0 & & M & 0 \end{bmatrix} . (12)$$

Note that M is nonsingular, since the A_k matrices are also nonsingular.

We now show by contradiction that the matrices X_k are full rank if (9) holds. Assume e.g. that X_1 does *not* have full column rank and that the columns of the matrix V span $\mathcal{V} := KerX_1$. Then it follows from (12) that $X_2 M V = S_2 X_1 V = 0$ and hence $M\mathcal{V} \subset KerX_2$. By similar arguments one shows that $M^k\mathcal{V} \subset KerX_{k+1}$ for $k = 2, \cdots, K-1$ and finally, $M^K\mathcal{V} \subset KerX_1 = \mathcal{V}$. So $\mathcal{V}$ is an invariant subspace of M^K and it must contain an eigenvector v corresponding to some eigenvalue ξ of M^K. Since the eigenvalues of M^K are the K-th powers of those of M, there exists a value $\lambda \in \sigma(M)$ such that $\xi = \lambda^K$. The assumption (9) then implies that $(M - \lambda I)v = 0$ and hence v is also an eigenvector of M. Since $\lambda \neq 0$ we found a vector v which is in the kernel of all matrices X_k and hence also of X. So we proved that $rank(X) = d$ implies $rank(X_1) = d$. The same proof also holds for all other matrices X_k since a circulant permutation of column and row blocks in (12) can bring any block X_k to the first position.

Observing (10) and using that E_k is nonsingular, it is obvious that Y_k is also of full column rank d. This completes the proof. □

Next, we want to link the system of equations (7) with one which only involves X_k rather than X_k and Y_k. For this, we first rewrite (7) in a slight more general form, which we will retrieve also later on:

$$\begin{cases} E_1 X_1 = Y_1 \bar{E}_1 \\ E_2 X_2 = Y_2 \bar{E}_2 \\ E_3 X_3 = Y_3 \bar{E}_3 \end{cases} ; \qquad \begin{cases} A_1 X_3 = Y_1 \bar{A}_1 \\ A_2 X_1 = Y_2 \bar{A}_2 \\ A_3 X_2 = Y_3 \bar{A}_3 \end{cases} , (13)$$

or

$$\begin{bmatrix} \lambda E_1 & & -A_1 \\ -A_2 & \lambda E_2 & \\ & -A_3 & \lambda E_3 \end{bmatrix} \mathcal{X} = \mathcal{Y} \begin{bmatrix} \lambda \bar{E}_1 & & -\bar{A}_1 \\ -\bar{A}_2 & \lambda \bar{E}_2 & \\ & -\bar{A}_3 & \lambda \bar{E}_3 \end{bmatrix} (14)$$

where the regularity of $\lambda\mathcal{E} - \mathcal{A}$ implies the regularity of

$$\lambda\bar{\mathcal{E}} - \bar{\mathcal{A}} = \begin{bmatrix} \lambda \bar{E}_1 & & -\bar{A}_1 \\ -\bar{A}_2 & \lambda \bar{E}_2 & \\ & -\bar{A}_3 & \lambda \bar{E}_3 \end{bmatrix},$$

provided the X_k and Y_k matrices are of full column rank d. Therefore, we have for all k:

$$\text{rank} \underbrace{[\bar{E}_k \ \bar{A}_k]}_{2d} = d.$$

And hence there exists a full column rank right null space, which we partition as follows:

$$[\bar{E}_k \ \bar{A}_k] \begin{bmatrix} \tilde{A}_k \\ -\tilde{E}_k \end{bmatrix} = 0. \quad (15)$$

Notice that for invertible matrices $\bar{E}_k, \tilde{E}_k$ this is like " swapping " factors since (15) implies

$$\bar{E}_k^{-1} \bar{A}_k = \tilde{A}_k \tilde{E}_k^{-1}.$$

Multiplying the left equations of (13) by $\tilde{A}_k$ and the right ones by $\tilde{E}_k$ and equating corresponding terms yields finally:

$$\begin{cases} E_1 X_1 \tilde{A}_1 = A_1 X_3 \tilde{E}_1 \\ E_2 X_2 \tilde{A}_2 = A_2 X_1 \tilde{E}_2 \\ E_3 X_1 \tilde{A}_3 = A_3 X_2 \tilde{E}_3 \end{cases} , \quad (16)$$

which does not involve Y_k anymore. To go from (16) to (13) again, we rewrite (16) as

$$[E_k X_k \quad A_k X_{k-1}] \begin{bmatrix} \tilde{A}_k \\ -\tilde{E}_k \end{bmatrix} = 0,$$

which indicates that $[E_kX_k \quad A_kX_{k-1}]$ is in the left null space of $\begin{bmatrix} \tilde{A}_k \\ -\tilde{E}_k \end{bmatrix}$. But a basis for that is given by $\begin{bmatrix} \bar{E}_k & \bar{A}_k \end{bmatrix}$. So there exists a matrix Y_k such that

$$[E_kX_k \quad A_kX_{k-1}] = Y_k \left[\bar{E}_k \quad \bar{A}_k\right], \qquad (17)$$

which yields again (13). From (14) it follows then that the Y_k must be full rank or otherwise we would have

$$\dim(\mathcal{E}\mathcal{V} + \mathcal{A}\mathcal{V}) < \dim \mathcal{V},$$

for

$$\mathcal{V} = Im \begin{bmatrix} X_1 & & \\ & X_2 & \\ & & X_3 \end{bmatrix},$$

which would mean $\lambda\mathcal{E} - \mathcal{A}$ is singular. Finally we point out that (17) can be expressed geometrically as the condition

$$\dim(E_k\mathcal{V}_k + A_k\mathcal{V}_{k-1}) = d = \dim \mathcal{V}_k,$$

for $\mathcal{V}_k = Im(X_k)$, because $rank\left[\bar{E}_k \quad \bar{A}_k\right] = d$.

To go from (13) to (7), we replace X_k with $\tilde{X}_k$ and Y_k with $\tilde{Y}_k$ in (13):

$$\begin{cases} E_1\tilde{X}_1 = \tilde{Y}_1\bar{E}_1 \\ E_2\tilde{X}_2 = \tilde{Y}_2\bar{E}_2 \\ E_3\tilde{X}_3 = \tilde{Y}_3\bar{E}_3 \end{cases}; \quad \begin{cases} A_1\tilde{X}_3 = \tilde{Y}_1\bar{A}_1 \\ A_2\tilde{X}_1 = \tilde{Y}_2\bar{A}_2 \\ A_3\tilde{X}_2 = \tilde{Y}_3\bar{A}_3 \end{cases}. \quad (18)$$

Notice that the $\bar{E}_k$ and $\bar{A}_k$ matrices are nonsingular, since the $E_k, A_k, \tilde{X}_k$, and $\tilde{Y}_k$ matrices are all full rank. Define

$$\tilde{M}_k = \bar{E}_k^{-1}\bar{A}_k, \quad \text{for} \quad k = 1, 2, 3.$$

And let M be a K-th root (K=3) of nonsingular matrix $\tilde{M}_3\tilde{M}_2\tilde{M}_1$. Then (18) induces (7) with

$$\begin{cases} X_k = \tilde{X}_k\left(\tilde{M}_k \cdots \tilde{M}_1 M^{-k}\right) \\ Y_k = \tilde{Y}_k\bar{E}_k\left(\tilde{M}_k \cdots \tilde{M}_1 M^{-k}\right), \text{ for } k = 1, 2, 3, \end{cases}$$

since then

$$\begin{aligned} E_kX_k &= E_k\tilde{X}_k\left(\tilde{M}_k \cdots \tilde{M}_1 M^{-k}\right) \\ &= \tilde{Y}_k\bar{E}_k\left(\tilde{M}_k \cdots \tilde{M}_1 M^{-k}\right) \\ &= Y_k; \end{aligned}$$

and

$$\begin{aligned} A_kX_{k-1} &= A_k\tilde{X}_{k-1}\left(\tilde{M}_{k-1} \cdots \tilde{M}_1 M^{-(k-1)}\right) \\ &= \tilde{Y}_k\bar{A}_k\left(\tilde{M}_{k-1} \cdots \tilde{M}_1 M^{-(k-1)}\right) \\ &= \tilde{Y}_k\left(\bar{E}_k\tilde{M}_k\right)\left(\tilde{M}_{k-1} \cdots \tilde{M}_1 M^{-(k-1)}\right) \\ &= \tilde{Y}_k\bar{E}_k\left(\tilde{M}_k \cdots \tilde{M}_1 M^{-k}\right) \cdot M \\ &= Y_kM, \end{aligned}$$

which yield (7). Moreover, in the extraction of M such that $M^3 = \tilde{M}_3\tilde{M}_2\tilde{M}_1$, we can take any complex number λ that satisfies $\lambda^3 = \xi$ (here ξ is an eigenvalue of $\tilde{M}_3\tilde{M}_2\tilde{M}_1$) as an eigenvalue of M (see (Gantmacher, 1959)). Thus, in particular, M can be chosen to satisfy the assumption of Theorem 1. That is, for each eigenvalue ξ of $\tilde{M}_3\tilde{M}_2\tilde{M}_1$, we extract a fixed number λ from the set of cubic roots of ξ as the only candidate for entering into the set of eigenvalues of M, even if there are multiple Jordan blocks corresponding to ξ.

With M satisfying the assumption of Theorem 1, all X_k and Y_k matrices in (7) are of full column rank d. Now we construct unitary matrices Q_k and Z_k such that

$$Z_k^*X_k = \hat{X}_k = \begin{bmatrix} R_k \\ 0 \end{bmatrix} \}d$$

and

$$Q_k^*Y_k = \hat{Y}_k = \begin{bmatrix} S_k \\ 0 \end{bmatrix} \}d,$$

where the R_k and S_k matrices are square invertible. Putting these transformations in block form :

$$\mathcal{Q} \doteq \begin{bmatrix} Q_1 & & & \\ & Q_2 & & \\ & & \ddots & \\ & & & Q_K \end{bmatrix}, \quad \mathcal{Z} \doteq \begin{bmatrix} Z_1 & & & \\ & Z_2 & & \\ & & \ddots & \\ & & & Z_K \end{bmatrix},$$

we apply the block transformation to the cyclic pencil

$$\mathcal{Q}^*(\lambda\mathcal{E} - \mathcal{A})\mathcal{Z} = \lambda\hat{\mathcal{E}} - \hat{\mathcal{A}},$$

and we find in the new coordinate system that

$$\begin{bmatrix} \lambda\hat{E}_1 & & -\hat{A}_1 \\ -\hat{A}_2 & \lambda\hat{E}_2 & \\ & -\hat{A}_3 & \lambda\hat{E}_3 \end{bmatrix} \hat{\mathcal{X}} = \hat{\mathcal{Y}} \begin{bmatrix} \lambda I & & -M \\ -M & \lambda I & \\ & -M & \lambda I \end{bmatrix},$$

with

$$\hat{\mathcal{X}} \doteq \begin{bmatrix} \hat{X}_1 & & \\ & \hat{X}_2 & \\ & & \hat{X}_3 \end{bmatrix}, \quad \hat{\mathcal{Y}} \doteq \begin{bmatrix} \hat{Y}_1 & & \\ & \hat{Y}_2 & \\ & & \hat{Y}_3 \end{bmatrix}.$$

This indicates that in this coordinate system the $\hat{E}_k$ and $\hat{A}_k$ matrices are upper block triangular:

$$\hat{E}_k = \left[\begin{array}{c|c} S_kR_k^{-1} & * \\ \hline 0 & * \end{array}\right], \hat{A}_k = \left[\begin{array}{c|c} S_kMR_{k-1}^{-1} & * \\ \hline 0 & * \end{array}\right].$$

So (6) induces a block triangular periodic Schur decomposition, provided M satisfies the assumption of Theorem 1.

We thus closed the following set of equivalence relations for pencils with invertible matrices E_k, A_k.

R1 $$\begin{bmatrix} \lambda E_1 & & -A_1 \\ -A_2 & \lambda E_2 & \\ & -A_3 & \lambda E_3 \end{bmatrix} \begin{bmatrix} X_1 \\ X_2 \\ X_3 \end{bmatrix} = \begin{bmatrix} Y_1 \\ Y_2 \\ Y_3 \end{bmatrix} (\lambda I - M)$$

with M satisfying the assumption of Theorem 1, X_k and Y_k matrices having full column rank d.

R2 $\begin{bmatrix} \lambda E_1 & & -A_1 \\ -A_2 & \lambda E_2 & \\ & -A_3 & \lambda E_3 \end{bmatrix} \tilde{\mathcal{X}} = \tilde{\mathcal{Y}} \begin{bmatrix} \lambda \bar{E}_1 & & -\bar{A}_1 \\ -\bar{A}_2 & \lambda \bar{E}_2 & \\ & -\bar{A}_3 & \lambda \bar{E}_3 \end{bmatrix}$,
where $M^3 = \bar{E}_3^{-1}\bar{A}_3\bar{E}_2^{-1}\bar{A}_2\bar{E}_1^{-1}\bar{A}_1$,
$\mathrm{Im}\left(\tilde{X}_k\right)=\mathrm{Im}(X_k)$, and $\mathrm{Im}(Y_k)=\mathrm{Im}(\bar{Y}_k)$,
for $k = 1, 2, 3$.

R3 $\begin{cases} E_1\tilde{X}_1\bar{A}_1 = A_1\tilde{X}_3\bar{E}_1 \\ E_2\tilde{X}_2\bar{A}_2 = A_2\tilde{X}_1\bar{E}_2 \\ E_3\tilde{X}_3\bar{A}_3 = A_3\tilde{X}_2\bar{E}_3 \end{cases}$,
where $\bar{E}_k\bar{A}_k = \bar{A}_k\bar{E}_k$, for $k = 1, 2, 3$.

R4 $\dim (E_k\mathcal{V}_k + A_k\mathcal{V}_{k-1}) = d = \dim \mathcal{V}_k$,
where $\mathcal{V}_k = Im\,(X_k) = Im\left(\tilde{X}_k\right)$,
for $k = 1, 2, 3$.

R5 There exist a block triangular periodic Schur decomposition with leading $d \times d$ blocks.

We can also show that relations (R3), (R4), and (R5) are still valid for E_k, A_k arbitrary for as long as $\lambda\mathcal{E} - \mathcal{A}$ is regular. In this case (R1) and (R2) only apply to the invariant subspaces with finite nonzero reduced spectrum.

Example 2 The pencil $(n = 1, K = 3)$

$$\lambda \begin{bmatrix} 1 & 0 & 0 \\ 0 & 1 & 0 \\ 0 & 0 & 1 \end{bmatrix} - \begin{bmatrix} 0 & 0 & 0 \\ 1 & 0 & 0 \\ 0 & 1 & 0 \end{bmatrix}$$

is clearly regular with triple eigenvalue 0. It is already in triangular periodic Schur form, but (R1) with $M = 0$ $(d = 1)$ has

$$X_1 = X_2 = 0 \quad \text{and} \quad Y_1 = Y_2 = 0.$$

So (R1) is not equivalent to (R5) in the case, neither is (R2). □

Finally, this also allows to give a new definition of eigenvalue/eigenvector pairs for periodic pencils. Proofs of the validity of these definition are given in the full paper.

Definition 1 Let $((A_k, E_k))_{k=1}^K$ be regular periodic $n \times n$ matrix pairs. If there exist complex numbers $\alpha_1, \cdots, \alpha_K$ and $\beta_1, \cdots, \beta_K$ such that $\left(\prod_{j=1}^K \alpha_j, \prod_{j=1}^K \beta_j\right) \neq (0, 0)$ and

$$\begin{bmatrix} \alpha_1 E_1 & & & -\beta_1 A_1 \\ -\beta_2 A_2 & \alpha_2 E_2 & & \\ & \ddots & \ddots & \\ & & -\beta_K A_K & \alpha_K E_K \end{bmatrix} \qquad (19)$$

is singular, then we say $< \prod_{j=1}^K \alpha_j, \prod_{j=1}^K \beta_j >$ is an eigenvalue of $(A_k, E_k)_{k=1}^K$. The set of eigenvalues of $(A_k, E_k)_{k=1}^K$ is denoted as $\sigma\,(A_k, E_k)_{k=1}^K$. □

Definition 2 Let $(A_k, E_k)_{k=1}^K$ be regular periodic $n \times n$ matrix pairs. If there exist complex numbers $\alpha_1, \cdots, \alpha_K, \beta_1, \cdots, \beta_K$, and *nonzero* vectors $x_1, \cdots, x_K$ such that

$$\beta_k A_k x_{k-1} = \alpha_k E_k x_k, \text{ for } k = 1, 2, \cdots, K \quad (20)$$

with $(\prod_{j=1}^K \alpha_j, \prod_{j=1}^K \beta_j) \neq (0, 0)$, we say that $(x_k)_{k=1}^K$ is an eigenvector sequence of $((A_k, E_k))_{k=1}^K$ with eigenvalue $< \prod_{j=1}^K \alpha_j, \prod_{j=1}^K \beta_j >$. □

ACKNOWLEDGMENTS

This paper presents research supported by NSF contract CCR-97-96315 and by the Belgian Programme on Inter-university Poles of Attraction, initiated by the Belgian State, Prime Minister's Office for Science, Technology and Culture. The scientific responsibility rests with its authors.

REFERENCES

A. Bojanczyk, G. Golub, and P. Van Dooren, "The periodic Schur decomposition. Algorithms and applications", in *Proceedings of the SPIE Conference*, San Diego, Vol. 1770, July 1992, pp. 31-42.

W. R. Ferng, W.-W. Lin, and C.-S. Wang, "On computing the stable deflating subspaces of cyclic symplectic matrix pairs", submitted for publication, 1998.

F. R. Gantmacher, *The Theory of Matrices* Vol. I, Chelsea, New York, 1959.

J. Sreedhar, P. Van Dooren, "A Schur approach for solving some periodic matrix equations," in *Systems and Networks : Mathematical Theory and Applications*, pp. 339-362, Eds. U. Helmke, R. Mennicken, J. Saurer, *Mathematical Research*, Vol 77, Akademie Verlag, Berlin 1994.

J. Sreedhar, P. Van Dooren, "Periodic descriptor systems : solvability and conditionability", *IEEE Trans. Aut. Contr.*, pp. 310-313, February 1999.

J. Sreedhar, P. Van Dooren, "Forward/backward decomposition of periodic descriptor systems and two point boundary value problems", in *Proceedings of the European Control Conf.*, paper FR-A-L7, 1997.

P. Van Dooren and J. Sreedhar, "When is a periodic discrete-time system equivalent to a time-invariant one ?", *Linear Algebra & Applications*, Vol. 212/213, pp. 131-152, 1994.

Copyright © IFAC Periodic Control Systems,
Cernobbio-Como, Italy, 2001

ON THE PERIODIC REALISATION OF TRANSFER MATRICES

D. C. McLernon and D. A. Wilson

The School of Electronic and Electrical Engineering
The University of Leeds
Leeds LS2 9JT, UK

Abstract: It is understood how to transform a linear periodically time-varying (LPTV) filter/difference equation into an equivalent multiple-input/multiple-output (MIMO) structure, or transfer matrix, but no published method exists for the reverse operation. This paper now presents a technique to return from the MIMO structure to the original LPTV difference equation. This new result is then used to 'approximate' a MIMO structure with a single LPTV filter, represent parallel and cascade connections of LPTV systems as single LPTV filters, implement order reduction of a LPTV difference equation, and finally obtain a LPTV difference equation representation that is equivalent to a LPTV state-space structure. *Copyright© 2001 IFAC*

Keywords: Periodic structures, time-varying systems, linear filters, multi-input/multi-output, order reduction, state-space realisation.

1. INTRODUCTION

While the modern origins of research into linear periodically time-varying (LPTV) continuous-time filters/systems go back over forty years (Jury and Mullin, 1958; Jury and Mullin, 1959), the first investigation of LPTV digital filters was a short paper (Fjällbrant, 1970) - not widely referenced. While these filters can be either one- or two-dimensional (McLernon, 1999; McLernon and King, 1990), the growing interest in one-dimensional LPTV filters/systems has been driven by applications in communications where signals with cyclostationary statistics are often observed and processed (Gelli *et al.*, 1998). Even H_∞, a framework traditionally used by control engineers to address robustness issues, has recently been employed in conjunction with LPTV filters (Xie *et al.*, 2000).

In (McLernon, 1999), all the interrelationships between the various equivalent structures for a LPTV filter were derived, and examples were given to show how this could assist the analysis of LPTV cascade, parallel and inverse filtering configurations. One well-known equivalent structure for a LPTV filter is known as the multiple-input/multiple-output (MIMO) realisation, $\mathbf{H}(z)$, or the *transfer matrix*. But what has not been understood before, is how to then go backwards from $\mathbf{H}(z)$ to the equivalent LPTV difference equation. This new result will now be derived, and its implication in analysing and simplifying different LPTV configurations will also be presented. It should be mentioned that other researchers, including (Lin and King, 1993; Colaneri and Longhi, 1995), have tackled the problem of obtaining the equivalent minimal LPTV state-space representation, when one is given the elements of $\mathbf{H}(z)$.

2. THE PROBLEM – HOW TO GO FROM MIMO MODEL TO LPTV DIFFERENCE EQUATION?

Consider the general LPTV difference equation (period-N) for the direct-form-I structure (as opposed to direct-form-II, lattice, etc.):

$$y(n) = \sum_{i=0}^{M_1} a_i(n)x(n-i) + \sum_{j=1}^{M_2} b_j(n)y(n-j) \qquad (1)$$

$$a_i(n+N) = a_i(n), b_j(n+N) = b_j(n)$$

Now to simplify nomenclature, if we let

$$x_k(n) \triangleq x(nN+k),\ y_k(n) \triangleq y(nN+k)$$
$$a_{ik} \triangleq a_i(nN+k), b_{jk} \triangleq b_j(nN+k)$$
$$-\infty \le n \le \infty, 0 \le k \le N-1$$

then after taking the z-transform and some manipulation, we can write:

$$\mathbf{Y}(z) = \mathbf{B}^{-1}(z)\mathbf{A}(z)\mathbf{X}(z) = \mathbf{H}(z)\mathbf{X}(z) \tag{2}$$

where we have the following:

$$\mathbf{Y}(z) \triangleq [Y_0(z)Y_1(z) \cdots Y_{N-1}(z)]^{\mathrm{T}}$$
$$\mathbf{X}(z) \triangleq [X_0(z)X_1(z) \cdots X_{N-1}(z)]^{\mathrm{T}}$$
$$y_k(n) \leftrightarrow Y_k(z),\ x_\ell(n) \leftrightarrow X_\ell(z)$$
$$\mathbf{A}(z) \triangleq \{A_{k\ell}(z)\}_{k,\ell=0,0}^{N-1,N-1},\ \mathbf{B}(z) \triangleq \{B_{k\ell}(z)\}_{k,\ell=0,0}^{N-1,N-1}$$
$$\mathbf{H}(z) \triangleq \{H_{k\ell}(z)\}_{k,\ell=0,0}^{N-1,N-1} = \{N_{k\ell}(z)/D(z)\}_{k,\ell=0,0}^{N-1,N-1}$$

Now $A_{k\ell}(z)$ is a polynomial in z^{-1} with coefficients formed from $\{a_{ij}\}$, $B_{k\ell}(z)$ is a polynomial in z^{-1} with coefficients formed from $\{b_{ij}\}$, $N_{k\ell}(z)$ is a polynomial in z^{-1} with coefficients formed from $\{a_{ij}\}$ and $\{b_{ij}\}$, and $D(z) = 1 + \sum_{k=1}^{M_2} \beta_k z^{-k}$ with each coefficient β_k formed from $\{b_{ij}\}$. So we have gone from the LPTV difference equation in (1) to the MIMO equivalent structure $\mathbf{H}(z)$ in (2). That much is well known. But what is not understood is, if we know the elements of $\mathbf{H}(z)$ in (2), how do we find the LPTV coefficients $\{a_{ij}\}$ and $\{b_{ij}\}$ in (1)?

Consider first the obvious approach. Let the LPTV filter in (1) be second-order with period $N = 2$. Then from (2) we get

$$\mathbf{H}(z) = \mathbf{B}^{-1}(z)\mathbf{A}(z) =$$
$$\begin{bmatrix} 1-b_{20}z^{-1} & -b_{10}z^{-1} \\ -b_{11} & 1-b_{21}z^{-1} \end{bmatrix}^{-1} \begin{bmatrix} a_{00}+a_{20}z^{-1} & a_{10}z^{-1} \\ a_{11} & a_{01}+a_{21}z^{-1} \end{bmatrix}$$
$$= \begin{bmatrix} H_{00}(z) & H_{01}(z) \\ H_{10}(z) & H_{11}(z) \end{bmatrix} \tag{3}$$

where $H_{k\ell}(z) = N_{k\ell}(z)/D(z)$, with

$$N_{00}(z) = \sum_{i=0}^{2} \alpha_i^{00} z^{-i} = a_{00} + (a_{20} - a_{00}b_{21} + a_{11}b_{10})z^{-1} - a_{20}b_{21}z^{-2}$$
$$N_{01}(z) = \sum_{i=0}^{2} \alpha_i^{01} z^{-i} = (a_{10} + a_{01}b_{10})z^{-1} + (a_{21}b_{10} - a_{10}b_{21})z^{-2} \tag{4}$$
$$N_{10}(z) = \sum_{i=0}^{2} \alpha_i^{10} z^{-i} = (a_{11} + a_{00}b_{11}) + (a_{20}b_{11} - a_{11}b_{20})z^{-1}$$
$$N_{11}(z) = \sum_{i=0}^{2} \alpha_i^{11} z^{-i} = a_{01} + (a_{21} - a_{01}b_{20} + a_{10}b_{11})z^{-1} - a_{21}b_{20}z^{-2}$$
$$D(z) = \sum_{i=0}^{2} \beta_i z^{-i} = 1 - (b_{20} + b_{21} + b_{10}b_{11})z^{-1} + b_{20}b_{21}z^{-2}$$

So if we are given $\mathbf{H}(z)$ (i.e. the equivalent MIMO coefficients $\{\alpha_i^{k\ell}; i = 0:M_1, k\ell = 0,0:N-1,N-1\}$ and $\{\beta_i; i = 1:M_2\}$), one approach would be to attempt to solve 12 non-linear equations in 10 unknowns ($\{a_{ij}\}$ and $\{b_{ij}\}$) – i.e.

$$\begin{gathered} a_{00} = \alpha_0^{00} \\ (a_{20} - a_{00}b_{21} + a_{11}b_{10}) = \alpha_1^{00} \\ \vdots \\ -(b_{20} + b_{21} + b_{10}b_{11}) = \beta_1 \\ b_{20}b_{21} = \beta_2 \end{gathered} \tag{5}$$

But even for only a second-order LPTV filter with period $N = 2$, using various non-linear optimisation algorithms is cumbersome and global convergence is not guaranteed. But for higher-order filters and larger N, it becomes totally impractical. So is there a closed-form solution? The answer is yes.

3. PROPOSED SOLUTION

We are given the MIMO transfer matrix $\mathbf{H}(z)$ in (2) and we wish to find the corresponding LPTV coefficients in (1). In order to convey the basic principle, the derivation in this section will be for the particular case of a second-order LPTV filter in (1) with period $N = 2$. The general result (which is a simple extension) has been derived, but cannot be presented here due to lack of space.

So let us expand $\mathbf{A}(z)$, $\mathbf{B}(z)$ and $\mathbf{H}(z)$ in (2) and (3) as matrix polynomials in z^{-1}:

$$\mathbf{A}(z) = \begin{bmatrix} a_{00}+a_{20}z^{-1} & a_{10}z^{-1} \\ a_{11} & a_{01}+a_{21}z^{-1} \end{bmatrix}$$
$$= \begin{bmatrix} a_{00} & 0 \\ a_{11} & a_{01} \end{bmatrix} + \begin{bmatrix} a_{20} & a_{10} \\ 0 & a_{21} \end{bmatrix} z^{-1} = \mathbf{A}_0 + \mathbf{A}_1 z^{-1} \tag{6}$$

$$\mathbf{B}(z) = \begin{bmatrix} 1-b_{20}z^{-1} & -b_{10}z^{-1} \\ -b_{11} & 1-b_{21}z^{-1} \end{bmatrix}$$
$$= \begin{bmatrix} 1 & 0 \\ -b_{11} & 1 \end{bmatrix} + \begin{bmatrix} -b_{20} & -b_{10} \\ 0 & -b_{21} \end{bmatrix} z^{-1} = \mathbf{B}_0 + \mathbf{B}_1 z^{-1} \tag{7}$$

$$\mathbf{H}(z) = \mathbf{B}^{-1}(z)\mathbf{A}(z) = \frac{\mathbf{N}(z)}{D(z)}$$
$$= \frac{\begin{bmatrix} \alpha_0^{00} & \alpha_0^{01} \\ \alpha_0^{10} & \alpha_0^{11} \end{bmatrix} + \begin{bmatrix} \alpha_1^{00} & \alpha_1^{01} \\ \alpha_1^{10} & \alpha_1^{11} \end{bmatrix} z^{-1} + \begin{bmatrix} \alpha_2^{00} & \alpha_2^{01} \\ \alpha_2^{10} & \alpha_2^{11} \end{bmatrix} z^{-2}}{1 + \beta_1 z^{-1} + \beta_2 z^{-2}} \tag{8}$$
$$= \frac{\mathbf{N}_0 + \mathbf{N}_1 z^{-1} + \mathbf{N}_2 z^{-2}}{1 + \beta_1 z^{-1} + \beta_2 z^{-2}}$$

Then from (6) to (8)

$$\mathbf{H}(z)=\mathbf{B}^{-1}(z)\mathbf{A}(z)$$
$$\Rightarrow(\mathbf{A}_0+\mathbf{A}_1z^{-1})(1+\beta_1z^{-1}+\beta_2z^{-2}) \quad (9)$$
$$=(\mathbf{B}_0+\mathbf{B}_1z^{-1})(\mathbf{N}_0+\mathbf{N}_1z^{-1}+\mathbf{N}_2z^{-2})$$

and by equating powers of z^{-1} we can partition the resulting linear eqns in terms of the unknown block matrices $[\mathbf{A}_0\ \mathbf{A}_1]$ and $[\mathbf{B}_0\ \mathbf{B}_1]$ to get:

$$\begin{aligned}[\mathbf{A}_0\ \mathbf{A}_1]\begin{bmatrix}\mathbf{I} & \beta_1\mathbf{I}\\ \mathbf{0} & \mathbf{I}\end{bmatrix}&=[\mathbf{B}_0\ \mathbf{B}_1]\begin{bmatrix}\mathbf{N}_0 & \mathbf{N}_1\\ \mathbf{0} & \mathbf{N}_0\end{bmatrix}\\ [\mathbf{A}_0\ \mathbf{A}_1]\begin{bmatrix}\beta_2\mathbf{I} & \mathbf{0}\\ \beta_1\mathbf{I} & \beta_2\mathbf{I}\end{bmatrix}&=[\mathbf{B}_0\ \mathbf{B}_1]\begin{bmatrix}\mathbf{N}_2 & \mathbf{0}\\ \mathbf{N}_1 & \mathbf{N}_2\end{bmatrix}\end{aligned} \quad (10)$$

Solving for $[\mathbf{B}_0\ \mathbf{B}_1]$ gives

$$\mathbf{BM}=[\mathbf{B}_0\ \mathbf{B}_1]\mathbf{M}=\begin{bmatrix}1 & 0 & -b_{20} & -b_{10}\\ -b_{11} & 1 & 0 & -b_{21}\end{bmatrix}\mathbf{M}=\mathbf{0} \quad (11)$$

where

$$\mathbf{M}=\left(\begin{bmatrix}\mathbf{N}_0 & \mathbf{N}_1\\ \mathbf{0} & \mathbf{N}_0\end{bmatrix}\begin{bmatrix}\mathbf{I} & \beta_1\mathbf{I}\\ \mathbf{0} & \mathbf{I}\end{bmatrix}^{-1}\begin{bmatrix}\beta_2\mathbf{I} & \mathbf{0}\\ \beta_1\mathbf{I} & \beta_2\mathbf{I}\end{bmatrix}-\begin{bmatrix}\mathbf{N}_2 & \mathbf{0}\\ \mathbf{N}_1 & \mathbf{N}_2\end{bmatrix}\right) \quad (12)$$

After re-arranging this produces (in general) the following eight eqns in four unknowns

$$\begin{bmatrix}\mathbf{r}_4^T & \mathbf{r}_3^T & \mathbf{0} & \mathbf{0}\\ \mathbf{0} & \mathbf{0} & \mathbf{r}_1^T & \mathbf{r}_4^T\end{bmatrix}\begin{bmatrix}b_{10}\\ b_{20}\\ b_{11}\\ b_{21}\end{bmatrix}=\begin{bmatrix}\mathbf{r}_1^T\\ \mathbf{r}_2^T\end{bmatrix}\Rightarrow\mathbf{P}_b\mathbf{b}=\mathbf{b}_1 \quad (13)$$

where $\mathbf{r}_i$ is the i-th row of $\mathbf{M}$. From the Appendix, let the solution be

$$\mathbf{b}=\mathbf{P}_b^+\mathbf{b}_1 \quad (14)$$

where $\mathbf{P}_b^+$ represents the pseudo-inverse of $\mathbf{P}_b$.

Now consider (11). The two rows of $\mathbf{B}$ in $\mathbf{BM}=\mathbf{0}$ are two linearly independent left eigenvectors of $\mathbf{M}$ corresponding to two zero eigenvalues. This implies that the rank of $\mathbf{M}$ must be at least two less than its maximum possible rank. Since $\mathbf{M}$ is 4x4 we can say that if (11) is true then

$$\text{rank}(\mathbf{M})\le 2 \quad (15)$$

So (15) is a *necessary* (*but not sufficient*) condition[1] for an arbitrary second-order MIMO $\mathbf{H}(z)$ to be realised by a second-order LPTV difference eqn. in (1) with the feedback coefficients calculated from (14). Conversely, if (15) does not hold, then (1) cannot model $\mathbf{H}(z)$.

[1]The fact that (15) is only necessary and '*not sufficient*' for $\mathbf{H}(z)$ to be realised by the LPTV difference eqn. in (1), follows on two counts. Firstly, if a LPTV filter in (1) produces the MIMO structure $\mathbf{H}(z)$, this implies that (11) holds – not the other way around. Secondly, if (11) holds this implies that $\text{rank}(\mathbf{M})\le 2$. There may however be cases where $\text{rank}(\mathbf{M})\le 2$ but one cannot find suitable coefficients so that (11) holds with the appropriate structure for $[\mathbf{B}_0\ \mathbf{B}_1]$. Finally, in practice, (15) is necessary and almost sufficient – see the Conclusions.

Now using this solution for $\mathbf{b}$ from (14) in (10), we can then solve for $[\mathbf{A}_0\ \mathbf{A}_1]$ (eight eqns in six unknowns):

$$\begin{bmatrix}a_{00} & 0 & a_{20} & a_{10}\\ a_{11} & a_{01} & 0 & a_{21}\end{bmatrix}\left(\begin{bmatrix}\mathbf{I} & \beta_1\mathbf{I}\\ \mathbf{0} & \mathbf{I}\end{bmatrix}+\begin{bmatrix}\beta_2\mathbf{I} & \mathbf{0}\\ \beta_1\mathbf{I} & \beta_2\mathbf{I}\end{bmatrix}\right) \quad (16)$$
$$=\begin{bmatrix}1 & 0 & -b_{20} & -b_{10}\\ -b_{11} & 1 & 0 & -b_{21}\end{bmatrix}\left(\begin{bmatrix}\mathbf{N}_0 & \mathbf{N}_1\\ \mathbf{0} & \mathbf{N}_0\end{bmatrix}+\begin{bmatrix}\mathbf{N}_2 & \mathbf{0}\\ \mathbf{N}_1 & \mathbf{N}_2\end{bmatrix}\right)$$

$$\Rightarrow\begin{bmatrix}\hat{\mathbf{r}}_1^T & \hat{\mathbf{r}}_4^T & \hat{\mathbf{r}}_3^T & \mathbf{0} & \mathbf{0} & \mathbf{0}\\ \mathbf{0} & \mathbf{0} & \mathbf{0} & \hat{\mathbf{r}}_2^T & \hat{\mathbf{r}}_1^T & \hat{\mathbf{r}}_4^T\end{bmatrix}\begin{bmatrix}a_{00}\\ a_{10}\\ a_{20}\\ a_{01}\\ a_{11}\\ a_{21}\end{bmatrix}=\begin{bmatrix}\tilde{\mathbf{r}}_1^T\\ \tilde{\mathbf{r}}_2^T\end{bmatrix}$$

$$\Rightarrow\mathbf{P}_a\mathbf{a}=\mathbf{a}_1 \quad (17)$$

where $\hat{\mathbf{r}}_i$ is the i-th row of $\left(\begin{bmatrix}\mathbf{I} & \beta_1\mathbf{I}\\ \mathbf{0} & \mathbf{I}\end{bmatrix}+\begin{bmatrix}\beta_2\mathbf{I} & \mathbf{0}\\ \beta_1\mathbf{I} & \beta_2\mathbf{I}\end{bmatrix}\right)$, and $\tilde{\mathbf{r}}_i$ the i-th row of $\begin{bmatrix}1 & 0 & -b_{20} & -b_{10}\\ -b_{11} & 1 & 0 & -b_{21}\end{bmatrix}\left(\begin{bmatrix}\mathbf{N}_0 & \mathbf{N}_1\\ \mathbf{0} & \mathbf{N}_0\end{bmatrix}+\begin{bmatrix}\mathbf{N}_2 & \mathbf{0}\\ \mathbf{N}_1 & \mathbf{N}_2\end{bmatrix}\right)$.

Again from the Appendix, let the solution to (17) be

$$\mathbf{a}=\mathbf{P}_a^+\mathbf{a}_1 \quad (18)$$

What are the implications of (14) and (18)? Firstly, if we have the elements of the MIMO model $\mathbf{H}(z)$ that we *know* originated from a LPTV filter representation in (1), then (14) and (18) will give us the exact LPTV coefficients in (1). Secondly, if we are given the elements of an arbitrary MIMO model $\mathbf{H}(z)$, then we can immediately check if an exact LPTV representation might exist by performing a rank examination as in (15). If $\mathbf{H}(z)$ fails this test, then perhaps (14) and (18) would still give us coefficients for a LPTV filter in (1) that might in some way 'approximate' $\mathbf{H}(z)$. Thirdly, using (14) and (18) with the recent results from (McLernon, 1999), we can now replace cascade and parallel combinations of any different LPTV structures with a single equivalent LPTV difference eqn. as in (1). Finally, order reduction and LPTV state-space to LPTV difference eqn. transformation can be carried out. All these points will now be illustrated.

4. EXAMPLES

Let the period N=2 in all the following examples.

Example 1: Exact LPTV Realisation of a MIMO System, $\mathbf{H}(z)$

Consider an arbitrary second-order MIMO structure $\mathbf{H}(z)$ as in (3) and (4). The problem is to examine whether this MIMO filtering operation can be more efficiently carried out using a LPTV filter structure as in (1). So let $\mathbf{H}(z)$ be

$$N_{00}(z)=1+9.2z^{-1}-1.2z^{-2}, N_{01}(z)=6.4z^{-1}+8.0z^{-2},$$
$$N_{10}(z)=4.5-9.3z^{-1}, N_{11}(z)=4+1.8z^{-1}-6.4z^{-2}, \quad (19)$$
$$D(z)=1+0.45z^{-1}+0.32z^{-2}$$

From (12) $\operatorname{rank}(\mathbf{M})=2$, and so from (15) $\mathbf{H}(z)$ *may* have an exact second-order LPTV filter realisation. So from (14) and (18) we get the following LPTV coefficients in (1):

$$\begin{aligned}[a_{00} \quad a_{10} \quad a_{20} \quad b_{10} \quad b_{20}]&=[1 \quad 2 \quad 3 \quad 1.1 \quad 0.8] \\ [a_{01} \quad a_{11} \quad a_{21} \quad b_{11} \quad b_{21}]&=[4 \quad 6 \quad 8 \quad -1.5 \quad 0.4]\end{aligned} \quad (20)$$

From (3) the LPTV coefficients in (20) are equivalent to a MIMO $\hat{\mathbf{H}}(z)$, which if calculated is identical to $\mathbf{H}(z)$ in (19). So (20) is the LPTV filter equivalent to $\mathbf{H}(z)$ in (19). This example corresponds to condition 1 in the Appendix.

Example 2: Approximate LPTV Realisation of a MIMO System, $\mathbf{H}(z)$

Repeat example 1 where now we have

$$N_{00}(z)=1+2z^{-1}+3z^{-2}, N_{01}(z)=1+5z^{-1}+3z^{-2},$$
$$N_{10}(z)=6+4z^{-1}+7z^{-2}, N_{11}(z)=5+7z^{-1}+8z^{-2}, \quad (21)$$
$$D(z)=1+1.7z^{-1}+0.9z^{-2}$$

From (12) $\operatorname{rank}(\mathbf{M})=4$, and so from (15) $\mathbf{H}(z)$ will not have an exact second-order LPTV filter realisation. But from (14) and (18) we still get the following LPTV coefficients for use in (1):

$$\begin{aligned}&[a_{00} \quad a_{10} \quad a_{20} \quad b_{10} \quad b_{20}] \\ &=[1.893 \quad 7.147 \quad 3.764 \quad -0.8096 \quad 1.0632] \\ &[a_{01} \quad a_{11} \quad a_{21} \quad b_{11} \quad b_{21}] \\ &=[-1.212 \quad -1.443 \quad -6.039 \quad 2.814 \quad 0.479]\end{aligned} \quad (22)$$

From (3), the LPTV coefficients in (22) are equivalent to a MIMO $\hat{\mathbf{H}}(z)$, which if calculated is (as expected) <u>not identical</u> to $\mathbf{H}(z)$ in (21). But perhaps these coefficients for the more efficient LPTV filter implementation might give some sort of 'approximation' to the original MIMO structure, $\mathbf{H}(z)$? While no claim of optimality is made, consider an arbitrary input $x(n)=\sin(n\frac{2\pi}{100})$ and examine the outputs from both the MIMO structure of (21) and the LPTV approximation of (1) and (22). In this particular case there is a good approximation by the LPTV filter, but this cannot always be guaranteed. This example corresponds to condition 3 in the Appendix.

Example 3: A Single LPTV Filter Equivalent of Two LPTV Systems in Cascade

What are the coefficients in (1) for an equivalent single LPTV filter representation of two LPTV filters in cascade (LPTV1 followed by LPTV2 – the order is important)? Consider two first-order LPTV filters with the following coefficients:

$$\text{LPTV1:} \quad \begin{aligned}[a_{00} \quad a_{10} \quad b_{10}]&=[1 \quad 2 \quad 0.8] \\ [a_{01} \quad a_{11} \quad b_{11}]&=[-3 \quad -4 \quad 1.2]\end{aligned} \quad (23)$$

$$\text{LPTV2:} \quad \begin{aligned}[a_{00} \quad a_{10} \quad b_{10}]&=[6 \quad 7 \quad 0.5] \\ [a_{01} \quad a_{11} \quad b_{11}]&=[-8 \quad -9 \quad 1.8]\end{aligned} \quad (24)$$

The equivalent second-order LPTV filter has a MIMO representation $\mathbf{H}(z)=\mathbf{H}_2(z)\mathbf{H}_1(z)$. So from (3) calculate $\mathbf{H}_1(z)$ and $\mathbf{H}_2(z)$ for LPTV1 and LPTV2 respectively, and thus derive $\mathbf{H}(z)$. We note from (12) that for $\mathbf{H}(z)$ $\operatorname{rank}(\mathbf{M})=2$, and so $\mathbf{H}(z)$ *may* have an exact second-order LPTV filter realisation. In fact it is not difficult to show that for this cascade situation, $\operatorname{rank}(\mathbf{M})\leq 2$ is both a *necessary and sufficient* condition for a single second-order LPTV filter realisation. Then from $\mathbf{H}(z)$, (14) and (18), the equivalent second-order LPTV filter coefficients in (1) are:

$$\begin{aligned}&[a_{00} \quad a_{10} \quad a_{20} \quad b_{10} \quad b_{20}] \\ &=[6 \quad -5.1290 \quad -22.837 \quad -0.2613 \quad 1.3703] \\ &[a_{01} \quad a_{11} \quad a_{21} \quad b_{11} \quad b_{21}] \\ &=[24 \quad 20.9661 \quad -22.0678 \quad 0.5390 \quad 0.6305]\end{aligned} \quad (25)$$

This example corresponds to condition 1 in the Appendix.

Example 4: A Single LPTV Filter Equivalent of Two LPTV Systems in Parallel

Let the same two LPTV filters in example three now be in parallel. The equivalent second-order LPTV filter has a MIMO representation $\mathbf{H}(z)=\mathbf{H}_1(z)+\mathbf{H}_2(z)$. Since $\operatorname{rank}(\mathbf{M})=2$, $\mathbf{H}(z)$ will have an exact LPTV filter realisation, as in example three. So from (14) and (18) the equivalent second-order LPTV filter coefficients in (1) are:

$$\begin{aligned}&[a_{00} \quad a_{10} \quad a_{20} \quad b_{10} \quad b_{20}] \\ &=[7 \quad 1.5 \quad -9 \quad -0.1 \quad 1.08] \\ &[a_{01} \quad a_{11} \quad a_{21} \quad b_{11} \quad b_{21}] \\ &=[-11 \quad -2.4 \quad 13.2 \quad 0.2 \quad 0.8]\end{aligned} \quad (26)$$

This example corresponds to condition 1 in the Appendix.

Example 5: Order Reduction for a LPTV Filter

Consider a second-order LPTV filter as in (1) with the following coefficients:

$$\begin{aligned}&[a_{00} \quad a_{10} \quad a_{20} \quad b_{10} \quad b_{20}] \\ &=[1 \quad 3.3260 \quad 1.7680 \quad 0.3580 \quad 0.3978] \\ &[a_{01} \quad a_{11} \quad a_{21} \quad b_{11} \quad b_{21}] \\ &=[3 \quad 4.5488 \quad 1.0976 \quad 0.3512 \quad 0.4390]\end{aligned} \quad (27)$$

Can we replace this second-order LPTV filter with a lower-order filter? From (3) and (4) we get the second-order MIMO equivalent $\mathbf{H}(z)$ where:

$$\begin{aligned} N_{00}(z) &= 1+2.9574z^{-1}-0.7762z^{-2} \\ N_{01}(z) &= 4.4z^{-1}-1.0672z^{-2} \\ N_{10}(z) &= 4.9-1.1885z^{-1} \\ N_{11}(z) &= 3+1.0723z^{-1}-0.4366z^{-2} \\ D(z) &= 1-0.9626z^{-1}+0.1746z^{-2} \end{aligned} \quad (28)$$

Checking the zeros of the numerator and denominator terms of $\mathbf{H}(z)$ in (28) shows a common factor of $1-0.2425z^{-1}$, which we can cancel out to give a new $\hat{\mathbf{H}}(z)$ with:

$$\begin{aligned} &\hat{N}_{00}(z) = 1+3.2z^{-1}, \hat{N}_{01}(z) = 4.4z^{-1}, \\ &\hat{N}_{10}(z) = 4.9, \hat{N}_{11}(z) = 3+1.8z^{-1}, \\ &\hat{D}(z) = 1-0.72z^{-1} \end{aligned} \quad (29)$$

So from the first-order[2] equivalent of (13) and (17) (i.e. configured for a first-order $\hat{\mathbf{H}}(z)$ and not a second-order MIMO structure to avoid anomaly one – see Conclusions) then rank(**M**)=1, and we get the first-order (see (1)) LPTV filter equivalent of (27):

$$\begin{aligned} \begin{bmatrix} \hat{a}_{00} & \hat{a}_{10} & \hat{b}_{10} \end{bmatrix} &= \begin{bmatrix} 1 & 2 & 0.8 \end{bmatrix} \\ \begin{bmatrix} \hat{a}_{01} & \hat{a}_{11} & \hat{b}_{11} \end{bmatrix} &= \begin{bmatrix} 3 & 4 & 0.9 \end{bmatrix} \end{aligned} \quad (30)$$

Remembering that rank(**M**)=1 satisfies only the necessary condition for a first-order LPTV realisation to exist, then from (3) we can confirm that (30) does indeed realise $\hat{\mathbf{H}}(z)$. This corresponds to condition 1 in the Appendix.

Example 6: From a LPTV State-Space Structure to the Equivalent LPTV Difference Equation

Consider the following SISO LPTV state-space structure:

$$\begin{aligned} \mathbf{w}(n+1) &= \mathbf{A}(n)\mathbf{w}(n)+\mathbf{B}(n)x(n) \\ y(n) &= \mathbf{C}(n)\mathbf{w}(n)+D(n)x(n) \\ \mathbf{A}(n) &= \mathbf{A}(n+2), \text{ etc., where} \end{aligned} \quad (31)$$

$$\mathbf{A}(0)=\begin{bmatrix} 0 & 1 \\ -0.7 & -0.2 \end{bmatrix}, \mathbf{B}(0)=\begin{bmatrix} 1 \\ 1 \end{bmatrix}, \mathbf{C}(0)=\begin{bmatrix} -3 & 4 \end{bmatrix}, D(0)=2$$

$$\mathbf{A}(1)=\begin{bmatrix} 0 & 1 \\ -0.5 & -0.9 \end{bmatrix}, \mathbf{B}(1)=\begin{bmatrix} -2 \\ -4 \end{bmatrix}, \mathbf{C}(1)=\begin{bmatrix} 1 & 1 \end{bmatrix}, D(1)=-4$$

[2] For example, (11) and (12) for a first-order LPTV filter now become:

$$\begin{bmatrix} 1 & 0 & 0 & -b_{10} \\ -b_{11} & 1 & 0 & 0 \end{bmatrix}\mathbf{M}=\mathbf{0}$$

$$\mathbf{M}=\left(\begin{bmatrix} \mathbf{N}_0 & \mathbf{N}_1 \\ \mathbf{0} & \mathbf{N}_0 \end{bmatrix}\begin{bmatrix} \mathbf{I} & \beta_1\mathbf{I} \\ \mathbf{0} & \mathbf{I} \end{bmatrix}^{-1}\begin{bmatrix} \beta_1\mathbf{I} & \mathbf{0} \\ \mathbf{I} & \beta_1\mathbf{I} \end{bmatrix}-\begin{bmatrix} \mathbf{N}_1 & \mathbf{0} \\ \mathbf{N}_0 & \mathbf{N}_1 \end{bmatrix}\right)$$

It is well known (Lin and King, 1993) how to obtain $\mathbf{H}(z)$:

$$\begin{aligned} N_{00}(z) &= 2-6.56z^{-1}-2.5z^{-2} \\ N_{01}(z) &= -10z^{-1}-16.72z^{-2} \\ N_{10}(z) &= 2+0.22z^{-1} \\ N_{11}(z) &= -4-5.88z^{-1}-4.76z^{-2} \\ D(z) &= 1+1.02z^{-1}+0.35z^{-2} \end{aligned} \quad (32)$$

and so from (14) and (18) we get the coefficients for the equivalent LPTV difference eqn. in (1):

$$\begin{bmatrix} a_{00} & a_{10} & a_{20} & b_{10} & b_{20} \end{bmatrix} = \begin{bmatrix} 2 & -174.4 & 57.5 & -41 & 8.05 \end{bmatrix}$$

$$\begin{bmatrix} a_{01} & a_{11} & a_{21} & b_{11} & b_{21} \end{bmatrix} = \begin{bmatrix} -4 & 1.5565 & 0.5913 & 0.2217 & 0.0435 \end{bmatrix}$$

5. CONCLUSIONS

It could be argued that the origins of LPTV systems go back nearly two hundred years (Richards, 1983). But modern research in the subject began over forty years ago, in both control and signal processing, but with little overlap between the two disciplines. In a recent paper (McLernon, 1999) a summary of many LPTV filter/system applications was given, and all the interrelationships between the various equivalent structures for a LPTV filter were derived. But one question remained unsolved - how to return from the MIMO structure $\mathbf{H}(z)$ to the equivalent LPTV difference eqn. A solution has now been derived for any order of LPTV filter with any period N, and a number of examples have been presented to show how this result may be applied in practice.

In addition, this result also allows both the simplification and the analysis of interconnections of various linear time invariant (LTI) and LPTV structures within a consistent mathematical framework, where issues of complexity, sensitivity, etc., may now be dealt with. The LTI structures may be represented as LPTV systems with period N, but where the coefficients do not alter. Even the LPTV structures can each also have different periods (say N_1 to N_Q), and analysis can now be carried out over the whole interconnection using period $N=\hat{N}$, where $\hat{N}$ is the lowest common multiple of $\{N_i\}_{i=1}^{Q}$.

Further work remains to be done on the rank test in (15), and under what conditions it is both *necessary and sufficient* for exact LPTV filter realisation. Also, the topic of *minimality* has not been fully explored in the context of a LPTV difference eqn. representation. The general issue of 'approximating' a MIMO structure with a LPTV filter (as in example two) has only been touched upon, as has what we can say about its optimality. Finally, three anomalies have been observed.

Firstly, consider a first-order LPTV filter as in (1), and via (3) generate the first-order equivalent MIMO

H(z). If we now use H(z) in (14) and (18), configured for a second-order LPTV filter in (1), then we get a set of second-order LPTV coefficients for (1), and not the original first-order filter coefficients. But once again via (3), this second-order LPTV filter can be shown to produce a second-order $\hat{\mathbf{H}}(z)$ which is identical (through pole/zero cancellation) to the original **H**(z). This is intuitively what we would expect to happen. Finally, the whole process can be repeated ad infinitum to generate a set of second-order LPTV filters equivalent to the original first-order LPTV filter.

Now for the second anomaly. Let us suppose that we have a second-order LPTV filter in (1) and we generate an equivalent MIMO **H**(z) via (3). If we subsequently use (14) and (18) to return to a LPTV filter realisation, we get (as expected) the original second-order coefficients of (1). Rare exceptions have been observed where we do not return to the original coefficients for (1), and this new LPTV filter also does not have the same 'impulse responses' as the original LPTV filter. One such exception is for the coefficients below (albeit unstable, but this is irrelevant):

$$\begin{bmatrix} a_{00} & a_{10} & a_{20} & b_{10} & b_{20} \end{bmatrix} = \begin{bmatrix} 1 & 2 & 3 & 4 & 5 \end{bmatrix}$$

$$\begin{bmatrix} a_{01} & a_{11} & a_{21} & b_{11} & b_{21} \end{bmatrix} = \begin{bmatrix} 6 & 7 & 8 & 9 & 10 \end{bmatrix}$$

Note also that $\mathbf{P}_a$ in (18) is now unusually not full rank. Also, note that rank(**M**) = 2 (see (15)). This is unusual as (15) also has been observed to be not only necessary, but also *almost sufficient,* for exact LPTV realisation, although here the feedback (not the feedforward) LPTV coefficients are actually correctly estimated. Strangely, just changing the coefficient 10 to 10.0001, and the algorithm performs correctly, re-generating the original LPTV coefficients.

Finally, the third anomaly. In the case of example two we considered a stable MIMO **H**(z), and derived the coefficients for an 'approximate' LPTV equivalent filter in (1). It is impossible to predict in advance of applying the algorithm whether this 'approximate' LPTV filter will be stable or unstable, although in simulations it is usually stable.

REFERENCES

Colaneri, P. and S. Longhi (1995). The realisation problem for linear periodic systems. *Automatica,* **31**, 775-779.

Fjällbrant, T. (1970). Digital filters with a number of shift sequences in each pulse repetition interval. *IEEE Trans. on Circuit Theory*, **17**, 452-455.

Gelli, G., L. Paura and A.M. Tulino (1998). Cyclostationarity-based filtering for narrowband interference suppression in direct-sequence spread spectrum systems. *IEEE J. Select. Areas Commun.,* **16**, 1747-1755.

Jury, E.I. and F.J. Mullin (1958). A note on the operational solution of linear difference equations. *J. Franklin Institute*, **266**, 189-205.

Jury, E.I. and F.J. Mullin (1959). The analysis of sampled-data control systems with a periodically time-varying sampling rate. *IRE Trans. Auto. Control*, **4**, 15-21.

Lin, C.-A. and C.-W. King (1993). Minimal periodic realisations of transfer matrices. *IEEE Trans. on Auto. Control*, **38**, 462-466.

McLernon, D.C. and R.A. King (1990). A multiple-shift, time-varying, two-dimensional filter. *IEEE Trans. Circuits and Systems,* **37**, 120-127.

McLernon, D.C. (1999). One-dimensional linear periodically time-varying structures: derivations, interrelationships and properties. *IEE Proc: Vis. Image Signal Process.,* **146**, 245-252.

Richards, J.A. (1983). *Analysis of periodically time-varying systems*. Springer-Verlag. New York.

Xie, L., S. Wang, C. Du and S. Zhang (2000). H_∞ deconvolution of periodic channels. *Signal Processing*, **80**, 2365-2378.

APPENDIX

Why use $\mathbf{P}_b^+$ *and* $\mathbf{P}_a^+$ *in (14) and (18)?*

Consider $\mathbf{P}_b \mathbf{b} = \mathbf{b}_1$ in (14), with the following discussion, where appropriate, also applying to (18). This represents an over-determined set of linear eqns, and from simulations we have observed three conditions that can and do occur. But condition 2, while observed in simulations, has not been represented in this paper.

Condition 1: Consistent set of eqns with unique solution. $\mathbf{P}_b$ is full rank and $rank[\mathbf{P}_b] = rank[\mathbf{P}_b \mid \mathbf{b}_1]$. *Choose unique solution.*

Condition 2: Consistent set of eqns with infinite number of solutions. $\mathbf{P}_b$ is not full rank and $rank[\mathbf{P}_b] = rank[\mathbf{P}_b \mid \mathbf{b}_1]$. *Choose minimum-norm solution.*

Condition 3: Inconsistent set of eqns with unique least-squares solutions. $\mathbf{P}_b$ is full rank and $rank[\mathbf{P}_b] < rank[\mathbf{P}_b \mid \mathbf{b}_1]$. *Choose unique least-squares solution.*

To accommodate the above three scenarios, use the pseudo-inverse and write $\mathbf{b} = \mathbf{P}_b^+ \mathbf{b}_1$. Finally, from simulations it appears that $\mathbf{P}_a$ is rarely rank deficient – see Conclusions.

Copyright © IFAC Periodic Control Systems,
Cernobbio-Como, Italy, 2001

LPV PREDICTIVE CONTROL OF THE STALL AND SURGE FOR JET ENGINE [1]

P. Falugi [*,1] **L. Giarré** [**,1] **L. Chisci** [*] **G. Zappa** [*,1]

[*] *Dipartimento di Sistemi e Informatica*
Università di Firenze, Firenze, Italy.
chisci,falugi,zappa@dsi.unifi.it *39-(055)-4796-569,359,263*
[**] *Dipartimento di Ingegneria Automatica e Informatica*
Università di Palermo, Palermo, Italy.
giarre@ias.unipa.it *39-(091)-481119*

Abstract: Predictive control of constrained LPV systems is applied to the model of the stall and surge control for jet engine compressors. The objective of the used technique is to optimize nominal performance while guaranteeing robust stability and constraint satisfaction. This is achieved by exploiting invariant sets and a receding horizon optimization procedure which provides on-line a non-linear correction to a gain-scheduled linear feedback designed off-line. A comparison with a contractive gain-scheduling control technique is also shown. *Copyright ©2001 IFAC*

Keywords: LPV models, Stall and Surge control, Predictive control, Gain scheduling, Robust control

1. INTRODUCTION

Predictive control of constrained LPV systems (Chisci *et al.* 2001*b*) will be applied to the stall and surge control for jet engine compressors. The objective of the used technique is to optimize nominal performance while guaranteeing robust stability and constraint satisfaction. The latter goal is achieved by exploiting invariant sets and a receding horizon optimization procedure which provides on-line a non-linear correction to a gain-scheduled linear feedback designed off-line.

Gain Scheduling is a two-step procedure where: 1) linear controllers are locally designed based on linearization of the nonlinear system at several different equilibria (operating conditions); 2) a global nonlinear controller for the nonlinear system is obtained by *interpolating* or *scheduling* among the local operating point designs, see (Rugh 1991) and (Shamma and Athans 1991).

A useful paradigm to study gain-scheduling is the one of LPV systems. The terminology Lin-

[1] Research supported by MURST Grant ex-40%

ear Parameter Varying (LPV) systems has been introduced in (Shamma 1996).

A discrete-time LPV system is represented in state space as

$$\begin{aligned} x(t+1) &= A(p(t))x(t) + B(p(t))u(t) \\ y(t) &= C(p(t))x(t) + D(p(t))u(t) \end{aligned}$$

The time evolution of the exogenous parameter $p(t)$ is assumed *a priori* unknown. However, it can be measured or estimated upon operation of the system; bounds on the magnitude on the scheduling parameter and rate of change can also be given a-priori. LPV systems can be exploited as alternative description of nonlinear systems. Rather than modeling the dynamical evolution of a particular variable, one can treat it as an exogenous variable and the bounds on the parameter evolution are obtained by analyzing the dynamics of the corresponding variable. In this way a Quasi-LPV system is obtained, see (Shamma and Athans 1991), (Shamma 1996), the recent survey (Rugh and Shamma 2000) and references therein.

A possible control design for LPV systems is to design robust controllers around each operating point and to switch between controllers according to some gain-scheduling policy. This is a good compromise between performance and robustness; moreover the stability question of the switching control has been solved in (Shamma and Athans 1991), (Blanchini 2000).

Model Predictive Control (MPC) can be an effective tool to design a gain scheduling policy for LPV systems. In fact, bounds in parameter variations can be explicitly considered in the optimization step of MPC, guaranteeing stability, robustness and performances.

Stability and robustness of predictive control have been thoroughly investigated (Keerthi and Gilbert 1988, Kouvaritakis *et al.* 1999). Typical robust constrained predictive control algorithms minimize either a worst-case performance index (Lee and Yu 1997) or an upper bound of it (Kothare *et al.* 1996), under suitable stability constraints. The present paper like (Badgwell 1997, Kouvaritakis *et al.* 1999) turns aside from the objective of worst-case performance optimization for a twofold reason. First, min-max optimization may be too computationally demanding. Secondly, the paradigm of optimizing performance for the worst-case may be unrealistic in the common situation where a nominal (most likely) model is available. For the above reasons, a more sensible approach seems to minimize the deviation from the input, expressed by a gain-scheduling linear control, which provides the "nominal" performance, while robustly guaranteeing stability and constraint satisfaction.

Accordingly we consider hereafter a novel predictive control algorithm for LPV systems with state and control constraints. It postulates a control sequence, along an infinite prediction horizon, consisting of the linear gain scheduled feedback plus N free control moves. The receding-horizon controller selects the control at sample time t as the first element of the control sequence which minimizes the energy of the control moves subject to appropriate state-dependent linear constraints. In particular it is imposed that after N steps the state enters a polytopic set, which is feasible and $\lambda-$contractive under the gain scheduling control. Such a set can be computed by exploiting an appropriate version of the controlled invariance kernel algorithm, see for instance (Blanchini 1999, Shamma and Xiong 1999). It can be shown that the algorithm provides asymptotic stability if the initial state is feasible. The feasibility region can be computed by solving LP problems.
This approach has some connection with the algorithm presented in (Sznaier 1999) which however considers unconstrained continuous time systems.

The effectiveness of the algorithm is shown in the surge and stall control in jet engine compressors based on the Moore and Greitzer model (Moore and Greitzer 1986). In particular, it is compared with a Lyapunov function based control developed in (Tu and Shamma 1998) which exploits the LPV approximation of the nonlinear dynamics.

2. PROBLEM FORMULATION

Following (Shamma and Xiong 1999) we consider the discrete-time LPV system

$$\begin{aligned} x(t+1) &\in \mathcal{F}(p(t)) \begin{bmatrix} x(t) \\ u(t) \end{bmatrix}, \; x(0) = x_0 \\ p(t+1) &\in Q(p(t)) \qquad\qquad p(0) = p_0 \end{aligned} \tag{1}$$

where $x(t) \in \mathbb{R}^n$ is the state vector and $u(t) \in \mathbb{R}^m$ the control input, $p(t)$ is a time-varying parameter which belongs to a discrete-set $P = \{p_1, \cdots, p_l\}$ and evolves according to a set-valued map $Q : P \rightsquigarrow P$. Finally the map $\mathcal{F} : P \rightsquigarrow \mathbb{R}^{n\times(n+m)}$ is also set-valued in order to represent additional uncertainty in the system dynamics.

Assumption 1 - For any $p_j \in P$, $\mathcal{F}(p_j)$ is a closed convex polytope P_j, i.e :

$$\begin{aligned} x(t+1) &= A(t)x(t) + B(t)u(t) \\ \text{for some} &\quad [A(t), B(t)] \in P_j \end{aligned} \tag{2}$$

Moreover the system (1) is subject to pointwise-in-time control and state constraints

$$u(t) \in \mathcal{U}, \; x(t) \in \mathcal{X} \qquad \forall t \geq 0 \tag{3}$$

for some appropriate polyhedra $\mathcal{U} \subset \mathbb{R}^m$, $\mathcal{X} \subset \mathbb{R}^n$ containing the origin as an interior point.

Assumption 2 - For each value p_j of the time-varying parameter there exists a linear feedback control law $u(t) = F_j x(t)$ such that the closed loop polytopic system

$$x(t+1) \in \mathcal{F}(p_j) \begin{bmatrix} x(t) \\ F_j x(t) \end{bmatrix} \tag{4}$$

is absolutely asymptotically stable.

In this respect, it is important to investigate: (1) if the linear gain scheduling control law $u(t) = F_j x(t)$ stabilizes also the LPV system and (2) how to design a non-linear state-feedback regulator

$$u(t) = g(x(t), p(t)) \tag{5}$$

which improves the performance of the linear one. Notice that, because of constraints (3), stabilization may be guaranteed in a possibly small neighborhood of the origin . The non linear feedback (5) could enlarge the stability domain in a significant way.

Let Σ_0 denote the set of initial extended states $s(0) = [x'(0),\ p'(0)]' \in \mathbb{R}^n \times P$ for which the plant state is asymptotically steered to the origin under the gain-scheduled linear feedback (4) without violating the constraints. Further consider the LPV system

$$\begin{aligned} x(t+1) &\in \mathcal{F}(p(t)) \begin{bmatrix} x(t) \\ F(p(t))x(t) + c(t) \end{bmatrix} \\ p(t+1) &\in Q(p(t)) \\ F(p) &= F_j, \quad \text{if } p = p_j \end{aligned}$$

for which the actual input sequence turns out to be the sum of the gain-scheduled linear feedback (4) plus an open-loop parameter independent signal $c(\cdot)$. Let S_N be the set of vectors $[s'(0), c'(0), \ldots, c'(N-1)]'$ such that $\{c(0), c(1), \ldots, c(N-1)\}$ steers the initial state s_0 to Σ_0 in N steps. For the calculation of the invariant polytopes S_N , see (Chisci *et al.* 2001*a*) and (Chisci *et al.* 2001*b*). In the latter reference, the following predictive control algorithm has been proposed.

Parameter Varying - Predictive Control (PV-PC) Algorithm At each sample time t, given $s(t) = [x(t)',\ p(t)]' \in \mathbb{R}^n \times P$, find

$$\hat{\underline{c}}(t) = \arg\min \|\underline{c}(t)\|^2, \tag{6}$$

subject to

$$\begin{bmatrix} s(t) \\ \underline{c}(t) \end{bmatrix} \in S_N. \tag{7}$$

Then apply to the plant the control signal

$$u(t) = F_j x(t) + \hat{c}_1(t) \tag{8}$$

where $\underline{c}(t)^T = [\hat{c}_1^T, \cdots, \hat{c}_N^T]$ and j is such that $p(t) = p_j$.
The above algorithm selects, at time t, among all admissible sequences $\underline{c}(t)$ the one with minimum l_2 norm . Since S_N is a collection of convex polytopes (6)-(7) amounts to a *Quadratic Programming (QP)* problem. As far as stability is concerned, the following result holds (Chisci *et al.* 2001*a*).

Theorem - Provided that the initial state $s(0)$ is feasible, the receding-horizon control (6)-(8) guarantees that (i) the constraints (3) are satisfied and (ii) $\lim_{t\to\infty} x(t) = \mathbf{0}$.

The PV-PC algorithm ensures, therefore, asymptotic stability with a certain domain of attraction $\underline{\Sigma}_N$, where $\underline{\Sigma}_N$ is the projection of S_N onto the space $\mathbb{R}^n \times P$.

2.1 *An alternative LPV-based scheme*

An alternative control scheme based on LPV models can be considered along the lines of (Shamma and Xiong 1999) and (Blanchini 1994). Let Σ be a polytopic λ-contractive controlled invariant set in the extended space $\mathbb{R}^n \times P$ for the LPV system (1). This means that $\Sigma = \{\Sigma^1, \Sigma^2, \ldots, \Sigma^l\}$ is a collection of polytopes Σ^j, one for each possible parameter value $p_j \in P$, such that if $x(t) \in \Sigma^j$ and $p(t) = p_j$ there exists $u(t) \in \mathcal{U}$ which ensures $x(t+1) \in \lambda\Sigma^i$ for all i such that $p_i \in Q(p(t))$. Such sets can be computed by an appropriate set recursion (Shamma and Xiong 1999). Then the following on-line control scheme, referred to hereafter as *Parameter Varying - Contractive Control* (PV-CC) can be introduced

$$\begin{gathered} u^*(x(k), p(k)) = \min_{\lambda \geq 0, u \in \mathcal{U}} \lambda \\ \text{subject to} \\ \mathcal{F}(p(k)) \begin{bmatrix} x(k) \\ u \end{bmatrix} \in \bigcap_{p_j \in Q(p(k))} \lambda\Sigma^j \end{gathered} \tag{9}$$

Given the current state and scheduling parameter, the optimization procedure selects the input control vector u, giving the best possible contraction of the state under the given constraints. Since the map $\mathcal{F}(p)$ is polytopic for each $p \in P$ and the sets Σ^j, $j = 1, 2, \ldots, l$, are all polytopes, the optimization (9) is an LP problem.

The two procedures PV-PC and PV-CC will be compared on the control problem described in the next section.

3. THE STALL AND SURGE CONTROL PROBLEM

Rotating stall and surge phenomena in compressors of jet engines represent critical operating conditions that limit the stability region at low mass flow in a compressor map. In particular, rotating stall occurs when a circumferential flow pattern is disturbed and the local region of stalled flow propagates around the compressor annulus at a fraction of the rotor speed. Surge, instead, is an axisymmetrical oscillation of mass flow and pressure which affects the whole compression system. In most cases this may lead to limit cycles in the compressor map. The essential differences between rotating stall and surge are that the average flow in pure rotating stall is steady in time, but the flow has circumferentially nonuniform mass deficit, while in pure surge the flow is unsteady but circumferentially uniform. Physically, these aerodynamics flow instabilities may cause a rapid heating of the blades and increase the exit temperature of the compressor. Also, additional periodic loads, blade vibrations and fatigue are responsible for material durability reduction such as severe damages to the machine.

Let us consider a simplified, lumped compression system model that describes rotating stall and surge instabilities, based on (Moore and Greitzer 1986):

$$\begin{aligned} \dot{\Omega} &= \frac{1}{l_c}[(-\Xi + \Xi_C(\Omega) - 3R(\Omega - 1))] \\ \dot{\Xi} &= \frac{1}{4l_cB^2}(\Omega - \Omega_T) \\ \dot{R} &= 3\mu R(2\Omega - \Omega^2 - R) \end{aligned} \qquad (10)$$

where: Ω is the compressor mass flow; Ξ is the pressure rise of the compressor; R is the amplitude square of the first harmonic mode of the rotating stall disturbance; Ω_T is the input viz. the throttle mass flow. Moreover $\Xi_C(\Omega)$ is the steady-state characteristic of the compression system (Moore and Greitzer 1986).

The objective of rotating stall and surge control is to achieve the maximum compression efficiency while eliminating stall, assuming that the state variable is available for feedback. In order to design a gain-scheduling control, it is convenient to model the compression system dynamics into an LPV form.

According to (Tu and Shamma 1998), we define $\omega(k) = \Omega(k) - \Omega_{eq}(k)$, $\omega_T(k) = \Omega_T(k) - \Omega_{T_{eq}}(k)$ and $\xi(k) = \Xi(k) - \Xi_{eq}(k)$, where we choose the equilibrium $\Omega_{eq} = \Omega_{T_{eq}} = 2$ and $\Xi_{eq} = 4 - \frac{3}{2}\omega^2 - \frac{1}{2}\omega^3$ so as to give maximum efficiency. In addition, we select $p(k) = \omega(k)$ as *scheduling* variable. Then, by Euler discretization of (10) with sampling time T, we get the following discrete-time Quasi-LPV model, using the relaxation auxiliary variable $v(k)$:

$$\mathbf{x}(k+1) = A(p(k))\mathbf{x}(k) + B_1 u(k). \qquad (11)$$

with state $\mathbf{x}(k) = [v(k), \quad \xi(k)), \quad R(k)]'$, input $u(k) = \omega_T(k)$, and matrices

$$A(p) = \begin{bmatrix} 1 & -\frac{T}{l_c} & -\frac{3T}{l_c}(p+1) \\ 0 & 1 - \frac{T}{l_c}(3p + \frac{3}{2}p^2) & -\frac{3T}{l_c}(3p + \frac{3}{2}p^2)(p+1) \\ 0 & 0 & 1 - 3\mu T(2p + p^2) - 0.9\mu T\rho \end{bmatrix} \qquad (12)$$

$$B_1 = \begin{bmatrix} 0 \\ -\frac{T}{4l_cB^2} \\ 0 \end{bmatrix},$$

where $\rho = \rho(k)$ is a bounded disturbance ($|\rho(k)| \leq 1$) taking into account some nonlinear terms and $p = p(k)$ is the scheduling parameter.

The bounds on the scheduling parameter dynamics are $|p(k)| \leq 0.5$, $|p(k+1) - p(k)| \leq T$. The following constraints on the input and state variables will be also considered:

$$\begin{gathered} |u(k)| \leq 1 \\ |v(k)| \leq 0.5, \quad |v(k+1) - v(k)| \leq T \\ \frac{T}{l_c}|\xi(k) + 3(p(k)+1)R(k)| \leq T \\ |R(k)| \leq 0.3 \end{gathered} \qquad (13)$$

According to **Assumption 2**, a linear feedback control law has been designed, solving an LQ optimal control, considering a finite set of possible values for the parameters. In particular we have chosen a set of 16 possible values p_j as follows :$\{-0.15, -0.13, -0.11, \ldots -0.01, 0.01, 0.03, \ldots, 0.15\}$. Let F_j be the feedback evaluated in $\frac{p_j + p_{j+1}}{2}$, that has been applied for any measured value $p(k)$: $[p_{j-1} \quad p_j]$. This feedback is such that it guarantees stability for the system under the varying parameter $p(k+1) \in Q(p(k))$, where the parameter dynamics is

$$Q(p) = \begin{cases} \{p_1, p_2\}, & p = p_1 \\ \{p_{j-1}, p_j, p_{j+1}\}, & p \in \{p_2, p_3, \ldots, p_{l-1}\} \\ \{p_{l-1}, p_l\}, & p = p_l \end{cases}$$

From this feedback law we computed the set S_N for PV-PC control. Moreover the invariant set $\Sigma = \{\Sigma^1, \Sigma^2, \ldots, \Sigma^l\}$ of the PV-CC control has been computed by considering all the extended states which can be driven in N steps to Σ_0 by exploiting a control sequence $u(k) = g(x(k), p(k))$ (see for details (Chisci *et al.* 2001*a*)) which explicitly takes into account the state evolution.

In the simulations we considered the parameter values $B = 0.3$, $l_c = 1$, $\mu = 0.6$ and a sampling time $T = 0.1$. Open-loop simulation of the transient response with initial state in the instability region, $x(0) = [-0.15 \quad 0 \quad 0]'$, shows the presence

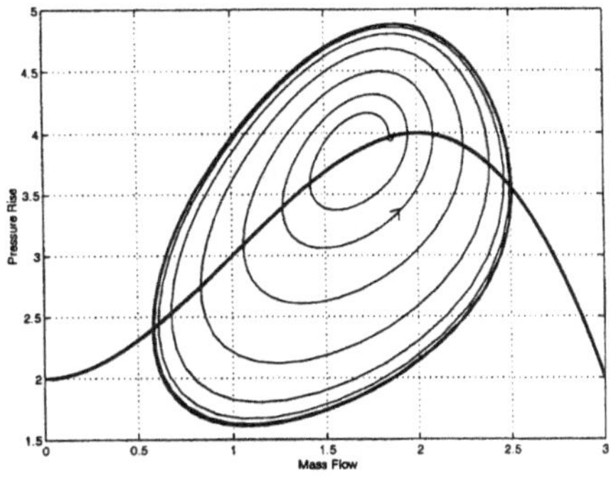

Fig. 1. Steady axisymmetric compressor characteristic: transient response with classic surge cycle.

of a limit cycle, i.e. the surge phenomenon appears (see fig. 1). The gain-scheduling control described in section 2, implemented with control horizon $N = 2$, guarantees that the surge phenomenon is avoided, as can be seen in Fig. 2.

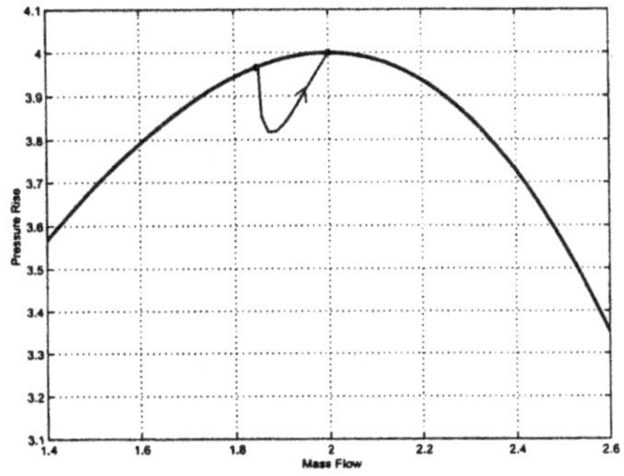

Fig. 2. Steady axisymmetric compressor characteristic: controlled transient response .

Under the condition of stall the gain-scheduled control achieving constrained regulation has been designed according to our procedure in order to design the robust overall stabilizing controller. Moreover, starting with state variable $x(0) = [-0.15,\ \ 0,\ \ 0.3]'$, the inception of stall can be robustly regulated to zero.

For predictive control horizon N equal to $2, 3, 4$ the controlled state variables are shown in Figs. 3, 4 and 5. The constraints are all satisfied and the rotating stall goes to zero; further, it is evident that there is no significant improvement beyond $N = 2$.

3.1 *Comparison with the PV-CC algorithm*

Under the condition of rotating stall, starting with the following initial conditions in the instability region: $x_1(0) = -0.15$, $x_2(0) = 0$, $x_3(0) = 0.1$; the two algorithms PV-PC and PV-CC have been applied and the resulting behaviour is shown in Figs. 6,7 and 8. It is evident how PV-PC exhibits a smoother responses as well as a faster recovery from stall.

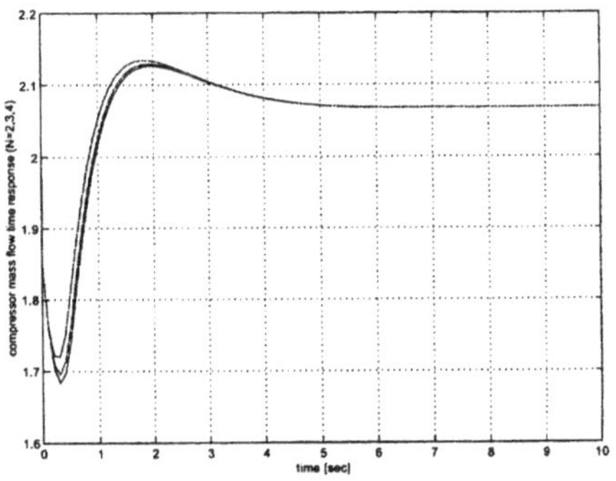

Fig. 3. Compressor mass flow time response ($N = 2, 3, 4$) .

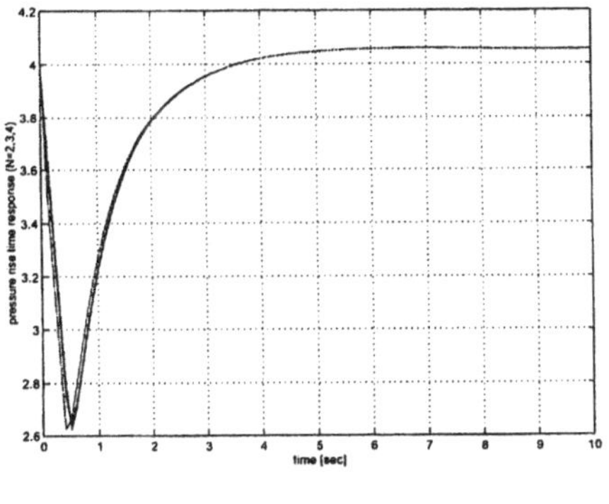

Fig. 4. Pressure rise time response ($N = 2, 3, 4$).

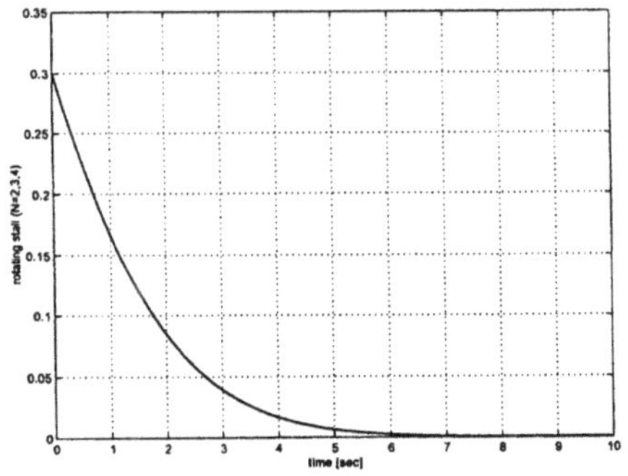

Fig. 5. Rotating stall ($N = 2, 3, 4$).

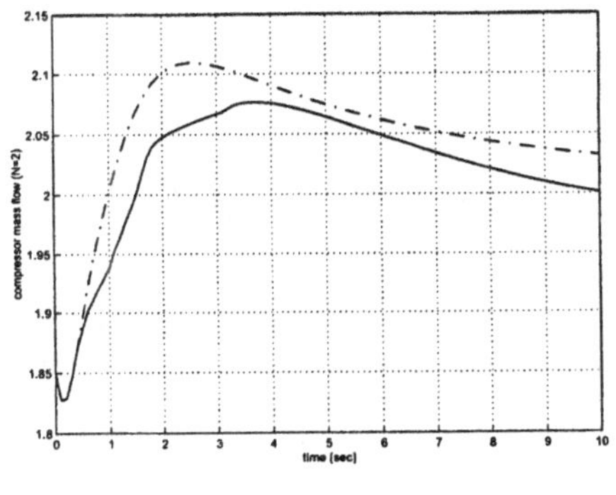

Fig. 6. Compressor mass flow time response (dash-dot line: PV-PC; solid line: PV-CC) .

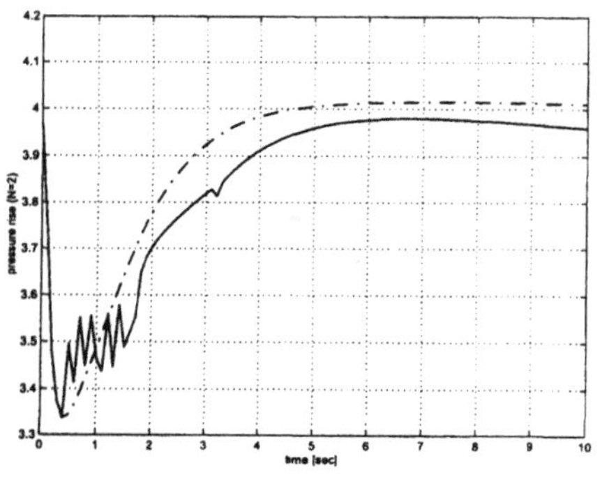

Fig. 7. Pressure rise time response (dash-dot line: PV-PC; solid line: PV-CC) .

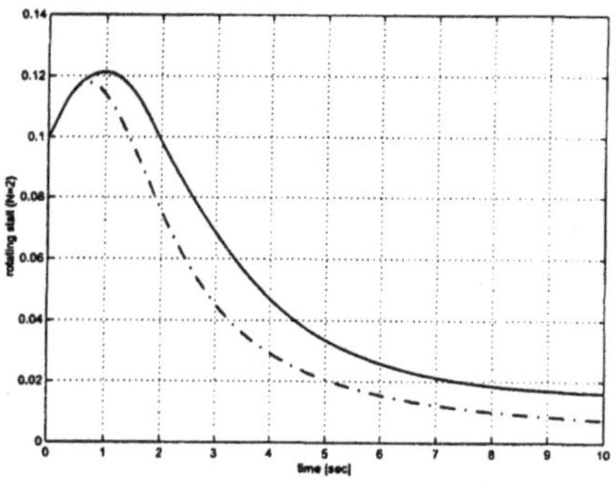

Fig. 8. Rotating stall (dash-dot line: PV-PC; solid line: PV-CC).

4. CONCLUSIONS

Predictive control of constrained LPV systems is applied to the model of the stall and surge control for jet engine compressors, in order to optimize nominal performance while guaranteeing robust stability and constraint satisfaction. This is achieved by exploiting invariant sets and a receding horizon optimization procedure which provides on-line a non-linear correction to a gain-scheduled linear feedback designed off-line. A comparison with a contractive gain-scheduling control scheme is also shown. The simulation results show that the presented method is robust in presence of rotating stall and surge.

5. REFERENCES

Badgwell, T.A. (1997). Robust model predictive control. *International Journal of Control* **68**, 797–818.

Blanchini, F. (1994). Ultimate boundedness control for discrete-time uncertain system via set-induced lyapunov function. *IEEE Trans. on Automatic Control* **39**(2), 428–433.

Blanchini, F. (1999). Set invariance in control. *Automatica* **35**, 1747–1768.

Blanchini, F. (2000). The gain scheduling and the robust state feedback stabilization problems. *IEEE Trans. on Automatic Control* **45**(11), 2061–2071.

Chisci, L., P. Falugi and G. Zappa (2001*a*). Predictive control for constrained LPV systems. *In Proc. of the European Control Conference*, 3074–3078.

Chisci, L., P. Falugi and G. Zappa (2001*b*). Predictive control for constrained systems with polytopic uncertainty. *In Proc. of the American Control Conference*, 3073–3078.

Keerthi, S.S. and E.G. Gilbert (1988). Optimal infinite-horizon feedback laws for a general class of constrained discrete-time systems: stability and moving-horizon approximations. *JOTA* **57**, 265–293.

Kothare, M.V., V. Balakrishnan and M. Morari (1996). Robust constrained model predictive control using linear matrix inequalities. *Automatica* **32**, 1361–1379.

Kouvaritakis, B., J.A. Rossiter and J. Schuurmans (1999). Efficient robust predictive control. *IEEE Trans. on Automatic Control* **45**(8), 1545–1549.

Lee, J.H. and Z. Yu (1997). Worst case formulations of model predictive control for systems with bounded parameters. *Automatica* **33**, 763–781.

Moore, F. and E. Greitzer (1986). A theory of pos-stall transients in axial compression systems: Part I-development of equations. *Journal of Engineering for Gas Turbines and Power* **1**, 68–76.

Rugh, W.J. (1991). Analytical framework for gain-scheduling. *IEEE Control systems Magazine* **11**, 79–84.

Rugh, W.J. and J.S. Shamma (2000). Research on gain scheduling. *Automatica* **36**, 1401–1425.

Shamma, J. (1996). Linearization and gain scheduling. In: *Control Handbook* (W. Levine, Ed.), 388–396. CRC Press.

Shamma, J. and M. Athans (1991). Guaranteed properties of gain scheduled control for linear parameter varying plants. *Automatica* **27**, 559–564.

Shamma, J. S. and D. Xiong (1999). Set-valued methods for linear parameter varying systems. *Automatica* **35**, 1081–1089.

Sznaier, M. (1999). Receding horizon: an easy way to improve performance in lpv systems. *In Proc. of the American Control Conference* **4**, 2262–2266.

Tu, K.H. and J.S. Shamma (1998). Nonlinear gain-scheduled control design using set-valued methods. *In Proc. of the American Control Conference*, 1195–1999.

Copyright © IFAC Periodic Control Systems,
Cernobbio-Como, Italy, 2001

MULTIVARIABLE CONTROL FOR A GAS TURBINE USING PERIODIC OUTPUT FEEDBACK

Arup Chakrabarti and Bijnan Bandyopadhyay

Interdisciplinary Group in Systems and Control Engineering
Indian Institute of Technology Bombay, Powai, Mumbai-400076,
India.

fax: (91)-(22) 572 3480
e-mail: bijnan@ee.iitb.ac.in

Abstract: The advent of electronic controllers in the field of gas turbine control, has allowed the implementation of sophisticated control algorithms. This paper proposes one such method, to design a controller using piecewise constant periodic output feedback, which has multifarious advantages, because the states are not required for feedback, complete pole-assignability is guaranteed and the method is easily implementable. *Copyright © 2001 IFAC*

Keywords: Gas turbines, Engine control, Linearization, State-space models, Reduced order models, Output feedback, Multivariable control, Simultaneous stabilization, Robust control

1. INTRODUCTION

Feedback control has been an essential part of gas turbine control from the very beginning. However, it is only recently that computers have been used to implement engine controls. Many engines even today, use hydromechanical controllers, commonly referred to as HMUs, which implement the desired control strategy in terms of cams and mechanical integrators.

The changeover to electronic controllers began in the 1970s, which led to higher engine operating efficiencies by allowing tighter engine control, through the use of higher loop gains and improved strategies to reduce the transient overshoot or undershoot. It also allowed implementation of control algorithms which would be difficult to implement mechanically.

Various control techniques like state feedback control, linear quadratic control, etc., have been proposed by numerous authors including De Hoff, *et al.* (1977) and Watts, *et al.* (1992). The problem with these control techniques is that the state feedback controller requires the availability of the entire state vector or needs an estimator, which may not be feasible in the case of gas turbines.

On the other hand, static output feedback requires only the measurement of system output. It is one of the most investigated problems in control theory and application, since it represents the simplest closed loop control that can be realised in practice. Secondly, many problems involving synthesizing dynamic controllers, can be formulated as static output feedback problems involving augmented plants. However, complete pole-assignment and guaranteed closed loop stability, has still not been obtained using static output feedback.

The periodic output feedback problem, which has been considered in this paper, has the features of static output feedback and also guarantees complete pole-assignability. Since the gains in periodic output feedback are piecewise constant, the method is easily implementable and compared to static output feedback, a much larger class of systems can be stabilized.

The linear model (state-space representation) data, required for most of the control applications was available for an F100 turbofan engine, from the work by De Hoff, *et al.* (1977). However, to arrive at this linear representation, one has to go through the necessary steps of modeling and simulation, followed by the generation of the linear models at the selected flight points. These aspects have been dealt with in brief, in the first few sections of the paper.

Gas turbines in general, are higher- order systems, and the cost and complexity involved in controller design is also high. In this work, a well known result is utilized to reduce the system order by retaining the dominant eigenvalues and then arriving at an output injection gain, which stabilizes the reduced order closed loop system. This when aggregated and applied to the higher order system, guarantees closed loop stability, which is an advantage of this approach. The output injection gains obtained for the various linear models are utilized to realize a simultaneously stabilizing periodic output feedback gain sequence.

The concept of periodic output feedback, the procedure followed to arrive at the robust feedback gain K, and the results have been presented in the later sections.

2. MODELING AND SIMULATION

The F100 engine has various static components, namely the inlet, fan, compressor, burner, HP turbine, LP turbine, mixer, afterburner and nozzle. Each of the components has a set of nonlinear equations forming it's mathematical model. A sample model for a compressor is presented here, with the suffixes 'in' and 'out', denoting the inlet and outlet of the compressor, respectively. The equations for the other components can be found in numerous texts on gas turbine theory, for example, Cohen, *et al.* (1972). A complete mathematical model for an automotive gas turbine is presented in the paper by Winterbone, *et al.* (1973).

2.1 *Analytical Model of the Compressor*

The compressor pressure ratio is given by

$$\frac{P_{out}}{P_{in}} = f\left(\frac{\dot{W}_C\sqrt{T_{in}}}{P_{in}}\right). \quad (1)$$

The compressor temperature rise is

$$\Delta T_C = \frac{T_{in}}{\eta_c}\left[\left(\frac{P_{out}}{P_{in}}\right)^{\frac{\gamma-1}{\gamma}} - 1\right]. \quad (2)$$

Temperature after compressor is

$$T_{out} = T_{in} + \Delta T_C. \quad (3)$$

The compressor power is given as

$$L_C = \left(\frac{\dot{W}_C\sqrt{T_{in}}}{P_{in}}\right)\left(\frac{P_{in}}{\sqrt{T_{in}}}\right).C_P.\Delta T_C. \quad (4)$$

2.2 *F100 Nonlinear simulation*

The nonlinear digital simulation of the F100 engine, includes both steady-state component descriptions (called component maps) as well as dynamic elements to represent gas volume, thermal capacitance and rotor inertial effects.

The steady-state operating characteristics (maps) are tied together by dynamic or time-varying relationships based on the conservation of mass and energy. Point-to-point calculations are made, using the rotor inertial effects, enclosed volume capacitive effects and transient heat transfer effects. The rotor dynamics, pressure dynamics and transient heat transfer effects, account for the sixteen states, which are described later.

The steady-state maps and dynamic relationships, combine to form a set of simultaneous, nonlinear differential equations. For e.g.., from the power balance equation, we can get an error variable,

$$E_1 = \dot{W}_C\Delta h_C + J\frac{dN}{dt} - \dot{W}_T\Delta h_T \quad (5)$$

which can be solved digitally, using recursive techniques. A detailed description of the simulation techniques has been discussed by Seller and Danielle (1975).

3. LINEAR MODEL GENERATION AND SELECTION

Most of the analytical design approaches used by control designers, require linear descriptions of steady-state and transient characteristics of the engine. Linear approximations to nonlinear systems can be obtained at a particular operating point (a particular equilibrium condition of power setting and flight condition) for small excursions about that point.

3.1 *Model generation*

A system with a number of states, inputs and outputs, when linearized about an operating point, can be described by the following linear state-space equations:

$$\begin{aligned} \Delta\dot{x} &= A\Delta x + B\Delta u, \\ \Delta y &= C\Delta x + D\Delta u, \end{aligned} \quad (6)$$

where,the elements of A, B, C, D are represented as follows:

$a_{ij} = \frac{\partial f_i}{\partial x_j}$; $b_{ik} = \frac{\partial f_i}{\partial u_k}$; $c_{lj} = \frac{\partial y_l}{\partial x_j}$; $d_{lk} = \frac{\partial y_l}{\partial u_k}$;

where i and j, vary from 1 to maximum number of state variables, k varies from 1 to maximum number of control variables, and l varies from 1 to maximum number of output variables.

The nonlinear simulation program is used to generate the finite difference approximations of these partial derivatives, by perturbing the steady-state level of each variable and system inputs independently, in both positive and negative directions, while holding other system and control variables fixed. For each step change, corresponding values of state derivatives/outputs are calculated. Geyser (1978), has proposed a method to calculate the required values.

3.2 *Model selection*

For an F100 engine, a number of altitude and mach number flight points have to be selected, with emphasis on extreme conditions of pressure, temperature, and at locations where control system limitations exist. Variations in power settings at these flight points, account for a larger number of linear models, as shown by Hackney and Miller (1977).

4. BRIEF INTRODUCTION TO PERIODIC OUTPUT FEEDBACK

The approach we use here is based on the result by Chammas and Leondes (1979) and further work carried out by Werner and Furuta (1995). The poles of a discrete system can be arbitrarily assigned (with the natural restriction that they appear in conjugate pairs) using a piecewise constant periodic output feedback. The idea is to sample the output at time instants $t = k\tau$, where $k = 0, 1, \ldots$.The control signal is generated according to

$$\begin{aligned} u(t) &= K_l y(k\tau), k\tau + l\Delta \leq k\tau + (l+1)\Delta, \\ K_{l+N} &= K_l, \end{aligned} \quad (7)$$

for $l = 0, 1, \ldots$,where a sampling interval τ is divided into N sub-intervals, $\Delta = \frac{\tau}{N}$.

Let the system sampled at the rate $\frac{1}{\Delta}$, be $\{\Phi, \Gamma, C\}$.

Now the closed loop system is

$$x(k\tau + \tau) = \left(\Phi^N + \mathbf{\Gamma} K C\right) x(k\tau), \quad (8)$$

where,

$$\mathbf{\Gamma} = \left[\Phi^N \Gamma \ldots\ldots\ldots. \Gamma\right], \quad (9)$$

and

$$K^T = \left[K_0^T \ldots\ldots\ldots\ldots. K_{N-1}^T\right]. \quad (10)$$

The sufficient condition for the existence of the gain matrix K, such that the closed loop system has any arbitrary set of poles, is that we choose N to be greater than or equal to the controllability index ν of the system.

The problem has now taken the form of a static output feedback problem. Eq.(8) suggests, that we find an output injection gain matrix G, such that,

$$\rho\left(\Phi^N + GC\right) < 1, \quad (11)$$

where, $\rho()$ denotes the spectral radius.
By observability, one can choose an output injection gain G, to achieve any desired self-conjugate set of eigenvalues for the closed loop matrix $(\Phi^N + GC)$, and from $N \geq \nu$, it follows that one can find a periodic output feedback gain, which realizes the output injection gain G, by solving

$$\mathbf{\Gamma} K = G \quad (12)$$

for K.

The controller obtained from Eq.(12) will give the desired behaviour, but might require excessive control action. To reduce this effect, we relax the condition that K exactly satisfy the above linear equation and include a constraint on the gain K. Thus we arrive at the equation

$$||\mathbf{\Gamma} K - G|| < \rho. \quad (13)$$

This can be formulated in the framework of Linear Matrix Inequalities as follows:

$$\begin{bmatrix} -\rho^2 I & (\mathbf{\Gamma} K - G) \\ (\mathbf{\Gamma} K - G)^T & -I \end{bmatrix} < 0. \quad (14)$$

The periodic output feedback controller obtained by the above method requires only constant gains and hence is easier to implement.

5. DESIGN OF THE CONTROLLER

The procedure used for designing the controller is as follows:

5.1 *Description of plants and control law*

Six linear models were available at the flight points / power settings, shown in Table 1. Two models were generated at the same settings (20/0/0).

Table 1: Index of linear models

Case	PLA (deg)	Alt (m)	M No
1	83	0	0
2	24	0	0
3	20	0	0
4	20	0	0
5	83	0	1.2
6	83	13,500	0.9

Two inputs have been considered here for the multivariable case, with one being the 'Main burner fuel flow,' which is the natural control parameter for any gas turbine. The other input is the 'High variable stator position(compressor)'. The outputs are the 'Compressor speed' and the 'Turbine inlet temperature,' both being critical parameters for the operation of the gas generator (core) portion of the engine. The linear models, for the above case, were found to be completely controllable and observable. The description of the engine states, inputs and outputs, is as follows:

Engine states

x_1–Fan speed;

x_2–Compressor speed;

x_3–Augmentor pressure;

x_4–Mass flow through the compressor;

x_5–Compressor discharge pressure;

x_6–Inter-turbine volume pressure;

x_7–Fan inside diameter discharge temperature;

x_8–Duct temperature;

x_9–Compressor discharge temperature;

x_{10}–Burner exit fast response temperature;

x_{11}–Burner exit total temperature;

x_{12}–Fan turbine inlet fast response temperature;

x_{13}–Fan turbine inlet slow response temperature;

x_{14}–Fan turbine exit temperature;

x_{15}–Duct exit temperature;

x_{16}–Nozzle temperature.

Engine inputs

u_1–Main burner fuel flow;

u_2–High variable stator position.

Engine outputs

y_1–Compressor speed;

y_2–Turbine inlet temperature.

5.2 *Model reduction*

Numerous model reduction techniques have been discussed by Mahmoud (1981). However, the approach we follow here, is based on the method proposed by Lamba and Rao (1974).

Consider a linear continuous time-invariant higher-order system, represented by:

$$\begin{aligned} \dot{x} &= Ax + Bu, \\ y &= Cx. \end{aligned} \tag{15}$$

If we sample this system at τ instants, we get the discrete representation as:

$$\begin{aligned} x(k+1) &= \Phi_\tau x(k) + \Gamma_\tau u(k), \\ y(k) &= Cx(k). \end{aligned} \tag{16}$$

The adjoint or the dual for the above system would be:

$$\begin{aligned} \hat{x}(k+1) &= \Phi_\tau^T \hat{x}(k) + C^T \hat{u}(k), \\ \hat{y}(k) &= \Gamma_\tau^T \hat{x}(k). \end{aligned} \tag{17}$$

Assuming that the eigen values of Φ_τ^T are real and distinct, there exists a transformation V, such that,

$$\hat{x} = V\hat{z}, \tag{18}$$

which transforms the dual system into the modal form:

$$\begin{aligned} \hat{z}(k+1) &= \Phi \hat{z}(k) + C\hat{u}(k), \\ \hat{y}(k) &= \Gamma \hat{z}(k), \end{aligned} \tag{19}$$

where,

$$\Phi = \begin{bmatrix} \lambda_1 & 0 & \cdots & 0 \\ 0 & \lambda_2 & \cdots & 0 \\ \vdots & \vdots & \ddots & \vdots \\ 0 & 0 & \cdots & \lambda_n \end{bmatrix}. \tag{20}$$

The eigenvalues of Φ are arranged in their order of dominance.

We now extract an rth order model, retaining the r dominant eigenvalues, by truncating the above system. Using Eqs.(19) and (20), we get,

$$\begin{aligned} \hat{z}(k+1) &= \begin{bmatrix} \Phi_1 & 0 \\ 0 & \Phi_2 \end{bmatrix} \hat{z}(k) + \begin{bmatrix} C_1 \\ C_2 \end{bmatrix} \hat{u}(k), \\ \hat{y}(k) &= \begin{bmatrix} \Gamma_1 & \Gamma_2 \end{bmatrix} \hat{z}(k). \end{aligned} \tag{21}$$

On truncation, we get,

$$\hat{z}_r(k+1) = \Phi_1 \hat{z}_r(k) + C_1 \hat{u}(k), \qquad (22)$$
$$\hat{y}_r(k) = \Gamma_1 \hat{z}_r(k).$$

5.3 *Output injection and periodic output feedback gains*

Using the **dlqr** command in MATLAB, one can get an output injection gain S_r , which stabilizes the reduced closed loop system $(\Phi_1 + C_1 S_r)$. Now,

$$\hat{z}_r = \left[I_r : 0_{r*(n-r)}\right] \hat{z} \qquad (23)$$
$$= \left[I_r : 0_{r*(n-r)}\right] V^{-1} \hat{x}.$$

Therefore, we get,

$$\hat{u}(k) = S_r \left[I_r : 0_{r*(n-r)}\right] V^{-1} \hat{x} \qquad (24)$$
$$= S \hat{x}.$$

which stabilizes the closed loop system $\left(\Phi_\tau^T + C^T S\right)$. Therefore,

$$\left(\Phi_\tau^T + C^T S\right)^T = \left(\Phi_\tau + S^T C\right) \qquad (25)$$

will also be stable.

Thus $S^T \equiv G$ is the output injection gain for the system in Eq.(16). Then, $\Gamma K = G$, can be solved for K, using LMI framework.

In this case, six LMIs are solved for the six plants, to get a common K, which would stabilize and give a good response for all the plants.

5.4 *Simulation results*

The open loop impulse response of the plants, indicate that the system is inherently stable with the eigenvalues lying inside the unit circle.

As discussed in the earlier section, the periodic output feedback gain K is arrived at, using the output injection gain G. For low values of τ, it is observed that the gain K is very high. However, by manipulation of the Q and R weighing matrices, one can get suitable values of K with good closed loop response.

The dominating time constant for the plants, is found to vary in the range of 0.1 to 1.67. A robust K in terms of stability and performance, has been obtained at $\tau = 0.15$ in this case. The gain values are low and realizable.

Sample closed loop response of the plants, along with the control efforts, are shown in Figs. 1 to 4.

Fig. 1 indicates the closed loop responses to the combined inputs in case of Plant 2, with the outputs being the Compressor speed and the Turbine entry temperature. The responses are stable and settle early. However, the initial fluctuations in the Turbine entry temperature is slightly high, attributable to the combined control effort.

Fig. 2 indicates the control efforts in case of Plant 2, with the inputs being the Main burner fuel flow and the High variable stator position. The initial fluctuations are slightly high, but again the settling time is very low, thus making it practically acceptable.

Fig. 3 indicates the closed loop responses to the combined inputs in case of Plant 5. The results are similar as in Fig. 1. Fig. 4 indicates the control efforts in case of Plant 5. Here again, the results are similar as in Fig. 2.

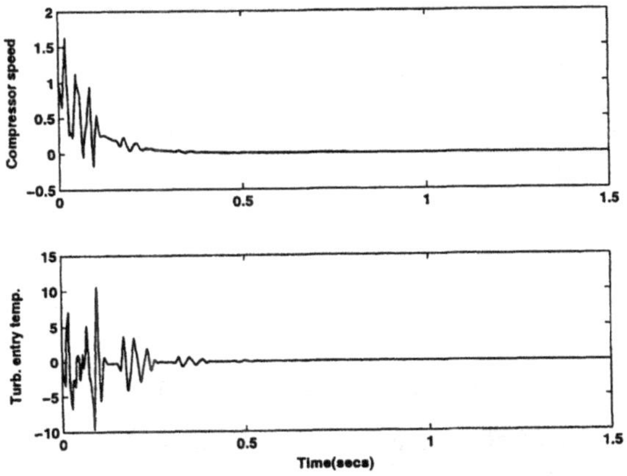

Fig. 1. Closed loop responses of Plant 2

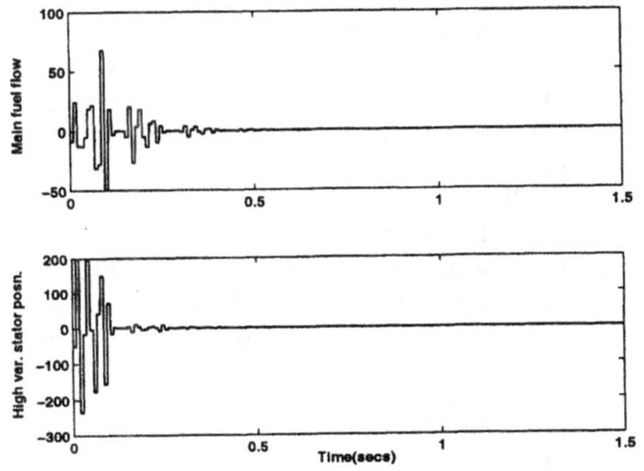

Fig. 2. Control efforts in case of Plant 2

6. CONCLUSION

The controller design for an F100 turbofan engine using piecewise constant periodic output feedback, has been discussed in this paper, with a brief

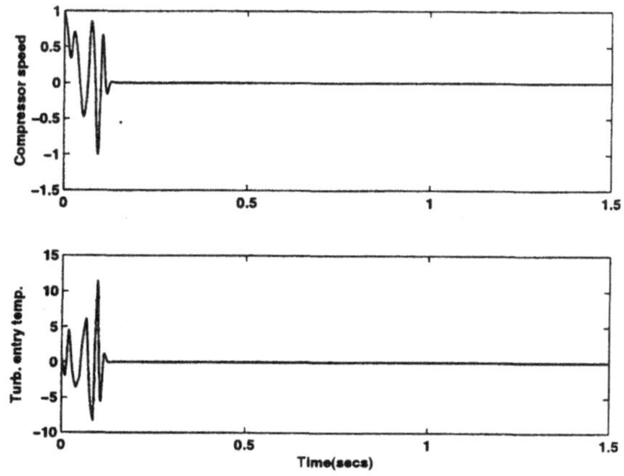

Fig. 3. Closed loop responses of Plant 5

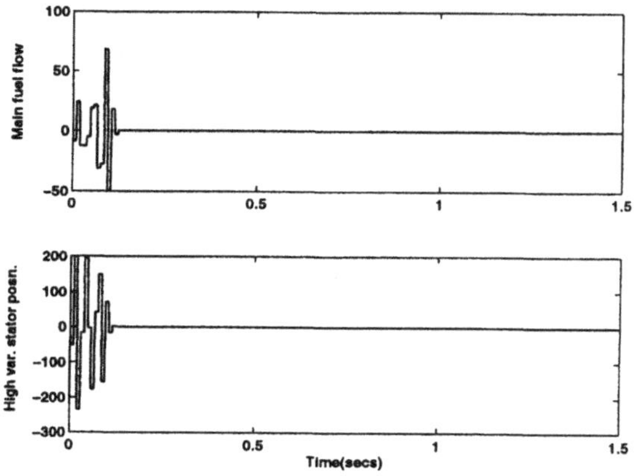

Fig. 4. Control efforts in case of Plant 5

mention of the modeling, simulation and linear model generation and selection aspects.

In practice, not all of the states are available for measurement. In this case, the optimal control law requires to design the state observer. This technique increases the implementation cost and reduces the reliability of the control system. Hence it is desirable to go for a periodic output feedback design which has the advantages of static output feedback and also guarantees complete pole-assignability. With the added features of ease of implementation and the ability to stabilize a larger class of systems, this technique appears to be a viable method for the design of a multivariable controller for Gas turbines.

The periodic output feedback gain K, arrived at via a reduced order model, is observed to perform satisfactorily.

Acknowledgement

The authors express their sincere gratitude towards Prof S. Sane of the Aerospace department, IIT Bombay, for his inspiring and encouraging guidance.

References

Chammas, A.B. and C.T. Leondes (1979). Pole-assignment by piecewise constant output feedback. *Int. Journal of control,* **29**, (1), 31-38.

Cohen, H., G.F.C. Rogers and H.I.H. Saravanamuttoo (1972). *Gas turbine theory.* Longman Group Limited, England.

De Hoff, R.L., W.E. Hall, R.J. Adams and N.K. Gupta (1977). F 100 multivariable control synthesis. *Systems control Inc.,* **AFAPL-TR-77-35**, Vol I and II.

Geyser, L.C. (1978). DYGABCD-A program for calculating linear A, B, C and D matrices from a nonlinear dynamic engine simulation. *Lewis research centre,* NASA.

Hackney, R.D., and R.J. Miller (1977). *Engine criteria and models for multivariable control system design,* AFAPL, Ohio.

Lamba, S.S. and S.V. Rao (1974). On suboptimal control via the simplified model of Davison. *IEEE Trans. Automat. Control,* **AC-19**, 448-450.

Mahmoud, M.S. and M.G. Singh (1981). *Large scale systems modeling.* Pergamon press, UK.

Seller, J.F. and C.J. Danielle (1975). DYNGEN-A program for calculating steady-state and transient performance of turbo-jet and turbo-fan engines. *Lewis research centre,* NASA.

Watts, J.W., T.E. Dwan and C.G. Brockus (1992). Optimal state-space control of a gas turbine engine. *ASME Trans. Engg. for GT and Power,* **114**, 763-767.

Werner, H. and K. Furuta (1995). Simultaneous stabilization based on output measurement. *Kybernetica,* **31**, (4), 395-411.

Winterbone, D.E., N. Munro and P.M.G. Lourtie (1973). A preliminary study of the design of a controller for an automotive gas turbine. *ASME Trans.Engg.for Power,* 244-256.

Copyright © IFAC Periodic Control Systems,
Cernobbio-Como, Italy, 2001

PERIODIC OUTPUT FEEDBACK CONTROL OF A LARGE NUCLEAR REACTOR

Chandan Nene * B. Bandyopadhyay * A. P. Tiwari **

* *Systems and Control Engineering Group, Department of Electrical Engineering, Indian Institute of Technology Bombay, Powai, Mumbai-400 076, INDIA.*
** *Reactor Control Division, Bhabha Atomic Research Centre, Trombay, Mumbai-400 085, INDIA.*

Abstract: The paper presents the design of piece-wise constant periodic output feedback control for a discrete three time-scale system resulting from the discretization of a continuous-time standard three time-scale system. By a suitable linear transformation of state variables the given continuous-time system model is converted into a block diagonal form in which the three time scale system is decoupled into a slow, a fast and a faster subsystem, respectively. The periodic output feedback gain for the slow subsystem is then calculated from the output injection gain computed from the slow subsystem and the same for the fast subsystems are set equal to zero. Finally the periodic output feedback gain for the composite system is obtained using all the above mentioned periodic output feedback gains. The method has been applied to a large Pressurized Heavy Water Reactor (PHWR) for control of Xenon-induced spatial oscillations and the efficacy of control has been demonstrated by simulation of transient behaviour of the nonlinear model of the PHWR. *Copyright ©2001 IFAC.*

Keywords: Nuclear Reactor, Space–time Kinetics, Singular perturbations and time–scale methods, Optimal Control.

1. INTRODUCTION

In a large Pressurized Heavy Water Reactor, such as in the 500 MWe PHWR, the neutronic coupling among different core–regions is so week that any deliberate attempt to operate the reactor with flattened axial and radial flux distribution is also accompanied with the xenon–induced spatial oscillations, which require control. If the oscillations in the power distribution are left uncontrolled, the power density and the time rate of change of power at some locations in the core may exceed the respective limits which may cause fuel failure. In several heavy water reactors, operational in different parts of the world, means for automatic control of power distribution have been devised. Apparently, these control systems have been designed utilizing the usual static output feedback concept. However, since late 80s new design trends, which would render better closed loop response characteristics, have emerged. Hence, there is a renewed interest in looking at the problem of spatial control.

The design of automatic power distribution control systems for nuclear reactors is a complicated task because the representation of the spatial effects in a nuclear reactor requires a large number of equations and the model is characterised by the simultaneous presence of both the slow and fast varying modes. While applying the design methods, special care must be taken to overcome the stiffness and ill–conditioning problems. In this paper, the singularly perturbed structure of the nuclear reactor model is exploited to decompose it into a slow subsystem, which is unstable, and two fast subsystems which are stable. An output injection gain is then designed for the slow subsystem only and the output injection gains for the fast subsystems are set equal to zero. An ingenious method is suggested whereby the periodic output feedback gains for the original system are determined without any significant difficulty. It is shown that the periodic output feedback control yields a satisfactory closed loop response in the case of the 500 MWe PHWR. The proposed design method eliminates the problems of stiffness and ill-conditioning.

2. MODEL OF THE PHWR

The space–time kinetic behaviour in a nuclear reactor is usually described by multigroup diffusion equations(Duderstadt and Hamilton, 1983; Henry, 1976; Stammler and Abbate, 1983). However, in the case of control design problems, the state equations are obtained by suitable approximation of the neutron diffusion equations and associated equations describing the dynamics of delayed neutron precursors, for the treatment of the fast transients. In spatial control problems involving slow transients, equations describing the dynamics of iodine and xenon are also needed. For a realistic modeling, the thermal reactivity feedback should also be considered. However, this effect is very small in PHWRs and it may be ignored while dealing with the spatial control problem. Also a two group formulation of the diffusion equations with one effective group of delayed neutrons is usually sufficient.

Employing the nodal approximation of the two group diffusion equations, the following dynamic equations can be derived for the 500 MWe PHWR(Nene, 2001):

$$l_{1i}\frac{d\phi_{1i}}{dt} = -\left[1+\frac{\omega_{1ii}+\Sigma_{1\to 2i}}{\Sigma_{a1i}}\right]\phi_{1i} + \sum_{j=1}^{N}\frac{\omega_{1ij}}{\Sigma_{a1i}}\phi_{1j} + \frac{(1-\beta)\left[\nu\Sigma_{f1i}\phi_{1i}+\nu\Sigma_{f2i}\phi_{2i}\right]}{\Sigma_{a1i}} + \frac{1}{\Sigma_{a1i}}\lambda C_i, \quad (1)$$

$$l_{2i}\frac{d\phi_{2i}}{dt} = -\frac{\omega_{2ii}}{\Sigma_{a2i}}\phi_{2i} + \sum_{j=1}^{N}\frac{\omega_{2ij}}{\Sigma_{a2i}}\phi_{2j} + \frac{\Sigma_{1\to 2i}}{\Sigma_{a2i}}\phi_{1i} - \left[1+\frac{\sigma_{ax}}{\Sigma_{a2i}}X_i + \frac{\kappa_i}{\Sigma_{a2i}}H_i\right]\phi_{2i}, \quad (2)$$

$$\frac{dC_i}{dt} = \beta\left[\nu\Sigma_{f1i}\phi_{1i}+\nu\Sigma_{f2i}\phi_{2i}\right] - \lambda C_i, \quad (3)$$

$$\frac{dI_i}{dt} = \gamma_I\left[\Sigma_{f1i}\phi_{1i}+\Sigma_{f2i}\phi_{2i}\right] - \lambda_I I_i, \quad (4)$$

$$\frac{dX_i}{dt} = \gamma_X\left[\Sigma_{f1i}\phi_{1i}+\Sigma_{f2i}\phi_{2i}\right] + \lambda_I I_i - \left[\lambda_X + \sigma_{ax}\phi_{2i}\right]X_i, \quad (5)$$

$$\frac{dH_i}{dt} = -mq_i, \quad (6)$$

$$P_i = E_{eff}V_i\left[\Sigma_{f1i}\phi_{1i}+\Sigma_{f2i}\phi_{2i}\right], \quad (7)$$
$$l_{1i} = \frac{1}{v_{1i}\Sigma_{a1i}}, \quad l_{2i} = \frac{1}{v_{2i}\Sigma_{a2i}},$$
$$(i = 1, 2, \ldots, N)$$

where N denotes the number of zones in the reactor, ϕ_{1i}, ϕ_{2i}, C_i, I_i, X_i and P_i denote respectively the fast group flux, thermal group flux, effective one group delayed neutron precursor concentration, iodine concentration, xenon concentration and power level in the zone i; V_i denotes the volume of zone i; E_{eff} denotes the average energy liberated in a fission reaction; λ, λ_I and λ_X denote the decay constants respectively for the effective one group delayed neutron precursor, iodine and xenon; ν denotes the average neutron yield in thermal fission, β the delayed neutron fraction and γ_I and γ_X the fission yields of iodine and xenon respectively; v_{1i} and v_{2i} denote respectively the average speed of fast and thermal neutrons in the reactor; ω_{1ii}, ω_{1ij}, ω_{2ii} and ω_{2ij} are nodal coupling coefficients characterizing the leakages of fast and thermal neutron fluxes among zones; Σ_{a1i} and Σ_{a2i} denote respectively the fast and thermal absorption crosssection, Σ_{f1i} and Σ_{f2i} the fission crosssections, and $\Sigma_{1\to 2i}$ the slowing down crosssection for fast neutrons; σ_{ax} denotes the microscopic thermal neutron absorption crosssection of xenon; H_i denotes the water level in the zone controller, q_i the signal to the inflow control valve of the ith zone, m the rate of change of water level in the zone controllers for unit change in the input signal to the control valve and κ_i denotes the change in thermal group neutron absorption crossection per unit change in the water level(Tiwari *et al.*, 1996; Tiwari *et al.*, 2000).

Accuracy of the model shall be in general better for large N. However, in case of the 500 MWe PHWR, satisfactory accuracy is obtained for $N = 14$. Thus there are 84 differential equations needed to describe the space–time behaviour of the reactor.

The above equations can be linearized around the steady state operating point to obtain the following standard state space representation:

$$\begin{bmatrix} \dot{z}_1 \\ \dot{z}_2 \\ \dot{z}_3 \end{bmatrix} = \begin{bmatrix} A_{11} & A_{12} & A_{13} \\ A_{21} & A_{22} & A_{23} \\ A_{31} & A_{32} & A_{33} \end{bmatrix} \begin{bmatrix} z_1 \\ z_2 \\ z_3 \end{bmatrix} + \begin{bmatrix} B_1 \\ 0 \\ 0 \end{bmatrix} \delta Q, \quad (8)$$

$$y = \begin{bmatrix} 0 & M_2 & M_3 \end{bmatrix} \begin{bmatrix} z_1 \\ z_2 \\ z_3 \end{bmatrix} \quad (9)$$

where $A_{11} \ldots A_{33}$, B_1, M_2 and M_3 are matrices of suitable dimensions,

$$z_1 = \left[\frac{\delta I_1}{I_{10}} \cdots \frac{\delta I_N}{I_{N0}} \frac{\delta X_1}{X_{10}} \cdots \frac{\delta X_N}{X_{N0}} \frac{\delta C_1}{C_{10}} \cdots \right.$$
$$\left. \cdots \frac{\delta C_N}{C_{N0}} \delta H_1 \ldots \delta H_N \right]^T,$$

$$z_2 = \left[\frac{\delta \phi_{21}}{\phi_{210}} \cdots \frac{\delta \phi_{2N}}{\phi_{2N0}} \right]^T;$$

$$z_3 = \left[\frac{\delta \phi_{11}}{\phi_{110}} \cdots \frac{\delta \phi_{1N}}{\phi_{1N0}} \right]^T,$$

$$y = \left[\frac{\delta P_1}{P_{10}} \cdots \frac{\delta P_N}{P_{N0}} \right]^T,$$

$$\delta Q = \left[\delta q_1 \ldots \delta q_N \right]^T,$$

and ϕ_{1i0}, ϕ_{2i0}, C_{i0}, I_{i0}, X_{i0} and P_{i0} denote the steady state values of the respective quantities.

It is worth to note that the states z_1 consists of quantities varying slowly with time, z_2 consists of quantities varying rapidly and z_3 of those varying very rapidly. The state space model corresponding to this particular grouping of the states makes the application of time–scale methods, convenient(Kokotovic *et al.*, 1986). The eigenvalues of the system matrix in the state equation (8), are seen to lie in three widely spaced clusters indicating that the PHWR model possesses the simultaneous presence of both the fast and slow phenomena, which is known to cause ill–conditioning and severe computational problems in control design and simulation. The singular perturbation approach is particularly suited in such a case. A particular property of the standard singularly perturbed model of special interest is that the model given by (8) and (9), can be transformed into the following block diagonal form(Kokotovic *et al.*, 1986):

$$\begin{bmatrix} \dot{z}_s \\ \dot{z}_f \\ \dot{z}_{ff} \end{bmatrix} = \begin{bmatrix} A_s & 0 & 0 \\ 0 & A_f & 0 \\ 0 & 0 & A_{ff} \end{bmatrix} \begin{bmatrix} z_s \\ z_f \\ z_{ff} \end{bmatrix} + \begin{bmatrix} B_s \\ B_f \\ B_{ff} \end{bmatrix} \delta Q, \quad (10)$$

$$y = \begin{bmatrix} M_s & M_f & M_{ff} \end{bmatrix} \begin{bmatrix} z_s \\ z_f \\ z_{ff} \end{bmatrix}. \quad (11)$$

Note that the original state equation having three widely separated clusters of eigenvalues is decoupled into a slow subsystem (A_s, B_s) of order 56, having eigenvalues of very small magnitude, a fast subsystem (A_f, B_f) of order 14 having eigenvalues of large magnitude and a faster subsystem (A_{ff}, B_{ff}) of order 14 having eigenvalues of very large magnitude. Moreover, the fast and faster subsystems are stable and have eigenvalues with large negative real parts.

3. CONTROL OF THE PHWR

For control of the 500 MWe PHWR, an observer based state feedback can be suggested. However, the observer based controller has generally very complex structure(Tiwari and Bandyopadhyay, 1998). Moreover, even a small variation in the model parameters and the reactor parameters, may result into significant degradation of the closed loop performance. In contrast to this, the control of the reactor based on the periodic output feedback is simpler and the variation of the parameters does not pose much problem as only the feedback of outputs is required. A brief description of the technique and its extension to multi–time–scale systems is given in the following.

3.1 *Periodic Output Feedback*

It has been established by Chammas and Leondes (1979) that the poles of a controllable and observable system discretized at the output sampling rate can be arbitrarily assigned by a piecewise constant periodic output feedback, provided the number of changes in the gain during one output sampling interval is greater than or equal to the controllability index of the system. The approach can be briefly described as follows (Werner and Furuta, 1995).

Consider the linear time invariant model

$$\dot{z} = Az + Bu, \quad (12)$$
$$y = Mz. \quad (13)$$

The discrete model corresponding to the above state equation, for sampling at the rate $\frac{1}{\tau}$, called as the output sampling rate, is

$$z[(k+1)\tau] = A_\tau z[k\tau] + B_\tau u[k\tau], \quad (14)$$
$$y[k\tau] = Mz[k\tau], \quad (15)$$

where

$$A_\tau = e^{A\tau}, \tag{16}$$

$$B_\tau = \int_0^\tau e^{At} B dt. \tag{17}$$

Now, let the output sampling interval, τ, be divided into N_c subintervals, $\Delta = \frac{\tau}{N_c}$. Then, the model corresponding to the sampling at the rate $\frac{1}{\Delta}$ will be

$$z[(l+1)\Delta] = A_\Delta z[l\Delta] + B_\Delta u[l\Delta], \tag{18}$$

$$y[l\Delta] = Mz[l\Delta], \tag{19}$$

where

$$A_\Delta = e^{A\Delta}, \tag{20}$$

$$B_\Delta = \int_0^\Delta e^{At} B dt. \tag{21}$$

Further, from (16) and (20) $A_\tau = A_\Delta^{N_c}$. Now let G be an output injection gain that yields the required performance for the system given by (14) and (15), *i.e.* the eigenvalues of $(A_\tau + GM)$ are at the desired locations in the circle of unit radius. Then there exists a sequence of gains

$$K_L = \begin{bmatrix} K_0^T & K_1^T & \dots & K_{N_c}^T \end{bmatrix}^T \tag{22}$$

such that when the control input is applied according to the rule,

$$u(t) = K_l y[k\tau] \tag{23}$$

where, $k\tau + l\Delta \le t < k\tau + (l+1)\Delta$ and,

$$K_{l+N} = K_l, \quad l = 0, 1, \dots N_c - 1,$$

the closed loop system corresponding to (14) and (23) given by,

$$z[(k+1)\tau] = A_\Delta^{N_c} z[k\tau] + \Gamma_L K_L M z[k\tau] \tag{24}$$

where

$$\Gamma_L = \begin{bmatrix} A_\Delta^{N_c-1} B_\Delta & A_\Delta^{N_c-2} B_\Delta & \cdots & B_\Delta \end{bmatrix} \tag{25}$$

is stable. Assuming that the system given by (14) and (15) is observable, it is possible to determine the output injection gain, G, such that $(A_\tau + GM)$ has desired performance characteristics. Hence, a periodic output feedback gain sequence K_L, determined from the relation

$$\Gamma_L K_L = G \tag{26}$$

would realize the output injection gain G. However, it is found that the time–varying gain K_L causes excessive variation of input during an output sampling interval though it meets the performance specifications at the output sampling instants. Hence, it is necessary to improve the closed loop behaviour of the system in some way. For this, the approach of Werner and Furuta (1995) is used. One begins by selecting a performance index which consists of a quadratic term in state, z, and a quadratic term in input, u, of the model given by (18) and a term that relaxes, to some extent, the condition that K_L achieves the same right hand side as in (26). The consideration leads to a two point boundary value problem which is solved to obtain K_L, that optimizes the closed loop system behaviour, during the output sampling intervals.

3.2 *Periodic Output Feedback Control for Three Time Scale Systems*

As the 500 MWe PHWR is a three time scale system, the attention is now turned to the determination of periodic output feedback gain sequence, K_L, for system as one described by equations (10) and (11). Here we have to consider the state z, in (12) as composite state vector of slow, fast and faster states, *i.e.*, $z = \begin{bmatrix} z_s^T & z_f^T & z_{ff}^T \end{bmatrix}^T$. The task of finding K_L from (26) for a G may seem trivial albeit the problems of ill–conditioning, usually encountered in case of multi–time scale systems. To overcome this difficulty, the approach in which all the slow, fast and faster phenomenon are separately considered is required. To begin with, consider the following discrete model for the system described by (10) corresponding to the sampling at the rate $\frac{1}{\Delta}$:

$$\begin{bmatrix} z_s[(l+1)\Delta] \\ z_f[(l+1)\Delta] \\ z_{ff}[(l+1)\Delta] \end{bmatrix} = \begin{bmatrix} A_{\Delta_s} & 0 & 0 \\ 0 & A_{\Delta_f} & 0 \\ 0 & 0 & A_{\Delta_{ff}} \end{bmatrix} \begin{bmatrix} z_s[l\Delta] \\ z_f[l\Delta] \\ z_{ff}[l\Delta] \end{bmatrix} + \begin{bmatrix} B_{\Delta_s} \\ B_{\Delta_f} \\ B_{\Delta_{ff}} \end{bmatrix} u \tag{27}$$

and the output equation corresponding to the sampling at the rate $\frac{1}{\tau}$

$$y[k\tau] = \begin{bmatrix} M_s & M_f & M_{ff} \end{bmatrix} \begin{bmatrix} z_s[k\tau] \\ z_f[k\tau] \\ z_{ff}[k\tau] \end{bmatrix}. \tag{28}$$

As all the modes are decoupled, it would be appropriate to consider the output injection matrix as,

$$G = \begin{bmatrix} G_s^T & G_f^T & G_{ff}^T \end{bmatrix}^T \tag{29}$$

where, G_s, G_f and G_{ff} denote the output injection matrices for the slow, fast and faster subsystems, respectively. Applying this G to the discrete system corresponding to the sampling at the rate $\frac{1}{\tau}$, the closed loop system matrix is obtained as

$$A_\tau + GM = A_\Delta^{N_c} + GM \tag{30}$$

Since both the fast and faster subsystems are stable, the output injection gains for both of these subsystems may be set equal to zero *i.e.*, $G_f = 0$ and $G_{ff} = 0$. This consideration simplifies (30) to

$$A_\Delta^{N_c} + GM = \begin{bmatrix} A_{\Delta_s}^{N_c} + G_s M_s & G_s M_f & G_s M_{ff} \\ 0 & A_{\Delta_f}^{N_c} & 0 \\ 0 & 0 & A_{\Delta_{ff}}^{N_c} \end{bmatrix}. \tag{31}$$

Using (22) and (25), the equation (26) can also be written as

$$\begin{bmatrix} A_{\Delta_s}^{N_c-1} B_{\Delta_s} & A_{\Delta_s}^{N_c-2} B_{\Delta_s} & \cdots & B_{\Delta_s} \\ A_{\Delta_f}^{N_c-1} B_{\Delta_f} & A_{\Delta_f}^{N_c-2} B_{\Delta_f} & \cdots & B_{\Delta_f} \\ A_{\Delta_{ff}}^{N_c-1} B_{\Delta_{ff}} & A_{\Delta_{ff}}^{N_c-2} B_{\Delta_{ff}} & \cdots & B_{\Delta_{ff}} \end{bmatrix} \times \begin{bmatrix} K_0 \\ K_1 \\ \vdots \\ K_{N_c-1} \end{bmatrix} = \begin{bmatrix} G_s \\ 0 \\ 0 \end{bmatrix}. \tag{32}$$

If sampling period, τ, and the number of gain changes, N_c, are chosen appropriately, the discrete system (18) would also possess three time scale property(Naidu and Rao, 1985), *i.e.*, the eigenvalues of A_{Δ_f} and $A_{\Delta_{ff}}$ will be very small. Hence, $A_{\Delta_f}, A_{\Delta_f}^2, \cdots, A_{\Delta_f}^{N_c-1}$ and $A_{\Delta_{ff}}, A_{\Delta_{ff}}^2, \cdots, A_{\Delta_{ff}}^{N_c-1}$ will be so small that they can be neglected in (32) to obtain the simplified equation

$$[\, A_{\Delta_s}^{N_c-1} B_{\Delta_s} \; A_{\Delta_s}^{N_c-2} B_{\Delta_s} \; \cdots \; A_{\Delta_s} B_{\Delta_s} \,] \times \begin{bmatrix} K_0 \\ K_1 \\ \vdots \\ K_{N_c-2} \end{bmatrix} = G_s \tag{33}$$

which can be solved for K_0, K_1, ..., K_{N_c-2} and set $K_{N_c-1} = 0$. It should be observed that (33) involves only the slow subsystem matices.

4. RESULTS

The proposed method is applied for designing the piecewise constant periodic output feedback control for the 500 *MWe* PHWR. The output sampling period, τ, and the input sampling period, Δ, should be chosen in such a way that the system given by (18) exhibits the three time-scale behaviour while N_c is larger than or equal to the controllability index of the system. Also the value of τ so selected should not cause loss of controllability and observability, which might happen due to sampling(Kuo, 1992). The fastest variable amongst the slow variables is the delayed neutron precursor concentration. The numerically largest eigenvalue of A_s was found to be 8.2138×10^{-2} *rad/sec*. In accordance with the considerations discussed above, the output sampling interval, τ, is chosen as 5 *s*. Further, let $N_c = 5$ which yields $\Delta = \tau/N_c = 1$ *s* for which the discrete system (18) also possesses 3 time scale behaviour. The output injection gain matrix G_0 which stabilizes the slow subsystem was obtained by the optimal control technique and the corresponding periodic output feedback gains, K_L, by the optimization method of Werner and Furuta (1995). It is seen that the periodic output feedback gains K_L determined using the proposed approximations also stabilizes the composite system.

To demonstrate the efficacy of the controller when placed in closed loop with the reactor, the behaviour of the system may be studied from simulation of nonlinear equations (1)–(6), for transient variation of reactivity which was simulated by perturbing the thermal neutron absorption cross section in zone 1, Σ_{a2_1}. The perturbation consisted of increasing Σ_{a2_1} linearly by 1.4% in 2.5 *s* and then reducing it back to original value in the next 2.5 *s*. The simulation was carried out using SIMULINK and the *ode15s* solver (for stiff equations) was used for computations. The variation of global

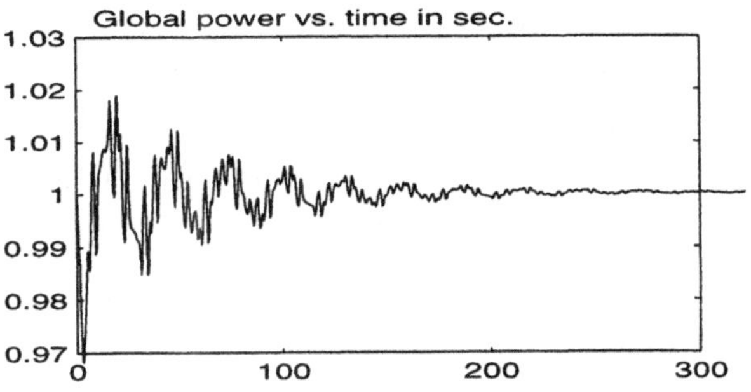

Fig. 1. Variation of normalized global power.

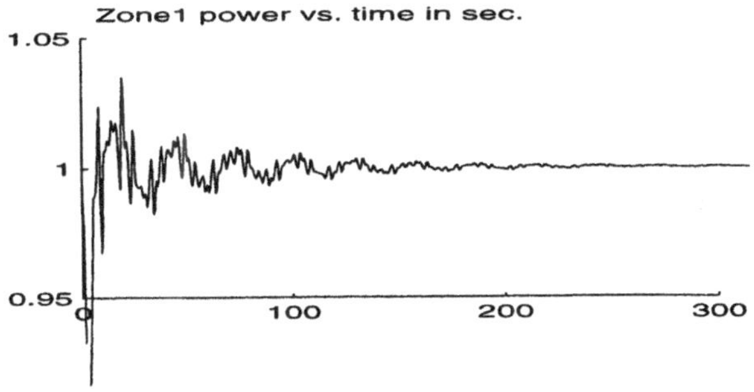

Fig. 2. Variation of normalized power in zone–1.

power and power level, xenon concentration and iodine concentration in zone–1, is shown in Fig. 1 to 4. It is seen that the global power settles quickly following the transient. The zonal power levels also settle to their respective equilibrium values in about 200 s. The changes in xenon and iodine concentrations is also small although it takes very large time for these variables to settle at the respective equilibrium values. Such a response is considered satisfactory for the operation of the 500 MWe PHWR.

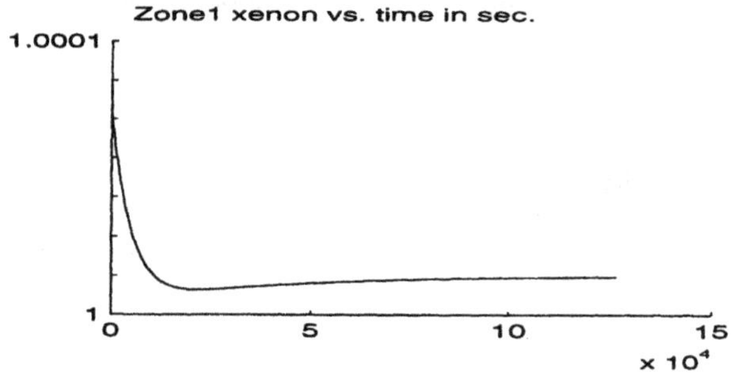

Fig. 3. Variation of normalized xenon concentration in zone–1.

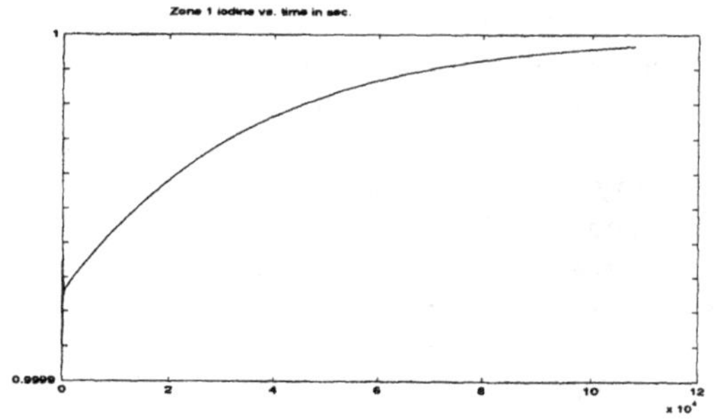

Fig. 4. Variation of normalized iodine concentration in zone–1.

5. CONCLUSION

It is well known that models of large scale physical systems are characterised by the presence of fast–decaying dynamic modes which generally do not require control. This usual property is exploited here and an approach for designing piecewise constant periodic output feedback control for multi–time–scale continuous–time systems is formulated. The method can also be applied to systems which are directly represented by standard singularly perturbed discrete–time model. It would be only necessary to transform the given model into the block diagonal form, in which fast subsystems are decoupled. Because Δ is permitted to be several times large compared to the largest time constant of the fast subsystem, speed of actuators would generally not be a limit in achieving the desired performance. Moreover, the output sampling period, τ is also large *i.e.* sampling rate is not required to be very high. This renders a considerable reduction in the cost of associated processing hardware.

Here, the attention is focussed on controlling the xenon–induced spatial oscillations in a large PHWR though the method can be directly applied to other types of thermal reactors which are described by the nodal model. The method for computing the periodic output feedback gains can be applied directly to other large scale systems.

6. REFERENCES

Chammas, A. B. and C. T. Leondes (1979). Pole assignment by piecewise constant output feedback. *International Journal of Control* **29**, 31–38.

Duderstadt, James J. and Louis J. Hamilton (1983). *Nuclear Reactor Analysis.* John Wiley & Sons Inc.. New York.

Henry, A. F. (1976). *Nuclear Reactor Analysis.* The MIT Press. Cambridge.

Kokotovic, P. V., H. K. Khalil and John O'Reilly (1986). *Singular Perturbation Methods in Control Analysis and Design.* Academic Press. New York.

Kuo, B. C. (1992). *Digital Control Systems.* Saunders College Publ.. Ft. Worth.

Naidu, D. S. and A. K. Rao (1985). *Lecture Notes in Mathematics: Singular Perturbation Analysis of Discrete Control Systems.* pp. 50–62. Vol. 1154. Springer–Verlag. Berlin Heidelberg.

Nene, Chandan R. (2001). Modeling and control of a large nuclear reactor. M. Tech Dissertation. Indian Institute of Technology Bombay, India.

Stammler, R. J. J. and M. J. Abbate (1983). *Methods of Steady State Reactor Physics.* Academic Press.

Tiwari, A. P. and B. Bandyopadhyay (1998). Control of xenon induced spatial oscillations in a large PHWR. In: *Proceedings of the IEEE Region 10 Conference on Global Connectivity in Energy, Computer Communication and Control (TENCON–98).* New Delhi, India.

Tiwari, A. P., B. Bandyopadhyay and G. Govindarajan (1996). Spatial control of a large pressurized heavy water reactor. *IEEE Transactions on Nuclear Science* **43**, 2440–2453.

Tiwari, A. P., B. Bandyopadhyay and H. Werner (2000). Spatial control of a large PHWR by piecewise constant periodic output feedback. *IEEE Transactions on Nuclear Science* **47**, 389–402.

Werner, H. and K. Furuta (1995). Simultaneous stabilization based on output measurement. *Kybernetika* **31**(4), 395–411.

Copyright © IFAC Periodic Control Systems,
Cernobbio-Como, Italy, 2001

MANAGEMENT STRATEGY OF THE PERIODICALLY FLOODED NATURE PARK "KOPAČKI RIT" (CROATIA)

Franjo Jović[*], Melita Mihaljević[#], Željko Jagnjić[*], Janja Horvatić[$]

** - University of Osijek, Faculty of Electrical Engineering*
Laboratory for Artificial Intelligence
 # - Public Institution of Nature Park "Kopački rit"
$ - University of Osijek, Institute for Biology

Abstract: Sustainable development of the periodically flooded Nature park "Kopački rit" is based on the following components: knowledge management, control mechanisms and risk assessment. The basic force function for growth and disturbance is flooding (i)regularity. Connection of flooding regularity to the main management components was proposed. Identified models of periodic flooding show high determinism. *Copyright © 2001 IFAC*

Keywords: control actions, measured value, risk, control functions, devoloping countries

1. INTRODUCTION AND PROBLEM DEFINITION

There are still a few places in Western and Central Europe that have not been completely changed through human manipulation and usage. One of them is Nature park "Kopački rit" situated in Croatia, at the inlet of Drava river into the Danube. The surface of the protected area is about 230 square kilometers. The structure of complex ecosystem of the Kopački rit area was occasionally investigated (Mihaljević, 1999). In order to maintain biodiversity and productivity and a wise use of the area, a definite strategy should be proposed and applied favoring a multidisciplinary scientific approach to the problem. Due to multidisciplines, the management project is recognized as the optimum control problem in the distributed nonlinear hierarchical large-scale system. The project task can be divided as: identification and estimation of the ecosystem balance, management tools and techniques, and design of the minimum risk system. Using the wise coordination of all three subactivities a feasible strategy of sustainable development of the area can be expected. The natural dynamic of a yearly periodic water flooding is the target function for system identification and the basic system time constant.

2. ECOSYSTEM BALANCE

A complex system exhibits a dynamic balance. A static balance can seldom be expected. Such a system can be simplified and presented as a linear system partitioned into (Åström, 1998): S_{or} – observable and reachable part, $S_{or'}$ – observable not reachable part, $S_{o'r}$ – not observable but reachable part, and $S_{o'r'}$ – not observable and not reachable part, Fig. 1. The intention of the management is to introduce some type of feedback in order to improve the transient behavior of the system, decrease the sensitivity to parameter change in the open-loop system, and eliminate steady-state errors in the open-loop system. The error is given with the expression (Åström, 1998):

$$e(k) = \frac{u_c(k)}{1 + H_o(q)H(q)} \qquad (1)$$

where H_o is transfer function of the observed process, H(q) is the transfer function of the feedback path, and u_c is input to the system, Fig 1.

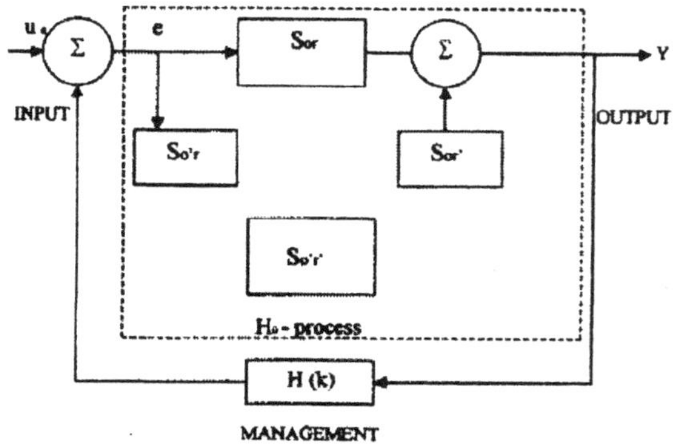

Fig.1. A general control scheme with concentrated parameters

The food chain consists of a whole hierarchy of subsystem processes and the applied scheme should be taken as a basic design scheme. The time ordered set of such models, where each order of magnitude is valid for each subsystem, make the basic structure of the system with subsystem interconnections by inputs or disturbances to the investigated subsystem.
Such subsystems are spatially nonhomogenious. The geospatial model takes into account the biodiversity character of the habitat in order to spatially model the disturbance regimes and to integrate the ground based on nonspatial data with the spatial characters of the landscape (Roy and Tomar, 2000).

3. MANAGEMENT TOOLS AND TECHNIQUES

The complex system can be managed by means of: power enforcement, money, or knowledge (Wilke, 1998). The dimensions of money as a control medium is given in Table 1

Table 1. Dimensions of economical and societal efficiency of the money controlled medium

Dimension	Economical Efficiency	Societal Efficiency
Material	Generalization of Choices	Idea of Growing
Social	Indifferential Options	Ordering of Indifferences
Temporal	Time scale of Options	Disposability of Options
Operational	Reflection of Options	Virtual Economy
Cognitive	Knowledge-Based Options	Monetarism of Knowledge

Money as all-societal activity contains shadow material (Wilke, 1998; Jović, 2000). Symbolically generated communication media are diabolically generated media (Luhmann, 1988). The expansion of the solution remains in the shadow of money, forever. The solution herein does not contain a balance between autonomy and interdependence. This does not lead to the sustainable solution of the development. The solution of management tools and media lies in the togetherness of interactions of specialistic functional systems of the modern society (Wilke, 1998). The indicators of sustainability can be obtained by different indirect measures that contain soil acidity, nutrient status and plant root physiology (Syers Hamblin and Pushparajah, 1995). Special care should be given to storage and decomposition of organic matter alongside main water flows (Delong and Brusven, 1993). Therefore, knowledge as the control medium has to be established, with the following features:

- Input side
    - quality of the input knowledge elements
    - minimum interproject delay between exact value estimation of each project contributing part
    - organizational motivation for such discipline
    - disposition of abstraction and abstract language of the team learning process
- Infrastructure side
    - adequate database system
    - semantic approach to the database material
    - thematic, conceptual, and user problem-oriented structure of the database
    - development of knowledge state indicators
    - precision of the information
    - pointing to internal and external experts for help procedures
    - availability of such experts
- User's side
    - active and routine usage of the knowledge base
    - evaluation of the knowledge usage
    - readiness of each user to enter into the expert role.

4. DESIGN OF THE RISK PREVENTION SYSTEM

Risk in complex systems can have many roots (Grabowski et. al., 2000): internal, external, human, organizational. Risk in a distributed system can migrate making its identification and mitigation difficult. Modeling risk in distributed systems is also difficult because it can have long incubation periods due to poor information flow between distributed subsystems, making risk analysis and identification of leading error chains difficult. The right combination of triggering events can awake a great risk possibility as well. Finally, addressing human and organizational errors in such distributed system is more complicated due to their dispersed physical oversight. The design of a risk prevention system is based on the risk modeling procedure. Risk, being basically the probability of accident multiplied with its cost, deserves formulation of the following steps in its design:

- establishment of the error chain because accidents do not happen "accidentally"(Vorko and Jović, 1992)

- modeling of the risk domain (historical performance, risk dynamic, risk domain, human and organizational error),
- risk assessment processes

Risk assessment processes, according to NCR's Committee on Risk Assessment identify the following general objectives as:

- *getting the science right*, means that risk analyses should follow the scientific standards in metric, analytic, data, plausibility and detachment procedures
- *getting the right science*, means that specific sciences should be addressed according to the type of risk events expected or eventually happened
- *getting the right participation*, means that decision-relevant number of people are involved
- *getting the participation right,* means that information, viewpoints, and concerns have been adequately represented.
- *development of an accurate, balanced and informative synthesis*, meaning that risk characterization should reflect relevant and state-of-the–art knowledge of the processes at risk.

The disturbances introduced into the food chain have been considered as the major risk of the system biodiversity of the national park Kopački rit (Horvatić et.al., 2000).

However the terrain organization enables very low vegetation gradients leading to the sensitive landscapes (Bridge and Johnson, 2000).

5. IDENTIFICATION OF THE YEARLY PERIODIC FLOODING PROCESS OF THE NATURE PARK "KOPAČKI RIT"

The Nature Park "Kopački rit", Fig.2. is uncontrollably flooded at extreme high and uncontrollably dried at medium or low water levels of the Drava and Danube river and semicontrollably flooded at medium high water levels (Tadić, 2001).

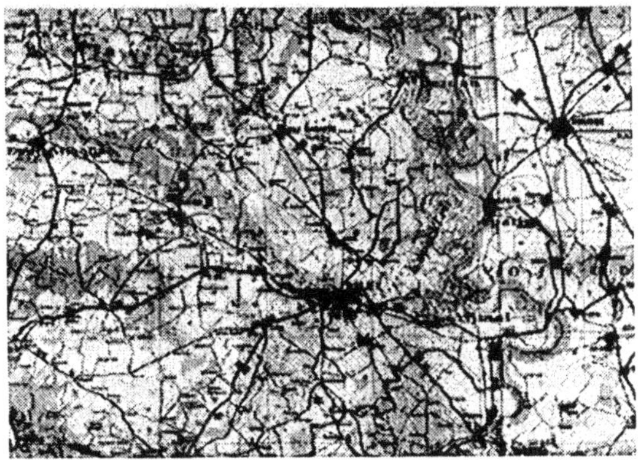

Fig.2. Water level measurement points in the area of "Kopački rit"

Two AI identification tools have been used, similar to neural networks but exhibiting algebraic explicit forms of the solution, based on the circular quantitative to qualitative information conversion (Jović, 1997; Jagnjić, 2001). Data from the six level measurement points on Drava river and Danube from 1951 to 1999 were taken during spring months: March, April and May as an mean daily levels (Hrvatske vode, 1999). Ten groups of solution have been found that follow the same or similar algebraic form and three of them are presented in Table 2, Table 3 and Table 4.

Table 2. Three groups of spring flooding models for the water level at Aljmaš site (Medusa 2000)

Group No.	Goal function	Medusa-2000 modeling algorithm
1.	Aljmaš51	Ap+0.08 Os/Bez.
1.	Aljmaš64	Ap+0.02 Os/Bez
1.	Aljmaš65	Ap
1.	Aljmaš85	Ap+0.06 DMih/Bez
2.	Aljmaš65	Ap
2.	Aljmaš51	Ap+0.08 Os/Bez
2.	Aljmaš70	Ap+0.35 Os/Bez
2.	Aljmaš73	Ap
3.	Aljmaš88	Ap+0.03 DMih/Ap
3.	Aljmaš52	Ap+0.05 Os/Bez
3.	Aljmaš55	Ap+0.01 Os/Bez
3.	Aljmaš71	Ap+0.04 Os/Ap

Table 3. Three groups of spring flooding models for the water level at Aljmaš site (QEks algorithm)

Group No.	Goal Function	QEks modeling algorithm
1.	Aljmaš51	Ap+0.06 DMih+0.01 Os
1.	Aljmaš64	Ap.+0.12 DMih+0.02 Os
1.	Aljmaš65	Ap
1.	Aljmaš85	Ap+0.13 Os
2.	Aljmaš65	Ap
2.	Aljmaš51	Ap+0.06 DMih+0.01 Os
2.	Aljmaš70	Ap+0.64 Os
2.	Aljmaš73	Ap
3.	Aljmaš88	Ap+0.5 Os
3.	Aljmaš52	Ap+0.17 Os
3.	Aljmaš55	Ap
3.	Aljmaš71	Ap+0.14 Os

Legend: Os and DMih designate water level of Drava river at Osijek and Donji Miholjac; Ap and Bez designate water level of Danube river at Bezdan and Apatin. Aljmaš51 designates water level of Drava and Danube summing point at Aljmaš during spring of the year 1951.

Table 4. Controllability of water level for three groups of spring floods

Group No.	Goal function	Controllability of water level in the park
1.	Aljmaš51	Medium
1.	Aljmaš64	Medium
1.	Aljmaš65	Low
1.	Aljmaš85	High
2.	Aljmaš65	Low
2.	Aljmaš51	Low
2.	Aljmaš70	Extremely low
2.	Aljmaš73	Medium
3.	Aljmaš88	Extremely low
3.	Aljmaš52	Low
3.	Aljmaš55	Low-medium
3.	Aljmaš71	High

Legend: Attribute of controllability according to time proportion of the water level state between 2 and 4 meters

Correlation coefficients were always higher than 0.99, or determinacy higher than 0.98. Goal function was chosen as the water level at the inlet of rivers at the Aljmaš site. The high determinacy of connection between water levels on Danube and Drava river and lake Sakadaš in the inside part of the "Kopački rit" Nature Park was proven earlier (Jović and Mihaljević, 1998).

The field validation of a "habitat hydraulics" approach should give basic directions in a mesoscale units of the channel bed description as "physical biotopes" leading to a pragmatic use of river flooding regularity (Newson and Newson, 2000).

6. DISCUSSION AND INDICATION OF FURTHER INVESTIGATION

From the standpoint of the Nature Park "Kopački rit" the uncontrollably high water levels were observed in spring periods of the years 1926,1965 and 1972 (Mihaljević, 1999). Models of the summing point at Aljmaš show for such occasions a complete correlation with the water level at Apatin. For medium high water levels, the water level at Aljmaš is partly controllable from the Drava river for the medium proportion of the spring flooding period. There are somehow bulk time intervals when any intervention is not possible because we cannot reach the state of control. Therefore the main task of the management team is to establish the distinction between reachable and nonreachable points of the spring flooding period. This is fortunately possible because of the high level of prediction of the majority of the observed spring flooding periods. Any manipulation during the reachable and controllable period of the flooding wave should involve a risk model regarding the food chain of the Nature Park. Some states of the food chain are not observable. The task of the biodiversity investigation is to establish the distinction among observable and unobservable states of the resilient system. Concerning the control scheme – their hierarchy, number of levels, and coordination methods have to be defined. The other problem might be spring waves that are not recognizable from previous observations. Obviously they have to be taken as uncontrollable periods.

The authors are thankful to the management of the Hrvatske Vode (Water Management Agency) in Osijek for their support and long term cooperation in the investigation process.

REFERENCES

Åström, K.J., Wittenmark B. (1984). *Computer Controlled Systems, Theory and Design,* Prentice Hall International, Inc. Englewood Cliffs, N.J.

Bridge, S.R.J., and Johnson, E.A. (2000). Geomorphyic Principles of Terrain Organization and Vegetation Gradients Source, Journal of Vegetation Science, 11(1)pp 57-70.

Delong, M.D. Brusven M.A. (1993). Storage and Decomposition of Organic Matter Along the Longitudinal Gradient of an Agriculturally-impacted Stream, Hydrobiologia. 262 (2) pp 77-88.

Grabowski M., Merrick J.R.W., Harrald J.R., Mazzuchi T.A., and van Dorp J.R. (2000). Risk Modelling in Distributed, Large-Scale Systems, *IEEE Transactions on Systems Man, and Cybernetics, Part A: Systems and Humans,* Vol.30, No.6., pp 651-660.

Horvatić, J., Jović, F. and Popović, Ž. (2000). Model of the Algal Growth Potential (AGP) of Chlorella kessleri FOT et NOV. in vitro Samples from Lake Sakadaš by Miniaturized Bioassay. In *Proceedings of the Danube Conference Limnological Reports of International Association for Danube Research of the International Association of the theoretic and applied limnology* (Horvatić, J. (Ed)), pp 81-88, Faculty of Education, Josip Juraj Strossmayer University of Osijek, Osijek

Hrvatske vode (1999). (Croatian Water Management) Data on water level measurement from Drava and Danube river from 1929 to 1999.

Jagnjić, Ž. (2001). Master Thesis, Faculty of Electrical and Computer Engineering, Zagreb, in procedure.

Jović, F. (1997). Qualitative Reasoning and a Circular Information Processing Algebra, *Informatica,* 21, pp. 31-47.

Jović, F., Mihaljević, M. (1998). A discrete water level model for water management in the 4D GIS concept of the Kopački rit wetland, In: *Proc. of the Wydzial Techniki Uniwersytetu Slaskiego, Pretwarzieni i ochrona olanych,* pp. 198-208, Katowice,

Jović F. (2000). Die Technik und das Unbewusste, Fachhochschule Albstadt Sigmaringen, November 2000. Vortrag für VDI/VDE. Author's publication.

Luhmann, N. (1988). *Die Wirtschaft der Gesellschaft*, Frankfurt: Suhrkamp

Mihaljević, M., et all (1999). *Kopački rit – Pregled istraživanja i bibliografija Kopački rit – (Surveillance of Investigation and Bibliography)*; Hrvatska akademija znanosti i umjetnosti, Zavod za znanstveni rad, Zagreb- Osijek,

Newson M.D. and Newson C.L. (2000(. Geomorphology, ecology and River Channel Habitata: Mesoscale Approaches to Basin-Scale Challenges, Progress in Physical Geography, 24 (2) pp 195-217.

Roy, P.S. and Tomar, S. (2000). Biodiversity Characterization at Landscape Level Using Geospatial Modelling Technique, Biological Conservation 95 (1) pp 95-109.

Syers, J.K. Hamblin, A. and Pushparajah, E. (1995). Indicators and Thresholds for the Evaluation of Sustainable Land Management, Canadial Journal of Soil Science, 75 (4), pp 423-428.

Vorko, A, and Jović, F. (1992). Macro Model Prediction of Elderly People's Injury and Death in Road Traffic Accidents in Croatia, *Accident Analysis and Prevention*, Vol 24, No. 6, pp 667-672

Wilke H. (1998). *Systemtheorie III: Steuerungstheorie,* 2. Auflage, Lucius&Lucius Verlag, Stuttgart.

Tadić Z. (2001). private communication

Copyright © IFAC Periodic Control Systems,
Cernobbio-Como, Italy, 2001

Studying a Basic Price Equation as a Periodic System

Teresa P. de LIMA

Member of Instituto de Sistemas e Robótica - Pólo de Coimbra, Portugal

Faculdade de Economia da Universidade de Coimbra
Av. Dias da Silva, 165, 3004-512, Coimbra, Portugal.
e-mail: tpl@sonata.fe.uc.pt

Abstract:

This paper is concerned with the study of the basic price equation of the dynamic Leontief model described by

$$F(k)\,p(k+1) = [1+r(k)]\,[F(k) - I]\,p(k) + Gp_0(k+1)\,a_0,\ k = 0, 1, \dots, L-1,$$

where $F(k) = I + (I-A^T)^{-1}B^T(k)$ and $G = (I-A^T)^{-1}$ considering both time-invariant and periodic coefficients, with particular emphasis on establishing some results in the periodic case where F(k) are commuting normal matrices. *Copyright © 2001 IFAC*

Keywords: descriptor discrete-time systems, periodic coefficients, economic models

1. INTRODUCTION

Usually named the basic price equation of the dynamic Leontief model, the class of descriptor linear discrete-time systems described by

$$F(k)\,p(k+1) = = [1+r(k)]\,[F(k) - I]\,p(k) + Gp_0(k+1)\,a_0, \quad (1)$$

$k = 0, 1, \dots, L-1$,

where $F(k) = I + (I-A^T)^{-1}B^T(k)$ and $G = (I-A^T)^{-1}$, is often discussed in some economic problems (Livesey, 1973; Luenberger and Arbel 1977; Campbell, 1980; Szyld, 1988; Amaral, 1991; Takayama, 1996).

Since in the economic model, $p(k)=[p_j]_{n\times 1}$ is the price vector in period k, $a_0 = [a_{0j}]_{n\times 1}$ where a_{0j} is the amount of labor necessary to produce one unit of the jth good, $p_0(k)\in\Re$ the price of labor ("wages") in period k, $r(k)\in[0,1]$ the interest rate which prevails in the economy throughout the period k, $A = [a_{ij}]_{n\times n}$ is the technique coefficient matrix and $B(k) = [b_{ij}(k)]_{n\times n}$ is the capital coefficient matrix (at the period k), it is usual to assume that:

(i) the matrices A, B(k) as well as the vectors p(k) and a_0 are nonnegative;

(ii) The matrix A satisfies to the following conditions:

(ii-a) if A is irreducible

(ii-a-1) $0\le a_{ij} < 1$, $i,j\in\{1, \dots, n\}$

(ii-a-2) $\sum_{i=1}^{n} a_{ij} \le 1$, $j\in\{1, \dots, n\}$

(ii-a-3) $\exists s\in\{1, \dots, n\}$: $\sum_{i=1}^{n} a_{is} < 1$;

(ii-b) if A is reducible (i.e., if $a_{ij}=0$, $i\in\mathbf{I}$, $j\in\mathbf{J}$, $\mathbf{I}\cup\mathbf{J}=\{1,\dots,n\}$, $\mathbf{I}\cap\mathbf{J}=\varnothing$):

(ii-b-1) $0\le a_{ij} < 1$, $i,j\in\{1, \dots, n\}$

(ii-b-2) $\sum_{i=1}^{n} a_{ij} \leq 1, j \in \{1, ..., n\}$

(ii-b-3) $\exists s \in \mathbf{J}: \sum_{i=1}^{n} a_{is} < 1.$

So, defining $I-A=[\bar{a}_{ij}]_{nxn}$ where $\bar{a}_{ij} = -a_{ij} \leq 0$, if $i \neq j$, and $\bar{a}_{ii} = 1 - a_{ii} > 0$, it is simple to verify that I-A is a Z-matrix satisfying the following conditions:

(ii-c) if I-A is irreducible

(ii-c-1) $\bar{a}_{jj} \geq \sum_{i \neq j} |\bar{a}_{ij}|, j \in \{1, ..., n\}$

(ii-c-2) $\exists s \in \{1, ..., n\}: |\bar{a}_{ss}| > \sum_{i \neq s} |\bar{a}_{is}|;$

(ii-d) if I-A is reducible (i.e., if $\bar{a}_{ij}=0$, $i \in \mathbf{I}$, $j \in \mathbf{J}$, $\mathbf{I} \cup \mathbf{I} = \{1,...,n\}$, $\mathbf{I} \cap \mathbf{J} = \emptyset$),

(ii-d-1) $\bar{a}_{jj} \geq \sum_{i \neq j} |\bar{a}_{ij}|, j \in \{1, ..., n\}$

(ii-d-2) $\exists\ s \in \mathbf{J}: \bar{a}_{ss} > \sum_{i \neq s} |\bar{a}_{is}|$.

Then, in conclusion, matrix A is convergent (and thus the inverse $(I-A)^{-1}$ exists and is nonnegative);

(iii) the matrices B(k) are singular and rankB(k)=m<n, k = 0, 1, ... , L - 1.

In previous work, when studying the output equation (which is the "dual" of the referred price equation) Borges and Lima (1996, 1998) have supposed that A and B are time-invariant, since in the original version of the static Leontief model no change in the technology is assumed over time. However, in the study of the dynamic version of the output model the above assumption has become less realistic than before (Livesey, 1973; Szyld, 1988; Cichocki and Wojciechovski, 1988; Amaral, 1991). Believing that this change is significant in that it motivates the study of time varying systems, this paper analyses a quite simple extension of the time-invariant case, i.e., discusses the case where the periodicity condition $B(k+N)=B(k)$ holds, for some $N \in Z^{+}$.

So, in what follows, both time-invariant and periodic basic price equations will be considered, with particular emphasis on the presentation of: a forward/backward decomposition of the invariant basic price equation; and an invariant formulation of the periodic basic price equation where F(k) are commuting normal matrices.

2. THE BASIC PRICE EQUATION OF THE DYNAMIC LEONTIEF MODEL

2.1 THE INVARIANT CASE

In this section the real Schur form of the matrix

$$F = I + (I-A^T)^{-1} B^T(k)$$

allows the statement of a forward/backward decomposition of the basic price equation.

As previously referred in the following equation

$$[I - A^T + B^T]p(k+1) = = (1+r)\, B^T p(k) + p_0(k+1)a_0, \quad (2)$$

$k = 0, 1, \ldots, L - 1.$

the tecnique coefficient matrix A is convergent and the capital coefficient matrix B is singular with rankB=rankF=m<n.

Let $H = I - A + B$. Since $\lambda H^T - (1+r)B^T$ is not a standard pencil [(1)] and taking into account that

$$\det[\lambda H^T - (1+r)B^T] \neq 0 \text{ for } \lambda = 1+r,$$

it is possible to transform the above regular pencil into a standard one, multiplying $\lambda H^T - (1+r)B^T$, on the left, by the matrix $(I - A)^{-1}$.

We obtain $\lambda \left[I + (I - A^T)^{-1} B^T \right] - (1+r)(I - A^T)^{-1} B^T$.

Similar operation can be carried out for the system (2) in order to get the following descriptor system

$$F\, p(k+1) = (1+r)\, [F - I]\, p(k) + G p_0(k+1)\, a_0, \quad k = 0, 1, \ldots, L - 1. \quad (3)$$

where $F = I + (I - A^T)^{-1} B^T$ and $G = (I-A^T)^{-1}$.

---

(1) Let E and F be two square matrices. If det(sE - F) doesn't vanish identically then sE - F is a regular pencil.
Let sE - F be a regular pencil. If there exist scalars α and β such that $\alpha E + \beta F = I$ then sE - F is a standard pencil and, consequently, EF = FE.

The descriptor system (3) is solvable [(2)] because $\det[\lambda F - (1+r)(F - I)] \neq 0$ for $\lambda = 1+r$.

According to [Horn and Johnson, 1985] there is an orthogonal matrix, $n \times n$, Q, such that

$$Q^T F Q = \begin{bmatrix} F_{11} & F_{12} & F_{13} \\ 0 & F_{22} & F_{23} \\ 0 & 0 & F_{33} \end{bmatrix},$$

where F_{11} is a triangular matrix with the real non null eigenvalues of F as the diagonal entries; F_{22} is a block upper triangular matrix with 2×2 blocks on the diagonal corresponding to the non real pairs of complex conjugate eigenvalues of F, and F_{33} is a nilpotent matrix corresponding to the null eigenvalues of F. The matrix $Q^T F Q$ is known as the real Schur form of F and it is not unique — the diagonal blocks of $Q^T F Q$ may be arranged in any prescribed order.

Then

$$Q^T (F - I) Q = \begin{bmatrix} F_{11} - I & F_{12} & F_{13} \\ 0 & F_{22} - I & F_{23} \\ 0 & 0 & F_{33} - I \end{bmatrix}$$

$$\text{and } Q^T G = \begin{bmatrix} G_{11} & G_{12} & G_{13} \\ G_{21} & G_{22} & G_{23} \\ G_{31} & G_{32} & G_{33} \end{bmatrix}.$$

By means of the coordinate transformation

$$Q^T p(k) = x(k) = \begin{bmatrix} x_1(k) \\ x_2(k) \\ x_3(k) \end{bmatrix}, \tag{4}$$

the system (3) is reduced to

$$\begin{bmatrix} F_{11} & F_{12} & F_{13} \\ 0 & F_{22} & F_{23} \\ 0 & 0 & F_{33} \end{bmatrix} \begin{bmatrix} x_1(k+1) \\ x_2(k+1) \\ x_3(k+1) \end{bmatrix} = \tag{5}$$

$$= (1+r) \begin{bmatrix} F_{11} - I & F_{12} & F_{13} \\ 0 & F_{22} - I & F_{23} \\ 0 & 0 & F_{33} - I \end{bmatrix} \begin{bmatrix} x_1(k) \\ x_2(k) \\ x_3(k) \end{bmatrix} +$$

$$+ \begin{bmatrix} G_{11} & G_{12} & G_{13} \\ G_{21} & G_{22} & G_{23} \\ G_{31} & G_{32} & G_{33} \end{bmatrix} \begin{bmatrix} u_1(k) \\ u_2(k) \\ u_3(k) \end{bmatrix}.$$

where $p_0(k+1)\, a_0 = u(k) = \begin{bmatrix} u_1(k) \\ u_2(k) \\ u_3(k) \end{bmatrix}$ and which can be decomposed into two subsystems

$$\begin{bmatrix} F_{11} & F_{12} \\ 0 & F_{22} \end{bmatrix} \begin{bmatrix} x_1(k+1) \\ x_2(k+1) \end{bmatrix} + \begin{bmatrix} F_{13} \\ F_{23} \end{bmatrix} x_3(k+1) = \tag{6.1}$$

$$= (1+r) \begin{bmatrix} F_{11} - I & F_{12} \\ 0 & F_{22} - I \end{bmatrix} \begin{bmatrix} x_1(k) \\ x_2(k) \end{bmatrix} +$$

$$+ \begin{bmatrix} F_{13} \\ F_{23} \end{bmatrix} x_3(k)) + \begin{bmatrix} G_{11} & G_{12} \\ G_{21} & G_{22} \end{bmatrix} \begin{bmatrix} u_1(k) \\ u_2(k) \end{bmatrix} +$$

$$+ \begin{bmatrix} G_{13} \\ G_{23} \end{bmatrix} u_3(k).$$

$$F_{33}\, x_3(k+1) = [1 + r(k)]\, (F_{33} - I)\, x_3(k) + \tag{6.2}$$

$$+ \begin{bmatrix} G_{31} & G_{32} & G_{33} \end{bmatrix} \begin{bmatrix} u_1(k) \\ u_2(k) \\ u_3(k) \end{bmatrix}.$$

As the initial assumptions imply the nonsingularity of the coefficients F_{11}, F_{22} and $F_{33} - I$ it is suitable to write the following forward/backward decomposition to the descriptor system (3)

$$\begin{bmatrix} F_{11} & F_{12} \\ 0 & F_{22} \end{bmatrix} \begin{bmatrix} x_1(k+1) \\ x_2(k+1) \end{bmatrix} = \tag{7.1}$$

$$= (1+r) \begin{bmatrix} F_{11} - I & F_{12} \\ 0 & F_{22} - I \end{bmatrix} \begin{bmatrix} x_1(k) \\ x_2(k) \end{bmatrix} +$$

$$+ \begin{bmatrix} F_{13} \\ F_{23} \end{bmatrix} \left[(1+r) x_3(k) + x_3(k+1) \right] +$$

$$+ \begin{bmatrix} G_{11} & G_{12} \\ G_{21} & G_{22} \end{bmatrix} \begin{bmatrix} u_1(k) \\ u_2(k) \end{bmatrix} + \begin{bmatrix} G_{13} \\ G_{23} \end{bmatrix} u_3(k).$$

$$x_3(k) = \frac{1}{(1+r)} (F_{33} - I)^{-1} F_{33}\, x_3(k+1) - \tag{7.2}$$

$$- \frac{1}{(1+r)} (F_{33} - I)^{-1} \begin{bmatrix} G_{31} & G_{32} & G_{33} \end{bmatrix} \begin{bmatrix} u_1(k) \\ u_2(k) \\ u_3(k) \end{bmatrix}.$$

So, given a final condition $x_3(L)$, the backward system (7.2) can be solved by iterating backwards in time. Similarly — after substituting $x_3(k)$ in (7.1) and specifying the initial conditions $x_1(0)$ and $x_2(0)$—, by iterating forwards in time, it is quite easy to solve the forward system (7.1).

---

(2) It is known that $\det[\lambda(I+F) - F] \neq 0$ is a necessary and sufficient condition to the existence of a solution to the descriptor system (3), (Campbell, 1980).

2.2 THE PERIODIC CASE

The next goal is to combine the simultaneous block diagonalization of a commuting family of N square normal matrices F(k), k = 0, 1, ... , N - 1, with a technique of stacked forms (Van Dooren et al., 97) in order to obtain an invariant formulation for the periodic basic price equation.

Therefore let's consider the periodic system

$$F(k)\,p(k+1) = \\ = [1+r(k)]\,[F(k) - I]\,p(k) + Gp_0(k+1)\,a_0, \qquad (8)$$

$k = 0, 1, \dots, L-1,$

where $F(k) = I + (I-A^T)^{-1}B^T(k)$, $G = (I-A^T)^{-1}$, and where the singular capital coefficient matrices, B(k), with constant rank, are periodic (with period $N \in Z^+$, such that $L = \mu N$, $\mu \in Z^+$).

Furthermore, imposing that the N matrices

$$F(k) = I + (I-A^T)^{-1}B^T(k),\ k=0, 1, \dots, N-1$$

are commutative and normal with rank$F(k) = m<n$ it is possible to assure (Horn and Johnson, 1985) the existence of an orthogonal matrix Q, of order n, such that

$$Q^TF(k)Q = \begin{bmatrix} F_1(k) & 0 & 0 & 0 \\ 0 & F_2(k) & \dots & 0 \\ 0 & 0 & 0 & F_{p(k)}(k) \end{bmatrix}, \qquad (9)$$

is a block diagonal matrix, where $F_j(k)$ is a real 1×1 matrix, or, a real 2×2 matrix, with a non real pair of complex conjugate eigenvalues of F(k).(3)

Since rank$B(k)$ = rank$F(k) = m<n$, we can assure that — in the diagonal of the block matrix $Q^TF(k)Q$ — there are n - m null matrices $F_j(k)$ of order one.

(3) Notice that, when j=1, ..., p(k), k=0, 1, ..., N-1, we have $F_j(k) = \lambda_j^k$ — where λ_j^k are the real eigenvalues of the matrix F(k) — or, $F_j(k) = \begin{bmatrix} \alpha_j^k & \beta_j^k \\ -\beta_j^k & \alpha_j^k \end{bmatrix}$ — where $\alpha_j^k \pm i\beta_j^k$ are the non-real complex conjugate eigenvalues of F(k), k=0, ..., N-1.

These facts allows for the study of the system (8) through the next periodic descriptor system

$$\begin{bmatrix} F_{11}(k) & F_{12}(k) & F_{13}(k) \\ 0 & F_{22}(k) & F_{23}(k) \\ 0 & 0 & F_{33}(k) \end{bmatrix} \begin{bmatrix} x_1(k+1) \\ x_2(k+1) \\ x_3(k+1) \end{bmatrix} =$$

$$= [1+r(k)] \begin{bmatrix} F_{11}(k)-I & F_{12}(k) & F_{13}(k) \\ 0 & F_{22}(k)-I & F_{23}(k) \\ 0 & 0 & F_{33}(k)-I \end{bmatrix} \begin{bmatrix} x_1(k) \\ x_2(k) \\ x_3(k) \end{bmatrix} +$$

$$+ \begin{bmatrix} G_{11} & G_{12} & G_{13} \\ G_{21} & G_{22} & G_{23} \\ G_{31} & G_{32} & G_{33} \end{bmatrix} \begin{bmatrix} u_1(k) \\ u_2(k) \\ u_3(k) \end{bmatrix} \Leftrightarrow$$

$$\Leftrightarrow H(k)x(k+1) = [1+r(k)]\,[H(k) - I]\,x(k) + Mu(k), \qquad (10)$$

$$k=0, 1, \dots, L-1,$$

where the N periodic matrices $H(k) = Q^TF(k)Q$ are of the form stated by (9), $M = Q^TG$, $x(k) = Q^Tp(k)$ and $u(k) = p_0(k+1)\,a_0$.

By stacking the equations (10) for k, k+1, ..., k+N-1, the following extended descriptor discrete-time linear periodic system is obtained

$$\overline{H}(k)\,\overline{x}(k+1) = \\ = [1+r(k)]\,(\overline{H-I})(k)\,\overline{x}(k) + \overline{M}\,\overline{u}(k), \qquad (11)$$

with $\overline{H}(k) = \text{diag}\big[H(k)\big]_{j=k}^{k+N-1} \in \Re^{Nn\times Nn}$,

$$(\overline{H-I})(k) = \text{diag}\big[H(k) - I\big]_{j=k}^{k+N-1} \in \Re^{Nn\times Nn},$$

$$\overline{M} = \text{diag}\big[M\big]_{j=k}^{k+N-1} \in \Re^{Nn\times Nn},$$

and

$$\overline{x}(k) = \text{col}\big[x(k)\big]_{j=k}^{k+N-1} \in \Re^{Nn},$$

$$\overline{u}(k) = \text{col}\big[u(k)\big]_{j=k}^{k+N-1} \in \Re^{Nn}.$$

The cyclic matrix, C_j, defined by

$$C_j = \begin{bmatrix} 0 & I_j \\ I_{(N-1)j} & 0 \end{bmatrix} \in \Re^{Nj\times Nj}.$$

satisfying the properties:

$C_j^{Nk} = I_j$, $\forall k \in Z$

$C_j^{-1} = C_j^{N-1}$,

will be useful in the statement of an invariant formulation to (8)

The multiplication by C_n^k , on the left, of each member of the system (14) and the following transformations

$$x_e(k) = C_j^{N-1}\, \overline{x}(k),\; u_e(k) = C_j^{N-1}\, \overline{u}(k), \qquad (12)$$

permit to write

$$C_n^k\, \overline{H}(k)\, C_n^{-k}\, x_e(k+1) = \qquad (13)$$
$$= [1+r(k)]C_n^k \overline{(H-I)}(k)\, C_n^{-k+1}\, x_e(k)+\overline{M}u_e(k).$$

A little manipulation — using some cyclic matrices properties as well as the periodicity of $\overline{H}(k)$ and $\overline{(I+H)}(k)$— shows that $C_n^{-1}\, \overline{H}(k)\, C_n = \overline{H}(k+1)$ and $C_n^{-1}\, \overline{(H-I)}(k)\, C_n = \overline{(H-I)}(k+1)$.

Thus the resulting conclusion is that

$$C_n^k\, \overline{H}(k)\, C_n^{-k} = \overline{H}(0)$$

and

$$C_n^k\, \overline{(H-I)}(k)\, C_j^{-k+1} = \overline{(H-I)}(0)\, C_n$$

are time-invariant matrices.

$$\text{Set } H_e = \overline{H}(0) = \begin{bmatrix} H(0) & \dots & \dots & \dots & 0 \\ 0 & H(1) & \dots & \dots & 0 \\ 0 & 0 & \dots & \dots & \dots \\ 0 & 0 & 0 & \dots & H(N-1) \end{bmatrix};$$

$(H-I)_e = \overline{(H-I)}(0) =$

$$= \begin{bmatrix} 0 & \dots & \dots & \dots & H(0)-I \\ H(1)-I & 0 & \dots & \dots & 0 \\ 0 & H(2)-I & \dots & \dots & 0 \\ \dots & \dots & \dots & \dots & \dots \\ 0 & 0 & \dots & H(N-1)-I & 0 \end{bmatrix}$$

and $M_e = \overline{M}$.

Therefore the initial problem can be mostly confined to the extended descriptor discrete-time system

$$H_e\, x_e(k+1) =$$
$$= [1+r(k)]\, (H-I)_e\, x_e(k) + M_e\, u_e(k). \qquad (14)$$

which is an invariant formulation of the periodic basic price equation (8).

Since H_e is a block diagonal matrix — where the blocks are the matrices $F_j(k)$ referred at (9) — we notice that in the diagonal of H_e there are $N(n-m)$ zeros

From (14), by means of a permutation matrix $P \in \Re^{Nn\times Nn}$ and the coordinate transformation

$$z_e(k) = \begin{bmatrix} z_{e1}(k) \\ z_{e2}(k) \end{bmatrix} = P^T x_e(k),$$

it is not hard to attain

$$PH_ePP^Tx_e(k+1) = [1+r(k)]\, P(H-I)_ePP^Tx_e(k) +$$
$$+ PM_e\, u_e(k),$$

which is equivalent to,

$$\begin{bmatrix} \hat{H}_e & 0 \\ 0 & 0 \end{bmatrix} \begin{bmatrix} z_{e1}(k+1) \\ z_{e2}(k+1) \end{bmatrix} =$$

$$= [1+r(k)] \begin{bmatrix} (H-I)_e^{11} & (H-I)_e^{12} \\ (H-I)_e^{21} & (H-I)_e^{22} \end{bmatrix} \begin{bmatrix} z_{e1}(k) \\ z_{e2}(k) \end{bmatrix} +$$

$$+ \begin{bmatrix} M_{e1} \\ M_{e2} \end{bmatrix} u_e(k),$$

or,

$$\hat{H}_e\, z_{e1}(k+1) =$$
$$= [1+r(k)]\left[(H-I)_e^{11}\, z_{e1}(k) + (H-I)_e^{12}\, z_{e2}(k)\right] + M_{e1}\, u_e(k)$$
$$0 = [1+r(k)]\left[(H-I)_e^{21}\, z_{e1}(k) + (H-I)_e^{22}\, z_{e2}(k)\right] +$$
$$+ M_{e2}\, u_e(k).$$

where, by construction, $\hat{H}_e$ is a nonsingular block diagonal matrix.

Then

$$z_{e1}(k+1) = [1+r(k)]\hat{H}_e^{-1}\, (H-I)_e^{11}\, z_{e1}(k) + \qquad (15.1)$$
$$+ [1+r(k)]\hat{H}_e^{-1}\, (H-I)_e^{12}\, z_{e2}(k)) +$$
$$+ \hat{H}_e^{-1}\, M_{e1}\, u_e(k)$$

$$0 = [1+r(k)]\,(H - I)^{21}_{e}\, z_{e1}(k) + \\ + [1+r(k)]\,(H - I)^{22}_{e}\, z_{e2}(k) + M_{e2}\, u_{e}(k). \qquad (15.2)$$

(15) is a state space system subject to an algebraic condition and is also an invariant formulation of the initial periodic basic price equation (8)

$$F(k)\, p(k+1) = [1+r(k)]\, [F(k) - I]\, p(k) + Gp_0(k+1)\, a_0, \quad k = 0, 1, \ldots, L-1,$$

CONCLUSIONS

The purpose of this paper is to call some attention to a special case of descriptor linear discrete-time systems which are related to some economic equations.

Nevertheless the presented study does not remove the most important of the difficulties — the causal indeterminacy [(4)] — because it doesn't assure the nonnegativity of the solutions.

There have been various attempts to rescue the dynamic Leontief models from this difficulty but there are almost no results about positive solutions of descriptor systems with variable coefficients.

REFERENCES

Amaral, J.F. (1991). *Curso Avançado de Análise Económica Multi-Sectorial*, Escher, Lisbon, Portugal.*(in Portuguese)*

Borges, A. and T. P. de Lima (1996). Nonsingular Formulation and Reachability for Leontief Dynamic Systems, pp.593-598. *Proc. 2nd Portuguese Conf. Aut. Control*, Porto, Portugal.

Borges, A. and T. P. de Lima (1998). An Invariant Formulation for Descriptor Discrete-Time Linear Periodic Systems,pp.53-56, *Proc. 3nd Portuguese Conf. Aut. Control*, Coimbra, Portugal.

Campbell, S.L. (1980). *Singular Systems of Differential Equations*, Pitman, vol I.

Cichocki. K. and W. Wojciechowski (1988) Investment Coefficient Matrix in Dynamic Input-Output Models: an Analysis and Prognosis. In: *Input-Output Analysis—current developments*, Ed. by M. Ciaschini, Int. Studies in Economic Models, Chapman and Hall.

Horn, R. A. and C. A. Johnson (1985). *Matrix Analysis*, Cambridge University Press.

Livesey, D.A. (1973). The Singularity Problem in the Dynamic Input-Ouput Model, *Int.J.Systems Sci.*, **Vol.4**, N°3, pp. 437-440.

Luenberger, D.G. and Arbel (1977). Singular Dynamic Leontief Systems, *Econometrica*, **Vol.45**, N°4, pp. 991-995.

Sreedhar, J., P. Van Dooren and B. Bamieh (1997). Computing H∞-Norm of Discrete Periodic Systems - A Quadratically Convergent Algorithm, *ECC*, June 1997.

Szyld, D.B. (1985). Conditions for the Existence of a Balanced Growth Solution for the Leontief Dynamic Input-Output, *Econometrica*, **vol.53**, N°6, pp.1411-1419.

Takayama, A. (1996). *Mathematical Economics*, 2nd ed, Cambridge University Press.

---

(4) That is, although the initial prices are nonnegative, it may happen that, for sufficiently large k, they become negative.

Copyright © IFAC Periodic Control Systems,
Cernobbio-Como, Italy, 2001

AN OVERVIEW OF PERIODIC TIME SERIES WITH EXAMPLES

Lynne Seymour
Department of Statistics
The University of Georgia
Athens, GA 30602-1952
seymour@stat.uga.edu

Abstract: Two environmental examples are used to demonstrate modeling with periodic autoregressive moving average (PARMA) processes. These two examples – average monthly temperatures in Big Timber, Montana, USA, and carbon dioxide exchange rates of flowering plants in a growth chamber – are clearly periodic in their first and second moments, and highlight the necessity as well as the pitfalls of PARMA modeling. *Copyright © 2001 IFAC*

Keywords: Modelling errors; Parameter estimation; Periodic ARMA models; Statistical inference; Time series analysis

1. INTRODUCTION

Periodicity occurs naturally in many environmental time series – hourly tide levels, daily stream flow, monthly average temperature, and carbon dioxide (CO_2) exchange of growing plants are but a few examples. This paper will focus on the last two examples in a demonstration of methods for analyzing periodically correlated (PC) time series. Specifically, a PC time series will tend to exhibit periodicity not only in its first moment, but also in its second moments. For modeling such phenomena, periodic autoregressive moving average (PARMA) models are discussed in some detail in Section 2.

The first example, shown in Figure 1, is of monthly average temperatures in degrees Celsius, measured over 86 years (1911 through 1996), in Big Timber, Montana, USA (Lund and Seymour, 2001). Beyond the obvious periodicity in the first moment, there is no other significant feature (*e.g.*, linear trend in the first moment of these data. There is also an obvious

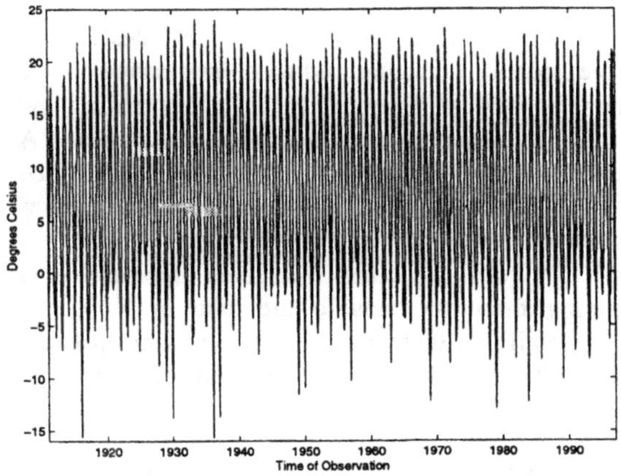

Figure 1. Monthly Average Temperatures (in °C) for Big Timber, Montana, USA, 1911-1996.

periodicity in the second moment of this data – the winter temperatures are more variable than the summer temperatures, which is true in general.

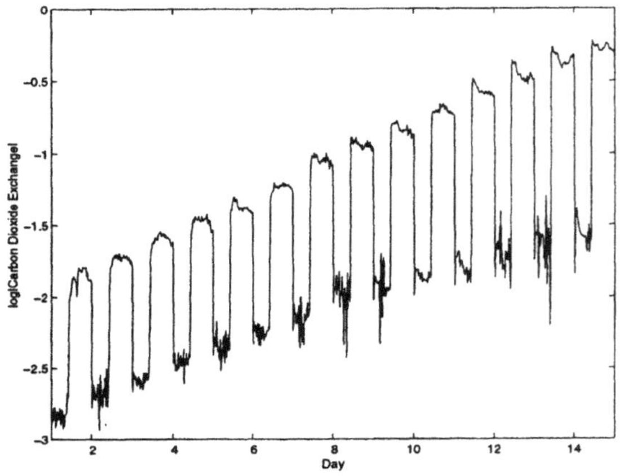

Figure 2. log| · |-Transformed CO_2 Exchange of a Group of Vinca Grown in a Growth Chamber.

Figure 2 shows the second example – the log| · |-transformed amount of CO_2 (originally in micromol/second) taken up by a group of thirty-two vinca [*Catharanthus roseus* (L.) G. Don.] plants growing inside a single growth chamber in which temperature and light intensity were controlled, and in which the only changes in CO_2 concentration would be due to the plants within the chamber (Seymour, 2001). The log| · | transform was used since the daylight data are all strictly positive while the darkness data are all strictly negative. The data were recorded every 20 minutes over 14 days, with 14 hours of daylight and 10 hours of darkness. There are 1008 observations: during any given 24-hour period, photosynthesis is occurring during the 42 daylight observations, but not during the 30 darkness observations. Like the temperature series, this series clearly exhibits periodic structure in both its first and second moments.

This paper continues in Section 2 with a description of methods for detecting and modeling periodic time series, as well as practical issues of estimating the covariance function of a periodic time series. A more detailed overview of this methodology may be found in Lund and Basawa (1999). Section 3.1 gives results of applying these methods to the data in Figure 1, while Section 3.2 gives the results of applying these methods to the data in Figure 2. More details on these results may be found in Lund *et al.* (2001) and Seymour (2001), respectively.

2. ESSENTIALS OF PERIODICALLY CORRELATED TIME SERIES

A process $\{X_t\}$ is said to be *periodically correlated with period* T (PC-T) if

$$E(X_{t+T}) = E(X_t) \qquad (1)$$
$$\text{Cov}(X_{s+T}, X_{t+T}) = \text{Cov}(X_s, X_t)$$

for all integers t and s. This property is also called *periodically stationary* or *cyclostationary*. A PC-T process is not (weakly) stationary; however, if a PC-T process is written as a T-variate process by grouping the data into "periods", then it is T-variate stationary (*cf.* Basawa and Lund, 2001). Such a process conveniently allows for a seasonal expression of the mean and covariance. Assume for simplicity that the series length N is evenly divisible by the period T, so that $d = N/T$ is an integer. Then for $\nu = 1, \dots, T$ and $n = 0, \dots, d-1$, the mean is a function of only season ν, and the covariance is a function of only season ν and lag h:

$$\mu_\nu = E(X_{nT+\nu}) \quad . \qquad (2)$$
$$\gamma_\nu(h) = \text{Cov}(X_{nT+\nu}, X_{nT+\nu+h}), h \geq 0$$

The covariance function $\gamma_\nu(h)$ is not symmetric in h, but $\gamma_\nu(-h) = \gamma_{\nu+h}(h)$. Note also that seasonal notation is used when emphasis upon the seasonality is necessary.

There are various tools for detecting periodicity, all of which are primarily visual. For detecting periodicity in the mean of a given time series $\{X_t\}$, besides visual inspection of $\{X_t\}$ itself, there is $\widehat{\rho}(h)$, the sample autocorrelation function (ACF) of $\{X_t\}$ at lag h (Brockwell and Davis 1996, or any introductory time series text). If $\{X_t\}$ has a substantial periodic component in its first moment, then $\widehat{\rho}(h)$ will have the same periodicity. However, a bit of caution is required in using the sample ACF to infer periodic behavior: a process for which $\widehat{\rho}(h)$ exhibits periodic behavior need not be periodic in its mean (Brockwell and Davis 1991).

Detecting periodicity in the second moments, or covariance structure, of a time series is more challenging. Perhaps the best tool to use is the average squared coherence (ASC) statistic of Goodman (1965), which involves the spectral representation of $\{X_t\}$ (see Bloomfield *et al.* (1994) or Lund *et al.* (1995) for the details). In essence, for a series of length N *from which the first moments have been removed*, the ASC statistic is given by

$$\overline{\kappa}_h = \frac{1}{N}\sum_{j=1}^{N-1}\left(\frac{\left|\sum_{m=0}^{M-1} I_{j+m}\overline{I_{j+h+m}}\right|^2}{\sum_{m=0}^{M-1}|I_{j+m}|^2 \sum_{m=0}^{M-1}|I_{j+h+m}|^2}\right) \qquad (3)$$

at spectral lags h and for a smoothing control M, where $I_j = \frac{1}{\sqrt{2\pi N}}\sum_{t=0}^{N-1} X_{t+1}\exp(-it\lambda_j)$, $i = \sqrt{-1}$ is the discrete Fourier transform of $\{X_t\}$ at the Fourier frequency $\lambda_j = 2\pi j/N$. A plot of $\overline{\kappa}_h$ versus h for all $1 \leq h \leq N/2$ then reveals any significant second-order periodicities: Small values of $\overline{\kappa}_h$ are statistical evidence that $\{X_t\}$ is stationary; large values of $\overline{\kappa}_h$ at some values of h that are integer multiples of $d = N/T$ is statistical evidence that

$\{X_t\}$ is PC-T; and if $\overline{\kappa}_h$ is large at values of h that are not all multiples of some common integer larger than one, this is statistical evidence that the series is neither stationary nor PC.

A well-developed class of models for mean zero stationary time series is the autoregressive moving average model of orders p and q, denoted ARMA(p, q) (Brockwell and Davis, 1991). An ARMA(p, q) series $\{Z_t\}$ is a solution to the linear difference equation

$$Z_t - \sum_{k=1}^{p} \phi_k^{\text{AR}} Z_{t-k} = \varepsilon_t + \sum_{k=1}^{q} \theta_k^{\text{MA}} \varepsilon_{t-k}, \quad (4)$$

where $\{\varepsilon_t\}$ is white noise with mean zero and variance σ_ε^2, denoted WN($0, \sigma_\varepsilon^2$). (Note: for the sake of parsimonious notation, $\{\varepsilon_t\}$ is used henceforth to denote a generic white noise process, while $\{Z_t\}$ is used to denote a generic ARMA(p, q) process.)

The most commonly-used model for seasonal time series based on the ARMA model in (4) is the seasonal autoregressive integrated moving average (SARIMA) model (Brockwell and Davis, 1996), in which an appropriately differenced series behaves as an ARMA process. SARIMA processes are not necessarily stationary; however, it is a simple exercise using (1) to verify that differencing an arbitrary PC-T process at lag T yields a PC-T process with mean zero. Hence the SARIMA models are inappropriate for PC time series.

The appropriate ARMA-type model for a PC-T series is the periodic autoregressive moving average (PARMA) model. The process $\{X_t\}$ is said to be a PARMA series if it is a solution to the periodic linear difference equation

$$(X_{nT+\nu} - \mu_\nu) - \sum_{k=1}^{p_\nu} \phi_{\nu,k}(X_{nT+\nu-k} - \mu_{\nu-k}) \quad (5)$$
$$= \sum_{k=0}^{q_\nu} \theta_{\nu,k}\, \varepsilon_{nT+\nu-k}$$

in which $n = 0, \ldots, d-1$, $d = N/T$, $\nu = 1, \ldots, T$, $\{\varepsilon_t\} \sim$ WN$(0, 1)$, μ_ν are the periodic means, and $\theta_{\nu,0}$ in particular are the periodic white noise variances. Recursive prediction and Gaussian likelihood evaluation for PARMA models is dealt with in Lund and Basawa (2000), but the main difficulty with these models is lack of parsimony. Fitting the PARMA model in (5) requires the estimation of $T + \sum_{\nu=1}^{T}(p_\nu + q_\nu + 1)$ parameters – a potentially formidable task. (For this reason, PARMA notation seldom includes parametric notation except in very simple cases.) A simple solution that has worked in many cases is to *layer*; that is, to fit first a very simple periodic model, and then to fit an ARMA model to the residuals from the first fit. Thus, $\{\varepsilon_t\}$ in (5) would be replaced by the ARMA(p, q) process $\{Z_t\}$ as given in (4).

One layering scheme which has been successfully used in climatology (Amato *et al.*, 1989; Lund and Seymour, 1999; Lund *et al.*, 2001) and in economics (Parzen and Pagano, 1979), and which is used here to model the temperature data in Figure 1, is to *seasonally standardize* the series by subtracting the periodic mean μ_ν, dividing by the periodic standard deviation σ_ν, and then fitting an ARMA(p, q) model to the series that remains – referred to henceforth as SS+ARMA (Bloomfield *et al.*, 1994; Lund *et al.*, 1995). Least-squares estimates of μ_ν and σ_ν are straightforward to calculate: simply average the observations occurring during a fixed season ν to obtain $\widehat{\mu}_\nu$, and compute a sample standard deviation of those observations to obtain $\widehat{\sigma}_\nu$. The seasonally standardized series is then given by $Z_{nT+\nu} = (X_{nT+\nu} - \widehat{\mu}_\nu)/\widehat{\sigma}_\nu$, on which ARMA modeling and diagnostics (Brockwell and Davis 1991) are then employed. The SS+ARMA layering scheme does in fact yield a parsimonious PARMA model with $p_\nu \equiv p$, $q_\nu \equiv q$, and

$$\phi_{\nu,k} = \frac{\sigma_\nu \phi_k^{\text{AR}}}{\sigma_{\nu-k}}, 1 \le k \le p \quad (6)$$
$$\theta_{\nu,k} = \theta_k^{\text{MA}} \sigma_\nu, 0 \le k < q, \theta_0^{\text{MA}} = 1$$

for $\nu = 1, \ldots, T$. Then the covariance matrix of $\{X_1, \ldots, X_N\}$ is then very easy to calculate. Let $\gamma_{\text{ARMA}}(h)$ be the autocovariance generating function of the model fitted in the ARMA layer. Then the covariance matrix $\Gamma_{\text{SS+ARMA}}$ of the layered SS+ARMA model is given by

$$\Gamma_{\text{SS+ARMA}} = \quad (7)$$
$$[\sigma_\nu \sigma_\lambda \gamma_{\text{ARMA}}((n-m)T + (\nu - \lambda))]_{nT+\nu, mT+\nu=1}^{N}$$

where σ_ν^2 is the seasonal variance for season $\nu = 1, \ldots, T$.

Another layering scheme used to model the data in Figure 2 (which has also worked well in climatology) is to fit first a periodic autoregressive process of order 1, denoted by PAR(1), followed by an ARMA(p, q) model – referred to hereafter as PAR+ARMA (Bloomfield *et al.*, 1994; Lund *et al.*, 1995). The process $\{X_t\}$ is said to be PAR(1) if it is a solution to

$$(X_{nT+\nu} - \mu_\nu) - \phi_\nu^{\text{PAR}}(X_{nT+\nu-1} - \mu_{\nu-1}), \quad (8)$$
$$= \delta_\nu \varepsilon_{nT+\nu}$$

in which $n = 0, \ldots, d-1$, $d = N/T$, $\nu = 1, \ldots, T$, $\{\varepsilon_t\} \sim$ WN$(0, 1)$, μ_ν are the periodic means, δ_ν are the periodic white noise variances, and ϕ_ν^{PAR} are the periodic autoregressive parameters. Such a process is PC-T if $|\prod_{\nu=1}^{T} \phi_\nu^{\text{PAR}}| < 1$ (Vecchia, 1985). Note that in this case, $p_\nu \equiv 1$ and $q_\nu \equiv 0$.

Approximate maximum likelihood estimates for the parameters in (8) are

$$\widehat{\phi}_\nu^{\text{PAR}} = \frac{\widehat{\gamma}_\nu(1)}{\widehat{\gamma}_\nu(0)}, \quad \widehat{\delta}_\nu = \widehat{\gamma}_\nu(0) - \widehat{\phi}_\nu^{\text{PAR}} \widehat{\gamma}_\nu(1) \quad (9)$$

where the sample covariance function of the series during season $\nu = 1, \dots, T$ at lag $h \geq 0$ is given by

$$\widehat{\gamma}_\nu(h) = \quad (10)$$
$$\frac{1}{d}\sum_{n=0}^{d-1}(X_{nT+\nu} - \widehat{\mu}_\nu)(X_{nT+\nu-h} - \widehat{\mu}_{\nu-h}),$$

where $\widehat{\mu}_\nu = \sum_{n=0}^{d-1} X_{nT+\nu}/d$ (Bloomfield *et al.*, 1994). If the PAR(1) model described in (8) is a good fit, then the residuals

$$\frac{(X_{nT+\nu} - \widehat{\mu}_\nu) - \widehat{\phi}_\nu^{\text{PAR}}(X_{nT+\nu-1} - \widehat{\mu}_{\nu-1})}{\widehat{\delta}_\nu} \quad (11)$$

should behave as a WN(0, 1) process. In the case of PAR+ARMA layering, (11) is an ARMA(p, q) process. This layering scheme results in a parsimonious PARMA model which will be used to successfully model the data in Figure 2.

Asymptotic results for PARMA parameter estimates have been recently studied by Basawa and Lund (2001). Taking an estimating equations approach, they establish the asymptotic normality of least squares estimates of the PARMA parameters, which also establishes the asymptotic normality of maximum likelihood estimates of the PARMA parameters under Gaussian assumptions.

The covariance matrix of a PAR+ARMA is complicated; indeed, the covariance function of a PARMA model is very complicated in general. There is a recursive method for calculating the covariance function $\gamma_\nu(h)$ for a PAR(1) (Lund and Basawa, 1999), but it is of no assistance in the PAR+ARMA case. However, a model-based approximation to the covariance matrix is possible based on the results in Lund and Basawa (1999). When the PARMA model is causal, solutions to (5) can be expressed uniquely (in mean square) in the form

$$X_{nT+\nu} = \sum_{k=0}^{\infty} \psi_{\nu,k}\varepsilon_{nT+\nu-k} \quad (12)$$

where $\sum_k |\psi_{\nu,k}| < \infty$ for $\nu = 1, \dots, T$. In the case of PAR+ARMA, the $\psi_{\nu,k}$ are given by

$$\psi_{\nu,k} = \sum_{l=0}^{\infty}\left(\psi_{k-l}^{\text{ARMA}}\delta_{\nu-l}\prod_{i=0}^{l-1}\phi_{\nu-i}^{\text{PAR}}\right), \quad (13)$$

where ψ_k^{ARMA} are the coefficients on the white noise terms in the causal representation of the ARMA process in (11) that is completely analogous to (12). Thus the covariance function of the PAR+ARMA process may be approximated by a finitely summed version of the general PARMA covariance

$$\gamma_\nu(h) = \sum_{k=0}^{\infty} \psi_{\nu,k+h}\psi_{\nu-h,k}. \quad (14)$$

One only needs to check that the convergence of the summands to zero is rapid enough that the tail sum becomes negligible after a point, which is reasonably easy via computer. The covariance matrix for $\{X_1, \dots, X_N\}$ is then given by

$$\underset{\sim}{\Gamma}_{\text{PAR+ARMA}} = \quad (15)$$
$$\left[\gamma_{\max(i,j)-T\lfloor\frac{\max(i,j)}{T}\rfloor}(|i-j|)\right]_{i,j=1}^{N}$$

where $\lfloor \cdot \rfloor$ is the floor function (*i.e.*, the greatest integer function). Computing the covariance matrix in this way is laborious, even on a computer. Currently, though, it is the only option.

Note that both of the layered estimation procedures given here are a combination of ordinary least squares (OLS) and maximum likelihood (ML): the regression parameters for the overall trend and the periodic mean fluctuations are estimated via OLS, while the PARMA parameters governing the covariance structure are estimated via ML.

To summarize, the PARMA modeling steps are:

1. Transform the data, if necessary.
2. Estimate and the trend function, including the periodic component.
3. Subtract the estimated trend from the data; that is, calculate the residuals.
4. Fit a PARMA model to the residuals obtained in step 3:
   a. Either seasonally standardize the residuals (simply by dividing by the seasonal standard deviation) or fit a PAR(1) to the residuals.
   b. Fit an ARMA model to the residuals from the fit obtained in step 4a.
5. Use the parameter estimates obtained in step 4 to calculate either $\widehat{\Gamma}_{\text{SS+ARMA}}$ or $\widehat{\Gamma}_{\text{PAR+ARMA}}$, based on the fit in step 4 using (13), (14), and (15).

The covariance matrix may then be used, for example, to compute standard errors for the trend estimates found in step 2.

3. RESULTS

In the following sections, the SS+ARMA layering scheme is applied to the temperature data, while the PAR+ARMA is applied to the CO_2 exchange data. Each model is the most parsimonious selection for

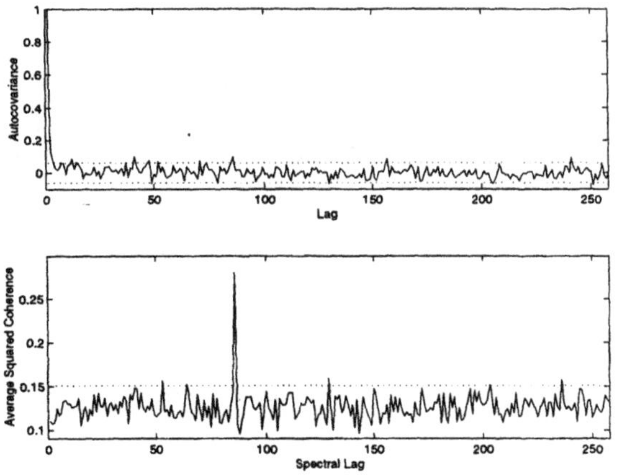

Figure 3. For the Temperature Data: a) ACF After Removal of All First Moments; b) ASC.

the data: in the first example, the PAR+ARMA model overparametrizes the temperature data, while in the second example, SS+ARMA does not account for all of the second-moment periodicity in the CO_2 exchange data.

3.1 SS+ARMA Applied to the Temperature Data.

Figure 3a shows the sample ACF of the temperature data depicted in Figure 1 *after* the removal of all first moments. The sample ACF, which is plotted along with a 99% confidence band for uncorrelated white noise, indicates that there is no further periodicity in the first moment. Figure 3b shows the ASC statistic (3) of the temperature series (recall that the ASC statistic must be calculated on zero-mean data). Note the one very large spike at lag $d = N/T = 1032/12 = 86$. This indicates a strong cycle of period $T = 12$ in the second moments of the series.

Observe the "confidence" line plotted in Figure 3b. The mean and variance of (3) are $E(\overline{\kappa}_h) = 1/M$ and $\mathrm{Var}(\overline{\kappa}_h) = c_M/N$, respectively, where M is a spectral smoothing parameter. The asymptotic normality of (3) is suspected and has been established in unpublished notes only for the case in which $\{X_t\}$ is a moving average of finite order (Bloomfield *et al.*, 1994; Lund *et al.*, 1995). Though (3) is suspected to be Gaussian (particularly in cases in which the data are approximately Gaussian), the constant c_M for spectral smoothing parameter M is still unknown. Calculations of the ASC used $M = 8$; thus the value $c_8 = .1277$, as given in a table of values for c_M simulated from Gaussian white noise (Lund *et al.*, 1995), was required. A "99% confidence" band is therefore $1/M + 2.33\sqrt{c_M/N}$. This value may not be appropriate in the present context. Nevertheless, it does provide a guideline against which spikes in the ASC plot may be evaluated.

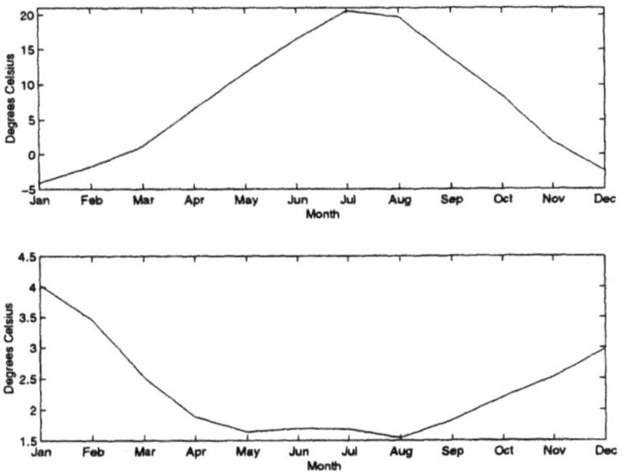

Figure 4. For the Temperature Data: a) Periodic Means; b) Periodic Standard Deviations.

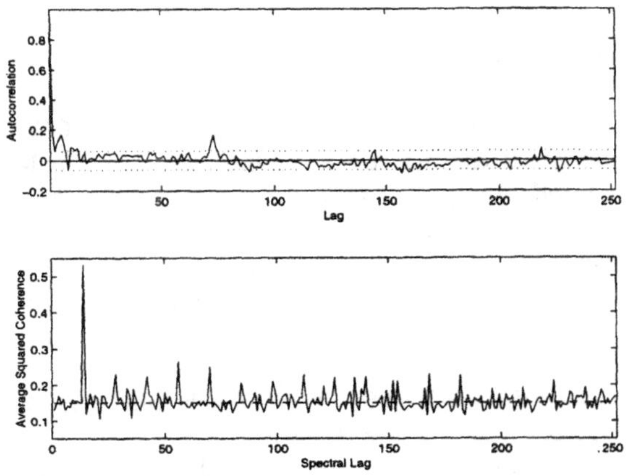

Figure 5. For the CO_2 Exchange Data: a) ACF After Removal of the First Moments; b) ASC.

The SS+ARMA layering approach gave an excellent fit to the data. Figure 4a depicts the seasonal mean temperature for each month, while Figure 4b shows the seasonal standard deviations. These values were used to seasonally standardize the temperature data. The seasonally standardized series was successfully fit with an ARMA(1,3). Hence a PAR+ARMA is a clear overparametrization, even though it may give an acceptable fit

3.2 PAR+ARMA Applied to the CO_2 Exchange Data.

Figure 5a shows the ACF of the CO_2 exchange data depicted in Figure 2. This plot indicates a possibility of seasonal autocorrelation of period $T = 72$. Figure 5b shows the ASC, with one very large spike at lag $d = N/T = 1008/72 = 14$ indicating a strong cycle of period $T = 72$ in the second moments of the series. Note that the "99% confidence" bound isn't

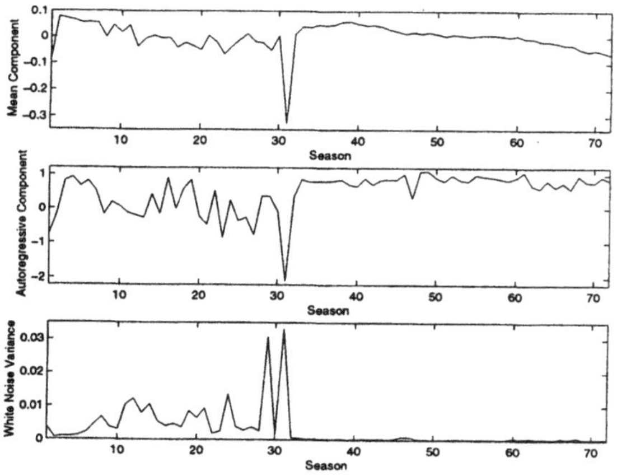

Figure 6. For the CO_2 Exchange Data: a) Periodic Means; b) Periodic Autoregressive Coefficients; c) Periodic White Noise Variances.

as clear as the one in Figure 3b.

Initially, the SS+ARMA model was tried on the CO_2 exchange series after removing the piecewise quadratic trend. In spite of seasonal standardization, the result still contained seasonality in its second moments, according to the ASC of the seasonally standardized series. Thus the SS+ARMA model was inadequate.

However, the PAR+ARMA approach did give a satisfactory model of the PC structure. A mean zero PAR(1) as in (8) was fitted to the detrended data. The fit is summarized in Figure 6: Figure 6a shows the periodic mean fluctuations $\widehat{\mu}_\nu$ about the trend, Figure 6b depicts the periodic autoregressive coefficients $\widehat{\phi}_\nu^{PAR}$, and Figure 6c gives the white noise variances $\widehat{\delta}_\nu$, each for $\nu = 1, \ldots, T = 72$.

The residuals from the PAR(1) fit were well-modeled by an ARMA(3,1). Thus the PAR+ARMA layering scheme gives a good fit to the CO_2 exchange data.

4. CONCLUSION

It is clear that PARMA models are handy tools for modeling periodic time series. Though parsimony remains a significant problem which is currently being addressed, the layering approaches outlined herein are practical solutions that will model many real time series.

REFERENCES

Amato, U., Cuomo, V., Fontana, F., and Serio, C. (1989). Statistical predictability and parametric models of daily ambient temperature and solar irradiance: An analysis in the Italian climate. *Journal of Applied Meteorology*, **28**, 711-721.

Basawa, I. V., and Lund, R. B. (2001). Large Sample Properties of Parameter Estimates for Periodic ARMA Models. *Journal of Time Series Analysis*. To Appear.

Bloomfield, P., Hurd, H. L., and Lund, R. B. (1994). Periodic Correlation in Stratospheric Ozone Data. *Journal of Time Series Analysis*, **15**, 127-150.

Brockwell, P. J., and Davis, R. A. (1991). *Time Series: Theory and Methods*. Springer-Verlag, New York.

Brockwell, P. J., and Davis, R. A. (1996). *Introduction to Time Series and Forecasting*. Springer-Verlag, New York.

Goodman, N. R. (1965). Statistical Tests for Stationarity Within the Framework of Harmonizable Processes. Rocketdyne Research Report No. 65-28.

Lund, R. B., and Basawa, I. V. (1999). Modeling and Inference for Periodically Correlated Time Series. In *Asymptotics, Nonparametrics, and Time Series* (Subir Ghosh (Ed.)), pp. 37-62. Marcel-Dekker, New York.

Lund, R. B., and Basawa, I. V. (2000). Recursive Prediction and Likelihood Evaluation for Periodic ARMA Models. *Journal of Time Series Analysis*, **21**, 75-93.

Lund, R. B., Hurd, H. L., Bloomfield, P., and Smith, R. L. (1995). Climatological Time Series with Periodic Correlation. *Journal of Climate*, **8**, 2787-2809.

Lund, R. B., and Seymour, L. (1999). Assessing Temperature Anomalies for a Geographical Region: A Control Chart Approach. *Environmetrics*, **10**, 163-177.

Lund, R. B., and L. Seymour (2001). Periodic Time Series. In: *Encyclopedia of Environmetrics* (A. El-Sharaawi and W. Piegorsch (Eds)). Wiley, New York. To Appear.

Lund, R. B., L. Seymour, and K. Kafadar (2001). Temperature Trends in the United States. *Environmetrics*. To Appear.

Parzen, E., and Pagano, M. (1979). An approach to modeling seasonally stationary time series. *Journal of Econometrics*, **9**, 137-153.

Vecchia, A. V. (1985). Maximum likelihood estimation for periodic autoregressive moving average models. *Technometrics*, **27**, 375-384.

Copyright © IFAC Periodic Control Systems,
Cernobbio-Como, Italy, 2001

MCMC METHODS FOR PERIODIC AR-ARCH MODELS

Wolfgang Polasek

Institute of Statistics and Econometrics, University of Basel
Bernoullistrasse 28, CH-4056 Basel
email: Wolfgang.Polasek@unibas.ch

Abstract: Many economic time series reveal periodic and seasonal patterns which can be modeled by periodic AR processes. Using a conjugate normal-gamma model we suggest to model seasonal data by a hierarchical prior distribution. We extend this approach to include periodic ARCH models which we call a PAR-GARCH model. A Metropolis-Hastings step is used for the ARCH part of the model and in case when the exact likelihood function is given. *Copyright ©2001 IFAC*
Keywords: Bayesian periodic AR-ARCH models, hierarchical models.

1. INTRODUCTION

Periodic time series have become increasingly important in many applied sciences and this has given new impact to develop different estimation methods. In this paper we will concentrate on the Bayesian estimation of time series models which is motivated by seasonal econometric applications. In a recent text book Franses (1998) has described the classical estimation of these models while an early Bayesian approach by hierarchical seasonal models can be found in Polasek (1984).
We will briefly describe the hierarchical model for seasonal time series models which follows the hierarchical Gaussian model and the embedding of period models into multivariate time series models. Also we show that this approach can be extended to periodic ARCH (autoregressive conditional heteroskedasticity) or GARCH (generalized ARCH, see Bollerslev (1986)) models and the estimation method we suggest is the MCMC approach which is now widely used in practice for hierarchical models. While we will concentrate on periodic AR (PAR) models it is readily seen that the model range can be extended to include ARMA processes, i.e. a moving average (MA) component.
Section 2 introduces the periodic AR model by specifying a hierarchical prior structure. Section 3 describes the estimation procedure and section 4 introduces periodic AR-ARCH processes. The last section concludes. The appendix contains all the technical details as how to run a Gibbs sampler for a hierarchical regression model and how to use MCMC methods to sample from the posterior distribution which is based on the exact likelihood function for AR models.

2. BAYESIAN PERIODIC AR MODELS

In this section we describe the Bayesian periodic AR model with period (or season) S and explain how we can can specify a prior distribution for the AR coefficients of a PAR model. We consider a time series with seasonal periodicity of length S and for each season we specify a Gaussian regression model. Several ways to introduce a hierarchical prior structure have been suggested over the last decades and for regression models many applications can be found in the literature (see e.g. Polasek (1984) for seasonal time series applications and Gelfand et al. (1990) for the first Gibbs sampling application for hierarchical models).
We will investigate two strategies for PAR models:

the first model is the hierarchical pooling model where we make the assumption that the PAR coefficients come from a common hyper-population.

$$\beta_s \sim N[\beta, \Sigma], \quad s = 1, ..., S. \tag{1}$$

The second model introduces smoothness restrictions between neighboring seasons, i.e.

$$\beta_s - \beta_{s-1} \sim N[0, \Sigma], \quad s = 1, ..., S \tag{2}$$

where we make a cyclical restriction on the seasonal regression coefficients $\beta_0 = \beta_S$.
To understand the 'periodic' approach consider the quarterly $PAR(1)$ model:

$$y_{1t} = \beta_{10} + \beta_{11} y_{4,t-1} + u_{1t}, \tag{3}$$
$$y_{2t} = \beta_{20} + \beta_{21} y_{1t} + u_{2t}, \tag{4}$$
$$y_{3t} = \beta_{30} + \beta_{31} y_{2t} + u_{3t}, \tag{5}$$
$$y_{4t} = \beta_{40} + \beta_{41} y_{3t} + u_{4t} \tag{6}$$

where we assume that the error term is homoskedastic iid or normally distributed: $u_{4t} \sim N[0, \sigma^2]$. Generalizing, the PAR(1) model with period S is defined as

$$y_{st} = x_{st}'\beta_s + u_{st}, \qquad s = 1, ..., S \tag{7}$$

where the regressors for the quarterly $PAR(1)$ model are defined by the vectors $x_{st} = (1, y_{s-1,t})'$, $s = 2, ..., S$ and $x_{1t} = (1, y_{4,t-1})'$.

3. THE HIERARCHICAL PAR MODEL

As we see from the quarterly $PAR(1)$ model, the periodic AR(p) model has more parameters than a non-periodic model and therefore it seems quite reasonable to make some similarity assumptions for the coefficients. A straightforward approach to do this is by specifying a Bayesian hierarchical structure. Assuming a Gaussian (normal) distribution for the data, the hierarchical PAR model is given by

$$y_{st} \sim N[x_{st}'\beta_s, \sigma_s^2 I_n], \qquad s = 1, ..., S, \quad t = 1, ..., T. \tag{8}$$

and the prior distributions are specified as

$$\beta_s \sim N[\beta, \Sigma] \tag{9}$$

$$\beta \sim N[b_*, H_*] \tag{10}$$

$$\Sigma^{-1} \sim W[(\gamma_* S_*)^{-1}, \nu_*] \tag{11}$$

$$\sigma_s^{-2} \sim Ga_2[\sigma_*^2, n_*], \tag{12}$$

where W stands for the Wishart distribution and Ga_2 for a gamma-2 distribution, i.e. stands for the re-parameterisation $Ga_2[a, b] = Ga[a/2, b/2]$. We adopt the convention that all hyper-parameters indexed with a '$*$' are known and have to be specified before the data are known. The estimation of this model is possible by Gibbs sampling and is described in Gelfand et al.(1990).

3.1 *The smoothness PAR model*

For the Gaussian smoothness PAR(p) model of order p we assume a diagonal regression system like the quarterly PAR(p) model which has the following form

$$\begin{pmatrix} y_{1t} \\ y_{2t} \\ y_{3t} \\ y_{4t} \end{pmatrix} = \begin{pmatrix} x_{1t}' & 0 & 0 & 0 \\ 0 & x_{2t}' & 0 & 0 \\ 0 & 0 & x_{3t}' & 0 \\ 0 & 0 & 0 & x_{4t}' \end{pmatrix} \begin{pmatrix} \beta_1 \\ \beta_2 \\ \beta_3 \\ \beta_4 \end{pmatrix} + \begin{pmatrix} u_{1t} \\ u_{2t} \\ u_{3t} \\ u_{4t} \end{pmatrix}$$

with $u_t \sim N[0, \Sigma]$. For general number of seasons S the PAR model is

$$y_t = X_t\beta + u_t \tag{13}$$

with $X_t' = (x_{1t}, ..., x_{St}), x_{st} = (1, x_{s,t-1}, ..., x_{s,t-p})'$ and $y_t : (S \times 1), X_t : (S \times Sk), \beta : (Sk \times 1), u_t : (S \times 1)$. Assuming Gaussian distributions the hierarchical model is given by

$$y_t \sim N[X_t\beta, \Omega] \tag{14}$$
$$A_k\beta \sim N[0, \sigma_*^2 I_S \otimes I_k] \tag{15}$$
$$\Omega \sim W[(\nu_*\Omega_*)^{-1}, \nu_*] \tag{16}$$

where the differencing matrix A_k is defined as

$$A = \begin{pmatrix} I_k & 0 & 0 & -I_k \\ -I_k & I_k & 0 & 0 \\ 0 & -I_k & I_k & 0 \\ 0 & 0 & -I_k & I_k \end{pmatrix} \tag{17}$$

and I_k is the identity matrix of order k. Note that the variance parameter σ_*^2 controls the tightness of the smoothness restriction in the prior distribution. a small σ_*^2 will enforce the equality of the β_s coefficients.

4. PERIODIC AR-ARCH MODELS

A further extension of PAR models covers volatile financial time series if we specify an ARCH structure or an GARCH (generalized ARCH) structure for the conditional variances. Using Gaussian distributions we assume a two-stage hierarchical model which is given by

$$y_{st} \sim N[x_{st}'\beta_s, z_{st}'\gamma_s] \tag{18}$$
$$\beta_s \sim N[b_*, H_*] \tag{19}$$
$$\gamma_s \sim N[\gamma_*, G_*]. \tag{20}$$

A 3-stage model in analogy to (2) would also be possible. In this model we assume a prior distribution which is the same for all seasons (we will call this also an equi-periodic prior distribution). In Polasek and Jin (1997) we have suggested a tightness prior structure for the AR coefficients which is the case if we assume $b_* = 0$ and $H_*^{-1} = diag(0, 1, 2, ..., k)$ for the prior of the AR coefficients and in similar way we assume for the ARCH coefficients $\gamma_* = 0.01$ and $G_* = H_*$. (The prior mean .01 can be varied with the dimension of the PAR system and the number of ARCH parameters.) The tightness structure is expressed through the variances which shrink in a linear way around the prior location, which is 0 for the AR coefficients and .01 for the ARCH coefficients.

The model specifies for each season s an univariate ARCH model where the prior distribution restricts the coefficients to the stationarity region and shrinks them to zero or a small positive value. The regression vector is $x'_{st} = (x_{st0}, ..., x_{stk})$ a vector of length $k+1$ and could consist of past time series data or other regression variables. In case of an AR(k) model the regression vector is given by $x'_{st} = (1, y_{s,t-1}, ..., y_{s,t-k})$ and z_{st} is a vector of length $p+q+1$ for the ARCH(p,q) coefficients, i.e.

$$z'_{st} = (1, h_{s,t-1}, ..., h_{s,t-p}, \varepsilon^2_{s,t-1}, ..., \varepsilon^2_{s,t-q}) \quad (21)$$

with $h_{st} = z'_{st}\gamma_s$ and $\varepsilon_{st} = y_{st} - x'_{st}\beta_s$.

For the MCMC algorithm we have to work out the full conditional distributions (f.c.d.'s). The f.c.d. for the AR coefficients are given by

$$p(\beta_s|\theta\backslash\beta_s, Y) \propto N[b_{s**}, \quad H_{s**}] \quad (22)$$

with the hyper-parameters

$$H_{s**}^{-1} = H_*^{-1} + X'_s D_s^{-1} X_s, \quad (23)$$

$$b_{s**} = H_{s**}(H_*^{-1} b_* + X'_s D_s^{-1} y_s), \quad (24)$$

where the weight matrix is $D_s = diag(z'_{s1}\gamma_s, ..., z'_{sT}\gamma_s)$ with $X'_s = (x_{s1}, ..., x_{sT})$ and $y'_s = (y_{s1}, ..., y_{sT})$.

The f.c.d. for the ARCH parameters has to be generated by a Metropolis step. The full conditional posterior distribution is given by

$$p(\gamma_s \mid \theta\backslash\gamma_s, Y) \propto \prod_{t=1}^{T} (z'_{st}\gamma_s)^{-1/2}$$

$$\exp[-\frac{1}{2}(y_{st} - x'_{st}\beta_s)^2 / z'_{st}\gamma_s] \cdot$$

$$\exp[-\frac{1}{2}(\gamma_s - \gamma_*)' G_*^{-1} (\gamma_s - \gamma_*)]. \quad (25)$$

We suggest to use as a proposal distribution the least squares regression from the squared residuals of the AR process on the variables in z_t:

$$\hat{\varepsilon}^2_{st} = z'_{st}\gamma_s + u_{st} \quad \text{with} \quad \hat{\varepsilon}_{st} = y_{st} - x'_{st}\beta_s \quad (26)$$

from where we get the parameters for the proposal distribution

$$p(\gamma_s) = N[\hat{\gamma}_s, \hat{G}_s] \quad (27)$$

with $\hat{\gamma}_s = (Z'_s Z_s)^{-1} Z'_s \hat{\varepsilon}_s$ and $\hat{G}_s = \hat{\tau}^2_s (Z'_s Z_s)^{-1}$ where $\hat{\tau}^2_s = \frac{1}{T}\sum_{t=1}^{T} \hat{u}^2_{st}$ and $Z'_s = (z_{s1}, ..., z_{sT})$.

Assuming a smoothness prior we can define a smoothness PAR-PARCH model. This would extend the 2-stage model in (??) to a 3-stage PAR(k)-ARCH(p,q) model:

$$y_{st} \sim N[x'_{st}\beta_s, z'_{st}\gamma_s] \quad (28)$$

$$\beta_s \sim N[b_*, H_*], \gamma_s \sim N[\gamma_*, G_*] \quad (29)$$

$$A_k\beta \sim N[0, \sigma_*^2 I_S \otimes I_k] \quad (30)$$

$$A_{p'}\gamma \sim N[0, \tau_*^2 I_S \otimes I_p] \quad (31)$$

where $\gamma' = (\gamma'_1, ..., \gamma'_S)$ and the differencing matrices A_k and $A_{p'}$ (of appropriate dimension $p' = p + q + 1$) are defined as in (17).

5. SUMMARY

The paper has given a brief introduction into MCMC methods for periodic AR models. We have described a hierarchical model for seasonal PAR models and we have shown how PAR models can be extended to PAR-GARCH models and the MCMC algorithm can be used as well. The MCMC approach is very flexible and can be extended to multivariate models or to models with outliers (see Polasek and Jin (1997)). A further open problem is a computational feasable approach to Bayesian model selection.

6. APPENDIX: THE METROPOLIS ALGORITHM

Consider a conditional distribution $p(\phi \mid y)$ which is specified up to a constant and cannot be sampled directly (also called target distribution). Then the Metropolis algorithm will create a sequence of samples which converge to the target distribution.

(1) Draw a point θ^0 from a starting distribution $p_0(\theta)$ so that $p(\theta|y) > 0$
(2) Sample a candidate θ^* from a jumping (or candidate) distribution at time t:
$$\theta^* \sim J_t(\theta^* \mid \theta^{t-1}) \quad (32)$$
which has to be symmetric
$$J_t(x \mid y) = J_t(y \mid x) \quad \text{for all} \quad x, y, t. \quad (33)$$
(3) Calculate the density ratio
$$r = p(\theta^* \mid y)/p(\theta^{t-1} \mid y). \quad (34)$$
(4) Accept the candidate with probability $min(r, 1)$
$$\theta^t = \begin{cases} \theta^* & \text{with prob.} \quad min(r,1) \\ \theta^{t-1} & \text{else.} \end{cases}$$

By cycling through step 1) to step 4) iterate for $t = 1, 2, ...$ until convergence. Note that if candidate draw is not accepted, then the sampler does not move: $\theta^{(t)} = \theta^{(t-1)}$. A good proposal is often difficult to obtain and in general the acceptance ratio should be as high as possible.

7. REFERENCES

Bollerslev, T. (1986). Generalized autoregressive conditional heteroskedasticity, *Journal of Econometrics,* **31**, 307-327.

Chib S. and E. Greenberg (1995). Understanding the Metropolis-Hastings Algorithm, *American Statistician,* **49** , 327-336.

Franses P.H. (1998). *Time Series Models for Business and Economic Forecasting,* Cambridge Univ. Press, UK.

Gelfand A.E, Hills S.E., Racine-Poon A. and A.F.M. Smith (1990). Illustration of Bayesian inference in normal data models using Gibbs sampling, *Journal of the Am. Statistical Association* **865**, 972-985.

Lund R. and I.V. Basawa (2000). Recursive Prediction and Likelihood Evaluation for Periodic ARMA Models, *J. of Time Series Analysis* **21**, 75-94.

Hamilton J. (1994). *Time Series Analysis,* Princeton University Press, NJ.

Polasek W.(1984) Multivariate Regression Systems: Estimation and Sensitivity Analysis for Two-Dimensional Data. In: *Robustness in Bayesian Statistics,* J. Kadane (ed.), 229-309, North-Holland.

Polasek W. and Jin S. (1996). Gibbs sampling in AR models with random walk priors. In: Gaul W. and D. Pfeifer (eds.) From Data to Knowledge, Springer Verlag, 86-93.

Polasek W. and Jin S. (1997). GARCH Models with Outliers. In: *Classification and Knowledge Organisation,* Klar R. and Opitz O. (eds.) Springer Verlag, 178-186.

Copyright © IFAC Periodic Control Systems,
Cernobbio-Como, Italy, 2001

APPLICATION OF CRONE CONTROL TO A SAMPLED TIME VARYING SYSTEM WITH PERIODIC COEFFICIENTS

Jocelyn Sabatier, Aitor Garcia Iturricha, Alain Oustaloup

LAP - ENSEIRB - Université Bordeaux 1 - Equipe CRONE – EP 2026 CNRS
351, Cours de la Libération, 33405 Talence, France
Tel : +33 (0)556 842 418   Fax : +33 (0)556 846 644
sabatier@lap.u-bordeaux.fr   www.lap.u-bordeaux.fr

Abstract: An application of CRONE control extended to discrete time varying systems with periodic coefficients is presented. The application is carried out in the frequency domain through the representation of considered systems using time varying z-transforms and time varying pseudo frequency responses. *Copyright ©2001 IFAC*

Keywords: Periodic systems, CRONE control, time varying frequency response, time varying z-transform

1. INTRODUCTION

CRONE Control (Oustaloup and Mathieu,1999) (CRONE is the French abbreviation of "Commande Robuste d'Ordre Non Entier") is a frequency robust control methodology based on fractional differentiation (Oustaloup, *et al.*, 2000). Recently, CRONE control has been extended to discrete time varying systems with periodic coefficients, also called discrete periodic systems (Sabatier, *et al.*, 2001; Sabatier and Garcia, 2000). This extension is possible through the use of time varying z-transform (TVZT) representations and time varying pseudo frequency responses (TVPFRs) (Sabatier, *et al.*, 2001) for discrete periodic system modelling, and through a generalisation of the Nyquist criterion (Sabatier, *et al.*, 2001).

With regard to the various representations found in the literature (Flamm, 1991; Misra, 1996; Grasselli, *et al.* 1995; Bittanti and Colaneri, 1996), the representation of periodic systems by time varying frequency responses (Zadeh, 1950) presents the following advantages :

- time varying character of the system is preserved, so the initial and final value theorems and others well-known theorems can be established easily (Garcia, 2002),
- several classes of time varying non-periodic systems can be represented.

The extension proposed in this paper, allows the synthesis of control laws which ensure :
- a stationary behaviour of the control loop for the nominal parametric state of the plant ;
- designer specified performances of the control loop for the nominal parametric state of the plant ;
- robustness of stability and of dynamic performances when the plant is reparametrated,
- some immunity to the time varying character of the plant.

These objectives are reached by the computation of an optimal open loop behaviour based on complex fractional differentiation (Oustaloup, *et al.*, 2000) which :
- respects the extended Nyquist criterion even with parametric variations of the plant ;
- minimises, at reparametration, variations of the resonance ratio of the stationary part of the TVPFR of the control loop, and the time varying part of this TVPFR.

In comparison with H_∞ type (Dahleh, et al, 1992; Cantoni, 1998) of LQG type (Wereley, 1991) control methods, the plant uncertainties are taken into account in a structured form with no overestimation, thus leading to control laws which are as little conservative as possible. Moreover, the fractional differentiation order, permits parametrization of the open loop transfer function with a small number of high-level parameters. The optimization of the control is thus reduced to only the search for the optimal values of these parameters.

In this study, this extension of CRONE control is applied to a testing bench with a DC motor.

Section 2 gives the definition of the time varying z-transform (TVZT). Section 3 proposes some tools which have been developed for the analysis of discrete periodic systems : procedures for the computation of TVPFRs of feedback systems and an extension of the Nyquist criterion. Section 4 summarises the extension of CRONE Control to discrete periodic systems. Section 5 presents the application of CRONE control to a real sampled periodic systems : a testing bench with a DC motor.

2. TIME VARYING Z-TRANSFORM

If $h(n, k)$ denotes the response at time nT_e (T_e being the sampling period) of a discrete time-varying system whose input is a Dirac function $\delta(t - kT_e)$, then, by analogy to the stationary case, the TVZT of this system can be defined by (Zadeh, 1950; Jury, 1964) :

$$H(n,z)=\mathcal{Z}[h(n,k)]=\sum_{r=0}^{\infty}h(n,n-r)z^{-r} \qquad z\in C, \tag{1}$$

or, using $k = n - r$ (assuming no input before time $kT_e = 0$) :

$$H(n,z)=\sum_{k=-\infty}^{n}h(n,k)z^{-n+k}=z^{-n}\sum_{k=0}^{n}h(n,k)z^{k}. \tag{2}$$

Using this representation, the output of the system at time nT_e, $y(n)$ is related to its input by (Jury, 1964) :

$$y(n)=\frac{1}{2\pi j}\int_{\Gamma}H(n,z)U(z)z^{n-1}\mathrm{d}z, \tag{3}$$

where $U(z)$ denotes the z-transform of the input, and Γ is a closed path in the z-plane which encircles counterclockwise the singularities of integral (3).

3. ANALYSYS OF TIME VARYING FEEDBACK SYSTEMS WITH PERIODIC COEFFICIENTS

We consider a discrete periodic system characterised by the state space description :

$$\begin{cases}x(n+1)=A(n)x(n)+B(n)u(n)\\ y(n)=C(n)x(n)\end{cases}, \tag{4}$$

where $u(n) \in \mathbb{R}$, $y(n) \in \mathbb{R}$, $x(n) \in \mathbb{R}^{q\times 1}$ and where coefficients $A(n)$, $B(n)$ and $C(n)$ are real-valued matrices of appropriate dimensions.
Matrices $A(n)$, $B(n)$ and $C(n)$ are also periodic functions of variable n, namely :

$$A(n)=A(n+T),\quad B(n)=B(n+T),\quad C(n)=C(n+T), \tag{5}$$

where period T represents the smallest value satisfying relation (5) and respects the following relation:

$$T=MT_e, \qquad M\in\mathbb{N}. \tag{6}$$

Matrices $A(n)$, $B(n)$ and $C(n)$ are respectively elements of $l_2^{q\times q}[0,T]$, $l_2^{q\times 1}[0,T]$ and $l_2^{1\times q}[0,T]$. Matrix $A(n)$ thus admits the following Fourier series expansions :

$$A(n)=\sum_{k=0}^{M-1}A_k e^{jk\frac{2\pi}{M}n}. \tag{7}$$

Similar series expansions are also possible for matrices $B(n)$ and $C(n)$, but using matrices B_k and C_k.
Given the previous comments, system (4) can be represented by a TVZT, $H(n, z)$, of the form (Sabatier, *et al.* 2001) :

$$H(n,z)=\sum_{k=0}^{M-1}H_k(z)e^{jk\frac{2\pi}{M}n}. \tag{8}$$

Theorem 1 (Sabatier, *et al.* 2001)

Transmittances $H_k(z)$ of relation (8) are given by :

$$\mathcal{H}=\mathcal{C}(\mathcal{N}-\mathcal{A})^{-1}\mathcal{B}, \tag{9}$$

in which vectors $\mathcal{H}$ and $\mathcal{B}$, and matrices $\mathcal{A}$ are respectively given by :

$$\mathcal{H}^T=[H_0(z)\ \ H_1(z)\ \ \dots\ \ H_{M-1}(z)], \tag{10}$$

$$\mathcal{B}^T=\left[B_0^T\ \ B_1^T\ \ \dots\ \ B_{M-1}^T\right], \tag{11}$$

$$\mathcal{A}=\begin{bmatrix}A_0 & A_{M-1} & A_{M-2} & \cdots & A_2 & A_1\\ A_1 & A_0 & A_{M-1} & \cdots & A_3 & A_2\\ A_2 & A_1 & A_0 & \ddots & \vdots & \vdots\\ \vdots & \vdots & \vdots & \ddots & A_{M-1} & A_{M-2}\\ A_{M-2} & A_{M-3} & A_{M-4} & \cdots & A_0 & A_{M-1}\\ A_{M-1} & A_{M-2} & A_{M-3} & \cdots & A_1 & A_0\end{bmatrix}. \tag{12}$$

Matrix $\mathcal{C}$ is defined as for matrix $\mathcal{A}$, but using C_k, and $\mathcal{N}=blkdiag\left(e^{jk2\pi/M}I_q\right)$, $k\in[0,\cdots,M-1]$, where I_q denotes the identity matrix of dimension q.

❑

Using bilinear transformation :

$$z=(w+1)/(1-w), \tag{13}$$

relation (13) becomes :

$$H(n,w)=\sum_{i=0}^{M-1}H_i(w)e^{ji\frac{2\pi}{M}n}, \tag{14}$$

where $H_i(w)$ is the w transform of transmittances $H_i(z)$. This permits a representation of considered systems using the time varying pseudo-frequency response (TVPFR):

$$H(n,jv)=\sum_{k=0}^{M-1}H_k(jv)e^{jk\frac{2\pi}{M}n}, \tag{15}$$

where pseudo-pulsation v is the imaginary part of w. Pseudo-pulsation v is also linked to the real pulsation ω by (Kuo, 1980) : $v=\tan(\omega T_e/2)$.

The standard control scheme of figure 1 is now considered, in which discrete periodic systems $\mathcal{C}$ and $\mathcal{P}$ can be defined by TVPFR $C(n, jv)$ and $P(n, jv)$ with :

$$C(n,jv)=\sum_{i=0}^{M-1}C_i(jv)e^{ji\frac{2\pi}{M}n},\ P(n,jv)=\sum_{i=0}^{M-1}P_i(jv)e^{ji\frac{2\pi}{M}n}. \tag{16}$$

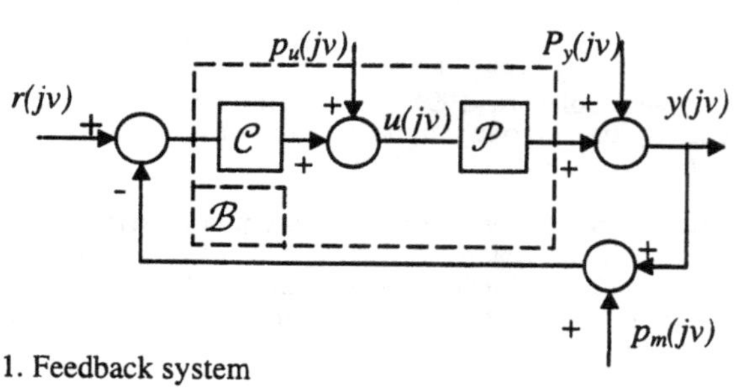

Fig. 1. Feedback system

Let $\beta(n, jv)$ be the TVPFR of the system resulting from the cascade connection of systems $\mathcal{C}$ et $\mathcal{P}$, with :

$$\beta(n, jv) = \sum_{i=0}^{M-1} \beta_i\,(jv) e^{ji\frac{2\pi}{M}n} . \qquad (17)$$

Theorem 2 (Sabatier, *et al.* 2001)

Pseudo frequency responses $\beta_i(jv)$ of relation (17) are given by :

$$\mathcal{B} = \hat{\mathcal{P}}\mathcal{C}, \qquad (18)$$

with :

$$\mathcal{B}^T = [\,\beta_0(jv)\,\beta_1(jv)\,\beta_2(jv)\ \ldots\ \beta_{M-1}(jv)], \qquad (19)$$

$$\mathcal{C}^T = [C_0(jv)\,C_1(jv)\,C_2(jv)\ \ldots\ C_{M-1}(jv)], \qquad (20)$$

$$\hat{\mathcal{P}} = \begin{bmatrix} P_0(jv_0) & \cdots & P_2(jv_{M-2}) & P_1(jv_{M-1}) \\ P_1(jv_0) & \cdots & P_3(jv_{M-2}) & P_2(jv_{M-1}) \\ \vdots & \ddots & \vdots & \vdots \\ P_{M-2}(jv_0) & \cdots & P_0(jv_{M-2}) & P_{M-1}(jv_{M-1}) \\ P_{M-1}(jv_0) & \cdots & P_1(jv_{M-2}) & P_0(jv_{M-1}) \end{bmatrix}, \qquad (21)$$

where :

$$jv_i = \frac{jv - 1 + e^{2\frac{ji\pi}{M}} + e^{2\frac{ji\pi}{M}}\,jv}{-\,jv + 1 + e^{2\frac{ji\pi}{M}} + e^{2\frac{ji\pi}{M}}\,jv}. \qquad (22)$$

❑

Let now $T(n, jv)$, $S(n, jv)$, $R(n, jv)$ and $PS(n, jv)$ denote respectively TVPFRs connecting :

- reference input $r(n)$ to output $y(n)$;
- output disturbance $p_y(n)$ to output $y(n)$;
- reference input $r(n)$ to input $u(n)$;
- plant input disturbance $p_u(n)$ to output $y(n)$.

The feedback system in figure 1 being a discrete periodic system, TVPFR $T(n, jv)$ is of the form :

$$T(n, jv) = \sum_{i=0}^{M-1} T_i\,(jv) e^{ji\frac{2\pi}{M}n} , \qquad (23)$$

where pseudo-frequency responses $T_i(jv)$ are given by the following theorem.

Theorem 3 (Sabatier, *et al.* 2001)

Transmittances $T_i(jv)$ of relation (23) are given by :

$$\mathcal{T} = \left(\hat{\mathcal{I}} + \hat{\mathcal{B}}\right)^{-1} \mathcal{B} \quad \text{if} \quad \det\left(\hat{\mathcal{I}} + \hat{\mathcal{B}}\right) \neq 0, \qquad (24)$$

where

$$\mathcal{T}^T = [\,T_0(jv)\,T_1(jv)\,T_2(jv)\ \ldots\ T_{M-1}(jv)]. \qquad (25)$$

Matrix $\hat{\mathcal{B}}$ is defined in the same way as matrix $\hat{\mathcal{P}}$ (relation (21)) but using functions $\beta_k(jv)$. Parameter jv_i is defined by relation (22).

Vector $\mathcal{B}$ is given by relation (19), and matrix $\hat{\mathcal{I}}$ is the identity matrix of dimension M.

❑

Similar theorems for the other TVPFRs, $S(n, jv)$, $PS(n, jv)$ and $R(n, jv)$, can be established in the same way (Garcia, 2002).

The representation of discrete periodic systems using TVPFRs, also permits an extension of the Nyquist criterion.

Theorem 4 : Nyquist criterion extension (Sabatier, *et al.* 2001)

If $\hat{\mathcal{B}}$ denotes matrix (parameter jv_i being given by relation (22)) :

$$\hat{\mathcal{B}} = \begin{bmatrix} \beta_0(jv_0) & \cdots & \beta_2(jv_{M-2}) & \beta_1(jv_{M-1}) \\ \beta_1(jv_0) & \cdots & \beta_3(jv_{M-2}) & \beta_2(jv_{M-1}) \\ \vdots & \ddots & \vdots & \vdots \\ \beta_{M-2}(jv_0) & \cdots & \beta_0(jv_{M-2}) & \beta_{M-1}(jv_{M-1}) \\ \beta_{M-1}(jv_0) & \cdots & \beta_1(jv_{M-2}) & \beta_0(jv_{M-1}) \end{bmatrix} \qquad (26)$$

and if system $\mathcal{B}$ has P_0 poles of modulus greater than 1 (poles of $\mathcal{B}$ are the eigenvalues of the monodromy matrix $\Phi(MT_e, 0)$, $\Phi(t, \tau)$ being the transition matrix of system $\mathcal{B}$), then the closed loop system of figure 1 is stable if and only if the Nyquist eigenvalue locus of matrix $\hat{\mathcal{B}}$ encircles the point $(-1, 0)$ P_0 times counterclockwise where $v \in [0, \sin(2\pi/M)/(\cos(2\pi/M)+1)[$ (supposing there are no hidden unstable modes).

❑

4. CRONE CONTROL OF SAMPLED PERIODIC PLANTS

4.1 Objectives

To extend CRONE control to sampled periodic plants (SPPs), an open loop behaviour for the nominal parametric state of the plant is required in the pseudo-continuous domain which :

- ensures a stationary behaviour of the closed loop system ;
- ensures performances set by the designer such as rapidity and the resonance ratio in tracking of the closed loop system ;
- takes into account the behaviour of the plant at the low and the high frequencies to ensure satisfactory accuracy of steady state, and immunity of the plant input to measurement noise ;
- takes into account the right half plane zeros of the plant which appear through the use of w transformation.

Whenever the plant is reparametrated, namely if the plant $\mathcal{P}$ is an element of the description family $\mathbb{P}$, this open loop must also ensure :

- robust closed loop stability and performances ;
- satisfactory immunity of the closed loop to the time varying character of the plant.

As for the stationary case, the behaviour thus defined can be described for stable SPPs, by transmittance based on frequency limited complex non integer integration (Oustaloup and Mathieu 1999; Oustaloup, *et al.*, 2000) :

$$\beta(w) = K \prod_{i=1}^{n_z}\left(1 - \frac{w}{z_i}\right)\left(1 + \frac{v'_b}{w}\right)^{n_b} \left(\frac{1+\frac{w}{v_b}}{1+\frac{w}{v'_b}}\right)\left(\frac{1+\frac{w}{v_h}}{1+\frac{w}{v_b}}\right)^{a}$$

$$\left(Re\left[\left(C_0\frac{1+\frac{w}{v_h}}{1+\frac{w}{v_b}}\right)^{ib}\right]_{C_j}\right)^{-sign(b)} \left(\frac{1+\frac{w}{v'_h}}{1+\frac{w}{v_h}}\right)^{2} \frac{1}{\left(1+\frac{w}{v'_h}\right)^{n_h}} \qquad (27)$$

where

$$C_0 = \left[\left(1+\frac{\nu_r^2}{\nu_b^2}\right)\Big/\left(1+\frac{\nu_r^2}{\nu_h^2}\right)\right]^{1/2} . \qquad (28)$$

K ensures the open loop unit gain pseudo-frequency ν_u set by the designer. ν'_b, ν_b, ν_h and ν'_h are transitional pseudo-frequencies. $n_b \in \mathbb{N}$ and $n_h \in \mathbb{N}$ are respectively the asymptotic behavior orders in open loop at the low ($\omega < \omega'_b$) and the high ($\omega > \omega'_h$) frequencies. $a \in \mathbb{R}^+$ and $b \in \mathbb{R}$ are the real and imaginary orders of integration. ν_r is the resonance frequency close to ν_u. n_z is the number of right half plane zeros z_i, $i \in [1, n_z]$, of the plant.

4.2 Optimisation of the open loop behaviour

The optimisation of the open loop behaviour consists in determining the seven optimal parameters of the nominal open loop transmittance $\beta(w)$:

- optimal real integration order a_{opt}, and optimal gain K_{opt}
- optimal imaginary integration order b_{opt}
- optimal transitional pseudo-frequencies ω'_{bopt}, ω_{bopt}, ω_{hopt} and ω'_{hopt}

The unit gain frequency and the tangency to an iso-overshoot contour are chosen by the designer, so only five independent parameters need to be considered.

The open loop behaviour which satisfies the objectives defined in section 4.1 can be computed by solving a constrained optimisation problem.
The performance criterion and constraints which provide optimal open loop behaviour respecting the objectives of section 4.1 thus comprise terms which guarantee :

- robustness of the stability degree of the control
- immunity of the control to the time varying character of the plant
- the performance objectives set by the designer.

This lead to minimisation of the criterion :

$$J = (Q_{max} - Q)^2 + (Q_{min} - Q)^2 + \delta \sup_{\substack{P\in\mathbb{P}, \nu\in\mathbb{R}^+ \\ 0<k<M,\ k\in\mathbb{N}}} |T_k(j\nu)| . \qquad (29)$$

With a judicious choice of weighting coefficient δ, the minimisation of criterion (29) ensures the minimisation of the resonance ratio variations of $T_0(j\nu)$ and the minimisation of the time varying part of the TVPFR $T(n, j\nu)$.

Minimisation of criterion (29) sometimes produce undesirable closed loop behaviours for one or more plant $\mathcal{P}$ of the description family $\mathbb{P}$. We thus define how each function $R(n, j\nu)$, $S(n, j\nu)$ and $T(n, j\nu)$ should be shaped to eliminate these behaviours.

The solicitation level of the plant input is taken into account through limitation of the TVPFR $R(n, j\nu)$:

$$\sup_{P\in\mathbb{P}, n\in\mathbb{N}} |R(n, j\nu)| < R_{adm}(\nu), \qquad \forall \nu \in \mathbb{R}^+ , \qquad (30)$$

where $R_{adm}(\nu)$ is the maximum admissible value of the modulus of $R(n, j\nu)$.

Also, in order to ensure a satisfactory rejection of plant output disturbances, function $S(j\omega, t)$ is bounded by the constraint :

$$\sup_{P\in\mathbb{P}, n\in\mathbb{N}} |S(n, j\nu)| < S_{adm}(\nu), \qquad \forall \nu \in \mathbb{R}^+ , \qquad (31)$$

where $S_{adm}(\nu)$ is the maximum admissible value of the modulus of $S(n, j\nu)$.

Finally, in order to cancel the effects of hauling on the step response in relation to its value in steady state and to ensure a satisfactory rejection of measurement noise, the three following constraints are introduced :

$$\sup_{P\in\mathbb{P}, 0<\nu<\nu_1} |T'_0(j\nu)| < h_1, \qquad \inf_{P\in\mathbb{P}, 0<\nu<\nu_2} |T'_0(j\nu)| > h_2,$$

and

$$\sup_{P\in\mathbb{P}, \nu_3<\nu} |T'_0(j\nu)| < h_3 \qquad h_1 \in \mathbb{R},\ h_2 \in \mathbb{R},\ h_3 \in \mathbb{R}. \qquad (32)$$

The optimisation of the open loop behaviour consists in determining the five optimal parameters of the nominal open loop transmittance $\beta(w)$ which minimise criterion (29) and satisfy the constraints (30), (31) and (32). The optimisation algorithm is based on the non linear simplex (Oustaloup and Mathieu, 1999).

4.3 Optimal controller

The optimal controller is computed so that the cascade connection of the controller and the nominal plant can be described by transmittance $\beta(j\nu)$. The resulting controller can be described by a TVPFR $C(n, j\nu)$ of the form :

$$C(n, j\nu) = C_0(j\nu) + \sum_{k=1}^{M-1} C_k(j\nu) e^{jk\frac{2\pi}{M}n}$$
$$= C_0(j\nu) + \sum_{k=1}^{n_c} Cc_k(j\nu)\cos\left(k\frac{2\pi}{M}n\right) + \sum_{k=1}^{n_s} Cs_k(j\nu)\sin\left(k\frac{2\pi}{M}n\right) \qquad (33)$$

with $n_c = M/2$ and $n_s = M/2-1$ if M even, or $n_c = n_s = (M-1)/2$ if M odd.

The synthesis of the controller thus consists in the approximation of transmittances $C_0(w)$, $Cc_k(w)$, $Cs_k(w)$ by transmittances of the form :

$$C(j\nu) = \sum_{i=0}^{n} b_i (j\nu)^i \Big/ \sum_{i=0}^{d} a_i (j\nu)^i , \qquad (34)$$

where degrees n and d are set by the designer. Two techniques can be used to determine coefficients a_i and b_i of relation (34). The first is a non iterative synthesis method based on the elementary symmetrical functions of Vietes roots (Oustaloup and Mathieu, 1999) and the second is based on the resolution of a linear programming problem (Oustaloup and Mathieu, 1999).
Z transformations of transmittances $C_0(w)$, $Cc_k(w)$ and $Cs_k(w)$ are obtained using inverse bilinear transformation $w = (z-1)/(z+1)$.

5. APPLICATION

5.1 Description of the plant

The extension of third generation CRONE control is applied to the speed control of a testing bench with a DC motor which rotates a disk on which equal weight loads are mounted. By varying the number of loads, inertia J_m driven by the motor varies.
For the testing bench to behave as a periodic system, the motor is supplied by a voltage of the form

$$V(t) = (A_0 + A_1\cos(\omega_0 t))U(t) \qquad (35)$$

with $A_0 = 0.05$, $A_1 = 0.5$ and $\omega_0 = 7$ rd/s.
The plant to be controlled thus admits state space description :

$$\begin{bmatrix}\dot{\Omega}(t)\\ \ddot{\Omega}(t)\end{bmatrix} = \begin{bmatrix} 0 & 1 \\ -\dfrac{0{,}42}{J_m} & \dfrac{-9{,}4.10^{-6} - J_m}{4{,}7.10^{-3} J_m}\end{bmatrix}\begin{bmatrix}\Omega(t)\\ \dot{\Omega}(t)\end{bmatrix} + \begin{bmatrix}0\\ \dfrac{5{,}44.10^2}{J_m}\end{bmatrix} V(t) \qquad (36)$$

Parametric variations of this plant result from variations of the number of loads and thus the inertia. Five parametric states have been considered :

- minimal load : $J_m = 0.108$ kg.m^2
- 25% of the maximal load : $J_m = 0.129$ kg.m^2
- 50% of the maximal load : $J_m = 0.150$ kg.m^2 (nominal plant)
- 75% of the maximal load : $J_m = 0.171$ kg.m^2
- maximal load : $J_m = 0.192$ kg.m^2

Control of the angular speed $\Omega(t)$ of the driving shaft must now be achieved.

5.2 Synthesis of the controller

Parameter M is fixed at 10. This leads to a sampling period $T_e = 2\pi/\omega_0 M = 0.0898$ s. To take into account right half plane zero $z_1 = 1$ of the nominal plant (given the use of w transformation), n_z is fixed at 1. The unit gain pseudo-frequency is fixed at $\nu_u = 0.135$ for the nominal behaviour of the plant. The asymptotic behaviour orders in open loop at low and at high frequencies are fixed at $n_b = 1$ and $n_h = 5$.

The optimal open loop behaviour which minimises the criterion

$$J = (Q_{\max} - 2.3)^2 + (Q_{\min} - 2.3)^2 + \sup_{\substack{P \in \mathbb{P}, \nu \in \mathbb{R}^+, \\ 0<k<M,\, k \in \mathbb{N}}} \left|T_k(j\nu)\right| , \quad (37)$$

(weighting coefficient δ was chosen equal to 1 after a few trails) is computed under constraints

$$\sup_{P \in \mathbb{P}, 0<\nu<\nu_1} \left|T_0(j\nu)\right| < h_1, \qquad \inf_{P \in \mathbb{P}, 0<\nu<\nu_2} \left|T_0(j\nu)\right| > h_2,$$

and (38)

$$\sup_{P \in \mathbb{P}, \nu_3<\nu} \left|T_0(j\nu)\right| < h_3,$$

with $h_1 = 5$ dB, $h_2 = -5$ dB, $h_3 = -10$ dB, $\nu_1 = 0.01$, $\nu_2 = 0.01$ and $\nu_3 = 50$.

This optimisation gives :

$$a = 1.472, \quad b = 0.2011, \quad K = 176.9,$$
$$\nu'_b = 0.00011, \quad \nu_b = 0.0033 \quad \nu_h = 67.36, \quad \nu'_h = 67.36,$$

and then permits the synthesis of an optimal controller of the form :

$$C(n,z) = C_0(z) + \sum_{k=1}^{5} Cc_k(z)\cos\left(k\frac{2\pi}{10}n\right) + \sum_{k=1}^{4} Cs_k(z)\sin\left(k\frac{2\pi}{10}n\right). \quad (39)$$

5.3 Results

Figure 2 shows the Bode diagrams (in the w-plane) of transmittance $C_0(w)$, $C_{ck}(w)$ $k \in [1, 5]$ and $C_{sk}(w)$ $k \in [1, 4]$.

Figure 3 shows the Nyquist eigenvalue locus of matrix $\hat{\mathcal{B}}$ with $\nu \in [0, \sin(2\pi/M)/(\cos(2\pi/M)+1)[$ for the five considered parametric states. Given that the plant is stable, this figure shows that the synthesised controller ensures closed loop stability for the five considered parametric states.

Figure 4 highlights the tangency of the Nichols locus of transmittances $\beta_0(w)$ for the nominal parametric state of the plant.
For the five considered parametric states, figure 5 shows gain diagrams of transmittances $T_0(w)$ and figure 6 shows gain diagrams of transmittances $T_k(w)$ for $k \in [1, 9]$.
These figures shows that the optimal open loop transmittance $\beta(w)$ obtained ensures : the minimisation of the resonance ratio variations of $T_0(j\nu)$, and the minimisation of the time varying part of the TVPFR $T(n, j\nu)$

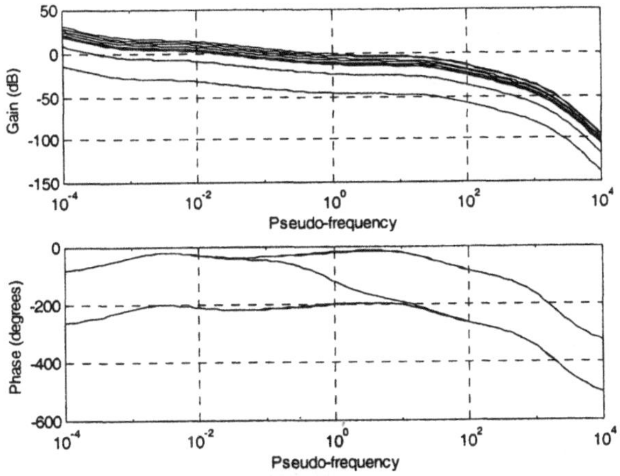

Fig. 2. Bode diagrams (in the w-plane) of transmittances $C_0(w)$, $C_{ck}(w)$ $k \in [1, 5]$ and $C_{sk}(w)$ $k \in [1, 4]$

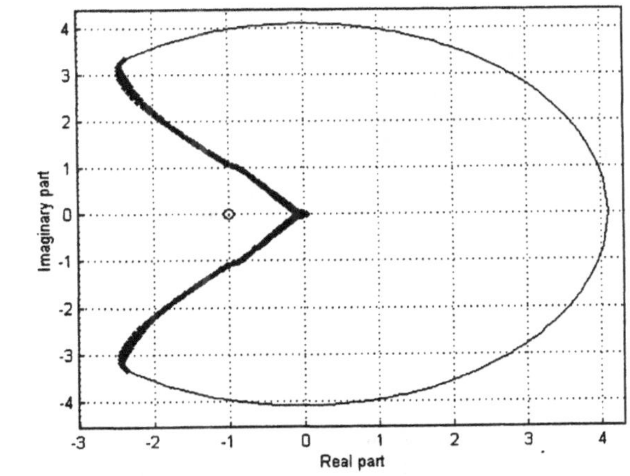

Fig. 3. Nyquist eigenvalue locus of matrix $\hat{\mathcal{B}}$ with $\nu \in [0, \sin(2\pi/M)/(\cos(2\pi/M)+1)[$ for the five considered parametric states

Figure 7 shows the responses of the closed loop with two extreme parametric states of the plant to the step function $r(t)=50\mathrm{H}(t-\tau)$ with $\tau = 0$, $\tau = 4T_e$, $\tau = 8T_e$ (H(t) denotes the Heaviside function) and demonstrates the efficiency of the synthesis method in spite of plant uncertainties and the time varying character of the plant.

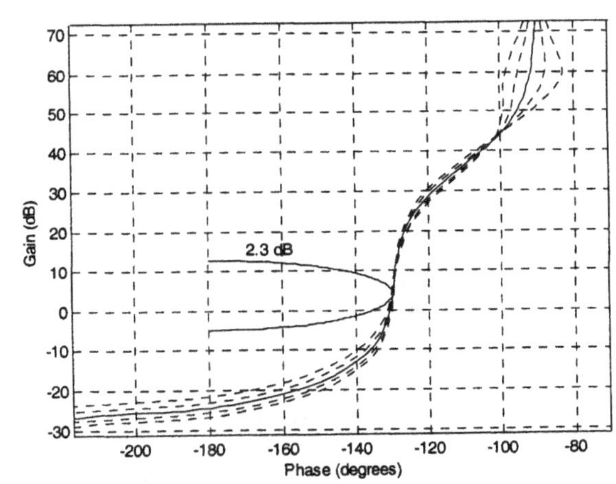

Fig. 4. Nichols locus of transmittance $\beta_0(w)$ for nominal (___) and reparametrated parametric state (- - -) of the plant.

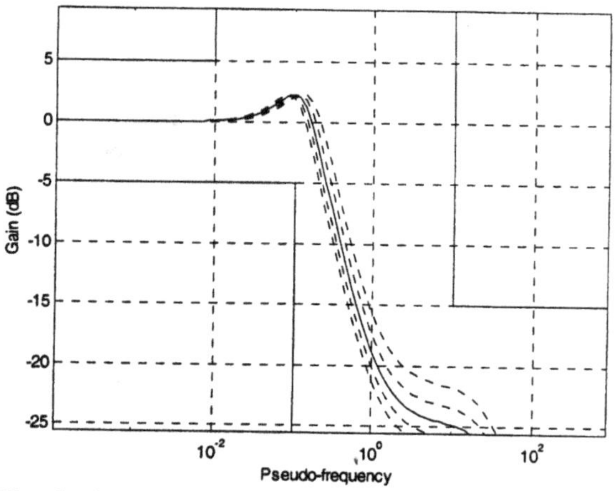

Fig. 5. Gain diagrams of transmittances $T_0(w)$ for the five parametric state of the plant and associated constraints

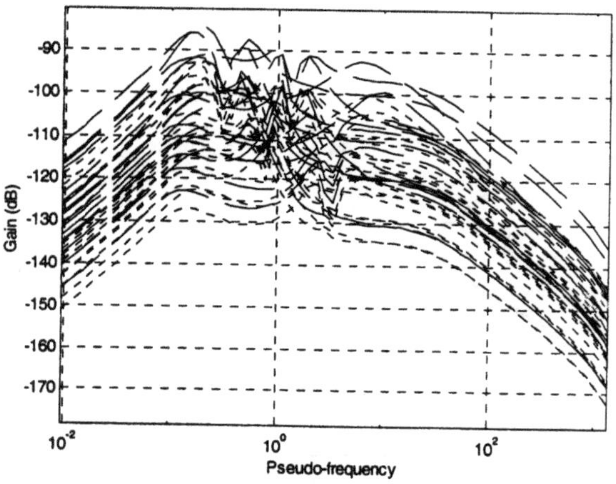

Fig. 6. Gain diagrams of transmittances $T_k(w)$ for $k \in [1, 9]$ and for the five parametric states of the plant.

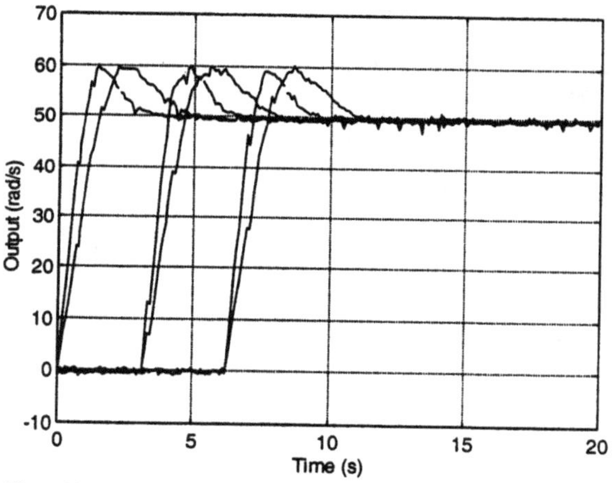

Fig. 7. Closed loop time responses, for each extreme parametric state of the plant, to step inputs $r(t)$=50H(t-τ), with $\tau=0$, $\tau=4T_e$, $\tau=8T_e$

6. CONCLUSION

An extension of CRONE control to discrete time varying systems with periodic coefficients is summarised. This robust control method is based on the computation of an optimal open loop behaviour which minimises the stability degree variations of the closed loop system when the plant is reparametrated. This extension is possible through the representation of considered systems using time varying z-transform.

Then, this extension is applied to a testing bench with a DC motor. This application proves the efficiency of CRONE control for discrete time varying systems with periodic coefficients.

It is demonstrated that this frequency method is very efficient for robust control of time varying systems. Thus, we now aim to extend the CRONE control to other classes of time varying systems using time varying frequency response representations.

REFERENCES

Oustaloup A., Mathieu.B. (1999). *La commande CRONE : du scalaire au multivariable.* HERMES Paris.

Oustaloup A., Levron F., Nanot F., Mathieu B. (2000). *Frequency band complex non integer differentiator : characterization and synthesis.* IEEE Trans. on Circ. and Syst., Vol 47, N° 1, p 25-40.

Sabatier J., Garcia Iturricha A., Oustaloup A. (2001). *Commande CRONE de systèmes linéaires non stationnaires échantillonnés à coefficients périodiques.* European Journal of Automation (JESA), Vol. 35, N° 1-2, pp 149-168,.

Sabatier J., Garcia Iturricha A. (2000). *Commande CRONE de systèmes non stationnaires à coefficients périodiques échantillonnés.* Proc. of the Conf. Intern. Francophone d'Automatique, CIFA 2000, Lilles, France, July 5-8.

Flamm D. S. (1991). *A new shift-invariant representation for periodic linear systems.* Systems & control letters Vol. 17, pp 9-14,.

Misra P. (1996). *Time-invariant representation of discrete periodic systems.* Automatica, Vol. 32, N°2, pp 267-272.

Grasselli O. M., Longhi S., Tornambe A. (1995). *System equivalence for periodic models and systems.* SIAM J. Contr. Optim, Vol. 33, N°2, pp 455-468.

Bittanti S. and Colaneri P. (1996). *Analysis of discrete-time periodic systems.* Control and Dynamic Systems : Digital Control and Signal Processing – Systems and techniques, C. T. Leondes Ed.

Garcia Iturricha A. (2002). *Commande CRONE de procédés non stationnaires.* Ph.D Thesis, Université Bordeaux I, to appear.

Dahleh M. A., Voulgaris P. G. and Valavani L. S. (1992). *Optimal and robust controllers for periodic and multirate Systems.* IEEE Trans. Autom. Contr., Vol. 37, N°1, pp 90-99.

Cantoni M. W. (1998). *Linear periodic systems : robustness analysis and sampled-data control.* PhD thesis, St John's College, Cambridge.

Wereley N. M. (1991). *Analysis and control of linear periodically time varying systems.* PhD thesis - Dept. of Aeronautics and Astronotics, M. I. T.

Zadeh A. L. (1950). *Frequency Analysis of Variable Networks,* Proc. of I.R.E., pp. 291-299.

E. I. Jury (1964). *Theory and Application of the z-Transfrom Method.* Robert E. Krieger Publishing Company, Malabar, Florida.

Kuo B. C. (1980). *Digital control systems.* Ed Holt, Rinehart and Winston .

Copyright © IFAC Periodic Control Systems,
Cernobbio-Como, Italy, 2001

PERIODICITY OF THE IDLE SPEED OF A DIESEL ENGINE

N. Kositza, Ch. Fleck, A. Schloßer and **H. Rake**

Institute of Automatic Control
Aachen University of Technology, 52056 Aachen, Germany
Phone: ++49-241-807481, Fax: ++49-241-8888296
E-Mail: N.Kositza@irt.rwth-aachen.de

Abstract: This paper presents studies of speed fluctuations of a diesel engine at idle speed. Especially fluctuations because of cylinder imbalance were taken into consideration. Therefore a model of the engine was developed. Simulated and measured speed were analysed by means of the fourier and wavelet transform. The aim was to point out possibilities for the design of a periodic controller.
Copyright ©2001 IFAC

Keywords: Automotive, Diesel Engines, Idle Speed Control, Modelling, Signal Analysis, Spectrum, Time-Frequency Representation

1. INTRODUCTION

The importance of high-speed direct-injection diesel engines has increased in the last few years. This is due to the low fuel consumption concomitant with excellent dynamic performance in vehicle movement. But further reduction of fuel consumption is still desired and needed. Vehicles driving in the city consume 30 % of their fuel while idling (Thornhill *et al.* 2000). Reducing the idle speed by 100 rpm for a vehicle with a fuel consumption of 10 l/100 km can save 4,25 % of fuel (Hrovat and Sun 1997).

There are several reasons which limit the reduction of idle speed: instability, friction and differences between single fuel injection pumps. This must be explained with regard to the injection system. The one of the diesel engine considered here consists of a unit pump system with a control rack. The injection map shows at lower speed a shape which leads to instability of the engine and can be noticed in speed fluctuations (Kamata *et al.* 1986). This phenomenon is of periodic behaviour and is called hunting. On the other hand if at low speed the energy supplied by the fuel injected is not sufficient to overcome the losses due to friction etc., the engine stops. This phenomenon is called lean misfire limit (LML). Furthermore differences between single fuel injection pumps lead to periodic disturbances. Especially these are regarded in this study. They can be reduced by cylinder balancing.

As customers want engines that operate reliably with low vibrations and are able to overcome load steps (e.g. by switching on the air conditioning system) the idle speed is set to a fairly high value. A reduction of these speed fluctuations would permit a reduction of idle speed and to fulfill both requests of the customer, namely low vibrations and low consumption. But before examining the idle speed control itself, it's important to understand the causes and to detect the periodic vibrations.

In this study different methods based on the speed measurement to permit the possibility of recognising defective pumps or analysing their influence on the speed are compared. Many methods exist like pattern recognition or correlation analysis, etc. The focus of this study has been the Fourier and the Wavelet Transform. Therefore the speed fluctuations of a diesel engine in the lower speed range have been examined based on a mathematical model of the engine and additional data obtained from a test bench. Especially the influence of cylinder imbalance was taken into consideration. Classical methods like Fourier transform are compared with newer ones like Wavelet transform. The use of an engine model is able to illustrate the operational behaviour of the engine and to give further information on the measured speed fluctuations. One aim was to point out possibilities for the design of a periodic controller which entails a further reduction of the average idling speed and takes comfort and consumption into consideration.

2. DIESEL ENGINE MODEL

In the literature many models of diesel engines can be found (Schmidt 1995, Kao and Moskwa 1995, Hild *et al.* 1999, Isermann *et al.* 1998, Zweiri *et al.* 1999). They differ in their independent variable (time or crankshaft angle), their accuracy, the parts of the engine regarded, the injection system etc. As already mentioned the injection system of the diesel engine considered here consists of a unit pump system with a control rack. Four or six single pumps driven by a camshaft produce the pressure in the fuel pipes, necessary for injection. A rack controls through its position the quantity of injected fuel. The resultant characteristic of the injection system has large influence on the fluctuations of the idle speed, especially on hunting.

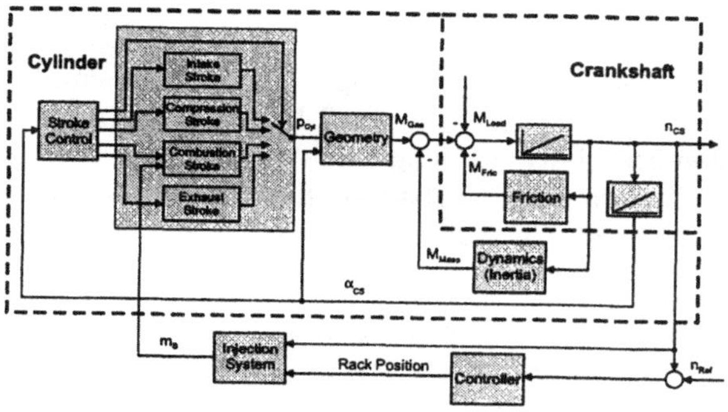

Fig. 1. Engine model

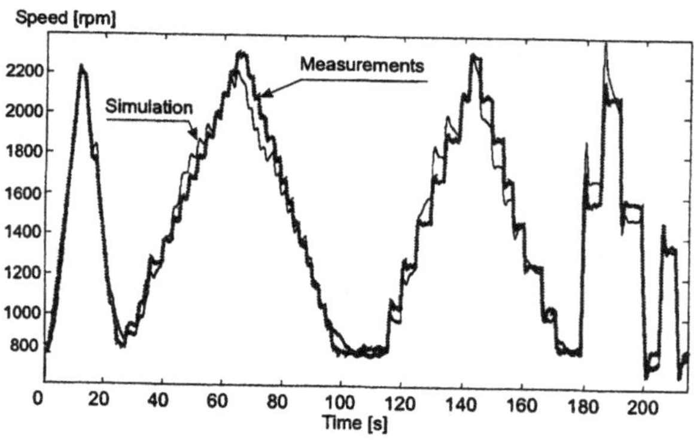

Fig. 2. Comparison of measurement and simulation for an engine cycle

In order to investigate the speed oscillations the crankshaft angle was chosen as independent variable of the engine model (Happe 2000). The resulting structure of the model for one cylinder is shown in figure 1.

Assuming an ideal combustion process the working process of the diesel engine is described by the thermodynamic characteristics of the basic engine cycles. The differential equation of the piston and crankshaft motion leads to the resulting rotational speed.

Figure 2 shows for comparison the measured and the simulated speed of one engine cycle of a 6 cylinder engine. The simulated signal is in sufficient accordance with the measured one.

3. FREQUENCY ANALYSIS OF SIMULATED SIGNALS

To simulate a defective pump a reduced amount of fuel was given to one cylinder of the model. Figure 3 shows the corresponding simulated speed of the engine model in idling operation.

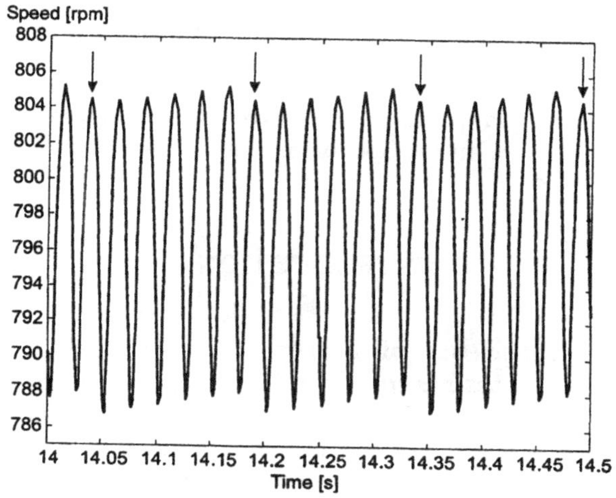

Fig. 3. Simulated idling speed with one cylinder receiving less fuel than others (see arrows)

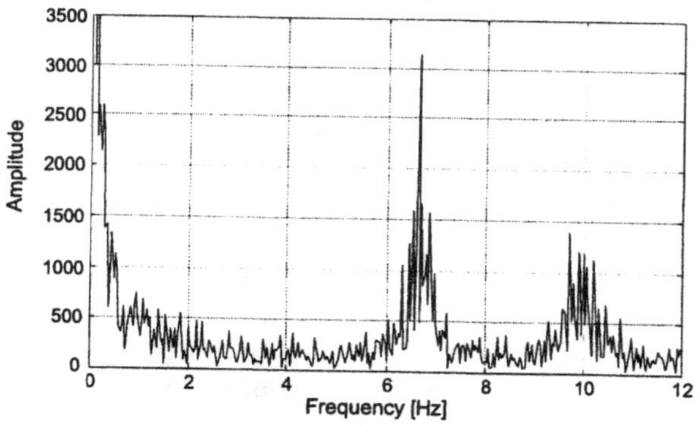

Fig. 4. Frequency spectrum of the simulated speed

Due to the cyclic combustion the engine speed is a periodic signal with a frequency spectrum that ideally includes only the speed and the combustion frequency. The defective pump however produces a disturbance frequency in the speed signal. This disturbance frequency ($f \approx 6.67$ Hz) depends on the number of cylinders (here: 6) and the average speed (here: 800 rpm) and can be clearly seen in figure 4.

4. FREQUENCY ANALYSIS OF MEASURED SIGNALS

4.1 *Fourier Transform*

The rotary-speed measurement is supplied at non-constant time intervals and the data acquisition system normally takes the values irregularly. For this reason the Discrete Fourier Transform (DFT) couldn't be applied directly to the measured data to calculate the frequency spectrum. Without changing the measurement system several approaches for a better interpretation of the measured signal have been realised. The speed data can be interpolated e.g. with cubic splines and then equidistantly sampled (Nahrath *et al.* 1999). Another possibility we used was to approximate the Fourier-coeffiecients by Trapezoidal or Simpson's Rules of the integrals. Both ways led to sufficient results.

The speed measurement system itself offers a third possibility: Though the speed n is measured at non-constant time intervals, the angel intervals are constant. The Fourier Transform

$$FT_{\alpha}(\omega) = \int_{-\infty}^{\infty} n(\alpha) \cdot \exp(-j\omega\alpha) \cdot d\alpha \qquad (1)$$

leads to a spectrum where the abscissa includes orders instead of frequencies. For a 4-stroke engine characterized by two alternate cycles the rotational speed always results in a peak at the second order.

Figure 5 shows the rotational speed, measured with 60 teeth per rotation, of a controlled 4-stroke diesel engine and the corresponding order spectrum. All cylinders behave almost the same. Beside the expected peaks at every 2nd order there can be noticed slight peaks at the order 0.15 and 1.5 and minimal ones at 0.5 and 1. Both slight peaks can also be found in the analyzed input signal of the engine, the diesel mass flow.

Figure 6 shows the controlled rotational speed and the corresponding order spectrum if one cylinder receives no fuel. The expected peak at order 0.5 can be clearly seen, but is not as high as the one at order 1.5. This leads to the assumption that the controller which works with larger sampling intervals creates the peak at order 1.5 by trying to balance the speed. The peak at order 0.15 exists no longer and has also not been remarked at several other speeds and measurements.

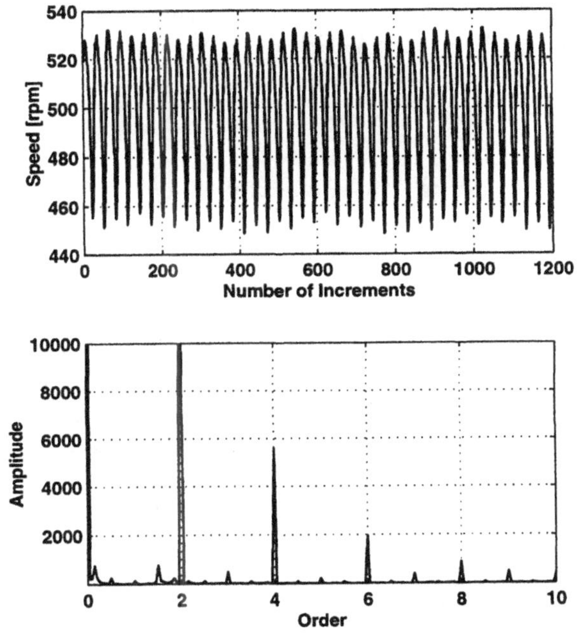

Fig. 5. Measured speed and corresponding order spectrum

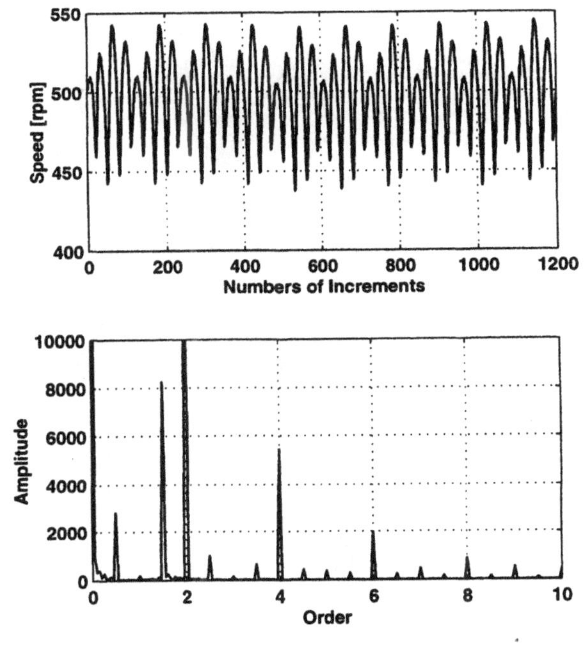

Fig. 6. Measured speed and corresponding order spectrum with one cylinder receiving no fuel

In the next sections only this disturbed measurement will be further examined.

4.2 *Short Time Fourier Transform*

The main disadvantage of the DFT is the impossibility to locate the occurence of specific frequencies in time. For example, if one pump is defective (less fuel at low speed), it's always defective at that speed. Therefore it would be useful to know, at what time interval and speed a special frequency occurs. One answer is the Short Time Fourier Transform (STFT)

$$STFT_\alpha(\alpha,\omega) = \int_{-\infty}^{\infty} n(\varphi)\cdot w(\alpha-\varphi)\cdot \exp(-j\omega\varphi)\cdot d\varphi \qquad (2)$$

which transforms by means of a so-called window w with constant size only one part of the signal at a time. The size of the window determines the frequency (order) resolution. The result for a rectangular window and a window size of one revolution is shown in figure 7. The order resolution is 1. To recognize differences, the corresponding order resolution must be at least twice the order wanted to be regarded. Therefore the window size should be four revolutions or longer. The result is shown in figure 8.

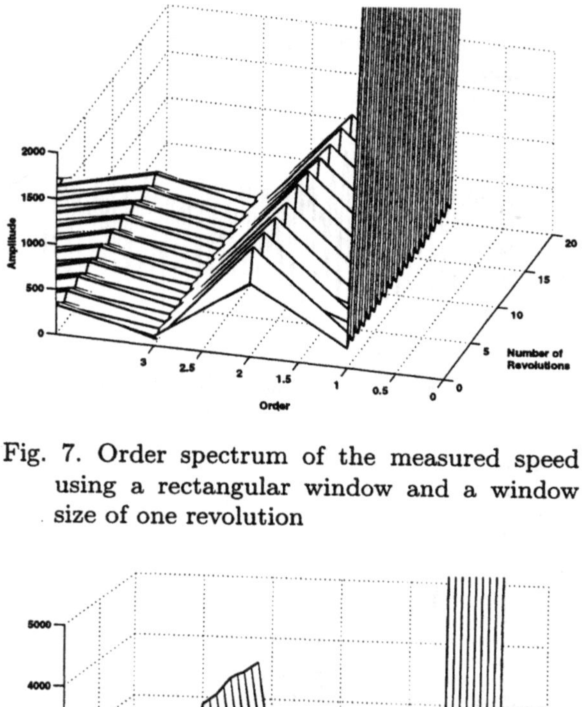

Fig. 7. Order spectrum of the measured speed using a rectangular window and a window size of one revolution

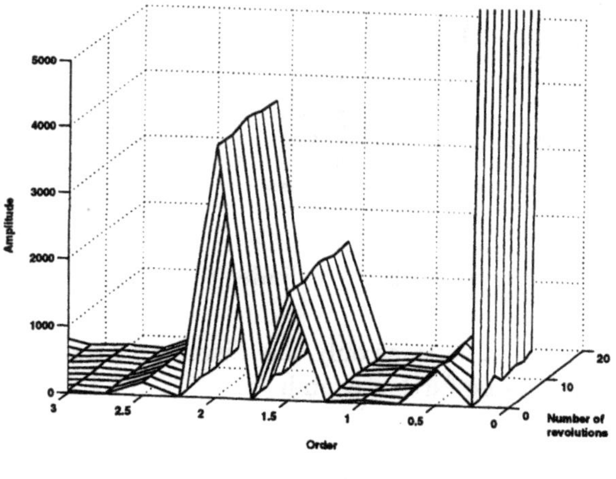

Fig. 8. Order spectrum of the measured speed using a rectangular window and a window size of four revolutions

As already seen in figure 6, the spectrum of the signal has three (lines of) peaks at orders 0.5, 1.5 and 2. The peak line at order 0.5 confirms the existence of a defective pump and consequently the need either for an exchange of the pump or for algorithms especially developed for cylinder balancing.

The disadvantage of the larger window is the loss of time information. Comparing figure 7 and 8 it can be clearly seen that the local maxima and minima at order 2 along the time axis (Number of Revolutions) got lost. One thing should at last be regarded: The achieved result depends also on the chosen window. Figure 9 shows the order spectrum using a Hanning window

$$w(k+1) = 0.5\cdot\left(1-\cos\left(2\pi\frac{k}{n-1}\right)\right)$$
$$k = 0,\ldots,n-1 \qquad (3)$$

Unfortunately most of the frequency accuracy got lost.

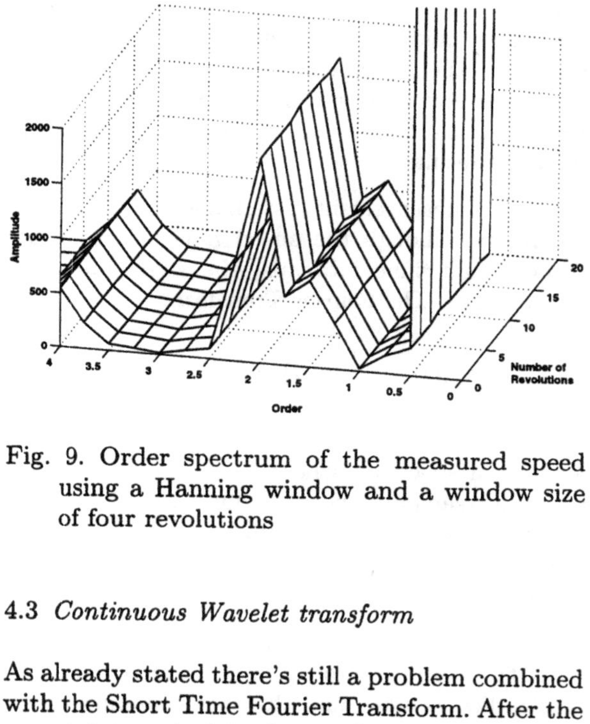

Fig. 9. Order spectrum of the measured speed using a Hanning window and a window size of four revolutions

4.3 *Continuous Wavelet transform*

As already stated there's still a problem combined with the Short Time Fourier Transform. After the uncertainty principle the resolution of frequency decreases with higher time resolution, which corresponds to the selection of a narrower window. That's why the selection of the constant window size is important. The selection should be adapted to the signal on the basis of a-priori-knowledge.

Another interesting possibility is the Continuous Wavelet Transform (CWT) (Misiti *et al.* 2000, Wickerhauser 1996)

$$WT_\alpha(\alpha,a) = \frac{1}{\sqrt{|a|}}\int_{-\infty}^{\infty} n(\varphi)\Psi(\frac{\alpha-\varphi}{a})d\varphi \quad (4)$$

which transforms the signal with variable-sized windows. The uncertainty principle restricts only the product of time and frequency intervals - therefore the wavelet transform provides precise informations about low frequencies by using a long time window and vice versa. Other transforms which use adapted window sizes are the Wigner distribution or the Choi-Williams Distribution (Meltzer 1999).

In contrast to the Short Time Fourier Transform the result of the Wavelet Transform doesn't depend on time and frequency, but on time and scale s. Instead of a shifted window Sinus-Function the wavelet transform uses scaled and shifted versions of a small wave - the wavelet Ψ. The scales can be converted to a so-called pseudo-frequency f_s

$$f_s = \frac{\Delta \cdot f_c}{s} \tag{5}$$

with the sampling period Δ and the center frequency f_c of the wavelet. Figure 10 shows the wavelet transform of the measured signal using a Morlet wavelet. The scales were transformed to (pseudo-)frequencies. The figure shows with perfect time resolution the known peaks indicating the rotational speed and the defective pump. Furthermore the transitions between the peaks can be clearly seen.

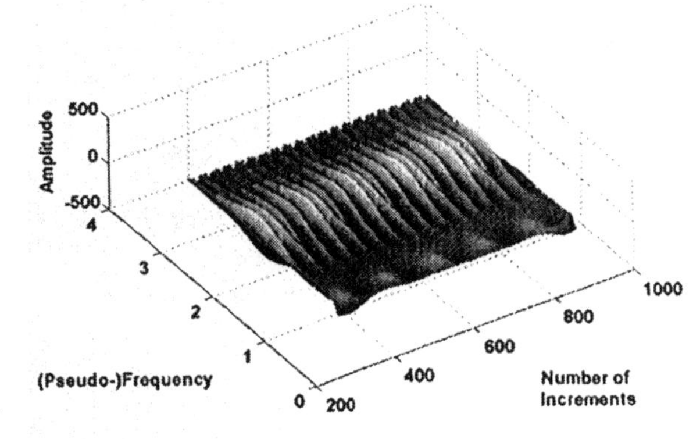

Fig. 10. Continuous Wavelet Transform of the measured signal using Morlet wavelets

4.4 *Discrete Wavelet transform*

The Discrete Wavelet Transform (DWT) corresponds to filtering and is therefore useful for signal analysis and the detection of failures. It can be developed of the CWT by using scales and positions based on powers of two - so-called dyadic scales and positions (Misiti *et al.* 2000). The original signal will be decomposed by a low and a high pass filter in *approximations* and *details*. The filters belong directly to the wavelets.

Figure 11 shows the 5 step decomposition of the measured signal using Daubechies Wavelets of order 5. In each decomposition step the approximation of the signal (s) is devided in the next approximation ($a_1 \dots a_5$) and the detail ($d_1 \dots d_5$). Included in the figure is the original signal and the detail of step 5.

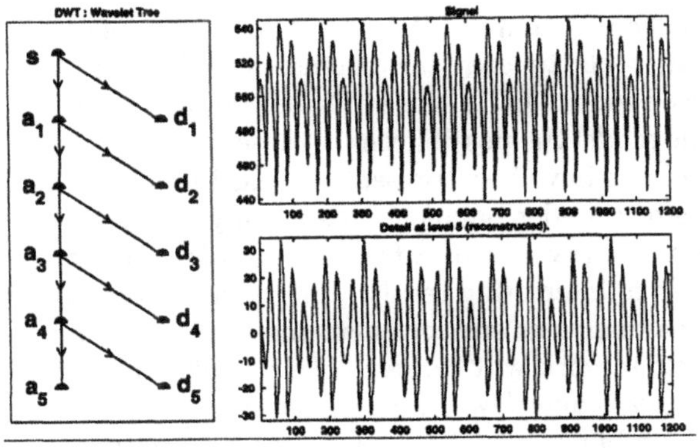

Fig. 11. Discrete Wavelet Transform of the measured signal using Daubechies wavelets

The DWT can therefore be used to denoise or compress a signal. A possibility to use this information for diagnosis would be to calculate the Crest-Factor (Stockmanns 2000) of the signal which is defined as

$$C = \frac{\max(|n_i|)}{\sqrt{\frac{1}{N}\sum_{i=1}^{k} n_i^2}} \tag{6}$$

The Crest-Factor is used to e.g. determine disturbances in gears.

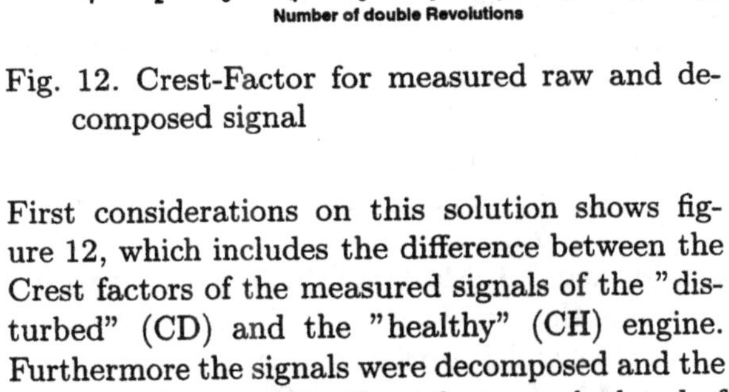

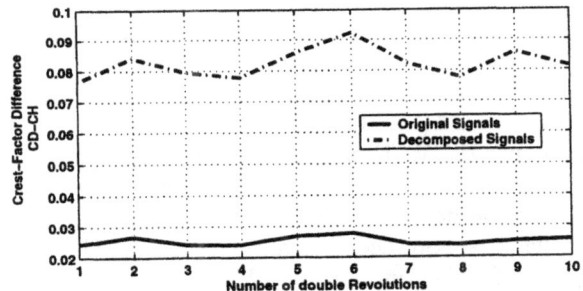

Fig. 12. Crest-Factor for measured raw and decomposed signal

First considerations on this solution shows figure 12, which includes the difference between the Crest factors of the measured signals of the "disturbed" (CD) and the "healthy" (CH) engine. Furthermore the signals were decomposed and the difference between the Crest factors calculated of the approximization of the third step. For calculation the factor was determined in blocks of two revolutions.

Principally the Crest Factor would be a good possibility to detect defective pumps. The decomposition increases the difference in the factor between the healthy and the defective engine. Further work will reinforce this idea.

5. CONCLUSION

The Fourier and the Wavelet Transforms are powerful tools to combine diagnostics with an idle speed controller. In this study beside the diesel engine model the possibilities of these signal analysis methods have been presented regarding the problem of cylinder balancing. Especially the Discrete Wavelet Transform, a fast and simple algorithm, leads to interesting results. First steps in developing a diagnosis and/or balancing tool were made. Further work will examine their applicability for diagnosis and develop an algorithm for balancing in combination with a periodic idle speed controller. One aspect will be a possible use in engine production control.

6. REFERENCES

Happe, J. (2000). Dynamische Modellierung und Simulation des Ladungswechsels eines NfZ-Dieselmotors zum Entwurf einer Leerlaufdrehzahlregelung. Master's thesis. RWTH Aachen.

Hild, O., A. Schloßer, K. Fieweger, St. Pischinger and H. Rake (1999). Die Regelstrecke eines PKW-Dieselmotors mit Direkteinspritzung im Hinblick auf Ladedruck- und Abgasrückführregelung. *Motortechnische Zeitschrift* **60**(3), 186–192.

Hrovat, D. and Jing Sun (1997). Models and Control Methodologies for IC Engine Idle Speed Control Design. *Control Engineering Practice* **5**(8), 1093–1100.

Isermann, R., M. Hafner, J. Schaffnit, M. Schüler and S. Sinsel (1998). Modellbildung und Simulation des statischen und dynamischen Verhaltens von Dieselmotoren mit Turbolader. In: *GMA-Kongress '98, Mess- und Automatisierungstechnik.* pp. 21–37. VDI Berichte 1397. VDI-Verlag. Düsseldorf.

Kamata, M., J. Nishihama and H. Sakai (1986). Improvement of idle speed stability of diesel engines by digital control. *JSAE Review* **7**(2), 16–22.

Kao, M. and J.J. Moskwa (1995). Turbocharged Diesel Engine Modeling for Nonlinear Engine Control and State Estimation. *Transaction of the AMSE - Journal of Dynamic Systems, Measurement and Control* **117**(1), 20–30.

Meltzer, G. (1999). Stand und Tendenzen der Schwingungsüberwachung und -diagnostik/ Innovative Diagnosetechnik. In: *Schwingungstagung '99, Schwingungsüberwachung und -diagnose von Maschinen und Anlagen.* VDI Berichte 1466. Düsseldorf. pp. 1–29.

Misiti, M., Y. Misiti, G. Oppenheim and J.-M. Poggi (2000). Wavelet Toolbox User's Guide: For Use with MATLAB. 2nd ed. The MathWorks, Inc.

Nahrath, Th., B. Bauer and A. Seeliger (1999). Vibration monitoring at unstable speeds. In: *COMADEM 99. 12th International Congress on Condition Monitoring and Diagnostic Engineering Management..* University of Sunderland, England. pp. 554–562.

Schmidt, C. (1995). Digitale kurbelwinkelsynchrone Modellbildung und Drehschwingungsdämpfung eines Dieselmotors mit Last. Fortschritt-Berichte VDI: Verkehrstechnik/ Fahrzeugtechnik. VDI-Verlag. Düsseldorf.

Stockmanns, G. (2000). Wavelet-Analyse zur Detektion von Zustandsänderungen. Fortschritt-Berichte VDI: Biotechnik/ Medizintechnik. VDI-Verlag. Düsseldorf.

Thornhill, M., S. Thompson and H. Sindano (2000). A comparison of idle speed control schemes. *Control Engineering Practice* **8**(5), 519–530.

Wickerhauser, M. V. (1996). Adaptive Wavelet-Analysis, Theorie und Software. Vieweg. Braunschweig/Wiesbaden.

Zweiri, Y.H., J.F. Whidborne and L.D. Seneviratne (1999). Dynamic simulation of a single-cylinder diesel engine including dynamometer modelling and friction. *Proceedings of the institution of mechanical engineers - Part D* **213**(4), 391–401.

Copyright © IFAC Periodic Control Systems,
Cernobbio-Como, Italy, 2001

PERIODIC CONTROL OF A PRESSURE SWING ADSORPTION PLANT

M. Bitzer [1] F. J. Christophersen M. Zeitz

*Institut für Systemdynamik und Regelungstechnik,
Universität Stuttgart, Pfaffenwaldring 9, D-70550 Stuttgart,
Germany*

Abstract: Pressure swing adsorption (PSA) plants are used for the separation of gas mixtures. Their periodic operation is achieved by a cyclic reversal of the flow conditions within the adsorbers. Based on the distributed parameter model of the considered two-bed PSA plant, the arising control problem is specified and a reduced model is used for the design of a periodic control strategy. The designed process control is evaluated with the simulated PSA plant model. *Copyright ©2001 IFAC.*

Keywords: periodic nonlinear system, distributed parameter system, variable sampling rate, adsorption process, nonlinear travelling concentration waves.

1. INTRODUCTION

Pressure swing adsorption (PSA) is a common process technique for the separation of gas mixtures (Ruthven *et al.*, 1994). The plants consist in general of several fixed-bed adsorbers and are operated as cyclic multi-step processes, i.e. the connections between the different adsorbers are changed by the switching of valves at the transition from one cycle step to the next. Thereby, a periodic operation is realized for the adsorption process.

As example, a two-bed pressure swing adsorption plant for the production of oxygen from air is considered. Its flowsheet is shown in Figure 1, see also (Bitzer and Zeitz, 2001). Each fixed-bed adsorber is described by a nonlinear model with distributed parameters (Unger, 1999). The implementation of the detailed PSA model within e.g. the simulation environment DIVA (Köhler *et al.*, 2001) enables its dynamical analysis.

A characteristic feature of PSA plants concerns the occurrence of nonlinear travelling concentration waves which are alternating their propagation direction as a consequence of the periodic process operation. In accordance with the cyclic coupling of the fixed-bed adsorbers, the occurring waves travel back and forth within the two adsorber beds and are thereby changing their shape, see Figure 1 (Box 1). The cycle time as well as the duration of the cycle steps do considerably affect the product concentration, because they determine the extent of breakthrough of a concentration front at the product end of the adsorber beds. The cycle step times are therefore considered the manipulating variables of the process.

If the adsorption plant is operated with fixed cycle step times, the plant approaches a cyclic steady state (CSS), i.e. the conditions at the end of each cycle are identical to those at its start. A numerical approach for the determination and the optimization of the CSS of adsorption processes was presented by Nilchan and Pantelides (1998).

During the operation, the PSA plant is subject to disturbances influencing its performance. In order to control the purity of the product, i.e. the averaged concentration in the oxygen tank, see Figure 1, a periodic operating point has to be stabilized by a process control scheme which is manipulating the duration of the cycle steps.

[1] Corresponding author: bitzer@isr.uni-stuttgart.de

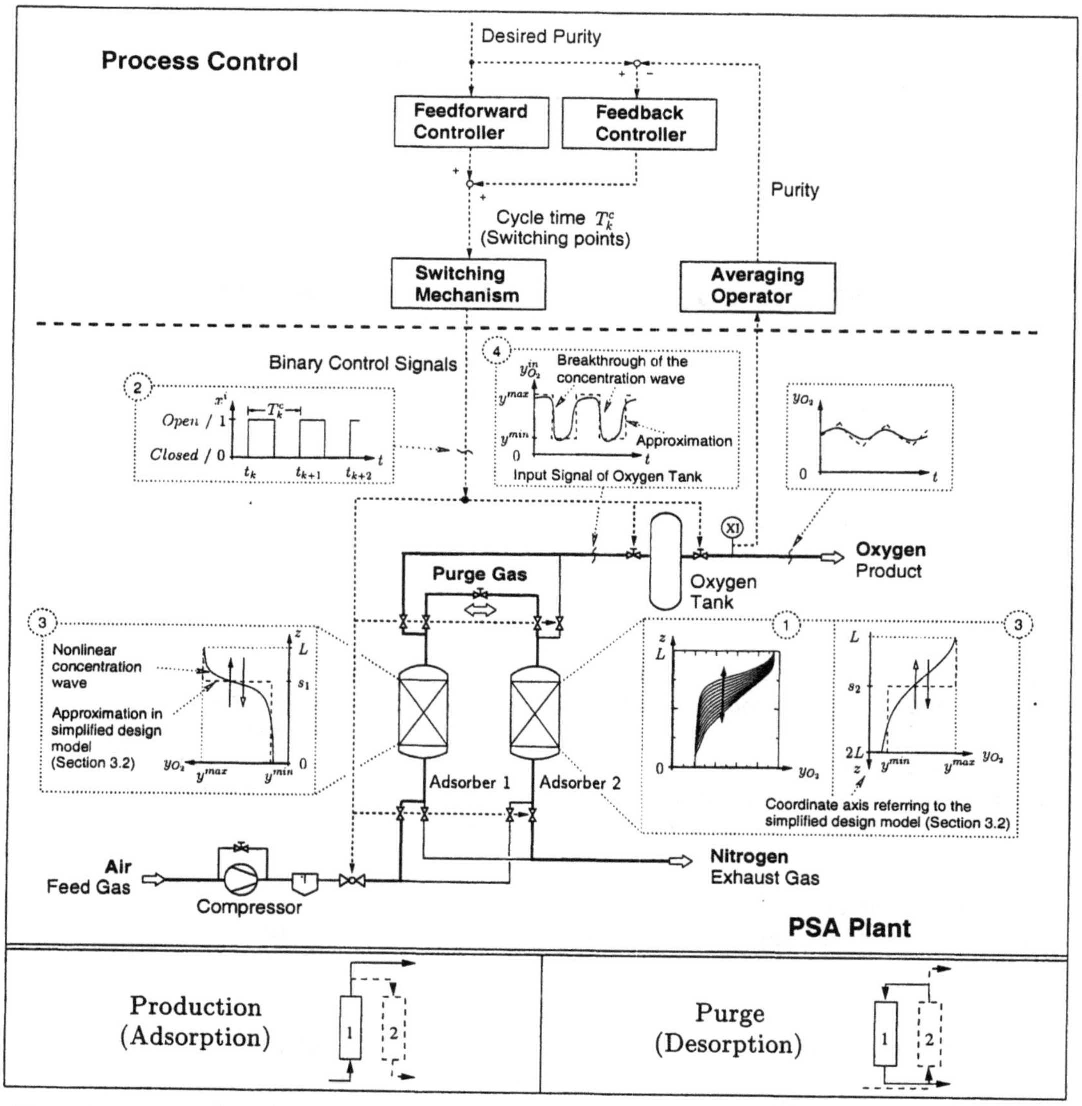

Fig. 1. Flowsheet of a two-bed pressure swing adsorption plant for oxygen production from air with the proposed process control strategy and the realized coupling schemes of the adsorbers during the cycle steps of production and purge.

Another control task is the realization of set-point changes or a start-up procedure. Up to now, there exists no model based process control strategy which is applicable to this type of plant and known to the authors. As a first contribution to such a process control concept, a nonlinear observer with distributed parameters was designed (Bitzer and Zeitz, 2001).

In this paper, a process control strategy is presented which comprises a feedforward and a superimposed feedback controller, see Figure 1. The paper is organized as follows: in the next section, the model of the two-bed pressure swing adsorption plant for the production of oxygen from air is introduced and the occurring control problem is specified. Then, the derivation of a reduced design model is explained and the process control strategy is presented. Finally, the effectiveness of the control concept is demonstrated by simulations with the detailed PSA model.

2. TWO-BED PRESSURE SWING ADSORPTION PLANT

The considered PSA plant shown in Figure 1 is used for the production of oxygen from air for medical purposes, see (Bitzer and Zeitz, 2001). The produced oxygen is stored in a tank from which it is taken by the consumer.

2.1 *PSA plant model*

The considered adsorption model[2], depicted in Table 1, treats air as a binary mixture of oxy-

[2] This model was derived by further simplifying the more detailed model of Unger (1999) based on physical considerations, e.g. by assuming an isothermal process and by neglecting the adsorption of oxygen. The aim was to get a model which is as simple as possible but still reveals all characteristic effects of PSA plants, i.e. the nonlinear concentration waves as well as their related breakthrough behavior.

Adsorber Beds $i \in \{1,2\}$:

Mass Balance
$$\frac{\epsilon}{RT}\frac{\partial p^i}{\partial t} + \rho_{Sch}\frac{\partial q^i_{N_2}}{\partial t} = -\frac{1}{A_{Sch}}\frac{\partial \dot{n}^i}{\partial z} \qquad z \in (0,L), \quad t > 0$$

Component Mass Balance (Gaseous Phase)
$$\frac{\epsilon}{RT}\left(y^i_{O_2}\frac{\partial p^i}{\partial t} + p^i\frac{\partial y^i_{O_2}}{\partial t}\right) = -\frac{1}{A_{Sch}}\frac{\partial \dot{n}^i_{O_2}}{\partial z} \qquad z \in (0,L), \quad t > 0$$

Component Mass Balance (Adsorbed Phase)
$$\frac{\partial q^i_{N_2}}{\partial t} = k^{LDF}_{N_2}\left(q^*_{N_2}(p^i, y^i_{O_2}) - q^i_{N_2}\right) \qquad z \in (0,L), \quad t > 0$$
$$\text{with} \quad q^*_{N_2} = q_{mon,N_2}\frac{K\,p^i\,y^i_{N_2}}{1 + K\,p^i\,y^i_{N_2}} \quad \text{and} \quad y^i_{N_2} = 1 - y^i_{O_2}$$

Ergun-Equation
$$0 = \frac{\partial p^i}{\partial z} + 150\frac{\eta}{c}\frac{(1-\epsilon_{Sch})^2}{\epsilon^3_{Sch} d_p^2 A_{Sch}}\dot{n}^i + 1.75\frac{M}{c}\frac{1-\epsilon_{Sch}}{\epsilon^3_{Sch} d_p A^2_{Sch}}(\dot{n}^i)^2 \qquad z \in (0,L), \quad t \geq 0$$

Initial Conditions
$$p^i(z,0) = p^i_0(z), \quad y^i_{O_2}(z,0) = y^i_{O_20}(z), \quad q^i_{N_2}(z,0) = q^i_{N_20}(z) \qquad z \in [0,L]$$

Boundary Conditions:

$$\dot{n}^i(0,t) = \dot{n}^i_{Feed}(\nabla p)\cdot x^i_0 + \dot{n}^i_{Exhaust}(\nabla p)\cdot(1 - x^i_0) \qquad \dot{n}^i(L,t) = \dot{n}^i_{Tank}(\nabla p)\cdot x^i_L - (-1)^i\dot{n}^i_{Purge}(\nabla p)$$

$$y^i_{O_2}(0,t) = \begin{cases} y^{in}_{O_2}(t) & \text{if } \dot{n}^i(0,t) > 0, \\ y^i_{O_2}(0^+,t) & \text{otherwise.} \end{cases} \qquad y^i_{O_2}(L,t) = \begin{cases} y^{i+\text{mod}(i,2)}_{O_2}(L,t) & \text{if } \dot{n}^i(L,t) < 0, \\ y^i_{O_2}(L^-,t) & \text{otherwise.} \end{cases}$$

Oxygen Tank:

Mass Balance
$$\frac{V}{RT}\frac{dp}{dt} = \sum_{i=1}^{2}\dot{n}^i_{Tank} - \dot{n}_{Product}, \qquad \dot{n}_{Product} = const. \qquad t > 0, \quad p(0) = p_0$$

Component Mass Balance
$$\frac{V}{RT}\left(y_{O_2}\frac{dp}{dt} + p\frac{dy_{O_2}}{dt}\right) = \sum_{i=1}^{2} y^i_{O_2}(L,t)\dot{n}^i_{Tank} - y_{O_2}\dot{n}_{Product} \qquad t > 0, \quad y_{O_2}(0) = y_{O_20}$$

Table 1. Model of the PSA plant with parameters from (Bitzer and Zeitz, 2001).

gen and nitrogen, and emanates from two phases, i.e. a gaseous phase and an adsorbed phase. The distributed parameter model with one space coordinate describing the adsorption process consists of a set of four quasilinear partial-differential-algebraic equations (PDAEs) for the pressure $p(z,t)$, oxygen mole fraction $y_{O_2}(z,t)$ in the gaseous phase, adsorbed amount $q_{N_2}(z,t)$ of nitrogen, and molar flux $\dot{n}(z,t)$ in dependence of space z and time t.

The boundary conditions depend on the connections between the adsorbers and on the gas flow through the tubing which is itself depending on the time-dependent pressure gradient ∇p. Therefore, the boundary conditions are described by algebraic nonlinear characteristic lines $\dot{n}^i = \dot{n}^i(\nabla p)$ and the binary control signals $x^i_{0/L}$ of the valves, see Figure 1 (Box 2). The model of the PSA plant is implemented in the simulation environment DIVA (Köhler *et al.*, 2001). Thereby, the model equations are spatially discretized according to the Method of Lines and their simulation is used for the testing of the developed control strategy.

2.2 *Control problem*

Operational cycles of PSA plants are generally consisting of several different steps (Ruthven *et al.*, 1994). For the sake of simplicity, a simplified 2-step cycle consisting of the production and the purge step is considered in order to develop a process control strategy. The two adsorbers are run in a phase-shifted manner in order to attain a quasi-continuous production, as shown in Figure 2. Thereby, the cycle time T^c_k is the manipulating variable of the process. The control variable is the purity P_k of the product, i.e. the oxygen mole fraction y_{O_2} in the product tank, averaged over one cycle

$$P_{k+1} := P(t_k + T^c_k) = \frac{1}{T^c_k}\int_{t_k}^{t_k+T^c_k} y_{O_2}(t)\,dt. \quad (1)$$

3. CONTROL DESIGN MODEL

For the design of a periodic control scheme, a simplified model is required representing appropriately the Input/Output behavior of the PSA plant. The modeling of the I/O behavior is based on the analysis of the CSS and of the transient

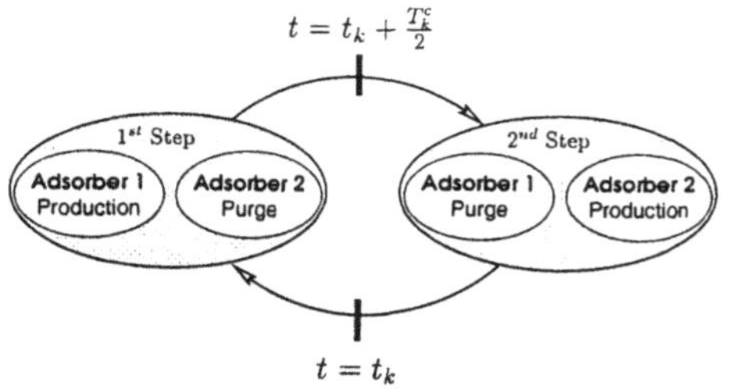

Fig. 2. Cyclic operation scheme of the PSA plant with the beginning of the kth cycle given by $t = t_k$ and its duration T_k^c.

behavior of the simulation model of the plant (see Table 1).

3.1 *Plant dynamics of the cyclic steady state (CSS)*

CSS is reached when the plant is operated in open-loop and the cycle period T_k^c is kept constant. In Figure 3, the oxygen mole fraction profiles $y_{O_2}(z, t_j)$ are given for the first cycle step of the CSS. The profiles of the second cycle step are mirror images of the first one. Due to the ad- and desorption of nitrogen, the occurring non-linear travelling waves are permanently changing their shape. A compression wave can be observed during the production step, and an expansion one during the purge step. For cycle-times $T_k^c > 40s$,

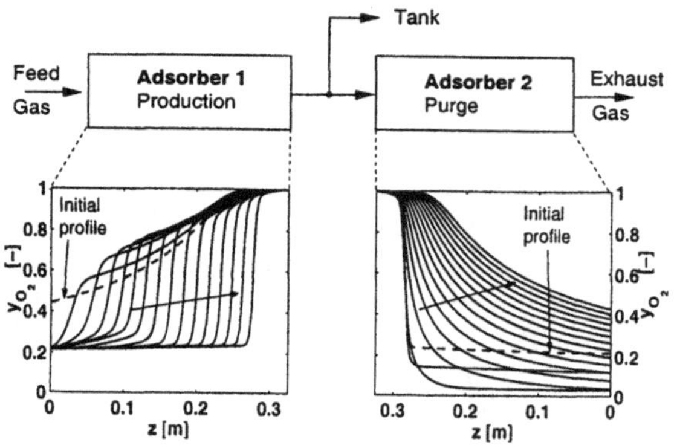

Fig. 3. Temporal evolution of the spatial oxygen mole fraction profiles $y_{O_2}(z, t_j)$ for the first cycle step of the CSS with $T_k^c = 30s$.

the entire concentration front breaks through. It is therefore evident that the purity P_k is mainly influenced by the breakthrough of the concentration front, because the amount of low concentrations entering the product tank is increasing, which leads to a decrease of the purity P_k. The graph shown in Figure 4 reflects the stationary I/O-behavior $P_k = f(T_k^c)$ of the plant operated in CSS.

3.2 *Derivation of the control design model*

The purity P_k of the product gas is mainly determined by the breakthrough concentration of each

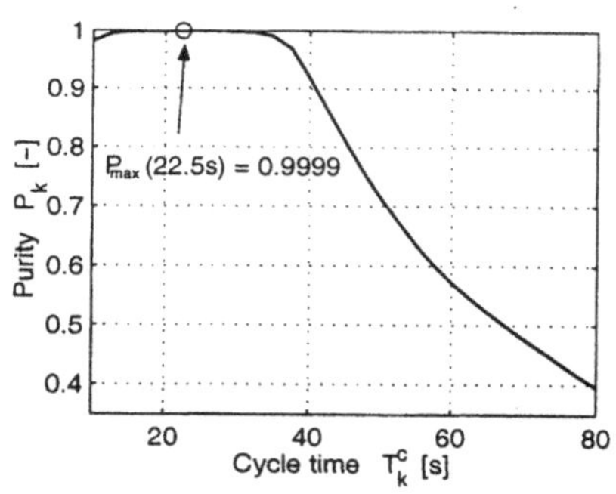

Fig. 4. Stationary I/O model $P_k = f(T_k^c)$ of the plant operated in CSS.

adsorber. Therefore, the control design model needs to reflect the occurring travelling waves and the influence of their breakthrough behavior. Additionally, the control design model has to describe the stationary I/O behavior of the plant. It would be desirable to derive such a low order model by extending the constant pattern wave concept of Kienle (2000). Such a lumped parameter model would contain position and shape of the dominating waves as new states.

But due to the complexity of this approach, the concentration profile is approximated by a simple rectangularly shaped profile as clarified in Figure 1 (Box 3). The control design model contains the front position s_1 and the oxygen mole fraction y_{O_2} in the tank as internal states, its signal flow diagram is given in Figure 5. The aim is to approximate the periodic input signal into the oxygen tank, see Figure 1 (Box 4). Since the low order design model is simplified to such an extent, the remaining parameters, e.g. the propagation velocity w, the distance b between the wave fronts s_1 and s_2, and the lower wave amplitude y^{min} are adjusted in such a way that the reduced design model approximates the breakthrough behavior of the concentration waves and that it reflects the stationary I/O behavior of the plant.

Finally, the tank is modelled by a linear first-order system. Similarly to the PSA plant, the input of the reduced design model is the cycle time T_k^c and its output the oxygen mole fraction y_{O_2} of the tank. The cyclic operation is realized by a periodic reversal of the propagation direction of the shock fronts according to the switching mechanism which is modelled using a hysteresis loop, see Figure 5.

4. PROCESS CONTROL STRATEGY

The process control has to accomplish both, the control of periodic operating points and the trajectory control for set-point changes. The design proceeds as follows: first the control of operating points is considered, then the strategy is extended for trajectory control tasks.

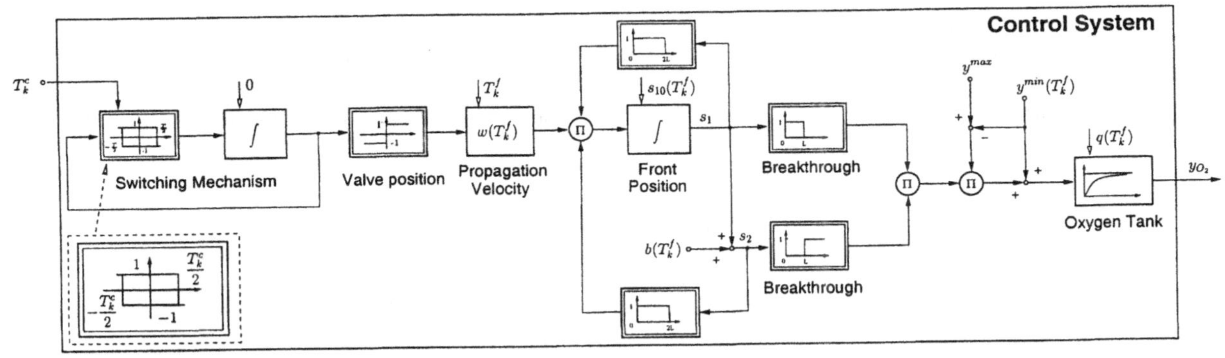

Fig. 5. Signal flow diagram of the design model for the concentration fronts with the cycle time T_k^c as input and the oxygen mole fraction y_{O_2} as output ($s_{1,2}$ - position of wave fronts, b - distance between wave fronts, y^{max}, y^{min} - upper and lower wave amplitude).

4.1 *Control of periodic operating points*

The process control strategy comprises a feedforward and a superimposed feedback controller which is stabilizing the desired purity P_k^d, see Figure 1. The feedforward controller for the feedforward part T_k^f of the cycle time is designed by inverting the sloping portion of the analyzed stationary I/O model $P_k = f(T_k^c)$ of the plant, see Figure 4. This leads to $T_k^f = f^{-1}(P_k^d)$. For the superimposed feedback controller a simple PID controller is used. For set-point changes, the feedforward controller is extended by an inverse I/O model $g^{-1}(\cdot)$ taking the transient behavior of the plant into account. In Figure 6, the blockdiagram of such a process control strategy is shown.

The PID controller is given as $\Delta T_k^{PID} = -(\Delta T_k^P + \Delta T_k^I + \Delta T_k^D)$. This means that the overall cycle time $T_k^c = T_k^f + \Delta T_k^{PID}$ is adapted after each cycle with respect to the actual control error $e_k = P_k^d - P_k$. The proportional, integral, and differential parts of the PID controller are given as $\Delta T_k^P = K^P e_k$, $\Delta T_k^I = \Delta T_{k-1}^I + K^I T_{k-1}^c e_k$ with $\Delta T_0^I = 0$, and $\Delta T_k^D = K^D \frac{e_k - e_{k-1}}{T_{k-1}^c}$ with the parameters K^P, K^I, $K^D \geq 0$.

4.2 *Trajectory control strategy*

Set-point changes can be accomplished by a trajectory control which considers the transient behavior of the plant. This task requires an inverse I/O model $g^{-1}(\cdot)$ of the process in order to design an appropriate feedforward controller. Such a model can be obtained using the reduced design model in Figure 5. Due to the time-discrete nature of the process, a discrete I/O model similar to a Poincaré-map, i.e. $P_{k+1} = g(P_k, P_{k-1}, \ldots, u_k)$, which is describing the transient open-loop behavior of the purity P_k is sufficient. Here, u_k represents the input variable. Simulations showed that even linear models $P_{k+1} = a_k P_k + a_{k-1} P_{k-1} + b_k u_k$ for the transient behavior of the purity P_k can be used.

The planning of trajectories implies the fixing of a time-horizon $t_{k_0}, t_{k_0+1}, \ldots, t_{k_0+N}$ during which the set-point change shall be accomplished. In this case, a certain number N of cycles has to be specified. Then, a desired trajectory P_k^d can be set up by using e.g. a sufficiently smooth polynomial. A feedforward control sequence P_k^f for the set-point change is then calculated using the inverse I/O model $u_k = g^{-1}(P_{k+1}, P_k, P_{k-1}, \ldots)$ and the desired trajectory P_k^d which leads to $P_k^f = g^{-1}(P_{k+1}^d, P_k^d, P_{k-1}^d, \ldots)$. The related cycle time T_k^f is again calculated using the stationary I/O model, i.e. $T_k^f = f^{-1}(P_k^f)$. For the control of operating points, it is evident that $P_k^f \equiv P_k^d$ has to be guaranteed.

5. SIMULATION RESULTS

The PID controller designed using the simplified design model was applied to the detailed model of the PSA plant (Table 1). Therefore, the parameters of the PID controller have been only slightly readjusted, which is astonishing considering the extremely simplified design model.

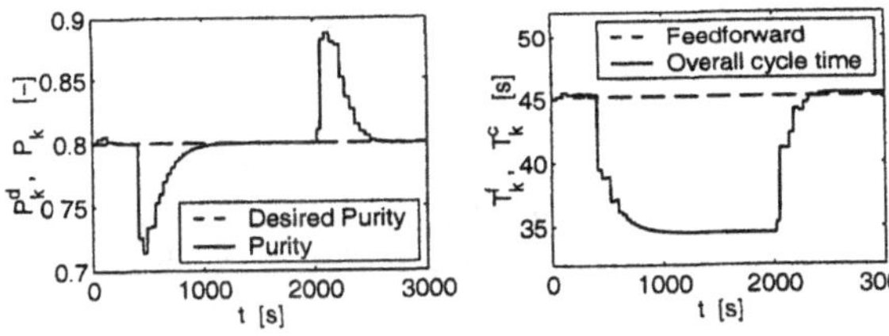

Fig. 7. Simulation of the controlled purity P_k subject to step disturbances.

Figure 7 shows the simulation of the controlled purity P_k subject to step disturbances. At $t = 200s$, the amount of product $\dot{n}_{Product}$ drawn off the tank is altered by +60%, which is a considerable disturbance. At $t = 2000s$ it is set back to its nominal value.

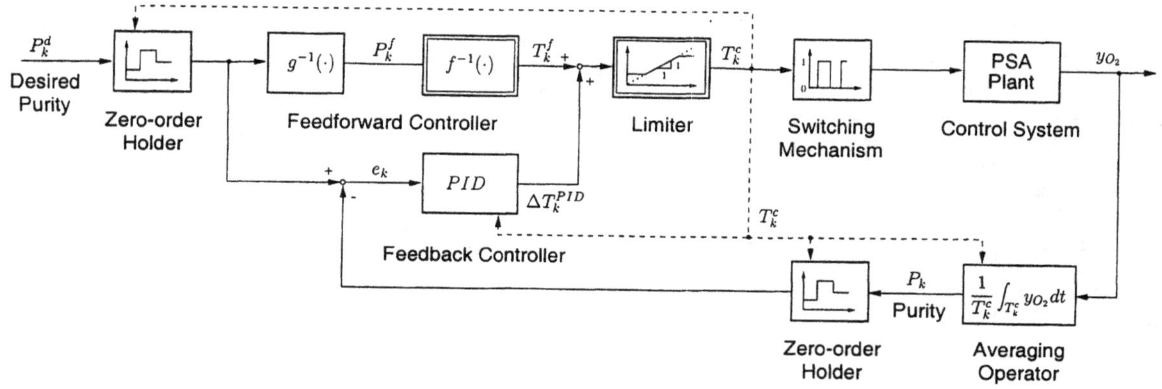

Fig. 6. Block diagram of the feedforward and the superimposed feedback controller for the purity P_k.

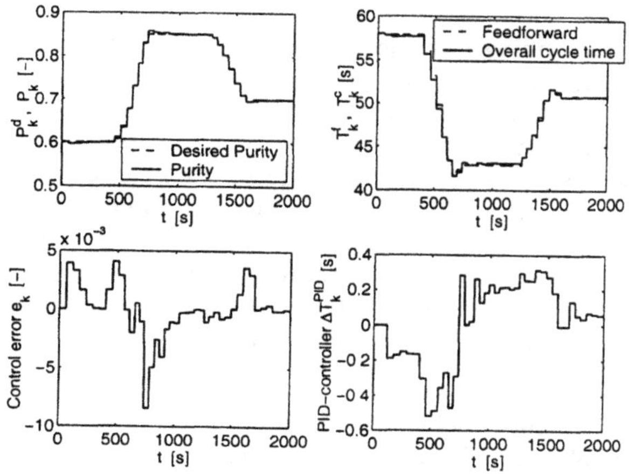

Fig. 8. Simulation of the trajectory control of the purity P_k with an increase and decrease of the desired purity P_k^d at $t = 400s$ and $t = 1250s$ respectively.

The simulation of the trajectory control of two consecutive set-point changes of the purity P_k is given in Figure 8. The desired trajectories P_k^d, $k = k_0, k_0+1, \ldots, k_0+8$ were set up using a 4th-order polynomial. It is stressed that the feedforward controller implies an inverse I/O model of an identified linear model of the transient behavior of the purity P_k of the PSA plant.

6. CONCLUSIONS AND OUTLOOK

A process control strategy consisting of a feedforward and a superimposed feedback controller has been proposed for a PSA plant. Its design is based on the analysis of the stationary I/O model as well as on the transient behavior of the plant. The performance of the control concept shows good results and encourages further research. A challenging task is the development of a more sophisticated reduced design model which provides a better representation of position and shape of the nonlinear waves and which includes the pressure dynamics as well as the space and time depending flow velocity of the gas within the adsorbers. The development of such a reduced design model is certainly crucial with respect to the examination of stability issues.

ACKNOWLEDGMENT

The authors gratefully acknowledge the cooperation with Prof. G. Eigenberger, W. Lengerer, and M. Stegmaier of the *Institut für Chemische Verfahrenstechnik* at the University of Stuttgart. The research work is supported within *Sonderforschungsbereich* 412 by *Deutsche Forschungsgesellschaft (DFG).*

7. REFERENCES

Bitzer, M. and M. Zeitz (2001). Design of a nonlinear distributed parameter observer for a pressure swing adsorption plant. *Journal of Process Control.*

Kienle, A. (2000). Low-order dynamic models for ideal multicomponent distillation processes using nonlinear wave propagation theory. *Chemical Engineering Science* **55**, 1817–1828.

Köhler, R., K. D. Mohl, H. Schramm, M. Zeitz, A. Kienle, M. Mangold, E. Stein and E. D. Gilles (2001). Method of lines within the simulation environment DIVA for chemical processes. In: *Adaptive Method of Lines* (A. Vande Wouwer, P. Saucez and W. Schiesser, Eds.). pp. 367–402. CRC Press. Boca Raton/USA.

Nilchan, S. and C. C. Pantelides (1998). On the optimisation of periodic adsorption processes. *Adsorption* **4**, 113–147.

Ruthven, D. M., S. Farooq and K. S. Knaebel (1994). *Pressure Swing Adsorption.* VCH Publishers, New York, Weinheim, Cambridge.

Unger, J. (1999). *Druckwechseladsorption zur Gastrennung - Modellierung, Simulation und Prozeßdynamik.* Fortschritt–Berichte Nr. 3/602. VDI-Verlag, Düsseldorf.

Copyright © IFAC Periodic Control Systems,
Cernobbio-Como, Italy, 2001

PERIODIC MODELLING OF POWER SYSTEMS

Henrik Sandberg * **Erik Möllerstedt** **

* *Department of Automatic Control, Lund Institute of Technology, Box 118, SE-221 00 Lund, Sweden*
henriks@control.lth.se
** *Decuma AB, IDEON Science Park, Ole Römers väg 12, SE-223 70 Lund, Sweden*
erik.mollerstedt@decuma.se

Abstract: This paper treats modelling of power systems with converters in a linear time-periodic framework. A power converter is a nonlinear switching device connecting an AC system to a DC system. The converter generates harmonics that might cause instabilities in systems of this kind. About a nominal periodic trajectory the power converter is well described by a periodic gain matrix, whereas the power grids often can be described by linear time-invariant models. Put together they form a linear time-periodic model. It is also shown in this paper how Integral Quadratic Constraints may be used for robustness analysis. To conclude an inverter locomotive is modeled with the described techniques. *Copyright ©2001 IFAC*

Keywords: linear, periodic, modelling, power systems, converters, inverters, harmonics, robustness

1. INTRODUCTION

The periodicity of currents and voltages ought to make AC power systems an ideal application for linear time-periodic system theory. These systems are driven by a voltage of defined frequency and amplitude. Since only relatively small deviations from this nominal voltage are allowed, the dynamics of these systems are well captured by models which are linearized about the nominal operating trajectory. This leads to linear time-periodic (LTP) models.

However, surprisingly little periodic modelling of power systems is found in the literature. The reason for this is the following: a traditional power system consists of a number of generators connected to the grid. The generators are rotating synchronously at fundamental frequency (generally 50 or 60 Hz). The heavy generators efficiently damp all other frequencies. This means that even though harmonics exist due to nonlinearities, they are not believed to have a significant effect on the dynamics of the system. For stability analysis it is then enough to work with linear time-invariant (LTI) models which capture the dynamics of the fundamental frequency component. Harmonics are considered as a static filtering problem.

The introduction of power electronics in power systems has dramatically changed the power systems during the last decades. Power electronic devices increase the flexibility and make more optimal utilization of the grid and improved load performance possible. New concepts and solutions have emerged, like high voltage DC (HVDC) transmission and distributed power generation (DPG). The deregulation of the electricity market has further helped to make these new concepts economically viable. To allow a more optimized operation of the system, accurate methods for analysis and control design are essential. However,

[1] Thanks to Markus Meyer at Adtranz Switzerland for kindly supporting us with models.

the switching nature of power electronics leads to systems that are much harder to analyze. Traditional analysis using LTI models is not sufficient to fully utilize the capacity of the power electronics.

Actively controlled power electronic devices like power converters, are very powerful actuators. Power flows can be changed in a fraction of a cycle. Since the grid itself is not low pass, the total system cannot be assumed to have slow dynamics. Furthermore, because of the switching dynamics, there is coupling between frequencies. Consequently, to fully utilize the possibilities brought by the power electronics, and to avoid overly conservative solutions, harmonics and frequency coupling must be considered. For reasons of simplicity and tradition, however, LTI models that only capture the dynamics of the fundamental frequency component are still often used for the analysis.

Power systems are very large and complex. It is therefore unrealistic to model complete systems as LTP. However, LTP models are very useful to understand the dynamics of power systems with switching power electronic components. In this paper it will be shown how periodic models of power systems can be used for improved analysis and control design. A power converter is used to illustrate the ideas. A related method is the so called dynamic phasors, (Stankovic *et al.*, 1999). Related approaches to steady state analysis of harmonic distortion in power systems are found in the literature under names like harmonic balancing, harmonic power flow studies etc. (Arrillaga *et al.*, 1994; Xu *et al.*, 1991). However, these methods do not capture the dynamics of the system and cannot be used for stability analysis.

2. CONVERTER MODELLING

A power converter is a nonlinear coupling between two electric systems. They are often built using GTO (Gate Turn Off)-thyristors with switching frequency up to 500Hz. IGBTs (Insulated Gate Bipolar Transistors) can also be used with switching frequency up to 10kHz. Most common is that the converter is used to connect an AC system to a DC system. The AC side and DC side dynamics can generally be captured with linear time-invariant models that are straightforward to derive. The problem is to obtain a good description of the coupling between the two sides, one that facilitates analysis and design of the complete system.

2.1 *An Ideal Converter*

An ideal single phase converter is shown in Fig. 1. The ideal converter has no losses and no energy

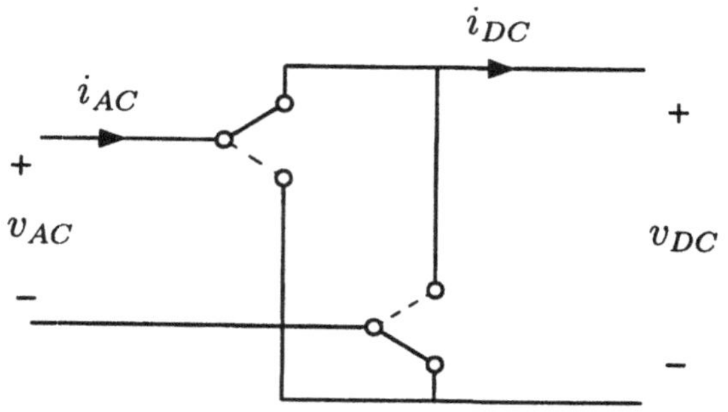

Fig. 1. An ideal converter. The switches are used to control the power flow through the converter, and the reactive power generated on the AC side.

storage. The basic goal of the converter control is to shape the AC voltage so that the desired power is fed through the converter. This is done by proper switching. From Fig. 1 it can be concluded that

$$v_{AC}(t) = s(t)v_{DC}(t) \qquad (1)$$

where the switch function $s(t)$ can be assigned the values 1, -1, and 0, since ideal switching is assumed. The desired AC voltage is smooth (sinusoidal), and must be approximated by using pulse width modulation, for instance.

The switch function also gives a relation between AC current and DC current:

$$i_{DC}(t) = s(t)i_{AC}(t). \qquad (2)$$

Since an ideal converter has no losses and no energy storage, the instantaneous power on the DC side and the AC side must be equal, that is,

$$P_{DC} = v_{DC}i_{DC} = v_{AC}i_{AC} = P_{AC}. \qquad (3)$$

The current relation (2) can also be derived from this power balance.

In (1) and (2) v_{DC} and i_{AC} are the chosen input quantities. It would be possible to instead choose v_{AC} and i_{DC} as inputs. Which pair to choose depends on the topology of the total system. As will be seen in section 3 the inputs to the converter are the outputs of the rest of the power system and thus the controlled variables. When modelling power systems using block diagrams the problem of choosing correct inputs and outputs always arises. It is vital that subsystems are modeled such that they are possible to interconnect. A behavioral modelling approach, as suggested by Willems (Willems, 1971), and used by equation based modelling languages like Modelica avoids this problem, (Elmqvist *et al.*, 1999).

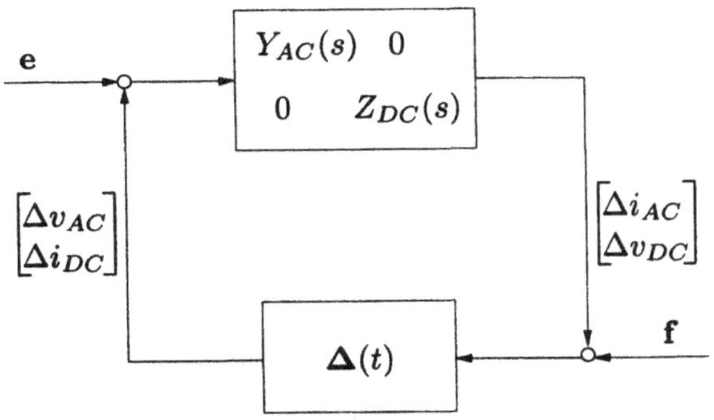

Fig. 2. A feedback model of a power converter, $\mathbf{\Delta}(t)$, that connects an AC power grid to a DC power grid. $\mathbf{\Delta}(t)$ is here a time-periodic gain matrix given by (5) or (7).

2.2 *Linearizing the Converter*

The local behavior of the converter in the neighborhood of a nominal periodic solution $\{v_{DC}^n(t), i_{AC}^n(t), s^n(t)\}$ is well described by a linear approximation of (1) and (2):

$$\begin{aligned}\Delta v_{AC}(t) &= s^n(t)\Delta v_{DC}(t) + \Delta s(t) v_{DC}^n(t)\\ \Delta i_{DC}(t) &= s^n(t)\Delta i_{AC}(t) + \Delta s(t) i_{AC}^n(t)\end{aligned} \quad (4)$$

This can be written

$$\begin{bmatrix}\Delta v_{AC}(t)\\ \Delta i_{DC}(t)\end{bmatrix} = \begin{bmatrix}0 & s^n(t) & v_{DC}^n(t)\\ s^n(t) & 0 & i_{AC}^n(t)\end{bmatrix}\begin{bmatrix}\Delta i_{AC}(t)\\ \Delta v_{DC}(t)\\ \Delta s(t)\end{bmatrix}$$

Since the nominal solution, around which the system is linearized, is periodic, the converter is represented by a periodic gain matrix. Ideally v_{DC}^n is constant and v_{AC}^n sinusoidal of fundamental frequency. From (1) it is seen $s^n(t)$ should also be sinusoidal. However, since $s(t)$ only takes discrete values $(\pm 1, 0)$ this can only be an approximation. For three-phase systems the same approach can be taken. The AC voltage and current, as well as the switch signal are replace by vectors of the three phase quantities, $v_{AC} = [v_a\, v_b\, v_c]$, etc. Three-phase AC systems are conveniently represented in rotating coordinates, the so called *dq*0-reference frame (Kundur, 1994). Due to the high degree of symmetry in balanced three-phase systems the effects of harmonics are often reduced.

2.3 *Stability of Interconnections*

Often the lines of the converter are connected to systems that are well described by linear time-invariant models (pure AC or DC systems), at least close to the nominal solution. Hence, Δi_{AC} and Δv_{DC} are obtained as

$$\begin{aligned}\Delta i_{AC} &= Y_{AC}(s)\Delta v_{AC},\\ \Delta v_{DC} &= Z_{DC}(s)\Delta i_{DC},\end{aligned}$$

where $Y_{AC}(s)$ is the admittance of the AC system, and $Z_{DC}(s)$ is the impedance of the DC system. The resulting closed-loop is shown in Fig. 2 with

$$\mathbf{\Delta}(t) = \mathbf{\Delta}(t+T) = \begin{bmatrix}0 & s^n(t)\\ s^n(t) & 0\end{bmatrix} \quad (5)$$

where T is the period of the nominal trajectory, and $\mathbf{e}$ and $\mathbf{f}$ are external noise signals. If the models of the AC/DC systems are finite dimensional, it is noted a periodic state-space model can be obtained. There are common examples of infinite dimensional systems, for example long transmission lines. However these are often approximated with finite dimensional systems.

The feedback interconnection in Fig. 2 is a well-posed problem if $Z_{DC}(\infty)Y_{AC}(\infty) < 1$ and $s^n(t) \leq 1$ for all t. It is of interest to determine if the nominal solution is asymptotically stable, i.e. if all unforced solutions will converge asymptotically to the nominal solution. Once the system is written in state-space form this can be determined by Floquet-analysis, for instance. If asymptotic stability is to be determined for different $s^n(t)$ various robustness criteria can be used, for example IQCs as described in the next section.

As the system trajectories are supposed to stay close to a nominal solution, which is different for different operating conditions, a stabilizing controller is generally needed. In the simplest case let us assume the switching signal is obtained from a differentiable static map $K(\cdot,\cdot)$:

$$s(t) = K(i_{AC}(t), v_{DC}(t)). \quad (6)$$

Then $\Delta s(t)$ in (4) is obtained by

$$\Delta s(t) = K_{i_{AC}}^n(t)\Delta i_{AC}(t) + K_{v_{DC}}^n(t)\Delta v_{DC}(t)$$

with $K_{i_{AC}}^n(t) = \partial K/\partial i_{AC}(i_{AC}^n(t), v_{DC}^n(t))$ and $K_{v_{DC}}^n(t) = \partial K/\partial v_{DC}(i_{AC}^n(t), v_{DC}^n(t))$. The feedback interconnection in Fig. 2 is still valid but the time-periodic block is updated by

$$\mathbf{\Delta}(t) = \begin{bmatrix} v_{DC}^n K_{i_{AC}}^n & s^n + v_{DC}^n K_{v_{DC}}^n\\ s^n + i_{AC}^n K_{i_{AC}}^n & i_{AC}^n K_{v_{DC}}^n\end{bmatrix}. (7)$$

Asymptotic stability and robustness of this feedback interconnection is, of course, a necessity for the power system. A static controller is a large restriction so it is vital to be able to use dynamic controllers with integral action, for instance. In the modelling example that is treated in section 3 it is shown how to include dynamic controllers and still keep much of the structure presented here.

2.4 *Stability and Robustness Analysis with IQCs*

To determine stability and check for robustness of the system in Fig. 2 the method of Integral

Quadratic Constraints (IQCs) is a valuable tool. For this analysis let us define

$$\mathbf{G}(s) = \begin{bmatrix} Y_{AC}(s) & 0 \\ 0 & Z_{DC}(s) \end{bmatrix} \quad (8)$$

and assume it is asymptotically stable. Let also $\mathbf{w} = \boldsymbol{\Delta}(t)\mathbf{v}$ with $\boldsymbol{\Delta}(t)$ defined as in (7). We would now like to prove stability of the interconnection for a set of different nominal periodic trajectories $\{v_{DC}^n(t), i_{AC}^n(t), s^n(t)\}$, that correspond to different actual operating conditions of the power system. One way to do this is to give constraints on the Fourier transforms of $\mathbf{w}$ and $\mathbf{v}$, denoted by $\hat{\mathbf{w}}$ and $\hat{\mathbf{v}}$, that emerges from these different $\boldsymbol{\Delta}(t)$. These constraints are formulated as

$$\int_{-\infty}^{\infty} \begin{bmatrix} \hat{\mathbf{v}}(j\omega) \\ \hat{\mathbf{w}}(j\omega) \end{bmatrix}^* \boldsymbol{\Pi}(j\omega) \begin{bmatrix} \hat{\mathbf{v}}(j\omega) \\ \hat{\mathbf{w}}(j\omega) \end{bmatrix} d\omega \geq 0, \quad (9)$$

where $\mathbf{v}, \mathbf{w} \in \mathbf{L}_2^2[0, \infty)$ are said to *satisfy the IQC defined by* $\boldsymbol{\Pi} : j\mathbb{R} \to \mathbb{C}^{4\times4}$, where $\boldsymbol{\Pi}$ is Hermitian. The main result is as follows: if the interconnection of $\mathbf{G}(s)$ and these $\boldsymbol{\Delta}(t)$ is well-posed and there exist $\epsilon > 0$ such that

$$\begin{bmatrix} \mathbf{G}(j\omega) \\ \mathbf{I} \end{bmatrix}^* \boldsymbol{\Pi}(j\omega) \begin{bmatrix} \mathbf{G}(j\omega) \\ \mathbf{I} \end{bmatrix} \leq -\epsilon\mathbf{I} \quad (10)$$

for all $\omega \in \mathbb{R}$, then the interconnection is asymptotically stable. As this is only a sufficient condition it is crucial that $\boldsymbol{\Pi}(j\omega)$ is well chosen, otherwise the result is very conservative.

In (Megretski and Rantzer, 1997) the theory of IQCs is given along with a list of $\boldsymbol{\Pi}$ for different types of $\boldsymbol{\Delta}(t)$. Of special interest here are those concerning time-periodic matrices. If $\boldsymbol{\Delta}(t)$ satisfies several IQCs they are readily combined into a single $\boldsymbol{\Pi}$ giving less conservative results. In (Möllerstedt *et al.*, 2000) some numerical experiments in this direction is done to study the robustness of Distributed Power Generators (DPGs) connected to a stiff power grid.

As the nominal system here is time-periodic it would be better to include the known periodicity in $\mathbf{G}$ to reduce conservatism in the analysis. The original theory of IQCs however require time-invariant models $\mathbf{G}(s)$. In (Jönsson *et al.*, 1999) an extension of the theory to handle linear time-periodic models is done. This should be utilized for the converter analysis.

Example 2.1. For uncertain periodic scalars there is an IQC that gives the result of Willems, (Willems, 1971). Signals $\mathbf{w} = s^n(t)\mathbf{I}\,\mathbf{v}$ with a T-periodic scalar $s^n(t) \leq 1$ satisfies the IQC given by

$$\boldsymbol{\Pi} = \begin{bmatrix} \mathbf{X}(j\omega) & \mathbf{Y}(j\omega) \\ \mathbf{Y}(j\omega)^* & -\mathbf{X}(j\omega) \end{bmatrix}$$

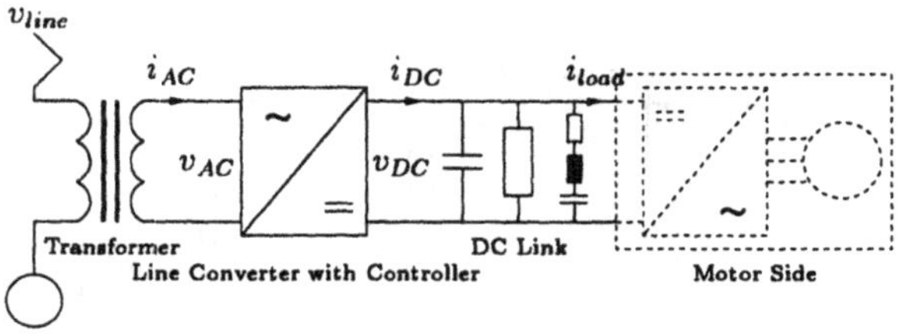

Fig. 3. A schematic of an inverter locomotive. The locomotive consists of a transformer, a line converter, a DC link, and the motor side. The construction opens new possibilities for control, and for operating the same locomotive on different power grids, which simplifies border crossing.

where

$$\mathbf{X}(j\omega) = \mathbf{X}(j(\omega + 2\pi/T)) = \mathbf{X}(j\omega)^* \geq 0$$
$$\mathbf{Y}(j\omega) = \mathbf{Y}(j(\omega + 2\pi/T)) = -\mathbf{Y}(j\omega)^*.$$

If (5) is used as $\boldsymbol{\Delta}(t)$ and we choose $\mathbf{X} = \text{diag}\{x_1, x_2\}$, and $\mathbf{Y} = 0$ the condition (10) becomes

$$\begin{bmatrix} x_2|Y_{AC}|^2 - x_1 & 0 \\ 0 & x_1|Z_{DC}|^2 - x_2 \end{bmatrix} \leq -\epsilon\mathbf{I} \quad (11)$$

for all $\omega \in \mathbb{R}$ and some $\epsilon > 0$. If we choose $x_1 = x_2 = 1$ this is just the small-gain theorem. If $Y_{AC}(s)$ and $Z_{DC}(s)$ are known we might however find $(j2\pi/T)$-periodic $x_1(j\omega)$ and $x_2(j\omega)$ that satisfies (11), and thereby proving stability of the interconnection in Fig. 2 for *every* nominal T-periodic $s^n(t)$.

3. PERIODIC MODELLING OF AN INVERTER LOCOMOTIVE

In this section it is shown that periodic modelling of a fairly complex power system is indeed possible, and also provides more information than a linear time-invariant approach would have done. Here an inverter locomotive is studied, schematically illustrated in Fig. 3, with two power converters connected via a DC-link, a so called back-to-back configuration. This is a common solution in modern variable-speed drives. It offers great flexibility. For example it can operate on different power grids and be used to minimize the reactive power on the grid. However this type of locomotive has been involved in instabilities on the power net, as in Zürich 1995 (Meyer, 1999), which makes it an interesting modelling object. Some of the problems are believed to be due to the creation of harmonics in the power converters of the locomotives. Such effects should be captured by a periodic model. In (Möllerstedt and Bernhardsson, 2000) the locomotive modelling was done in detail. Here those results will be summarized and it will be shown how the model fits into the framework of the former section.

The line converter is used to keep the DC-link voltage level constant and to draw a sinusoidal current of correct frequency and phase from the grid. By controlling the shape of i_{AC} the locomotive can be used to compensate for reactive power on the net. The equations (1) and (2) describe the relation between the transformer and the link. Ideally i_{DC} would be constant but from (2) it is seen that it equals the product of two sinusoids and thus a strong harmonic of twice the grid frequency is present. This frequency is damped by a passive filter in the link. The link also contains a large capacitor to stabilize the voltage level of the link. Here the link is modeled by a fourth-order impedance

$$Z_{DC}(s) = -\frac{C_f L_f s^2 + C_f R_f s + 1}{(C_f L_f s^2 + C_f R_f s + 1)(Cs + 1/R) + C_f s}$$

where C is the capacitance of the large capacitor, R is the resistance of the link, and C_f, L_f, and R_f belong to the passive filter. Now $v_{DC} = Z_{DC}(s)(i_{load} - i_{DC})$. The transformer that is connected to the other side of the converter is modeled with the admittance

$$Y_{AC}(s) = \frac{1}{sL_{tr} + R_{tr}}$$

with L_{tr} and R_{tr} being the inductance and resistance of the transformer. The relation is then $i_{AC} = Y_{AC}(s)(v_{line} - v_{AC})$.

The second converter generates three-phase AC of variable frequency to control an asynchronous engine. The symmetry in the three-phase arrangement reduces the harmonics from the engine. With the so called *dq*0-transformation, mentioned before, this side can be modeled as an LTI system. Here the motor side will be replaced with a current sink, i_{load} =constant. This is a fairly good approximation of a three-phase engine running in steady-state, as it then requires constant instantaneous power.

To get a linear time-periodic model a periodic trajectory to linearize about is needed. For more complex systems this is done by first simulating the entire system. Here the periodic trajectory will have a frequency of 16 + 2/3Hz (the grid frequency). $i^n_{AC}(t)$ and $s^n(t)$ are sinusoids, and v^n_{DC} and i_{load} are constant. The linearized model can be put together as in Fig. 4 with

$$\mathbf{\Delta}(t) = \begin{bmatrix} 0 & s^n(t) & v^n_{DC}(t) \\ s^n(t) & 0 & i^n_{AC}(t) \end{bmatrix} \qquad (12)$$

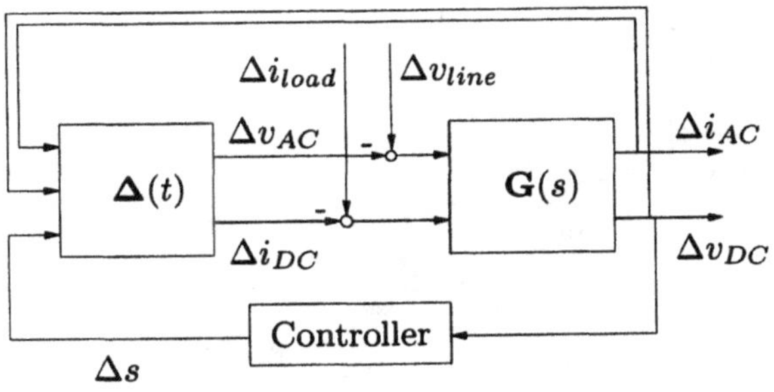

Fig. 4. A block diagram of the linearized inverter locomotive with a converter controller trying to keep v_{DC} constant. The converter model, a time-periodic matrix $\mathbf{\Delta}(t)$, is here given by (12).

and $\mathbf{G}(s)$ containing $Z_{DC}(s)$ and $Y_{AC}(s)$ as before. Notice $\mathbf{\Delta}(t)$ now has one more input that comes from a possibly dynamic controller. The loop now contains the normal control system blocks: the controller, the actuator (the converter), and the process (the transformer and the link).

As $\mathbf{G}(s)$ and $\mathbf{\Delta}(t)$ can be assembled into a linear time-periodic state-space model standard techniques can be used to design a local controller about the nominal trajectory. Here however a global controller is just taken from (Möllerstedt and Bernhardsson, 2000), is linearized and put into the loop to allow us to do some analysis. The controller contains a PI-controller, a model of the transformer and a simple model of the pulse width modulation used to get $s(t)$. With the Nyquist stability criteria for linear periodic systems presented in (Wereley, 1991) a gain margin for the PI-controller is obtainable. Simulations verify that this predicted stability margin actually is very close to the real one. This is valuable as before this, simulation has been the *only* tool to test stability, and then only a simple yes/no answer is given to the stability question. Now also some bounds on acceptable component parameters in the link can be obtained. This is useful as parameters of electrical components always are uncertain to some degree.

To study stability of interconnections of several locomotives the admittance of each locomotive is needed. Admittance is here defined to be the influence of the grid voltage (Δv_{line}) on the drawn current (Δi_{AC}). Normally this relation is considered to be time invariant. By plotting how harmonics interact in the above model we can see if this is a valid approximation. In Figure 5 the interaction is visualized for two different operating conditions of a locomotive. The main diagonal is essentially the amplitude part of a Bode plot. The sub-diagonals show the transfer from one frequency to another. The shift in these cases is twice the grid frequency, 33-1/3Hz. The sub-

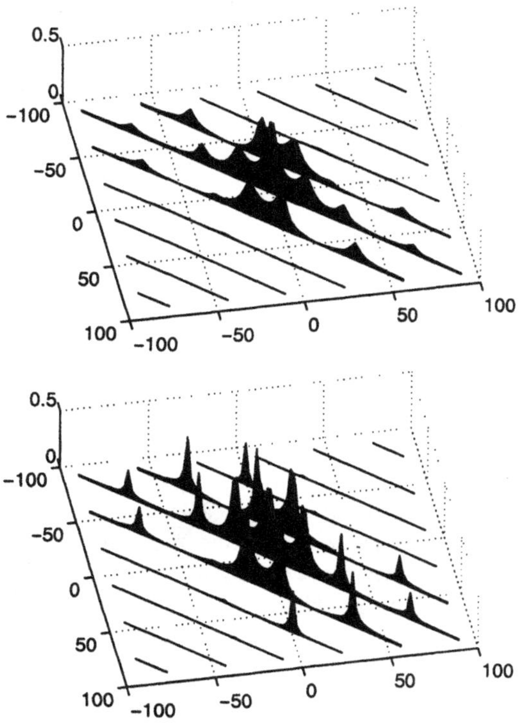

Fig. 5. Plots showing the amplitude (z-axis) of the coupling between the frequencies in Δv_{line} (x-axis) and Δi_{AC} (y-axis) for a single locomotive. The main diagonal is essentially the amplitude part of a Bode plot. The sub-diagonals show the transfer from one frequency to another. The load current is $i_{load} = 500$ A in the upper plot and $i_{load} = -500$ A in the lower illustrating the difference with the motor working as a load (driving) and as a generator (braking).

diagonals are considerably large in both cases and even contain resonance peaks. Thus to design controllers based only on time-invariant models (essentially the main diagonals) leads to either unsafe or overly conservative solutions. The analysis of interconnections of several locomotive models still remains to be done.

4. CONCLUSIONS AND FUTURE WORK

Some of the possibilities of using periodic modelling of power systems have been shown. A model of a power converter was developed. It turned out to be a periodic gain matrix. This was used to model an inverter locomotive. The locomotive had a considerable frequency coupling, something that normally is not taken into account during controller design and stability analysis.

It was also shown how IQCs may be used in the analysis of systems with power converters. This should be further studied. Also it would be interesting to look at larger problems. For example interconnections of several locomotives running on the same net. Such systems must be very large in order to be realistic. This should require model reduction of time-periodic models which also is an interesting topic to be further studied.

5. REFERENCES

Arrillaga, J., A. Medina, M.L.V. Lisboa, M. A. Cavia and P. Sánchez (1994). The harmonic domain. A frame of reference for power system harmonic analysis. *IEEE Trans. on Power Systems* **10**(1), 433–440.

Elmqvist, H., S.E. Mattsson and Martin Otter (1999). Modelica - a language for physical system modeling, visualization and interaction. In: *Proc. of the 1999 IEEE Symposium on Computer-Aided Control System Design.* Hawaii, USA.

Jönsson, U.T., C.-Y. Kao and A. Megretski (1999). Robustness analysis of periodic systems. In: *Proceedings of 38th IEEE conference on Decisions and Control.* Phoenix, Arizona, USA.

Kundur, P. (1994). *Power System Stability and Control.* McGraw-Hill Inc., USA.

Megretski, A. and A. Rantzer (1997). System analysis via integral quadratic constraints. *IEEE Transactions on Automatic Control* **42**(6), 819–830.

Meyer, M. (1999). Netzstabilität in grossen Bahnnetzen. *Eisenbahn-Revue* (7-8), 312–317.

Möllerstedt, E., A. Stothert and H. Sandberg (2000). Robust control of power converters. *To be submitted.*

Möllerstedt, E. and B. Bernhardsson (2000). Out of control because of harmonics - an analysis of the harmonic response of an inverter locomotive. *IEEE Control Systems Magazine* **20**(4), 70–81.

Stankovic, A.M., B.C. Lesieutre and T. Aydin (1999). Modeling and analysis of single-phase induction machines with dynamic phasors. *IEEE Transactions on Power Systems* **14**(1), 9–14.

Wereley, N.M. (1991). Analysis and Control of Linear Periodically Time Varying Systems. PhD thesis. Dept. of Aeronautics and Astronautics, MIT.

Willems, J. C. (1971). *The Analysis of Feedback Systems.* MIT Press. Cambridge, MA, USA.

Xu, W., J.R. Marti and H.W. Dommel (1991). A multiphase harmonic load flow solution technique. *IEEE Trans. on Power Systems* **6**(1), 174–182.

Copyright © IFAC Periodic Control Systems,
Cernobbio-Como, Italy, 2001

ACHIEVABLE H^∞ PERFORMANCE IN SAMPLED-DATA SMOOTHING: BEYOND THE $\|\check{D}_1\|$-BARRIER*

Leonid Mirkin[†] **Allan C. Kahane** **Zalman J. Palmor**

Faculty of Mechanical Eng., Technion — IIT, Haifa 32000, Israel

Abstract: The lifting technique is a powerful tool for handling the periodically time-varying nature of sampled-data systems. Yet all known solutions of sampled-data H^∞ problems are limited to the case when the feedthrough part of the lifted system, $\check{D}_1 : L^2[0,h] \mapsto L^2[0,h]$, satisfies $\|\check{D}_1\| < \gamma$, where γ is the required H^∞ performance level. While this condition is always necessary in feedback control, it might be restrictive in signal processing applications, where some amount of delay or latency between measurement and estimation can be tolerated. In this paper the sampled-data H^∞ fixed-lag smoothing problem with a smoothing lag of one sampling period is studied. The problem corresponds to the *a-posteriori* filtering problem in the lifted domain and is probably the simplest problem for which a smaller than $\|\check{D}_1\|$ H^∞ performance level is achievable. The necessary and sufficient solvability conditions derived in the paper are compatible with those for the sampled-data filtering problem. This result extends the scope of applicability of the lifting technique and paves the way to the application of sampled-data methods in digital signal processing.
Copyright © 2001 IFAC

1 INTRODUCTION

Since the early 90th, much attention has been paid to control and estimation of continuous-time systems using sampled measurements, see the book (Chen and Francis, 1995) and the references therein. This class of systems is referred to as *sampled-data systems*. An important breakthrough in the treatment of sampled-data systems was the introduction of "lifting," an operation that reduces periodically time-varying sampled-data problems to equivalent time-invariant, albeit inherently *infinite-dimensional*, discrete ones (Yamamoto, 1994; Toivonen, 1992; Tadmor, 1992; Bamieh and Pearson, 1992). The lifting technique enables unified and intuitively clear solutions to a wide spectrum of sampled-data (and other periodic) problems.

The key step in the use of the lifting technique is the reduction of infinite-dimensional H^∞ problems in the lifted domain to equivalent finite-dimensional discrete problems. Several different approaches to perform this step are currently available (Bamieh and Pearson, 1992; Hayakawa *et al.*, 1994; Toivonen and Sågfors, 1997; Mirkin *et al.*, 1999). Roughly, all of them are based on the assumption that the $(1,1)$ sub-block of the feedthrough part of the lifted generalized plant, say $\check{D}_1 : L^2[0,h] \mapsto L^2[0,h]$, satisfies $\|\check{D}_1\| < \gamma$ (here γ is the required H^∞ level). When the sampled-data controller/estimator is causal this assumption is not restrictive. Indeed, such a controller/estimator is always *strictly proper* in the lifted domain (Mirkin and Rotstein, 1997) and therefore the γ-suboptimal H^∞ sampled-data problem is solvable *only if* $\|\check{D}_1\| < \gamma$.

The assumption $\|\check{D}_1\| < \gamma$ does become restrictive when non-causal (smoothed) solutions are allowed. This is the case, for example, in signal processing applications, where some amount of delay or latency between measurement and estimation can be tolerated, see (Khargonekar and Yamamoto, 1996; Ishii *et al.*, 1999) for the application of the sampled-data H^∞ techniques to some problems in digital signal processing. Such

*This research was supported by the Israel Ministry of Science under contract no. 8573-1-98.

[†]Corresponding author, e-mail: *mirkin@tx.technion.ac.il*.

problems are reduced to the standard H^∞ control problem in the lifted domain with delayed regulated signal (or, equivalently, with a non-causal controller). When $\gamma > \|\breve{D}_1\|$, the latter problem can be solved exactly by the reduction to an equivalent discrete H^∞ problem. Yet the available approaches cannot be applied if further reduction of γ is required. To circumvent this difficulty, Khargonekar and Yamamoto (1996) used the rational approximation of the continuous-time delay, whereas Ishii *et al.* (1999) used the fast sampling approximation of $\breve{D}_1$. These approximations, however, might considerably increase the dimension of the problem. It is therefore important to develop new approaches, which can cope with the case of $\gamma \le \|\breve{D}_1\|$, see in this respect the discussion in (Yamamoto and Hara, 1999).

The purpose of this paper is to demonstrate, that the machinery introduced in (Mirkin and Palmor, 1999a) enables one to treat the case of interest in a straightforward manner with no need for approximations. To this end the sampled-data fixed-lag smoothing problem with the smoothing lag equal to the sampling period h is considered. In the lifted domain this problem corresponds to the *a-posteriori* filtering problem. This appears to be the simplest problem for which the H^∞ performance level smaller than $\|\breve{D}_1\|$ can be achieved. The problem is solved in two steps. First, the formal solution to the lifted *a-posteriori* filtering problem is just written down in terms of the operator-valued parameters of the lifted system. Second, the time-domain solvability conditions are "peeled-off" from their lifted counterparts. To this end the representation of the lifted parameters introduced in (Mirkin and Palmor, 1999a) plays a central role and enables the derivation of the solvability conditions to the sampled-data smoothing compatible with those for the sampled-data filtering (Sun *et al.*, 1993; Mirkin and Palmor, 1999b). In the later case the solvability conditions involve the existence of the solution to a differential Riccati equation over one sampling interval $[0, h]$. As shown in the paper, the solvability conditions for the smoothing problem are based upon the same differential Riccati equation, yet the smoothing problem may be solvable even when this equation does have a finite escape point in $[0, h]$.

Notation. In order to distinguish systems in the time domain from the corresponding transfer functions, the former are denoted by script capital letters, so $G(s)$ implies the transfer function of a continuous-time LTI system $\mathcal{G}$. The sampling operator, $\mathcal{S}_h$, acts on continuous signals so that $\bar{\zeta} = \mathcal{S}_h\zeta \iff \bar{\zeta}_k = \zeta(kh^-)$; the continuous-time delay operator $\mathcal{D}_\delta$ is defined as follows: $\zeta = \mathcal{D}_\delta\omega \iff \zeta(t) = \omega(t-\delta)$; and the backward unit shift operator, $\mathcal{U}$, acts on discrete-time (either real valued or functional space valued) signals so that $\breve{\zeta} = \mathcal{U}\breve{\omega} \iff \breve{\zeta}_k = \breve{\omega}_{k-1}$.

2 MAIN RESULT

In this section the sampled-data smoothing problem is posed and its solution is formulated. The next sections are devoted to the proofs.

2.1 *Problem formulation*

The problem of estimating a continuous-time signal z on the basis of sampled measurements $\bar{y} = \mathcal{S}_h y$ is considered. It is assumed that both z and y are driven by a "disturbance" signal w as follows:

$$z = \mathcal{G}_1 w \qquad \text{and} \qquad y = \mathcal{G}_2 w,$$

where the continuous-time LTI systems $\mathcal{G}_1$ and $\mathcal{G}_2$ are defined in terms of their state-space realizations:

$$\begin{bmatrix} G_1(s) \\ G_2(s) \end{bmatrix} = \left[\begin{array}{c|c} A & B \\ \hline C_1 & 0 \\ C_2 & 0 \end{array}\right].$$

Note that the fact that $G_2(s)$ is strictly proper implies in fact that pre-filtering by an anti-aliasing filter is provided. This is necessary to guarantee the boundedness of the sampling operation (Chen and Francis, 1995). $G_1(s)$ is assumed to be strictly proper to simplify the derivations and obtain more transparent results. It will also be assumed throughafter that

(A1): (C_2, A) is detectable and h is non-pathological with respect to A;

(A2): (A, B) has no uncontrollable modes on the imaginary axis.

These assumptions guarantee that the estimation problem is nonsingular.

When the estimation of $z(t)$ can be based on the measurements $\bar{y}_k$ available up to the time instance[1] $k = \lfloor \frac{t}{h} \rfloor$, i.e., the sampled-data estimator is *causal*, the estimation problem for z is the standard sampled-data filtering problem extensively studied in the literature, see (Sun *et al.*, 1993; Mirkin and Palmor, 1999b). In numerous signal processing application, however, *non-causal* estimation (smoothing) is permitted (Anderson, 1999). In other words, estimation of $z(t)$ can be based on $\bar{y}_k$ up to $k = \lfloor \frac{t}{h} \rfloor + \delta$, where a positive $\delta \in \mathbb{R}^+$ is called the *smoothing lag*. Such a problem can be equivalently formulated as the estimation of $\mathcal{D}_{\delta h} z$ by a causal estimator.

In this paper the problem of the achievable H^∞ performance in the sampled-data smoothing in the case of $\delta = 1$ will be studied. Such a problem can be formulated as follows:

[1] $\lfloor \cdot \rfloor$ stands for the "floor" (round towards $-\infty$) operation.

SP: Given $\mathcal{G}_1$ and $\mathcal{G}_2$ and the sampling period $h \geq 0$, determine whether there exists a causal operator (smoother) $\mathcal{K}$, which guarantees

$$\|\mathcal{D}_h\mathcal{G}_1 - \mathcal{K}\,\mathcal{S}_h\mathcal{G}_2\| < \gamma$$

for a given $\gamma > 0$ (here $\|\cdot\|$ stands for the L^2-induced operator norm).

To the best of our knowledge, no exact solutions to **SP** is available in the literature. The problems similar to **SP** with arbitrary smoothing lags were treated in (Khargonekar and Yamamoto, 1996; Ishii *et al.*, 1999). Yet only approximate solutions were obtained there, see Introduction. Thus this paper presents the first treatment of **SP**, where the effect of the previewed information on the achievable H^∞ performance in sampled-data estimation is accounted for in an exact manner.

2.2 *Problem solution*

The solution to **SP** is based on the following symplectic matrix function:

$$\Sigma(t) = \begin{bmatrix} \Sigma_{11}(t) & \Sigma_{12}(t) \\ \Sigma_{21}(t) & \Sigma_{22}(t) \end{bmatrix} \doteq \exp\left(\begin{bmatrix} A & BB' \\ -\frac{1}{\gamma^2}C_1'C_1 & -A' \end{bmatrix} t\right).$$

To simplify the notations, we simply write Σ instead of $\Sigma(h)$. Introduce also the matrices

$$M_l \doteq \begin{bmatrix} 0 & C_2 & 0 \\ \Sigma_{11}' & -\Sigma_{21}' & \Sigma_{11}'C_2' \\ -\Sigma_{12}' & \Sigma_{22}' & -\Sigma_{12}'C_2' \end{bmatrix}, \quad M_r \doteq \begin{bmatrix} 0 & 0 & 0 \\ I & 0 & 0 \\ 0 & I & 0 \end{bmatrix},$$

which constitute (Mirkin *et al.*, 1999) an extended symplectic matrix pair (M_l, M_r), i.e., the associated matrix pencil $M_l - \lambda M_r$ verifies *(a)* $\det(M_l - \lambda M_r) \not\equiv 0$, *(b)* If $\lambda \notin \{0, \infty\}$ is a generalized eigenvalue of $M_l - \lambda M_r$ of multiplicity r then so is $\frac{1}{\lambda}$, and *(c)* if 0 is an eigenvalue of M_l of multiplicity r then it is an eigenvalue of M_r of multiplicity $r + m$, where m is the row dimension of C_2. We say that $(M_l, M_r) \in \mathrm{dom}(\mathrm{Ric}_{\mathbb{D}})$ if there exist matrices $Y = Y'$ and L so that

$$M_r \begin{bmatrix} I \\ Y \\ L' \end{bmatrix} = M_l \begin{bmatrix} I \\ Y \\ L' \end{bmatrix} A_L$$

for some Schur A_L. It can be shown (Van Dooren, 1981) that whenever $(M_l, M_r) \in \mathrm{dom}(\mathrm{Ric}_{\mathbb{D}})$, the matrices Y and L as above are uniquely determined from (M_l, M_r) and it is thus possible to define the function $\mathrm{Ric}_{\mathbb{D}} : (M_l, M_r) \to (Y, L)$. Below, this function is denoted by the following notation: $(Y, L) = \mathrm{Ric}_{\mathbb{D}}(M_l, M_r)$.

Finally, let $C_2^\perp$ be any matrix such that $\begin{bmatrix} C_2 \\ C_2^\perp \end{bmatrix}$ has full column rank and $C_2^\perp C_2' = 0$. Then the main result of the paper is formulated as follows:

Theorem 1. *Let (A1) and (A2) hold. Then* **SP** *is solvable if, and only if,*

(a) $(M_l, M_r) \in \mathrm{dom}(\mathrm{Ric}_{\mathbb{D}})$;

(b) $Y \geq 0$, *where* $(Y, L) = \mathrm{Ric}_{\mathbb{D}}(M_l, M_r)$; *and*

(c) $M_S(t) \doteq \begin{bmatrix} C_2(\Sigma_{12}(t) + \Sigma_{11}(t)Y) \\ C_2^\perp(\Sigma_{22}(t) + \Sigma_{21}(t)Y) \end{bmatrix}$ *is nonsingular* $\forall t \in (0, h]$.

Note, that the only difference between the solution to the smoothing problem given by Theorem 1 and the solution to the sampled-data filtering problem in (Mirkin and Palmor, 1999b) is condition *(c)*. In the filtering case the nonsingularity of $M_F(t) \doteq \Sigma_{22}(t) + \Sigma_{21}(t)Y$ is required instead, see also (Sun *et al.*, 1993). It can be shown that the latter condition is equivalent to the absence of escape points in $[0, h]$ of the following differential Riccati equation:

$$\dot{Q} = AQ + QA' + BB' + \tfrac{1}{\gamma^2}QC_1'C_1Q, \quad Q(0) = Y.$$

Indeed, if $M_F(t)$ is nonsingular $\forall t \in [0, h]$, then $Q(t) = (\Sigma_{12}(t) + \Sigma_{11}(t)Y)M_F^{-1}(t)$.

Assume, that $M_F(t)$ becomes singular first time at $t = t_e > 0$. Then for all $t \in (0, t_e)$ $M_S(t)$ is nonsingular iff $\det(C_2Q(t)C_2') \neq 0$. Since $Q(t)$ is monotonically nondecreasing, the matrix $C_2Q(t)C_2' > 0$ for all $(0, t_e)$. This shows that condition *(c)* of Theorem 1 cannot be violated before its filtering counterpart is.

2.3 *Example*

Consider the following simple first-order system:

$$G_1(s) = G_2(s) = \tfrac{a}{s+a}$$

with $a > 0$. Obviously, $\gamma = 1 = \|G_1\|_\infty$ is achievable for any h in both the filtering and smoothing settings. We therefore consider only the case $\gamma < 1$. In this case (M_l, M_r) always belongs to $\mathrm{dom}(\mathrm{Ric}_{\mathbb{D}})$ with $Y = 0$ and then

$$M_F(t) = \cos\omega t + \tfrac{a}{\omega}\sin\omega t, \quad M_S(t) = \tfrac{a^2}{\omega}\sin\omega t$$

where $\omega \doteq a\sqrt{\gamma^{-2} - 1}$. Both these functions are depicted in Fig. 1(a) for the case of $a = 1$ and $\gamma = 0.3$. It is seen that the smoothing formulation enables one to reduce the requirement on the sampling rate to more than 2/3 of that in the filtering formulation. In general, for a given $\gamma < 1$ the maximal sampling period for which this γ is achievable in the filtering and smoothing cases are

$$h_F = \tfrac{\pi}{\omega} - \tfrac{1}{\omega}\arctan\tfrac{\omega}{a} \quad \text{and} \quad h_S = \tfrac{\pi}{\omega},$$

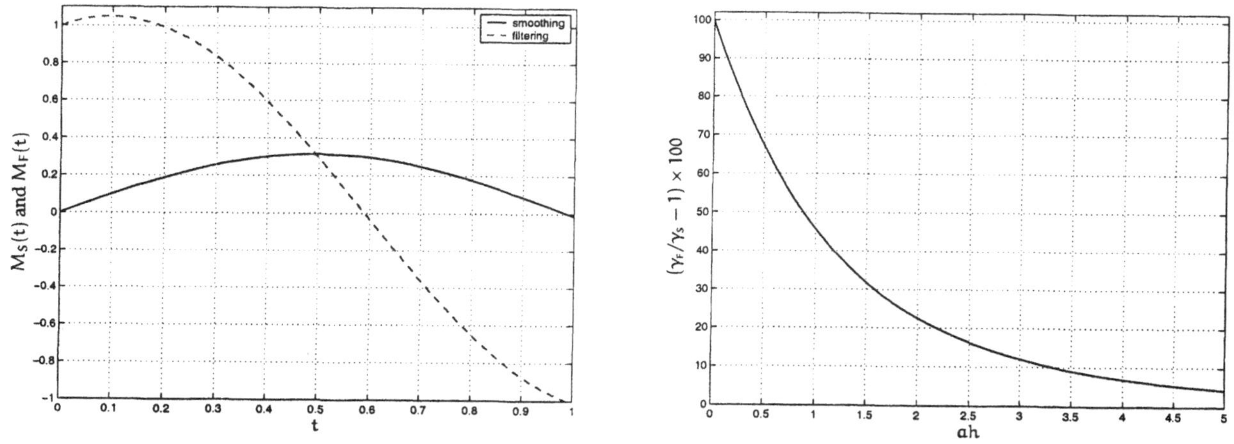

(a) $M_F(t)$ and $M_S(t)$ for $a = 1$ and $\gamma = 0.3$

(b) $\frac{\gamma_F}{\gamma_S} - 1$ vs. ah (in %)

Fig. 1: Comparison between smoothing and filtering in the Example

respectively. Thus, by allowing one additional measurement, $\bar{y}_{k+1}$, to be used to estimate $z(t)$, $t \in [kh, (k+1)h)$, an up to 2 times relaxation of the sampling rate can be achieved.

Alternatively, smoothing and filtering may be compared by the achievable performance for a given sampling period h. To this end, denote by γ_F and γ_S the achievable H^∞ performance in the filtering and smoothing cases, respectively. Fig. 1(b) shows the relative improvement of the latter with respect to the former, $\gamma_F/\gamma_S - 1$ (in percents) vs. the normalized sampling period, ah. One can see that for relatively fast sampling such an improvement reaches 100%.

3 REFORMULATION AND SOLUTION IN THE LIFTED DOMAIN

The solution of **SP** is based on its conversion to pure discrete-time time-invariant setting by the use of the so-called lifting technique. To this end, let $\mathcal{W}_h$ denotes the lifting operator converting real valued signals in continuous time into functional space valued sequences, see (Chen and Francis, 1995) for the exact definition and properties of $\mathcal{W}_h$. The lifted signals $\breve{w} \doteq \mathcal{W}_h w$ and $\breve{z} \doteq \mathcal{W}_h z$ are equivalent to their continuous-time counterparts w and z, respectively, in the sense that they contain the same information and their norms are equivalent. One thus can consider the lifted systems

$$\breve{\mathcal{G}}_1 \doteq \mathcal{W}_h \mathcal{G}_1 \mathcal{W}_h^{-1} : \breve{w} \mapsto \breve{z},$$
$$\acute{\mathcal{G}}_2 \doteq \mathcal{S}_h \mathcal{G}_2 \mathcal{W}_h^{-1} : \breve{w} \mapsto \bar{y}$$

instead of $\mathcal{G}_1 : w \mapsto z$ and $\mathcal{S}_h\mathcal{G}_2 : w \mapsto \bar{y}$, respectively. The main advantage of dealing with the lifted systems is that they are pure discrete and time invariant. The transfer functions of $\breve{\mathcal{G}}_1$ and $\acute{\mathcal{G}}_2$ are (Mirkin and Palmor, 1999a,b)

$$\breve{G}_1(z) = \left[\begin{array}{c|c} \bar{A} & \acute{B} \\ \hline \grave{C}_1 & \breve{D}_1 \end{array}\right],$$
$$\acute{G}_2(z) = z^{-1}\left[\begin{array}{c|c} \bar{A} & \acute{B} \\ \hline \bar{C}_2 & \acute{D}_2 \end{array}\right] \doteq z^{-1}\acute{G}_{2a}(z),$$

where

$$\begin{bmatrix} \bar{A} & \acute{B} \\ \grave{C}_1 & \breve{D}_1 \end{bmatrix} = \begin{bmatrix} \mathcal{J}_h^* \\ \grave{C}_1 \end{bmatrix} \left(\begin{array}{c|c} A & I \\ \hline I & 0 \end{array}\right) \begin{bmatrix} \mathcal{J}_0 & B \end{bmatrix}$$

and $[\,\bar{C}_2 \;\; \acute{D}_2\,] = C_2 [\,\bar{A} \;\; \acute{B}\,]$. Here the compact block notation $\breve{O} = \left(\begin{array}{c|c} A & B \\ \hline C & D \end{array}\right)$ is used to denote an operator $\breve{O} : L^2[0,h] \mapsto L^2[0,h]$ described by the following state equations:

$$\breve{O} : \begin{cases} \dot{x}(t) = Ax(t) + Bu(t), & x(0) = 0, \\ y(t) = Cx(t) + Du(t); \end{cases} \tag{1}$$

the *impulse operator* $\mathcal{J}_\theta$ transforms $\eta \in \mathbb{R}^n$ into a modulated δ-impulse as $(\mathcal{J}_\theta\eta)(t) = \delta(t-\theta)\eta$; and the *sampling operator*[2] $\mathcal{J}_\theta^*$ transforms $\zeta \in C_n[0,h]$ into a vector from $\mathbb{R}^n$ as $\mathcal{J}_\theta^*\zeta = \zeta(\theta)$. In the sequel, we shall also use operators $\breve{O} : L^2[0,h] \mapsto L^2[0,h]$, which are described by equations like (1) but with the *2-point boundary conditions* $\Omega x(0) + \Upsilon x(h) = 0$ for some square matrices Ω and Υ. Such operators, denoted as

$$\breve{O} = \left(\begin{array}{c|c|c} A & \Omega \rightleftharpoons \Upsilon & B \\ \hline C & & D \end{array}\right),$$

are well-posed iff $\Xi_0 \doteq \Omega + \Upsilon e^{Ah}$ is nonsingular.

Remark 3.1. To improve the readability of formulae when both finite and infinite dimensional input/output spaces are involved, the following

[2] It is worth stressing that $\mathcal{J}_\theta^*$ is not the adjoint of $\mathcal{J}_\theta$. Nevertheless, we will proceed with this abuse of notation for the reasons discussed in (Mirkin and Palmor, 1999a).

operator notation is used. A bar indicates an operator $\bar{O}$ with both input and output spaces finite dimensional; grave accent—$\grave{O}$, when the input space is finite dimensional and the output infinite dimensional one; acute accent—$\acute{O}$, when the input space is infinite dimensional and the output finite dimensional one; and finally breve—$\breve{O}$, when both input and output spaces are infinite dimensional.

With the definitions above, **SP** can be reformulated in the lifted domain as follows:

SP$_{\text{eq}}$: Given the systems $\breve{G}_1(z)$ and $\acute{G}_{2a}(z)$, determine whether there exists a proper $\grave{K}(z)$, which guarantees

$$\|\breve{G}_1(z) - \grave{K}(z)\acute{G}_{2a}(z)\|_\infty < \gamma$$

for a given $\gamma > 0$ (here $\|\cdot\|_\infty$ stands for the H^∞ induced norm).

SP$_{\text{eq}}$ is, in principle, a standard discrete *a-posteriori* filtering problem, solution to which is well understood (Hassibi *et al.*, 1999). The solvability condition can thus be expressed in terms of the parameters of $\breve{G}_1$ and $\acute{G}_{2a}$ in a straightforward manner. The only difference between the case treated in the literature and the one here is that $\breve{D}_1 \neq 0$; this, however, can easily be accounted for.

To formulate the solution to **SP$_{\text{eq}}$**, let us impose the following two standard assumptions upon $\acute{G}_{2a}(z)$:

(A3): $(\bar{C}_2, \bar{A})$ is detectable;

(A4): $\acute{G}_{2a}(e^{j\theta})$ is right invertible $\forall\theta \in [0, 2\pi)$;

and consider the H^∞ filtering Riccati equation

$$Y = \bar{A}Y\bar{A}' + \acute{B}\acute{B}^* - \acute{L}\breve{R}\acute{L}^*, \qquad (2)$$

where

$$\breve{R} = \begin{bmatrix} \breve{R}_{11} & \grave{R}_{12} \\ \acute{R}_{21} & \bar{R}_{22} \end{bmatrix}$$
$$\doteq \begin{bmatrix} \breve{D}_1 \\ \acute{D}_2 \end{bmatrix} [\breve{D}_1^* \ \acute{D}_2^*] + \begin{bmatrix} \grave{C}_1 \\ \bar{C}_2 \end{bmatrix} Y [\grave{C}_1^* \ \bar{C}_2^*] - \gamma^2 \begin{bmatrix} I & 0 \\ 0 & 0 \end{bmatrix}$$

and

$$\acute{L} = [\,\acute{L}_1 \quad \bar{L}_2\,]$$
$$\doteq -(\acute{B}[\,\breve{D}_1^* \quad \acute{D}_2^*\,] + \bar{A}Y[\,\grave{C}_1^* \quad \bar{C}_2^*\,])\breve{R}^{-1}.$$

The necessary and sufficient solvability conditions of **SP$_{\text{eq}}$** are then formulated as follows:

Theorem 2. *Let (A3) and (A4) hold, then* **SP$_{\text{eq}}$** *is solvable iff there exists a solution* Y *to* (2) *such that*

(a) $\bar{A} + \acute{L}_1\grave{C}_1 + \bar{L}_2\bar{C}_2$ *is stable (Schur);*

(b) $Y = Y' \geq 0$;

(c) $\breve{R}$ *and* $\begin{bmatrix} -I & \\ & I \end{bmatrix}$ *have the same inertia.*

Proof. A straightforward extension of (Hassibi *et al.*, 1999, Theorem 13.3.1). □

Theorem 2 yields the solution of **SP** in terms of the *infinite-dimensional* parameters of the lifted problem. It therefore is not readily verifiable and has to be translated back to the time domain ("peeled-off") and expressed in terms of finite-dimensional matrices.

Note that the difference between **SP$_{\text{eq}}$** and the lifted equivalent of the sampled-data filtering problem (Mirkin and Palmor, 1999b) is that in the latter case $\acute{K}$ is constrained to be *strictly* proper rather than just proper as above. In other words, the sampled-data smoothing reduces to an *a-posteriori* filtering in the lifted domain, whereas the sampled-data filtering reduces to an *a-priori* filtering. In terms of the lifted solution, this difference affects only condition *(c)* of Theorem 2. More precisely, in the smoothing case this condition is equivalent to $\bar{R}_{22} > 0$ and

$$\breve{R}_{11} - \grave{R}_{12}\bar{R}_{22}^{-1}\acute{R}_{21} < 0, \qquad (3)$$

while in the filtering case (3) is just replaced with

$$\breve{R}_{11} < 0. \qquad (3')$$

It is clear that whenever (3′) holds so does (3), but not vise versa. The smoothing problem can thus be solvable even when the filtering problem cannot. Yet (3′) requires $\breve{\Theta} \doteq \gamma^2 I - \breve{D}_1\breve{D}_1^* > 0$, while (3) requires neither $\breve{\Theta} > 0$ nor even $\breve{\Theta}$ to be nonsingular. This actually is the main source of difficulties in "peeling-off" the solvability conditions of Theorem 2.

4 "PEELING-OFF" THE LIFTED SOLUTION: OUTLINE

The purpose of this section is to prove the equivalence of the conditions of Theorems 1 and 2. Due to space limitations, all the proofs are only outlined. Moreover, items *(a)* and *(b)* of Theorem 1 can be proved in a similar fashion to their filtering counterparts in (Mirkin and Palmor, 1999b). Hence, only the proof of the equivalence between condition *(c)* of Theorem 1 and condition (3) will be presented.

We start with the following lemma which plays a key role in the sequel:

Lemma 1. *Define the (orthogonal) projection matrix* $\Pi_y \doteq C_2'(C_2C_2')^{-1}C_2$ *and the* $L^2[0,\tau] \mapsto L^2[0,\tau]$ *operator:*

$$\breve{O}_\tau \doteq \left(\begin{array}{cc|c|c} A & BB' & \multirow{2}{*}{$\begin{bmatrix} I & -Y \\ 0 & 0 \end{bmatrix} \rightleftharpoons \begin{bmatrix} 0 & 0 \\ \Pi_y & I-\Pi_y \end{bmatrix}$} & 0 \\ 0 & -A' & & -C_1' \\ \hline C_1 & 0 & & 0 \end{array}\right).$$

Then $\breve{O}_\tau$, $\tau \in (0, h]$, *is well-posed iff* $\bar{R}_{22} > 0$ *and, moreover,* $\breve{R}_{11} - \grave{R}_{12}\bar{R}_{22}^{-1}\acute{R}_{21} = \breve{O}_h - \gamma^2 I$.

Comparing Lemma 1 with Lemma 14 in (Mirkin and Palmor, 1999b), one can see that the only difference between the "state-space" representations of $\check{R}_{11}$ and $\check{R}_{11} - \grave{R}_{12}\bar{R}_{22}^{-1}\grave{R}_{21}$ is that the former requires $x_2(h) = 0$, while the latter requires $\Pi_y x_1(h) + (I - \Pi_y)x_2(h) = 0$. Thus the (finite-rank) term $\grave{R}_{12}\bar{R}_{22}^{-1}\grave{R}_{21}$ affects the result only by reshaping the final condition between two components of the state vector of $\check{O}_h$.

Having established Lemma 1, the "peeling-off" procedure for condition *(c)* of Theorem 2 consists of the following steps, proofs of which (except, perhaps, Step 1) are relatively straightforward:

Step 1: $\|\check{O}_\tau\|$ is a monotonically increasing function of τ and, moreover, $\lim_{\tau\to 0}\|\check{O}_\tau\| = 0$.

Step 2: Consequently, (3) holds iff $\gamma^2 I - \check{O}_\tau > 0$ for all $\tau \in (0, h]$.

Step 3: Therefore, (3) holds iff $(\gamma^2 I - \check{O}_\tau)^{-1}$ is well-posed for all $\tau \in (0, h]$.

Step 4: The latter is equivalent (see p. 4) to the invertibility of the matrix $\Pi_y(\Sigma_{12}(t) + \Sigma_{11}(t)Y) + (I - \Pi_y)(\Sigma_{22}(t) + \Sigma_{21}(t)Y)$, $\forall t \in (0, h]$, which, in turn, reduces to condition *(c)* of Theorem 1.

5 CONCLUDING REMARKS

This paper has studied the sampled-data H^∞ fixed-lag smoothing problem with a smoothing lag of one sampling period. This appears to be the simplest sampled-data problem for which the achievable performance level γ can be smaller than $\|\check{D}_1\|$, where $\check{D}_1$ is the system feedthrough term in the lifted domain. In this paper necessary and sufficient solvability conditions have been derived, which are compatible with those in the filtering case and require neither $\gamma^2 I > \check{D}_1\check{D}_1^*$ nor even $\gamma^2 I - \check{D}_1\check{D}_1^*$ to be invertible.

The technical machinery developed in this paper is expected to be applicable to more general sampled-data smoothing problems (with an arbitrary smoothing lag) as well. The main obstacle here is the absence of transparent solvability conditions for the standard discrete-time H^∞ fixed-lag smoothing problem to be used as a basis for the solution in the lifted domain. This is the subject of future research.

REFERENCES

Anderson, B. D. O. (1999). "From Wiener to hidden Markov models," *IEEE Control Systems Magazine*, **19**, no. 3, pp. 41–51.

Bamieh, B. and J. B. Pearson (1992). "A general framework for linear periodic systems with applications to H^∞ sampled-data control," *IEEE Trans. Automat. Control*, **37**, pp. 418–435.

Chen, T. and B. A. Francis (1995). *Optimal Sampled-Data Control Systems*, Springer-Verlag, London.

Hassibi, B., A. H. Sayed, and T. Kailath (1999). *Indefinite Quadratic Estimation and Control: A Unified Approach to H^2 and H^∞ Theories*, SIAM, Philadelphia.

Hayakawa, Y., S. Hara, and Y. Yamamoto (1994). "H^∞ type problem for sampled-data control systems—a solution via minimum energy characterization," *IEEE Trans. Automat. Control*, **39**, pp. 2278–2284.

Ishii, H., Y. Yamamoto, and B. A. Francis (1999). "Sample-rate conversion via sampled-data H^∞ control," in *Proc. 38th IEEE Conf. Decision and Control*, Phoenix, AZ, pp. 3440–3445.

Khargonekar, P. P. and Y. Yamamoto (1996). "Delayed signal reconstruction using sampled-data control," in *Proc. 35th IEEE Conf. Decision and Control*, Kobe, Japan, pp. 1259–1263.

Mirkin, L. and Z. J. Palmor (1999a). "A new representation of parameters of lifted systems," *IEEE Trans. Automat. Control*, **44**, pp. 833–840.

Mirkin, L. and Z. J. Palmor (1999b). "On the sampled-data H^∞ filtering problem," *Automatica*, **35**, pp. 895–905.

Mirkin, L. and H. Rotstein (1997). "On the characterization of sampled-data controllers in the lifted domain," *Syst. Control Lett.*, **29**, pp. 269–277.

Mirkin, L., H. Rotstein, and Z. J. Palmor (1999). "H^2 and H^∞ design of sampled-data systems using lifting, Part I: General framework and solutions," *SIAM J. Control Optim.*, **38**, pp. 175–196.

Sun, W., K. M. Nagpal, and P. P. Khargonekar (1993). "H^∞ control and filtering for sampled-data systems," *IEEE Trans. Automat. Control*, **38**, pp. 1162–1174.

Tadmor, G. (1992). "H^∞ optimal sampled-data control in continuous-time systems," *Int. J. Control*, **56**, pp. 99–141.

Toivonen, H. T. (1992). "Sampled-data control of continuous-time systems with an H^∞ optimality criterion," *Automatica*, **28**, pp. 45–54.

Toivonen, H. T. and M. F. Sågfors (1997). "The sampled-data H^∞ problem: a unified framework for discretization-based methods and Riccati equation solution," *Int. J. Control*, **66**, pp. 289–309.

Van Dooren, P. (1981). "A generalized eigenvalue approach for solving Riccati equations," *SIAM J. Scientific and Statistical Computing*, **2**, pp. 121–135.

Yamamoto, Y. (1994). "A function space approach to sampled data control systems and tracking problems," *IEEE Trans. Automat. Control*, **39**, pp. 703–713.

Yamamoto, Y. and S. Hara (1999). "Performance lower bound for a sampled-data signal reconstruction," in *Open Problems in Mathematical Systems and Control Theory* (V. D. Blondel, E. D. Sontag, M. Vidyasagar, and J. C. Willems, eds.), Springer-Verlag, London, pp. 277–279.

Copyright © IFAC Periodic Control Systems,
Cernobbio-Como, Italy, 2001

TWO APPLICATIONS FOR HYBRID $\mathcal{H}_\infty$-CONTROL: GENERALISED SAMPLED-DATA AND LOOP-SHAPING

A.-K. Christiansson * B. Lennartson ** H.T. Toivonen ***

* *Univ Trollh/Uddev, Box 957, SE-46129 Trollhättan, Sweden*
** *Chalmers Univ of Technology, SE-41296 Gothenburg, Sweden*
*** *Åbo Akademi University, FIN-20500 Åbo, Finland*

Abstract: A hybrid general $\mathcal{H}_\infty$-framework is presented, which can be applied to most linear control schemes, such as sampled-data control of time-varying systems, multirate sampling, mixed continuous- and discrete-time measurements and/or control signals as well as disturbance and performance signals. In infinite-time, the framework is specially emphasised on periodic systems. First necessary notations are introduced and the general hybrid framework is presented, and then the framework is applied to two different time-varying (periodic) systems. One is a generalised sampled-data system with no continuous-time measurements and control signals, however hybrid otherwise. The other corresponds to the well-known Glover-McFarlane loop-shaping design, with the advantage that there is no need for γ-iterations.
Copyright © 2001 IFAC

Keywords: H-infinity control, Periodic structures, Sampled-data control, Loop shaping, Hybrid systems

1. INTRODUCTION

$\mathcal{H}_\infty$-control of continuous-time and discrete-time systems are by now well established, see e.g. (Green and Limebeer, 1995). In the literature over all the continuous-time and discrete-time solutions are treated separately, and the discrete-time solutions seem to be more complicated than the continuous-time ones. Mostly the literature treats simplifying systems, e.g. $D_{11} = 0$, cross terms being zero, etc. In (Christiansson, 2000; Christiansson *et al.*, 2000) a hybrid $\mathcal{H}_\infty$-solution is presented for time-varying systems which considers both continuous-time and discrete-time solutions in the same framework, and no simplifying assumptions on the system is introduced. The pure continuous-time and discrete-time solutions are simply two different interpretations of the general result. This $\mathcal{H}_\infty$-solution can be applied to any linear system, even to multirate-sampling with mixed continuous/discrete-time performance. In this paper the general hybrid solution is first presented, and then applied to the sampled-data case and to a situation which does not need any γ-iteration. This corresponds to what mostly is referred to as Glover-McFarlanes loop-shaping method, see e.g. (Glover and McFarlane, 1989; McFarlane and Glover, 1992). A lifting procedure is generalised to take into consideration intersample behaviour for periodic situations. This approach is found in time-invariant sampled-data solutions in e.g. (Sun *et al.*, 1993; Toivonen and Sågfors, 1997).

The results in this paper are generalisations of known results in the literature on pure continuous-time, pure discrete-time, sampled-data and loop shaping situations, see e.g. (Green and Limebeer, 1995; Chen and Francis, 1991; McFarlane and Glover, 1992).

The paper is organised as follows: The unified hybrid output feedback $\mathcal{H}_\infty$-solution is presented in Section 2 starting with some useful notations introduced in Section 2.1. This general solution is then applied to the sampled-data case with no continuous-time measurements or control signals in Section 3, and to the loop shaping case in Section 4. Finally the results are summarised in Section 5. Detailed proofs are found in (Christiansson, 2000) as generalisations of known results in e.g. (Ravi *et al.*, 1991; Green and Limebeer, 1995; Toivonen and Sågfors, 1997).

2. THE HYBRID $\mathcal{H}_\infty$-SOLUTION

The general hybrid $\mathcal{H}_\infty$-solution was derived in (Christiansson, 2000), presented in (Christiansson *et al.*, 2000) and briefly below. In order to give a compact solution some hybrid notations are first introduced. In this paper the term hybrid is used for mixed continuous- and discrete-time signals and systems, and not for discrete-event systems, which nowadays often also is denoted hybrid.

2.1 *Hybrid notations*

Sampling time instants are considered at $t = t_k$, $k = 1, 2, \ldots$, and t_k^+ is the time instant just after the sampling at t_k. Similarly, t_{k+1}^- is the time instant just before t_{k+1}. The length of the sampling interval is $h_k = t_{k+1} - t_k$. Note that this allows for varying sampling intervals. Consider the general hybrid system parts with μ_c/μ_d as inputs and η_c/η_d as outputs, with $\mathcal{G}_c$ for $t \neq t_k$ and $\mathcal{G}_d$ for $t = t_k$:

$$\mathcal{G}_c : \begin{cases} \dot{x}(t) = A_c(t)x(t) + B_c(t)\mu_c(t) \\ \eta_c(t) = C_c(t)x(t) + D_c(t)\mu_c(t) \end{cases} \quad \text{(1a)}$$

$$\mathcal{G}_d : \begin{cases} x(t_k^+) = A_d(t_k)x(t_k^-) + B_d(t_k)\mu_d(t_k) \\ \eta_d(t_k) = C_d(t_k)x(t_k^-) + D_d(t_k)\mu_d(t_k) \end{cases} \quad \text{(1b)}$$

When no index $_c$ or $_d$ is used, the signals and matrices are assumed to be hybrid, i.e. either continuous-time when $t \neq t_k$, or discrete-time when $t = t_k$. Let the input signal μ be composed of the disturbance signal w and the control signal u as $\mu = [w' \;\; u']'$. Similarly, let the output signal η be composed of the performance signal z and the measured output y as $\eta = [z' \;\; y']'$. The system is now in the standard form for $\mathcal{H}_\infty$-applications. A hybrid signal is composed of both discrete-time and continuous-time contributions, and its size is expressed as (for z):

$$||z||^2_{[0,T_f]} = \int_0^{T_f} z_c'(t)z_c(t)dt + \sum_{t_k \in [0,T_f]} z_d'(t_k)z_d(t_k) \quad (2)$$

Note that z_c in (2) is not formally defined at times t_k. However, the time limits exist, and the integral can be considered as a sum of integrals. Variables, that are solutions to a mix of differential- and difference equations, appear as piece-wise continuous variables with jumps. To describe differential- and difference-equations for such variables similarly, a forward/backward notation is introduced as

$$\text{forward} \quad x^+ = \begin{cases} \dot{x}(t), & t \neq t_k \\ \lim\limits_{\epsilon \to 0} x(t_k + \epsilon) = x(t_k^+) \end{cases} \quad \text{(3a)}$$

$$\text{backward} \quad x^- = \begin{cases} -\dot{x}(t), & t \neq t_k \\ \lim\limits_{\epsilon \to 0} x(t_k - \epsilon) = x(t_k^-) \end{cases} \quad \text{(3b)}$$

where $\epsilon \in \mathbb{R}^+$. The systems in (1) can thus be considered as the hybrid system

$$\mathcal{G} : \begin{bmatrix} x^+ \\ z \\ y \end{bmatrix} = \begin{bmatrix} A & B_w & B_u \\ C_z & D_{zw} & D_{zu} \\ C_y & D_{yw} & D_{yu} \end{bmatrix} \begin{bmatrix} x \\ w \\ u \end{bmatrix} \quad (4)$$

where the B-, C- and D-matrices in (1) are partitioned accordingly. Note that the absence of time notations in systems and signals is just for simpler notations; the system may be time-varying throughout this paper. The state-transition matrix $\Pi_A(t,s)$ for the hybrid system $x^+ = Ax$ from time s to t, is defined by

$$\begin{cases} \dot{\Pi}_A(t,s) = A_c(t)\Pi_A(t,s), & t \neq t_k \\ \Pi_A(t_k^+, s) = A_d(t_k)\Pi_A(t_k^-, s) \end{cases} \quad (5)$$

The hybrid system A is exponentially stable, if there exist positive real numbers c_1, c_2, such that $||\Pi_A(t,s)|| \leq c_1 e^{-c_2(t-s)}, 0 \leq s \leq t$. This is also denoted as the system A being stable. The hybrid system $\mathcal{G}$ in (4) is said to be stabilisable (detectable), if there exist a bounded hybrid matrix L (K), such that the system $A - BL$ ($A - KC$) is stable. When the system is periodic with period T_p, the system matrices are repeated periodically, e.g. $A(t + \ell T_p) = A(t)$.

The $\mathcal{H}_\infty$-control problem can now be expressed: ***Find an output feedback controller $u = \mathcal{K}(y)$, such that the bound on the induced norm of the hybrid closed system from w to z***

$$||\mathcal{G}_{zw}||_{[0,T_f]} = \sup_{||w||_{[0,T_f]} \neq 0} \frac{||z||_{[0,T_f]}}{||w||_{[0,T_f]}} < \gamma \quad (6)$$

holds for a specified constant $\gamma > 0$. The supremum is taken over all w in $\mathcal{L}_{2[0,T_f]} \oplus l_{2[1,N]}$ such that $||w||_{[0,T_f]} \neq 0$. In infinite-time horizon, i.e. when $T_f \to \infty$, the controller shall also be stabilising.

Note that this problem statement implies that z and w can contain both continuous- and discrete-time signals and that she system (4) may be time-varying. In infinite-time horizon the solution is emphasised on periodic systems, such that a periodic pattern consisting of a number of sampling steps will be repeated. The sampling may be multirate, i.e. different signals may be sampled by different sampling periods, however as a subset of the periodically repeated sampling pattern.

There is a need for some more notations to be able to present the hybrid solution in a compact form. First define a hybrid matrix notation $\diamond$, suitable for Lyapunov- and Riccati equations, according to

$$A \diamond P = \begin{cases} A_c P + P A_c', & t \neq t_k \\ A_d P A_d', & t = t_k \end{cases} \tag{7}$$

Then introduce the notations δ_{t_k}, A_{t_k} that allow us to express some continuous- and discrete-time matrices in a unified form:

$$\delta_{t_k} = \begin{cases} 0, t \neq t_k \\ 1, t = t_k \end{cases}; \quad A_{t_k} = \delta_{t_k} A + (1 - \delta_{t_k}) I \tag{8}$$

See (Christiansson *et al.*, 2000) for more details on the hybrid notation.

2.2 *The hybrid output feedback $\mathcal{H}_\infty$-solution*

The design method is well known, based on the solutions of two Riccati equations, one $S(t)$ for the static feedback case, and one $P(t)$ for the filter. To achieve the compact hybrid solution, introduce a number of useful matrices. These are based on the hybrid system $\mathcal{G}$ in (4), the piece-wise continuous matrices S and P, the matrices $Q_\gamma = \mathrm{diag}(I_{n_w}, 0_{n_u}), R_\gamma = \mathrm{diag}(I_{n_z}, 0_{n_y})$, and the notations δ_{t_k}, A_{t_k}:

$$Q_\mu = D_z' D_z - \gamma^2 Q_\gamma + \delta_{t_k} B' S B \equiv \begin{bmatrix} Q_w & Q_{wu} \\ Q_{uw} & Q_u \end{bmatrix} \tag{9a}$$

$$Q_{\mu x} = D_z' C_z + B' S A_{t_k} \equiv \begin{bmatrix} Q_{wx} \\ Q_{ux} \end{bmatrix} \tag{9b}$$

$$R_\eta = D_w D_w' - \gamma^2 R_\gamma + \delta_{t_k} C P C' \equiv \begin{bmatrix} R_z & R_{zy} \\ R_{yz} & R_y \end{bmatrix} \tag{9c}$$

$$R_{x\eta} = B_w D_w' + A_{t_k} P C' \equiv \begin{bmatrix} R_{xz} & R_{xy} \end{bmatrix} \tag{9d}$$

where the system matrix partitions are obvious, e.g. $D_w = [D_{zw}' \;\; D_{yw}']'$. Note that the δ_{t_k}-notation implies that the matrices in (9) are different in continuous and discrete time. Furthermore, define the hybrid matrices

$$\hat{Q}_w = Q_{wu} Q_u^{-1} Q_{uw} - Q_w \tag{10a}$$

$$\hat{R}_z = R_{zy} R_y^{-1} R_{yz} - R_z \tag{10b}$$

To obtain the output feedback solution, use a signal transformation ending up in a system $\bar{\mathcal{G}}$:

$$\begin{bmatrix} x^+ \\ \bar{u} \\ y \end{bmatrix} = \begin{bmatrix} A + B_w L_w & \gamma B_w \hat{Q}_w^{-1/2} & B_u \\ Q_u^{1/2} \bar{L}_u & \gamma Q_u^{1/2} L_{uw} \hat{Q}_w^{-1/2} & Q_u^{1/2} \\ C_y + D_{yw} L_w & \gamma D_{yw} \hat{Q}_w^{-1/2} & D_{yu} \end{bmatrix} \begin{bmatrix} x \\ \bar{w} \\ u \end{bmatrix}$$
$$\equiv \begin{bmatrix} \bar{A} & \bar{B}_{\bar{w}} & B_u \\ \bar{C}_{\bar{u}} & \bar{D}_{\bar{u}\bar{w}} & Q_u^{1/2} \\ \bar{C}_y & \bar{D}_{y\bar{w}} & D_{yu} \end{bmatrix} \begin{bmatrix} x \\ \bar{w} \\ u \end{bmatrix} \tag{11a}$$

where

$$\bar{L}_u = L_u + L_{uw} L_w \tag{11b}$$

$$L_w = \hat{Q}_w^{-1}(Q_{wx} - Q_{wu} Q_u^{-1} Q_{ux}) \tag{11c}$$

$$L_u = Q_u^{-1} Q_{ux}, \quad L_{uw} = Q_u^{-1} Q_{uw} \tag{11d}$$

Define matrices for the system (11a) in analogy with (9c)-(9d):

$$\bar{R}_\eta = \bar{D}_{\bar{w}} \bar{D}_{\bar{w}}' - \gamma^2 R_\gamma + \delta_{t_k} \bar{C} \bar{P} \bar{C}' \equiv \begin{bmatrix} \bar{R}_{\bar{u}} & \bar{R}_{\bar{u}y} \\ \bar{R}_{y\bar{u}} & \bar{R}_y \end{bmatrix} \tag{12a}$$

$$\bar{R}_{x\eta} = \bar{B}_{\bar{w}} \bar{D}_{\bar{w}}' + \bar{A}_{t_k} \bar{P} \bar{C}' \equiv \begin{bmatrix} \bar{R}_{x\bar{u}} & \bar{R}_{xy} \end{bmatrix} \tag{12b}$$

Here $\bar{A}_{t_k}$ is defined for $\bar{A}$ as was A_{t_k} for A. The general hybrid solution is now summarised in a theorem. This is a generalisation of known results for different applications, see e.g. (Green and Limebeer, 1995; Ravi *et al.*, 1992; Sivashankar and Khargonekar, 1994).

Theorem 1. Consider the hybrid time-varying system (4) with $D_{zu}' D_{zu} > 0, D_{yw} D_{yw}' > 0$. Then

- there exists a hybrid controller $u = \mathcal{K}(y)$ on $t \in [0, T_f]$, which achieves the bound (6)

if and only if ($\Leftrightarrow$)

- there exist hybrid piece-wise symmetric continuous matrix functions $S \geq 0$, $P \geq 0$ satisfying the hybrid Riccati equations

$$S^- = A' \diamond S + C_z' C_z - L' Q_\mu L \tag{13a}$$

$$P^+ = A \diamond P + B_w B_w' - K R_\eta K' \tag{13b}$$

  where $L = Q_\mu^{-1} Q_{\mu x}, K = R_{x\eta} R_\eta^{-1}$ are found from (9). Necessary conditions for existence of bounded solutions S, P are $\hat{Q}_w > 0, Q_u > 0, \hat{R}_z > 0, R_y > 0$.
- the spectral radius $\rho(SP) < \gamma^2$

One controller, that achieves the bound is given by

$$\hat{x}^+ = \bar{A}\hat{x} + B_u u + \bar{K}_y (y - \hat{y}) \tag{14a}$$

$$\hat{y} = \bar{C}_y \hat{x} + D_{yu} u \tag{14b}$$

$$u = -\bar{L}_u \hat{x} - \bar{K}_{uy}(y - \hat{y}) \tag{14c}$$

where $\bar{A}$ and $\bar{C}_y$ are defined in (11a), $\bar{L}_u$ is defined in (11b), $\bar{K}_y = \bar{R}_{xy} \bar{R}_y^{-1}, \bar{K}_{uy} = Q_u^{-1/2} \bar{R}_{\bar{u}y} \bar{R}_y^{-1}$ with $\bar{R}_{xy}, \bar{R}_y$ and $\bar{R}_{\bar{u}y}$ defined in (12) and $\bar{P} = P(I - \gamma^{-2} S P)^{-1}$. □

Now pure continuous-time and discrete-time solutions are just different interpretations of this hybrid result. Note that the matrix $D_{zw} \neq 0$ here, but it is set to zero in most other presented solutions. Theorem 1 can be applied in infinite time horizon if additional detectability and stabilisability conditions are added, see (Christiansson *et al.*, 2000). Two applications of this general result are given below, however briefly, due to lack of space.

3. THE GENERALISED SAMPLED-DATA $\mathcal{H}_\infty$-SOLUTION

Compared to the general hybrid situation in Section 2.2 the generalised sampled-data case is the case with no continuous-time measurements y_c and no continuous-time control signals u_c, however there are no restrictions introduced in discrete time. This implies that some simplifications can be introduced. The intersample behavior is achieved by applying Theorem 1 in continuous time between the sampling instants. However this behaviour is preferably solved in a "lifted" (or discretised) system. For this introduce the continuous-time matrices

$$A_{\mu_c} = A_c + \gamma^{-2} B_{w_c}(I - \gamma^{-2} D'_{zw_c} D_{zw_c})^{-1} D'_{zw_c} C_{z_c}$$
$$B_{\eta_c} B'_{\eta_c} = B_{w_c}(I - \gamma^{-2} D'_{zw_c} D_{zw_c})^{-1} B'_{w_c}$$
$$C'_{\mu_c} C_{\mu_c} = C'_{z_c}(I - \gamma^{-2} D_{zw_c} D'_{zw_c})^{-1} C_{z_c}$$
$$\mathcal{H}_\mu = \begin{bmatrix} A_{\mu_c} & \gamma^{-2} B_{\eta_c} B'_{\eta_c} \\ -C'_{\mu_c} C_{\mu_c} & -A'_{\mu_c} \end{bmatrix} \quad (15)$$

As argued in (Christiansson, 2000), some matrices need to be nonsingular for solutions S and P to exist. In this special case this requires that $\rho(D'_{zw_c} D_{zw_c}) < \gamma^2$, which is always fulfilled for the often used simplification $D_{zw} = 0$. Let $\Pi(t,s)$ be the transition matrix associated with $\mathcal{H}_\mu$ from s to t, i.e.

$$\dot{\Pi}(t,s) = \mathcal{H}_\mu \Pi(t,s), \qquad \Pi(t,t) = I \quad (16)$$

Partition Π into sub matrices of size $n_x \times n_x$, and study two consecutive sampling instants t_k, t_{k+1}:

$$\Pi(t_k^+, t_{k+1}^-) = \begin{bmatrix} \Pi_{11}(t_k^+, t_{k+1}^-) & \Pi_{12}(t_k^+, t_{k+1}^-) \\ \Pi_{21}(t_k^+, t_{k+1}^-) & \Pi_{22}(t_k^+, t_{k+1}^-) \end{bmatrix} \quad (17)$$

Now a "lifting" theorem is presented that accounts for the intersample behaviour.

Theorem 2. Consider system (4) with no continuous-time measurements and no continuous-time control signal between the sampling instants t_k and t_{k+1}. The discretised system model

$$\begin{bmatrix} x(t_{k+1}^-) \\ \tilde{z}(t_k, h_k) \end{bmatrix} = \begin{bmatrix} \tilde{A} & \tilde{B}_{\tilde{w}} \\ \tilde{C}_{\tilde{z}} & 0 \end{bmatrix} \begin{bmatrix} x(t_k^+) \\ \tilde{w}(t_k, h_k) \end{bmatrix} \quad (18a)$$

where (time arguments in left-hand-side is (t_k, h_k))

$$\tilde{A} = \Pi_{11}^{-1}(t_k^+, t_{k+1}^-) \quad (18b)$$
$$\tilde{B}_{\tilde{w}} \tilde{B}'_{\tilde{w}} = -\gamma^2 \Pi_{11}^{-1}(t_k^+, t_{k+1}^-) \Pi_{12}(t_k^+, t_{k+1}^-) \quad (18c)$$
$$\tilde{C}'_{\tilde{z}} \tilde{C}_{\tilde{z}} = \Pi_{21}(t_k^+, t_{k+1}^-) \Pi_{11}^{-1}(t_k^+, t_{k+1}^-) \quad (18d)$$

generates the same static feedback and filter Riccati equations as the continuous-time system does at time instants t_k^+ and t_{k+1}^- respectively, see (13). These Riccati equations are

$$S(t_k^+) = \tilde{A}' S(t_{k+1}^-) \tilde{\Psi}_{\tilde{w}} \tilde{A} + \tilde{C}'_{\tilde{z}} \tilde{C}_{\tilde{z}} \quad (19a)$$
$$P(t_{k+1}^-) = \tilde{A} \tilde{\Psi}_{\tilde{z}} P(t_k^+) \tilde{A}' + \tilde{B}_{\tilde{w}} \tilde{B}'_{\tilde{w}} \quad (19b)$$

where

$$\tilde{\Psi}_{\tilde{w}}(t_k, h_k) = (I - \gamma^{-2} \tilde{B}_{\tilde{w}} \tilde{B}'_{\tilde{w}} S(t_{k+1}^-))^{-1} \quad (19c)$$
$$\tilde{\Psi}_{\tilde{z}}(t_k, h_k) = (I - \gamma^{-2} P(t_k^+) \tilde{C}'_{\tilde{z}} \tilde{C}_{\tilde{z}})^{-1} \quad (19d)$$

The intersample state transition, when the disturbance is the worst one w_c^*, is expressed by Π_{A-BL} for the static feedback solution, and Γ_{A-KC} for the filter:

$$\Pi_{A-BL}(t_{k+1}^-, t_k^+) = \tilde{\Psi}_{\tilde{w}}(t_k, h_k) \tilde{A}(t_k, h_k) \quad (20a)$$
$$\Gamma_{A-KC}(t_{k+1}^-, t_k^+) = \tilde{A}(t_k, h_k) \tilde{\Psi}_{\tilde{z}}(t_k, h_k) \quad (20b)$$

□

The discrete-time updates, reflecting the sampling instants, are achieved from applying Theorem 1 in discrete time, by simply adding indices $_d$ to all signals and systems. In discrete time there are no system restrictions introduced. Let the controller (14) in discrete time be expressed by the state-space model:

$$\mathcal{K}_d: \quad \begin{bmatrix} \hat{x}(t_k^+) \\ u_d(t_k) \end{bmatrix} = \begin{bmatrix} \hat{A}_d & \hat{B}_d \\ \hat{C}_d & \hat{D}_d \end{bmatrix} \begin{bmatrix} \hat{x}(t_k^-) \\ y_d(t_k) \end{bmatrix} \quad (21)$$

The results are summarised in a theorem:

Theorem 3. Consider the situation in Theorem 1, with neither control signals nor measurements in continuous time. Assume that the discretised Riccati equations reflecting the intersample behaviour together with the discrete-time Riccati equations at the sampling updates have symmetric solutions $S \geq 0, P \geq 0$. One sampled-data controller that achieves the bound (6) is

$$\hat{x}(t_{k+1}^-) = \Pi_{A-BL}(t_{k+1}^-, t_k^+) \hat{x}(t_k^+) \quad (22a)$$
$$\hat{x}(t_k^+) = \hat{A}_d \hat{x}(t_k^-) + \hat{B}_d y_d(t_k) \quad (22b)$$
$$u_d(t_k) = \hat{C}_d \hat{x}(t_k^-) + \hat{D}_d y_d(t_k) \quad (22c)$$

□

The discrete-time state transitions for the static feedback and filter cases can be expressed as

$$\Pi_{A-BL}(t_k^+, t_k^-) = A_d - B_d L_d \quad (23a)$$
$$\Gamma_{A-KC}(t_k^+, t_k^-) = A_d - K_d C_d \quad (23b)$$

The state transition between two sampling instants thus contains both the continuous- and the discrete-time transitions:

$$\Pi_{A-BL}(t_{k+1}^-, t_k^-) = \Pi_{A-BL}(t_{k+1}^-, t_k^+)(A_d - B_d L_d)$$
$$\Gamma_{A-KC}(t_{k+1}^-, t_k^-) = \Gamma_{A-KC}(t_{k+1}^-, t_k^+)(A_d - K_d C_d)$$

To assure stability in infinite-time horizon, additional demands must be put on the system, as in the general case, cf. (Christiansson, 2000). For a periodic case, with a period over r sampling steps, the transitions $\Pi_{A-BL}(t^-_{k+r}, t^-_k)$ and $\Gamma_{A-KC}(t^-_{k+r}, t^-_k)$ must have all their eigenvalues inside the open unit disc, where

$$\begin{aligned}
&\Pi_{A-BL}(t^-_{k+r}, t^-_k) = \\
&= \Pi_{A-BL}(t^-_{k+r}, t^-_{k+r-1}) \cdots \Pi_{A-BL}(t^-_{k+1}, t^-_k) \quad (24a)\\
&\Gamma_{A-KC}(t^-_{k+r}, t^-_k) = \\
&= \Gamma_{A-KC}(t^-_{k+r}, t^-_{k+r-1}) \cdots \Gamma_{A-KC}(t^-_{k+1}, t^-_k) \quad (24b)
\end{aligned}$$

Remark: When there is also a hold circuit at the controller output and the system is continuous-time and time-invariant, well-known results follow as in e.g. (Sun *et al.*, 1993; Toivonen, 1995; Toivonen and Sågfors, 1997).

4. A γ-FREE $\mathcal{H}_\infty$-SOLUTION

A system that can be described by the special structure

$$\mathcal{G}: \begin{bmatrix} x^+ \\ z \\ y \end{bmatrix} = \begin{bmatrix} A & [0\ B_u] & B_u \\ \begin{bmatrix} C_y \\ 0 \end{bmatrix} & \begin{bmatrix} I & 0 \\ 0 & 0 \end{bmatrix} & \begin{bmatrix} 0 \\ I \end{bmatrix} \\ C_y & [I\ 0] & 0 \end{bmatrix} \begin{bmatrix} x \\ \begin{bmatrix} e \\ v \end{bmatrix} \\ u \end{bmatrix} \quad (25)$$

turns out to give γ-free solutions. One reason for the simplifications that can be made lies in the way disturbances are introduced, i.e. $w = [e' \ v']'$. This is partly illustrated in Fig. 1. Furthermore the performance signal is composed as $z = [y' \ u']'$. Note that this implies that B_u is used for both u and w input signals, and C_y both for z and y output signals! For simpler notations, let $B = B_u, C = C_y$. Note that $D_{yu} = 0$ in (25) for simplicity. Furthermore, note that the notations B, C here are **not** identical to those in former sections (for simpler notations).

The same phenomenon, i.e. a γ-free solution, appears for the loop-shaping technique that was proposed by Glover and Mc Farlane, see (Glover and McFarlane, 1989; McFarlane and Glover, 1992), as an $\mathcal{H}_\infty$-method to avoid γ-iteration. The idea lies in the way weighting filters for achieving desired closed-loop properties are placed. For loop-shaping, the filters are placed *in* the loop, as opposed to standard methods, where they can be placed "anywhere".

Applying Theorem 1 on the hybrid system (25) results in Riccati equations, cf. (13), for S and P. Now introduce

$$\nu = 1 - \gamma^{-2}; \qquad S_\nu = \nu S \quad (26)$$

and the static feedback and filter Riccati equations can be expressed in hybrid form as

$$\begin{aligned}
S^-_\nu &= A' \diamond S_\nu + C'C - L'_u Q_u L_u \quad (27a)\\
P^+ &= A \diamond P + BB' - K_y R_y K'_y \quad (27b)
\end{aligned}$$

with Q_u, R_y in (9), L_u in (11d) and $K_y = R_{xy}R_y^{-1}$. Note that none of these equations are depending on γ, so there is no need for γ-iteration when solving the Riccati equations! The solution to the output feedback problem $\bar{P}$, cf. Theorem 1, can be calculated from these solutions. Introducing $\bar{P}_\nu = \nu^{-1}\bar{P}$ this solution can be expressed as

$$\bar{P}_\nu = P(I - \gamma^{-2}(I + S_\nu P))^{-1} \quad (28)$$

As can be seen in (28) this solution requires

$$\gamma^2 > 1 + \rho(S_\nu P) \quad (29)$$

Furthermore, introduce the hybrid gain matrices

$$\begin{aligned}
L_\nu &= (I + \delta_{t_k} B' S_\nu B)^{-1} B' S_\nu A_{t_k} \quad (30a)\\
K_\nu &= A_{t_k} \bar{P}_\nu C' (I + \delta_{t_k} C \bar{P}_\nu C')^{-1} \quad (30b)\\
K_{\nu y} &= \delta_{t_k} L_\nu \bar{P}_\nu C' (I + \delta_{t_k} \bar{P}_\nu C')^{-1} \quad (30c)
\end{aligned}$$

Now there are enough preliminaries to summarise the results:

Theorem 4. Consider the hybrid time-varying system (25). Then

- there exists a hybrid controller $u = \mathcal{K}(y)$ on $t \in [0, T_f]$, which achieves the bound $\|\mathcal{G}_{zw}\|_\infty < \gamma$

if and only if ($\Leftrightarrow$)

- there exist hybrid piece-wise continuous matrix functions $S_\nu \geq 0$, $P \geq 0$ satisfying the hybrid REs (27).
- the spectral radius $\rho(S_\nu P) < \gamma^2 - 1$.

One controller, that achieves the bound is given by

$$\begin{aligned}
\hat{x}^+_\nu &= A\hat{x}_\nu + Bu + K_\nu(y - \hat{y}) \quad (31a)\\
\hat{y} &= C\hat{x}_\nu \quad (31b)\\
u &= -L_\nu \hat{x}_\nu - K_{\nu y}(y - \hat{y}) \quad (31c)
\end{aligned}$$

where $L_\nu, K_\nu, K_{\nu y}$ are defined in (30) and $\bar{P}_\nu$ in (28). □

The performance problem for the system (25) can be considered as a robustness problem with multiplicative uncertainties in input and output respectively as in Fig. 1. The loop-shaping system discussed in (McFarlane and Glover, 1992) considered the situation in Fig. 2. When the plant is normalised coprime factored with $\mathcal{G} = M^{-1}N$ the problems are identical. Thus this paper claims that the presented solution is valid also for this loop shaping setup. In fact, for the time-invariant

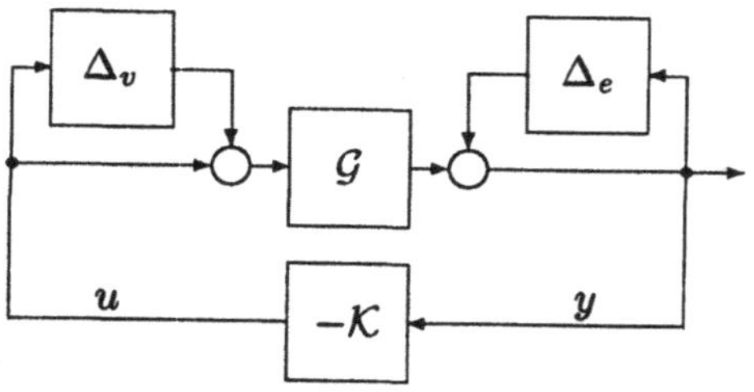

Fig. 1. Multiplicative uncertainties; $y = (I - \Delta_e)^{-1}\mathcal{G}(I + \Delta_v)u$

case it was mentioned in a remark in (McFarlane and Glover, 1990).

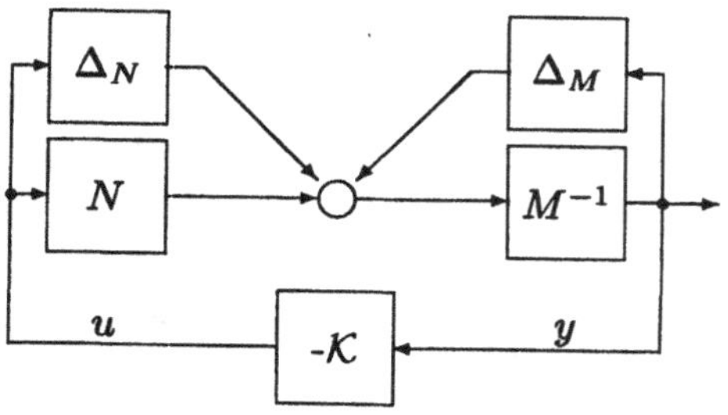

Fig. 2. Glover-McFarlane plant structure with additive uncertainties; $y = (M - \Delta_M)^{-1}(N + \Delta_N)u$

5. CONCLUSION

A general mixed continuous-time and discrete-time $\mathcal{H}_\infty$-solution for linear time-varying systems is presented. As opposed to the literature this shows that continuous-time and discrete-time solutions are closely related in that they are simply two different interpretations of the general result.

The general results are applied to the simpler generalised sampled-data and loop shaping cases. By generalised sampled-data is meant the situation when there is no continuous-time control signals and no continuous-time measurements, however there are no restrictions introduced to she system in discrete time. For the periodic case, state-transition matrices are derived covering the closed system behaviour over the whole period. This is essential for the evaluation of stability in the $\mathcal{H}_\infty$-synthesis procedure.

The loop shaping case is a special case that does not need any γ-iteration for achieving the (sub)optimal controller. It is shown how a multiplicative input/output uncertainty plant model gives γ-free Riccati equations, however a lowest achievable γ-value is presented. This method does not require any normalised coprime factorisation of the plant uncertainty model as is the case in the Glover-McFarlane method where additive uncertainties are used.

Acknowledgements The work is partly financed by the Swedish NFFP-programme.

6. REFERENCES

Chen, T. and B.A. Francis (1991). Input-output stability of sampled-data systems. *IEEE Trans on Automatic Control* **36**(1), 50–58.

Christiansson, A.-K. (2000). A General Framework for Hybrid $\mathcal{H}_\infty$-Control. Licentiate thesis. Chalmers University of Technology, Control and Automation Laboratory.

Christiansson, A.-K., B. Lennartson and H.T. Toivonen (2000). Continuous- discrete- and sampled-data $\mathcal{H}_\infty$ control - a unified framework. In: *CDC39*. Sydney, Australia. pp. 1936–1941.

Glover, K. and D.C. McFarlane (1989). Robust stabilization of normalized coprime factor plant description with $\mathcal{H}_\infty$-bounded uncertainty. *IEEE Trans. on Automatic Control* **34**(8), 821–830.

Green, M. and D.J.N. Limebeer (1995). *Linear Robust Control*. Prentice Hall, Englewood Cliffs, N. J.

McFarlane, D.C. and K. Glover (1990). *Robust Controller Design Using Normalized Coprime Factor Plant Descriptions*. Springer-Verlag.

McFarlane, D.C. and K. Glover (1992). A loop shaping design procedure using $\mathcal{H}_\infty$-synthesis. *IEEE Trans on Automatic Control* **37**(6), 759–769.

Ravi, R., A.M. Pascoal and P.P. Khargonekar (1992). Normalized coprime factorization for time-varying systems. *Systems and Control Letters* **18**, 455–465.

Ravi, R., K.M. Nagpal and P.P. Khargonekar (1991). $\mathcal{H}_\infty$ control of linear time-varying systems: A state-space approach. *SIAM J. Control Optim.* **29**(6), 1394–1413.

Sivashankar, N. and P.P. Khargonekar (1994). Characterization of the $\mathcal{L}_2$-induced norm for linear systems with jumps with applications to sampled-data systems. *SIAM J. Control Optim.* **32**(4), 1128–1150.

Sun, W., K.M. Nagpal and P.P. Khargonekar (1993). H_∞ control and filtering for sampled-data systems. *IEEE Trans. on Automatic Control* **38**(8), 1162–1175.

Toivonen, H.T. (1995). Digital control with $\mathcal{H}_\infty$ optimality criteria. In: *Control & Dynamic Systems* (C.T. Leondes, Ed.). pp. 215–262. Academic Press.

Toivonen, H.T. and M.F. Sågfors (1997). The sampled-data $\mathcal{H}_\infty$ problem: A unified framework for discretization based methods and Riccati-equation solution. *Int. J. Control* **66**, 289–309.

Copyright © IFAC Periodic Control Systems,
Cernobbio-Como, Italy, 2001

ROBUST STABILIZATION OF PERIODIC AND MULTIRATE SYSTEMS WITH GAP AND ν-GAP UNCERTAINTY

Kan Tan * **Li Chai** ** **Li Qiu** **,[1]

* *Department of Mechanical Engineering*
University of Houston, Houston, TX 77204 USA

** *Department of Electrical and Electronic Engineering*
Hong Kong University of Science and Technology
Clear Water Bay, Kowloon, Hong Kong, China

Abstract: In this paper we present a state space solution to the robust stabilization problem of general discrete-time periodic and multirate systems where the uncertainty is described in terms of the gap and ν−gap metrics. This robust stability problem is converted to a constrained $\mathcal{H}_\infty$ optimal control problem by using the lifting technique. Then the optimal robust stability margin is explicitly computed and a method is provided to get the controllers satisfying the optimal robust stabilization margin. The solution amounts to solving two discrete-time algebraic Riccati equations and an extended Parrot problem. *Copyright ©2001 IFAC*

Keywords: multirate systems, periodic systems, robust stabilization, gap metric, ν−gap metric

1. INTRODUCTION

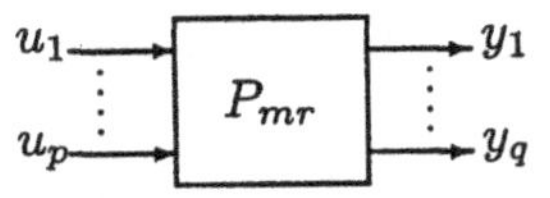

Fig. 1. A general multirate system.

A general multivariable discrete-time multirate system is depicted in Figure 1. Here the signals $u_1, \ldots, u_p$ and $y_1, \ldots, y_q$ are discrete-time signals with different sampling rates. Such a multirate system can result from sampling an analog system using multirate samplers and holds or can appear as it is in some special applications. In our study, we assume that this system is linear and causal, and satisfies certain periodic property. Because of this, it can be converted to an equivalent LTI system using the so-called lifting or blocking technique (Qiu and Chen, 1994; Ravi, *et al.*, 1990; Voulgaris, *et al.*, 1994). Hence the analysis and design techniques for LTI systems can be applied to such a multirate system. However, it has been known that the lifting results in a peculiar constraint on the equivalent LTI system, due to the causality of the original system. When solving a design problem, such as the robust stabilization problem considered in this paper, extra effort is needed to make sure that the designed multirate system is causal.

Our formulation of a multirate system includes a single rate periodic system as a special case. Periodic systems occur naturally in many applications and can be intentionally used to achieve something that time invariant system can not. See a recent survey paper (Bittanti and Colaneri, 2000) and the references therein. In this paper, periodic and multirate systems are treated in a unified framework by using the lifting technique. A related study was presented in (Iglesias, 2000) in which a method to design a strictly proper controller for the discrete-

[1] This work is supported by Hong Kong Research Grant Council under HKUST6054/99E. This work is completed when the third author was visiting the National Key Lab on Industrial Control Technology, Zhejiang University, HangZhou, Zhejiang, China.

time, normalized left-coprime factorization robust stabilization was given.

The first issue in robust control is the description of the uncertainty. The most natural way to describe system uncertainty is by using a metric in the set of all systems under consideration and an uncertain system is then simply a ball defined by this metric centered at a nominal system with certain radius. There are several metrics in the literature for this very purpose: gap metric (Georgiou and Khargonekar, 1987), pointwise gap metric (Qiu and Davison, 1992), ν-gap metric (Vinnicombe, 1993). In this paper, both the gap and ν-gap metrics are studied for multirate and periodic systems. We will see that the gap metric has an easy generization. However, it is not easy to extend the ν-gap metric to multirate and periodic systems following the method in (Vinnicombe, 1993). Here we generalize the treatment in (Wan and Huang, 2000) to define the ν-gap metric for multirate systems.

2. PRELIMINARY

Consider the multirate system P_{mr} showing in Figure 1. Assume that P_{mr} is linear and causal. Also assume that the signals $u_i, i = 1, 2, \cdots, p$, and $y_j, j = 1, 2, \cdots, q$, are synchronized at time zero. Let the sampling interval of u_i be $m_i h$ and that of y_j be $n_j h$, where h is a real number giving a time unit and m_i, n_j are integers. The overall input and output are denoted by

$$u = \begin{bmatrix} u_1 \cdots u_p \end{bmatrix}^T, y = \begin{bmatrix} y_1 \cdots y_q \end{bmatrix}^T.$$

respectively. Here u and y are vectors of signals with different sampling rates. Let l be a common multiple of m_i and n_j, $i = 1, \cdots, p$, $j = 1, \cdots, q$. Let $\bar{m}_i = l/m_i$ and $\bar{n}_j = l/n_j$. Denote the sets $\{m_i\}$ and $\{n_j\}$ by M and N respectively and the sets $\{\bar{m}_i\}$ and $\{\bar{n}_j\}$ by $\bar{M}$ and $\bar{N}$ respectively. Let U be the unit delay operator and let

$$\begin{aligned} U_{\bar{M}} &= \text{diag}\{U^{\bar{m}_1}, \dots, U^{\bar{m}_p}\}, \\ U_{\bar{N}} &= \text{diag}\{U^{\bar{n}_1}, \dots, U^{\bar{n}_q}\}. \end{aligned}$$

With the above notion, we finally assume that the system P_{mr} satisfies

$$P_{mr} U_{\bar{M}} = U_{\bar{N}} P_{mr}.$$

This property will be called the $(\bar{M}, \bar{N})$-shift-invariance. This multirate system is more general than what one can get from sampling an LTI analog system using a multirate sampling scheme. As an extreme case, if all m_i and n_j are equal to 1 and $l > 1$, then the multirate system is actually an l-periodic single-rate discrete-time system.

Define a lifting operator

$$L_m : \{\cdots | x(0), x(1), \cdots\} \mapsto$$
$$\left\{ \cdots \left| \begin{bmatrix} x(0) \\ x(1) \\ \vdots \\ x(m-1) \end{bmatrix}, \begin{bmatrix} x(m) \\ x(m+1) \\ \vdots \\ x(2m-1) \end{bmatrix}, \cdots \right. \right\}.$$

Its inverse L_m^{-1} is called a delifting operator. Let

$$\begin{aligned} L_{\bar{M}} &= \text{diag}\{L_{\bar{m}_1}, \dots, L_{\bar{m}_p}\} \\ L_{\bar{N}} &= \text{diag}\{L_{\bar{n}_1}, \dots, L_{\bar{n}_q}\}. \end{aligned}$$

Then the lifted system $P = L_{\bar{N}} P_{mr} L_{\bar{M}}^{-1}$ is an LTI system in the sense that $PU = UP$ since

$$\begin{aligned} PU &= L_{\bar{N}} P_{mr} L_{\bar{M}}^{-1} U = L_{\bar{N}} P_{mr} U_{\bar{M}} L_{\bar{M}}^{-1} \\ &= L_{\bar{N}} U_{\bar{N}} P_{mr} L_{\bar{M}}^{-1} = U L_{\bar{N}} P_{mr} L_{\bar{M}}^{-1} = UP. \end{aligned}$$

Hence it has transfer function $\hat{P}$ in λ-transform, i.e. replacing z in z-transform by $\frac{1}{\lambda}$. However, P is not an arbitrary LTI system, instead it is subject to a constraint that is resulted from the causality of P_{mr}. This constraint is best described using nest operators.

Let $\mathcal{X}$ be a finite dimensional vector space. A nest in $\mathcal{X}$, denoted $\{\mathcal{X}_i\}$, is a chain of subspaces in $\mathcal{X}$, including $\{0\}$ and $\mathcal{X}$, with the non-increasing ordering:

$$\mathcal{X} = \mathcal{X}_0 \supseteq \mathcal{X}_1 \supseteq \cdots \supseteq \mathcal{X}_{n-1} \supseteq \mathcal{X}_n = \{0\}.$$

Denote by $\mathcal{L}(\mathcal{X}, \mathcal{Y})$ the set of linear operators $\mathcal{X} \to \mathcal{Y}$ and abbreviate it as $\mathcal{L}(\mathcal{X})$ if $\mathcal{X} = \mathcal{Y}$. Assume that $\mathcal{X}$ and $\mathcal{Y}$ are equipped respectively with nest $\{\mathcal{X}_i\}$ and $\{\mathcal{Y}_i\}$ which have the same number of subspaces, say, $n+1$ as above. A linear map $T \in \mathcal{L}(\mathcal{X}, \mathcal{Y})$ is said to be a nest operator if

$$T\mathcal{X}_i \subseteq \mathcal{Y}_i, \; i = 0, 1, \cdots, n.$$

The set of all nest operators (with given nests) is denoted $\mathcal{N}(\{\mathcal{X}_i\}, \{\mathcal{Y}_i\})$ and abbreviated $\mathcal{N}(\{\mathcal{X}_i\})$ if $\{\mathcal{X}_i\} = \{\mathcal{Y}_i\}$.

Let us see how to characterize the causality constraint on P by using nest operators. Write $\underline{u} = L_{\bar{M}} u$, $\underline{y} = L_{\bar{N}} y$. Then

$$\begin{aligned} \underline{u}(0) &= [u_1(0) \cdots u_1((\bar{m}_1 - 1)m_1 h) \cdots \\ &\quad u_p(0) \cdots u_p((\bar{m}_p - 1)m_p h)]^T, \\ \underline{y}(0) &= [y_1(0) \cdots y_1((\bar{n}_1 - 1)n_1 h) \cdots \\ &\quad y_q(0) \cdots y_q((\bar{n}_q - 1)n_q h)]^T. \end{aligned}$$

Define for $k = 0, 1, \dots, l$,

$$\begin{aligned} \mathcal{U}_k &= \{\underline{u}(0) : u_i(rm_i h) = 0 \text{ if } rm_i h < kh\} \\ \mathcal{Y}_k &= \{\underline{y}(0) : y_j(rn_j h) = 0 \text{ if } rn_j h < kh\}. \end{aligned}$$

Then the lifted plant P will have

$$\hat{P}(0) \in \mathcal{N}(\{\mathcal{U}_r\}, \{\mathcal{Y}_r\}).$$

Now consider a linear causal $(\bar{M}, \bar{N})$-shift-invariant multirate system P_{mr}. The graph of P_{mr} is defined as $\mathcal{G}(P_{mr}) =$

$$\left\{ \begin{bmatrix} u \\ P_{mr}u \end{bmatrix} ; u \in \ell^2_+, P_{mr}u \in \ell^2_+ \text{ converges} \right\}.$$

Here ℓ^2_+ means possibly the direct sum of a collection of ℓ^2_+ signal spaces with different sampling rates. Clearly $\mathcal{G}(P_{mr})$ is a subspace of $\ell^2_+ \oplus \ell^2_+$. A subspace $\mathcal{G}$ of $\ell^2_+ \oplus \ell^2_+$ is called $(\bar{M}, \bar{N})$-shift-invariant if

$$\begin{bmatrix} U_{\bar{M}} & 0 \\ 0 & U_{\bar{N}} \end{bmatrix} \mathcal{G} \subset \mathcal{G}.$$

It is easy to see that the graph of P_{mr} is $(\bar{M}, \bar{N})$-shift-invariant. The gap between P_{mr} and $\tilde{P}_{mr}$ is defined by

$$\delta(P_{mr}, \tilde{P}_{mr}) = \|\Pi_{\mathcal{G}(P_{mr})} - \Pi_{\mathcal{G}(\tilde{P}_{mr})}\|,$$

where $\Pi_{\mathcal{G}(P_{mr})}$ and $\Pi_{\mathcal{G}(\tilde{P}_{mr})}$ are the orthogonal projections from $\ell^2_+ \oplus \ell^2_+$ onto $\mathcal{G}(P_{mr})$ and $\mathcal{G}(\tilde{P}_{mr})$ respectively. Since the lifting operators $L_{\bar{M}}$ and $L_{\bar{N}}$ are unitary operators, it is clear that

$$\delta(P_{mr}, \tilde{P}_{mr}) = \delta(P, \tilde{P}),$$

where $\tilde{P} = L_{\bar{N}} \tilde{P}_{mr} L_{\bar{M}}^{-1}$.

A subgraph of a linear causal $(\bar{M}, \bar{N})$-shift-invariant multirate system is defined as an $(\bar{M}, \bar{N})$-shift-invariant subspace of its graph. We denote the set of all subgraphs as $\mathcal{S}_{\mathcal{G}}(P_{mr})$. To define the ν−gap between two multirate systems, we need the notation of the index of a subgraph $\mathcal{V}$ with respective to $\mathcal{G}(P_{mr})$, defined as (Wan and Huang, 2000)

$$\text{ind}(\mathcal{V}) := \dim (\mathcal{G}(P_{mr}) \ominus \mathcal{V}).$$

The ν−gap between two plants P_{mr} and $\tilde{P}_{mr}$ is then defined by

$$\delta_\nu(P_{mr}, \tilde{P}_{mr}) = \inf_{\substack{\mathcal{V} \in \mathcal{S}_{\mathcal{G}}(P_{mr}) \\ \tilde{\mathcal{V}} \in \mathcal{S}_{\mathcal{G}}(\tilde{P}_{mr}) \\ \text{ind}(\mathcal{V}) = \text{ind}(\tilde{\mathcal{V}})}} \|\Pi_{\mathcal{V}} - \Pi_{\tilde{\mathcal{V}}}\|$$

where $\Pi_{\mathcal{V}}$ and $\Pi_{\tilde{\mathcal{V}}}$ are the orthogonal projections from $\ell^2_+ \oplus \ell^2_+$ onto $\mathcal{G}_s(P_{mr})$ and $\mathcal{G}_s(\tilde{P}_{mr})$ respectively. The ν-gap between two multirate systems can be computed from that between their equivalent LTI systems, where many efficient methods are available (Vinnicombe, 1993; Wan and Huang, 2000). Note that $\mathcal{V}$ is a subgraph of a multirate system P_{mr} if and only if $\begin{bmatrix} L_{\bar{M}} & 0 \\ 0 & L_{\bar{N}} \end{bmatrix} \mathcal{V}$ is a subgraph of $\mathcal{G}(P)$, where $P = L_{\bar{N}} P_{mr} L_{\bar{M}}^{-1}$. Then we have the following result:

Lemma 1. Let P_{mr} and $\tilde{P}_{mr}$ be two linear causal $(\bar{M}, \bar{N})$-shift-invariant multirate systems and their equivalent LTI systems are P and $\tilde{P}$ respectively, that is

$$P = L_{\bar{N}} P_{mr} L_{\bar{M}}^{-1}, \quad \tilde{P} = L_{\bar{N}} \tilde{P}_{mr} L_{\bar{M}}^{-1}.$$

Then we have $\delta_\nu(P_{mr}, \tilde{P}_{mr}) = \delta_\nu(P, \tilde{P})$.

The gap metric ball and ν-gap metric ball centered at P_{mr} with radius r are respectively given by

$$\mathcal{B}(P_{mr}, r) = \{\tilde{P}_{mr} : \delta(P_{mr}, \tilde{P}_{mr}) < r\}$$
$$\mathcal{B}_\nu(P_{mr}, r) = \{\tilde{P}_{mr} : \delta_\nu(P_{mr}, \tilde{P}_{mr}) < r\}.$$

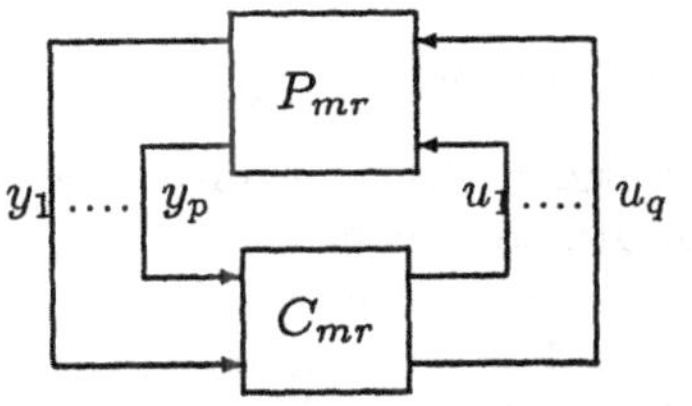

Fig. 2. A general multirate feedback control system.

Now consider the feedback system shown in Figure 2. Here we assume that P_{mr} is linear causal $(\bar{M}, \bar{N})$-shift-invariant and C_{mr} is linear causal $(\bar{N}, \bar{M})$-shift-invariant. An interesting problem is then to design C_{mr} for given P_{mr} so that the feedback system is stable and has optimal robust stability in the sense that it can tolerate the maximum amount of gap or ν-gap metric uncertainty in both P_{mr} and C_{mr}. Let $P = L_{\bar{N}} P_{mr} L_{\bar{M}}^{-1}$ and $C = L_{\bar{M}} C_{mr} L_{\bar{N}}^{-1}$. Then P and C are LTI and hence have transfer functions $\hat{P}$ and $\hat{C}$ respectively. For fixed P_{mr} and C_{mr}, the stability robustness of the feedback system is given by the following lemma (Vinnicombe, 1999; Qiu and Davison, 1992)

Lemma 2. Given a nominal plant P_{mr} and a stabilizing controller C_{mr}, assume that r_1 and r_2 are positive real numbers, then the feedback system with plant $\tilde{P}_{mr}$ and controller $\tilde{C}_{mr}$ is stable for all $\tilde{P}_{mr} \in \mathcal{B}(P_{mr}, r_1)$ (or $\tilde{P}_{mr} \in \mathcal{B}_\nu(P_{mr}, r_1)$) and all $\tilde{C}_{mr} \in \mathcal{B}(C_{mr}, r_2)$ (or $\tilde{C}_{mr} \in \mathcal{B}_\nu(C_{mr}, r_2)$)if and only if

$$\arcsin r_1 + \arcsin r_2 + \arccos b_{P,C} \le \frac{1}{2}\pi,$$

where

$$b_{P,C} = \| \begin{bmatrix} I \\ \hat{P} \end{bmatrix} (I - \hat{C}\hat{P})^{-1} \begin{bmatrix} I & -\hat{C} \end{bmatrix} \|_\infty^{-1}.$$

The quantity $b_{P,C}$ is defined as the robust stability margin. For a given P, the optimal robust stabilization problem is to maximize $b_{P,C}$ by choosing a stabilizing controller C.

Define

$$\hat{G} = \begin{bmatrix} \begin{bmatrix} I & 0 \\ \hat{P} & 0 \end{bmatrix} & \begin{bmatrix} I \\ \hat{P} \end{bmatrix} \\ \begin{bmatrix} \hat{P} & I \end{bmatrix} & \hat{P} \end{bmatrix}. \tag{1}$$

Then it is easy to see that

$$\begin{bmatrix} I \\ \hat{P} \end{bmatrix} (I - \hat{C}\hat{P})^{-1} \begin{bmatrix} I & -\hat{C} \end{bmatrix} = \mathcal{F}(\hat{G}, \hat{C})$$
$$:= \hat{G}_{11} + \hat{G}_{12}\hat{C}(I - \hat{G}_{22}\hat{C})^{-1}\hat{G}_{21} \tag{2}$$

Hence our optimal robust stabilization problem becomes a special discrete-time $\mathcal{H}_\infty$ optimal control problem. Since the causality of P_{mr} and C_{mr} is equivalent to

$$\hat{P}(0) \in \mathcal{N}(\{\mathcal{U}_r\}, \{\mathcal{Y}_r\}) \quad (3)$$

and

$$\hat{C}(0) \in \mathcal{N}(\{\mathcal{Y}_r\}, \{\mathcal{U}_r\}), \quad (4)$$

we should consider the causality constraint (4) and possibly utilize the causality constraint (3) in solving the special discrete-time $\mathcal{H}_\infty$ optimal control problem. The continuous-time counterpart of such an $\mathcal{H}_\infty$ optimal control problem (without causality constraint) has been explicitly solved in (Georgiou and Smith, 1990; Glover and McFarlane, 1989).

3. THE MAIN RESULTS

Suppose that P has a stabilizable and detectable state space realization $\left[\begin{array}{c|c} A & B \\ \hline C & D \end{array}\right]$. Then G in (1) has a state space realization

$$\hat{G} = \left[\begin{array}{c|c|c} A & \begin{bmatrix} B & 0 \end{bmatrix} & B \\ \hline \begin{bmatrix} 0 \\ C \end{bmatrix} & \begin{bmatrix} I & 0 \\ D & 0 \end{bmatrix} & \begin{bmatrix} I \\ D \end{bmatrix} \\ C & \begin{bmatrix} D & I \end{bmatrix} & D \end{array}\right]. \quad (5)$$

Let X and Y be the stabilizing solutions of Riccati equations

$$\begin{aligned} X &= A'XA + C'C - (A'XB + C'D) \\ &\quad (B'XB + I + D'D)^{-1}(B'XA + D'C), \quad (6) \\ Y &= AYA' + BB' - (AYC' + BD') \\ &\quad (CYC' + I + DD')^{-1}(CYA' + DB'). \quad (7) \end{aligned}$$

Denote

$$\begin{aligned} F &= -(B'XB + I + D'D)^{-1}(B'XA + D'C), \quad (8) \\ L &= -(AYC' + BD')(CYC' + I + DD')^{-1}. \quad (9) \end{aligned}$$

Here $(A + BF)$ and $(A + LC)$ are stable since X and Y are stablilizing solutions. Choose constant matrices $R \in \mathcal{N}(\{\mathcal{U}_r\}), S \in \mathcal{N}(\{\mathcal{Y}_r\})$ satisfying (Chen and Qiu,1994),

$$\begin{aligned} R'R &= B'XB + I + D'D, \quad (10) \\ SS' &= CYC' + I + DD'. \quad (11) \end{aligned}$$

Theorem 1. Given a nominal plant $\hat{P} = \left[\begin{array}{c|c} A & B \\ \hline C & D \end{array}\right]$ with $D \in \mathcal{N}(\{\mathcal{U}_r\}, \{\mathcal{Y}_r\})$, let X and Y be the stabilizing solutions of Riccati equations (6) and (7), and let F, L, R, S be defined as in (8)-(11). Then the optimal robust stabilization margin $\sup_C b_{P,C}$ is equal to

$$\left\{(\max_r \| \begin{bmatrix} I - \Pi_{\mathcal{U}_r} & 0 \\ 0 & I \end{bmatrix} H_T|_{(\mathcal{Y}_r \oplus \mathcal{R}^n)}) \|^2 + 1\right\}^{-\frac{1}{2}},$$

where $H_T =$

$$\begin{bmatrix} R'^{-1}D'S & R'^{-1}B'(X + XYX)^{\frac{1}{2}} \\ (Y^{-1} + X)^{-\frac{1}{2}}[(C + DF)'S - (A + BF)'XLS] & (Y^{-1} + X)^{-\frac{1}{2}}(A + BF)'(X + XYX)^{\frac{1}{2}} \end{bmatrix}. \quad (12)$$

Theorem 1 follows easily from Propositions 1, 2 and Lemma 3 which are to be established in the following.

Let

$$\hat{T} = \left[\begin{array}{c|c} \tilde{A} & \tilde{B} \\ \hline \tilde{C} & \tilde{D} \end{array}\right] = \left[\begin{array}{c|c} A + BF & BR^{-1} \\ \hline S'(C + DF) - A'L'X(A + BF) & S'DR^{-1} \end{array}\right] \quad (13)$$

Denote $\hat{T}^{\sim}(\lambda) = \hat{T}(\frac{1}{\lambda})'$ as the adjoint of $\hat{T}(\lambda)$.

Proposition 1.

$$\inf_{\hat{C}} \|\mathcal{F}(\hat{G}, \hat{C})\|_\infty^2 = \inf_{\hat{Q} \in \mathcal{RH}_\infty, \hat{Q}(0) \in \mathcal{N}(\{\mathcal{Y}_r\}, \{\mathcal{U}_r\})} \|\hat{T}^{\sim} + \hat{Q}\|_\infty^2 + 1. \quad (14)$$

Remark: This result is also given in (Georgiou and Smith, 1990) through an operator approach. The minimization problem in the right hand side of (14), if the causality constraint is removed, is a standard Nehari problem, which has an explicit solution (Al-Husari, *et. al.*, 1997).

Proof: Let X and Y be the stabilizing solutions of Riccati equations (6)-(7), then all stabilizing controllers of P satisfying the causality constraint are characterized by an LFT (Ravi, *et. al.*, 1990):

$$\hat{C} = \mathcal{F}(\hat{J}, \hat{Q}), \quad (15)$$

where $\hat{Q} \in \mathcal{RH}_\infty, \hat{Q}(0) \in \mathcal{N}(\{\mathcal{Y}_r\}, \{\mathcal{U}_r\})$,

$$\hat{J} = \left[\begin{array}{c|c} A + BF + LC + LDF & -L \quad B + LD \\ \hline F & \begin{bmatrix} 0 & I \\ I & -D \end{bmatrix} \\ -C - DF & \end{array}\right]. \quad (16)$$

Under this characterization, the closed loop transfer function is

$$\mathcal{F}(\hat{G}, \hat{C}) = \mathcal{F}[\hat{G}, \mathcal{F}(\hat{J}, \hat{Q})] = \hat{T}_{11} + \hat{T}_{12}\hat{Q}\hat{T}_{21},$$

where

$$\hat{T}_{11} = \left[\begin{array}{cc|c} A+BF & BF & \begin{bmatrix} B & 0 \end{bmatrix} \\ 0 & A+LC & -\begin{bmatrix} B+LD & L \end{bmatrix} \\ \hline \begin{bmatrix} F \\ C+DF \end{bmatrix} & \begin{bmatrix} F \\ DF \end{bmatrix} & \begin{bmatrix} I & 0 \\ D & 0 \end{bmatrix} \end{array}\right],$$

$$\hat{T}_{12} = \left[\begin{array}{c|c} A+BF & B \\ \hline \begin{bmatrix} F \\ C+DF \end{bmatrix} & \begin{bmatrix} I \\ D \end{bmatrix} \end{array}\right],$$

$$\hat{T}_{21} = \left[\begin{array}{c|c} A+LC & \begin{bmatrix} B+LD & L \end{bmatrix} \\ \hline C & \begin{bmatrix} D & I \end{bmatrix} \end{array}\right].$$

It follows (Chen and Francis, 1995) that

$$\hat{T}_{12}^{\sim}\hat{T}_{12} = B'XB + I + D'D, \tag{17}$$

$$\hat{T}_{21}\hat{T}_{21}^{\sim} = CYC' + I + DD'. \tag{18}$$

Now carry out matrix factorizations in (10) and (11) to get $R \in \mathcal{N}(\{\mathcal{U}_r\})$ and $S \in \mathcal{N}(\{\mathcal{Y}_r\})$. Define

$$U = \begin{bmatrix} R'^{-1}\hat{T}_{12}^{\sim} \\ I - \hat{T}_{12}R^{-1}R'^{-1}\hat{T}_{12}^{\sim} \end{bmatrix},$$

$$V = \begin{bmatrix} \hat{T}_{21}^{\sim}S'^{-1} & I - \hat{T}_{21}^{\sim}S'^{-1}S^{-1}\hat{T}_{21} \end{bmatrix}.$$

Then (17) and (18) imply $U^{\sim}U = I$ and $VV^{\sim} = I$. Hence we have

$$\begin{aligned} \|\mathcal{F}(\hat{G},\hat{C})\|_\infty &= \|\hat{T}_{11} + \hat{T}_{12}\hat{Q}\hat{T}_{21}\|_\infty \\ &= \|\hat{U}(\hat{T}_{11} + \hat{T}_{12}\hat{Q}\hat{T}_{21})\hat{V}\|_\infty \\ &= \left\| \begin{bmatrix} \hat{T}^{\sim} + R\hat{Q}S & -I \\ 0 & 0 \end{bmatrix} \right\|_\infty \\ &= (\|\hat{T}^{\sim} + \hat{Q}_R\|_\infty^2 + 1)^{\frac{1}{2}}, \end{aligned}$$

where $\hat{Q}_R = R\hat{Q}S$.
Notice that $\hat{Q}_R(0) \in \mathcal{N}(\{\mathcal{Y}_r\},\{\mathcal{U}_r\})$ if and only if $\hat{Q}(0) \in \mathcal{N}(\{\mathcal{Y}_r\},\{\mathcal{U}_r\})$. □

The constrained optimal distance problem in the right hand side of (14) is solved in (Georgiou and Khargonekar, 1987). The controllability Grammian Ψ and the observability Grammian Φ of system T are required. Ψ and Φ are the solutions of the following Lyapunov equations

$$\Psi = \tilde{A}\Psi\tilde{A}' + \tilde{B}\tilde{B}', \tag{19}$$

$$\Phi = \tilde{A}'\Phi\tilde{A} + \tilde{C}'\tilde{C}. \tag{20}$$

Lemma 3.

$$\begin{aligned} &\inf_{\hat{Q}_R \in \mathcal{RH}_\infty, \hat{Q}_R(0) \in \mathcal{N}(\{\mathcal{Y}_r\},\{\mathcal{U}_r\})} \|\hat{T}^{\sim} + \hat{Q}_R\|_\infty \\ &= \inf_{\hat{Q}_R(0) \in \mathcal{N}(\{\mathcal{Y}_r\},\{\mathcal{U}_r\})} \left\| \begin{bmatrix} \tilde{D}' + \hat{Q}_R(0) & \tilde{B}'\Phi^{\frac{1}{2}} \\ \Psi^{\frac{1}{2}}\tilde{C}' & \Psi^{\frac{1}{2}}\tilde{A}'\Phi^{\frac{1}{2}} \end{bmatrix} \right\| \\ &= \max_r \left\| \begin{bmatrix} I - \Pi_{\mathcal{U}_r} & 0 \\ 0 & I \end{bmatrix} \begin{bmatrix} \tilde{D}' & \tilde{B}'\Phi^{\frac{1}{2}} \\ \Psi^{\frac{1}{2}}\tilde{C}' & \Psi^{\frac{1}{2}}\tilde{A}'\Phi^{\frac{1}{2}} \end{bmatrix} |_{(\mathcal{Y}_r \oplus \mathcal{R}^n)} \right\|. \end{aligned}$$

We will see that Ψ and Φ are simple functions of X and Y.

Proposition 2.

$$\Psi = Y(I + XY)^{-1}, \tag{21}$$

$$\Phi = (I + XY)X. \tag{22}$$

Proof: The proof is parallel to the continuous-time case (Georgiou and McFarlane, 1989) and hence omitted here.

To characterize all controllers satisfying the robust stabilization margin given by Theorem 1, one way is to characterize all Q_R by solving the optimal distance problem in Lemma 3. This can be done by using the method proposed in (Georgiou and Khargonekar, 1987). Then Q is simply $R^{-1}Q_RS^{-1}$. We can also follow the coprime factorization approach proposed in (Georgiou and McFarlane, 1989).

Theorem 2. Given a nominal plant $\hat{P} = \left[\begin{array}{c|c} A & B \\ \hline C & D \end{array}\right]$ with $D \in \mathcal{N}(\{\mathcal{U}_r\},\{\mathcal{Y}_r\})$, let X and Y be the stabilizing solutions of Riccati equations (6) and (7), and let F, L, R, S be defined as in (8)-(11). Then the optimal robust stabilization margin is

$$\sup_C b_{P,C}^2 =$$

$$1 - \max_r \left\| \begin{bmatrix} I - \Pi_{\mathcal{U}_r} & 0 & 0 \\ 0 & I - \Pi_{\mathcal{Y}_r} & 0 \\ 0 & 0 & I \end{bmatrix} H_F|_{(\mathcal{Y}_r \oplus \mathcal{R}^n)} \right\|^2,$$

where

$$H_F = \begin{bmatrix} -D'S'^{-1} & -(B' + D'L')(X^{-1} + Y)^{-\frac{1}{2}} \\ S'^{-1} & L'(X^{-1} + Y)^{-\frac{1}{2}} \\ Y^{\frac{1}{2}}C'S'^{-1} & Y^{\frac{1}{2}}(A + LC)'(X^{-1} + Y)^{-\frac{1}{2}} \end{bmatrix}. \tag{23}$$

This theorem is straitforward after the Lemma 4, which is obtained by slightly modifying the result in (Georgiou and McFarlane, 1989), is introduced.

For a nominal plant P with $\hat{P}(0) \in \mathcal{N}(\{\mathcal{U}_r\},\{\mathcal{Y}_r\})$, there are normalized left coprime factorizations $P = \tilde{M}^{-1}\tilde{N}$ with $\hat{\tilde{N}}(0) \in \mathcal{N}(\{\mathcal{U}_r\},\{\mathcal{Y}_r\})$ and $\hat{\tilde{M}}(0) \in \mathcal{N}(\{\mathcal{Y}_r\})$. One particular realization of such factorization is

$$\begin{bmatrix} \hat{\tilde{N}} & \hat{\tilde{M}} \end{bmatrix} = \left[\begin{array}{c|cc} (A+LC) & B+LD & L \\ \hline S^{-1}C & S^{-1}D & S^{-1} \end{array}\right].$$

Lemma 4. Let $P = \tilde{M}^{-1}\tilde{N}$ be a normalized left coprime factorization with $\hat{\tilde{M}}(0) \in \mathcal{N}(\{\mathcal{Y}_r\})$ and $\hat{\tilde{N}}(0) \in \mathcal{N}(\{\mathcal{U}_r\},\{\mathcal{Y}_r\})$. The optimal robust stabilization margin is

$$\sup_C b_{P,C} = \left\{ 1 - \inf_{\substack{\hat{U},\hat{V} \in \mathcal{RH}_\infty \\ \hat{U}(0) \in \mathcal{N}(\{\mathcal{Y}_r\},\{\mathcal{U}_r\}) \\ \hat{V}(0) \in \mathcal{N}(\{\mathcal{Y}_r\})}} \left\| \begin{bmatrix} -\hat{\tilde{N}}^{\sim} \\ \hat{\tilde{M}}^{\sim} \end{bmatrix} + \begin{bmatrix} \hat{U} \\ \hat{V} \end{bmatrix} \right\|^2 \right\}^{\frac{1}{2}} \tag{24}$$

Let $\begin{bmatrix} \hat{U} \\ \hat{V} \end{bmatrix}$ be a solution of the minimization problem in (24), then an optimal robust stabilizing controller is given by $C = UV^{-1}$ with $\hat{C}(0) \in \mathcal{N}(\{\mathcal{Y}_r\},\{\mathcal{U}_r\})$.

The optimization problem in Lemma 4 can be solved in the same way as that in Lemma 3. Note that

$$\begin{bmatrix} -\hat{\tilde{N}}^{\sim} \\ \hat{\tilde{M}}^{\sim} \end{bmatrix}^{\sim} = \left[\begin{array}{c|cc} A+LC & -B-LD & L \\ \hline S^{-1} & -S^{-1}D & S^{-1} \end{array} \right].$$

Parallel to the continuous-time case, this system's controllability Grammian P_F and the observability Grammian Q_F are

$$P_F = Y, \quad Q_F = X(I+YX)^{-1},$$

where X, Y are the solutions of the Riccati equations (6-7). Now Theorem 2 is clear.

We can also get another formula for $\sup_C b_{P,C}$ symmetric to (23) by considering the normalized right coprime factorization of P. The optimization problem in Theorem 2 has a bigger size than that in Theorem 1. These two theorems actually give the same optimal robust stabilization margin.

In summary, the robust stabilization problem for a general multirate system can be solved as follows
1) Obtain the solutions X, Y of the two Riccati equations (6-7) to form the matrices in (12) or (23).
2) Compute the optimal robust stabilization margin by Theorems 1 or 2.
3) Use the algorithm in (Georgiou and Khargonekar, 1987) to get a solution of the minimization problem of (24), then the optimal robust stabilizing controller is $C = UV^{-1}$.

4. CONCLUSION

In this paper, we present a state space solution to the robust stabilization problem of discrete-time periodic and multirate systems. First, we show how the robust stabilization problem of multirate systems with gap and ν-gap metric uncertainty can be converted to a constrained $\mathcal{H}_\infty$ optimal control problem. The optimal robust stabilization margin is explicitly computed and a method is given to design an optimal controll. The computational burden is to solve two Riccati equations and an extended Parrot problem.

REFERENCES

Al-Husari, M. M. M., I. M. Jaimoukha and D. J. N. Limebeer (1997). A descriptor solution to a class of discrete distance problems. *IEEE Trans. Automat. Control*, 42, 1558-1564.

Bittanti, S. and P. Colaneri (2000). Invariant representations of discrete-time periodic systems. *Automatica*, 36, 1777-1793.

Chen, T. and B. A. Francis (1995). *Optimal Sampled-Data Control Systems.* Springer, London.

Chen, T. and L. Qiu (1994). $\mathcal{H}_\infty$ design of general multirate sampled-data control systems. *Automatica*, 30, 1139-1152.

Georgiou, T. T. and P. P. Khargonekar (1987). A constructive algorithm for sensitivity optimization of periodic systems. *SIAM J. Control and Optimization*, 25, 334-340.

Georgiou, T. T. and M. C. Smith (1990). Optimal robustness in the gap metric. *IEEE Trans. Automat. Control*, 35, 673-687.

Glover, K. and D. C. McFarlane (1989). Robust stabilization of normalized coprime factor plant descriptions with $\mathcal{H}_\infty$-bounded uncertainty. *IEEE Trans. Automat. Control*, 34, 821-830.

Iglesias, P. A. (2000). The strictly proper discrete-time controller for the normalized left-coprime factorization robust stabilization problem. *IEEE Trans. Automat. Control*, 45, 516-520.

Qiu, L. and T. Chen (1994). Multirate sampled-data systems: all $\mathcal{H}_\infty$ suboptimal controllers and the minimum entropy controller. *Proceedings of the 33rd IEEE Conference on Decision and Control*, 4, 3707-3712.

Qiu, L. and E. J. Davison (1992). Pointwise gap metric on transfer matrices. *IEEE Trans. Automat. Control*, 37, 770-780.

Qiu, L. and E. J. Davison (1992). Stability under simultaneous uncertainty in plant and controller. *Systems & control Letters*, 18, 9-22.

Ravi, R., P. P. Kargonekar, K. D. Minto, C. N. Nett (1990). Controller parametrization for time-varying multirate plants. *IEEE Trans. Automat. Control*, 35, 1259-1262.

Voulgaris, P. G., M. A. Dahleh and L. Valavani (1994). $\mathcal{H}^\infty$ and $\mathcal{H}^2$ optimal controllers for periodic and multirate systems. *Automatica*, 30, 251-263.

Vinnicombe, G. (1993). Frequency domain uncertainty and the graph topology. *IEEE Trans. Automat. Control*, 38, 1371-1383.

Vinnicombe, G. (1999). *Uncertainty and Feedback: $\mathcal{H}_\infty$ loop-shaping and the ν-gap metric.* Imperial College Press, 1999.

Wan, S. and B. Huang. Gap metric on the subgraphs of systems and the robustness problem. *IEEE Trans. Automat. Control*, 45, 1522-1526.

Copyright © IFAC Periodic Control Systems,
Cernobbio-Como, Italy, 2001

PERIODIC ATTITUDE CONTROL FOR SATELLITES WITH MAGNETIC ACTUATORS: AN OVERVIEW

Marco Lovera

Dipartimento di Elettronica e Informazione, Politecnico di Milano
32, Piazza Leonardo da Vinci, 20133 Milano, Italy
email: lovera@elet.polimi.it

Abstract: The problem of attitude control of a small spacecraft using magnetic actuators is considered and a review of the solutions available in the literature for the linear and the nonlinear control problems is proposed, together with a contribution to the design of robust control laws with respect to uncertainty in the knowledge of the geomagnetic field, based on H_∞ periodic control theory. *Copyright © 2001 IFAC*

Keywords: Attitude control, Optimal control, Robust control, Disturbance rejection.

1. INTRODUCTION

The use of magnetic coils for control purposes has been the subject of extensive study since the early years of satellite missions (see, e.g., (Stickler and Alfriend, 1976)). As is well known, the operation of magnetic actuators is based on the interaction with the geomagnetic field (Wertz, 1978; Sidi, 1997); the major consequence of this is that the torques which can be applied to the spacecraft for attitude control purposes are constrained to lie in the plane orthogonal to the magnetic field vector. In particular, 3 axis magnetic stabilisation is only possible if the considered orbit "sees" a variation of the magnetic field which is sufficient to guarantee the stabilisability of the spacecraft. Until recent years, only approximate solutions to the problem of dealing with such time varying actuators were available (see, e.g., (Stickler and Alfriend, 1976; Arduini and Baiocco, 1997)). In particular, while periodic control has been already proved successful in various applications (see, e.g., (Arcara *et al.*, 2000)), the feasibility of periodic techniques for the control of small satellites using magnetic actuators has become only recently a topic of active research (see, e.g., the recent works (De Marchi *et al.*, 1999; Wisniewski and Blanke, 1999; Wisniewski and Markley, 1999; Wang and Shtessel, 1999; Lovera *et al.*, 2001)).

In particular, both the linear and the nonlinear attitude control problems have been investigated; the former has been studied mostly in an LQ perspective, while the latter has been mainly investigated using Lyapunov methods and exploiting periodicity of the Earth magnetic field for the stability analysis. Finally, more recent contributions are available, aiming at solving the global magnetic stabilisation problem without resorting to the periodicity assumption.

The paper is organised as follows: in Section 2 a description of the dynamics of a spacecraft will be presented, and a linearised dynamic model derived; Section 3 and 4 offer an overview of the relevant results in linear and nonlinear control which are available in the literature for this problem.

2. SPACECRAFT DYNAMICS

2.1 *Attitude dynamics*

The attitude dynamics of a rigid spacecraft can be expressed by the well known Euler's equations, as follows (Wertz, 1978):

$$I\dot{\omega}(t) = -\omega(t) \wedge I\omega(t) + T_c(t) + T_d(t) \quad (1)$$

where $\omega \in \mathbb{R}^3$ is the vector of spacecraft angular rate, expressed in body frame, $I \in \mathbb{R}^{3\times3}$ is the inertia matrix, $T_c \in \mathbb{R}^3$ is the vector of control torques and $T_d \in \mathbb{R}^3$ the vector of external disturbance torques.

The most common parameterisation for spacecraft attitude is given by the four Euler parameters, which lead to the following representation for the attitude kinematics:

$$\dot{q}(t) = W(\omega(t))q(t) \tag{2}$$

where $q \in \mathbb{R}^4$ is the vector of unit norm Euler parameters and

$$W(\omega) = \frac{1}{2}\begin{bmatrix} 0 & \omega_z & -\omega_y & \omega_x \\ -\omega_z & 0 & \omega_x & \omega_y \\ \omega_y & -\omega_x & 0 & \omega_z \\ -\omega_x & -\omega_y & -\omega_z & 0 \end{bmatrix} \tag{3}$$

For the ACS of an Earth pointing spacecraft the following reference systems are adopted:

- Orbital Axes (X_O, Y_O, Z_O). The origin of these axes is in the satellite centre of mass. The X-axis points to the Earth's centre; the Y-axis points in the direction of the orbital velocity vector. The Z-axis is normal to the satellite orbit plane and completes the right-handed orthogonal triad.
- Satellite Body Axes (X, Y, Z). The origin of these axes is in the satellite centre of mass.

Under nominal pointing conditions, the Satellite Body Axes coincides with the Orbital Axes.

2.2 *Attitude control hardware*

On board small satellites the usual choice for the generation of external torques falls on a set of three magnetic coils, aligned with the spacecraft principal axes, which generate torques according to the law

$$T_{mag}(t) = m(t) \wedge b(t) = B(b(t))m(t) \tag{4}$$

where $m \in \mathbb{R}^3$ is the vector of magnetic dipoles for the three coils (which represent the actual control variables for the coils), $b \in \mathbb{R}^3$ is the vector formed with the components of the Earth's magnetic field in the body frame of reference and

$$B(t) = \begin{bmatrix} 0 & b_z(t) & -b_y(t) \\ -b_z(t) & 0 & b_x(t) \\ b_y(t) & -b_x(t) & 0 \end{bmatrix} \tag{5}$$

As is well known (Wertz, 1978; Sidi, 1997), spacecraft which corresponds to the so-called *momentum bias* configuration, run (usually only) one momentum wheel at constant speed in order to provide gyroscopic stiffness to one spacecraft axis. The presence of wheels also introduces additional gyroscopic terms in the system's dynamics, which must be modified as follows:

$$I\dot{\omega}(t) = -\omega(t) \wedge [I\omega(t) + J\nu(t)] + T_c(t) + T_d(t) \tag{6}$$

where $J \in \mathbb{R}^{3\times3}$ represents the overall wheels inertia matrix wrt the body frame.

As is often the case in practice, in this paper it will be assumed that an appropriate control loop has been designed in order to keep the momentum wheel at a constant angular velocity ν.

2.3 *Disturbance torques*

External disturbance torques occur naturally and have different sources, such as gravity gradient, aerodynamics, solar radiation and residual magnetic dipoles. Regardless of the physical mechanism giving rise to them (i.e., magnetic, aerodynamic, solar, gravity gradient), they can be separated into a secular component (i.e., the part with nonzero mean around each orbit) and a cyclic component (i.e., the zero mean, periodic part, with a period given by the orbit period). See, e.g., (Annoni *et al.*, 1999) for an overview of issues related to the simulation of spacecraft dynamics.

2.4 *Linearised dynamics*

Assuming a momentum bias configuration for the spacecraft (i.e., one momentum wheel, aligned with the body z axis) and defining the states for the linearised system as:

$$\delta q(t) = [\delta q_1(t)\ \delta q_2(t)\ \delta q_3(t)\ 0]^T \tag{7}$$

$$\delta\omega(t) = \omega(t) - [0\ 0\ -\Omega]^T \tag{8}$$

which corresponds to introducing small displacements from the nominal attitude quaternion $q_1 = q_2 = q_3 = 0$, $q_4 = 1$, and small deviations of the body rates from the nominal ones $\omega_x = \omega_y = 0, \omega_z = \Omega$, Ω being the angular frequency associated with the orbit period, the state vector for the linearised system can be defined as

$$\delta x = [\delta q_1\ \delta q_2\ \delta q_3\ \delta\omega_x\ \delta\omega_y\ \delta\omega_z]^T \tag{9}$$

The above equations (2) and (6) can be linearised and the local linear dynamics for the system defined as

$$\delta\dot{x}(t) = A\delta x(t) + \begin{bmatrix} 0 \\ I^{-1} \end{bmatrix}(B(b(t))m(t) + T_d(t)). \tag{10}$$

The complete expression of matrix A can be found in (Lovera *et al.*, 2001). Clearly, if the time variation of the magnetic field is fundamentally a periodic one, this model can be seen as a linear time-periodic one.

2.5 *Periodic approximation of the magnetic field*

A time history of the International Geomagnetic Reference Field (IGRF) model for the Earth's magnetic field (Wertz, 1978) along five orbits in Pitch-Roll-Yaw coordinates for a spacecraft in a circular, polar orbit (87° inclination, 450km altitude) is shown in Figure 1 (solid lines).

As can be seen, $b_x(t)$, $b_y(t)$ have a very regular and almost periodic behaviour, while the $b_z(t)$ component is much less regular. This behaviour can be easily interpreted by noticing that when the spacecraft is in the nominal attitude, the x and y body axes lie in the orbit plane while the z axis is normal to it. As a consequence, the x and y magnetometers sense only the variation of the magnetic field due to the orbital motion of the spacecraft (period equal to the orbit period) while the z axis sensor is affected by the variation of b due to the rotation of the Earth (period of 24h).

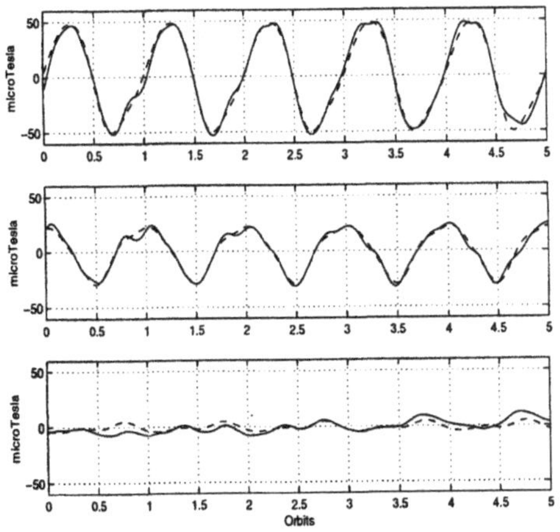

Fig. 1. Periodic approximation of the geomagnetic field in Pitch-Roll-Yaw coordinates, 87° inclination orbit, 450km altitude.

Combining the linearised dynamics derived in the previous Section with the periodic approximation of the magnetic field obtained herein, one gets a complete periodic model of the local dynamics of the spacecraft.

2.6 *Controllability issues*

As is clear from eq. (5), the B matrix in the equation for the magnetic control torque is structurally of rank two, so that the use of magnetic coils implies that the spacecraft is instantaneously underactuated, i.e., at each time instant it is not possible to apply a control torque to all three spacecraft axes in an independent manner. Full controllability is only guaranteed provided that the variations in the direction of the magnetic field experienced by the spacecraft along each orbit are such that

$$\bigcup_{t=0,T} span_{col}(B(t)) = \mathbb{R}^3$$

(see, e.g., (Wisniewski, 2000)). This condition depends considerably on the inclination of the selected orbit: it turns out that magnetic actuators give rise to almost uncontrollable dynamics in equatorial orbit; controllability characteristics tend to improve with orbit inclination.

In particular, when considering an ideal circular polar orbit and the simple dipole approximation for the geomagnetic field (see, e.g, (Wertz, 1978)), a closed form expression for the controllability Grammian for the spacecraft dynamics can be computed, and it can be showed that it has a nonzero determinant.

3. LINEAR DESIGN METHODS

The main issues in the design of a linear attitude controller based on magnetic actuators are the following:

(1) As showed in the previous Section, the local dynamics of a magnetically controlled spacecraft are (approximately) time periodic, therefore the stabilisation problem cannot be faced by means of conventional LTI design tools.
(2) The spacecraft is subject to external disturbances, which can be decomposed in a secular part (i.e., constant) and a cyclic part (i.e., periodic, with period equal to the orbit period $T = \frac{2\pi}{\Omega}$); the controller must try and attenuate the effect of such disturbances on the attitude.
(3) Robustness issues: the dynamics of the spacecraft are subject to a number of sources of uncertainty, the most important of which are the moments of inertia of the spacecraft and the actual behaviour of the magnetic field.
(4) Implementation issues: periodic gains are not easily implemented on board, as they give rise to a sensitive synchronisation problem. The design of a periodic controller which can exploit on board measurements of the magnetic field vector remains an open problem.

In this Section the main contributions given so far in the literature (dealing with points 1 and 2 in the above list) will be reviewed and the application of periodic H_∞ control theory will be proposed, in order to deal with point 3 in the above list.

3.1 *Optimal periodic control*

Assuming that the periodic approximation for the geomagnetic field is satisfactory, the resulting periodic dynamics for the spacecraft can be easily stabilised, either by state or output periodic feedback. A number of contributions have been given in the literature to the analysis of the LQ magnetic attitude control problem.

In (Pittelkau, 1993) both output feedback stabilisation and disturbance attenuation for a momentum biased spacecraft have been analysed; in particular, the disturbance issue has been addressed by means of an internal model principle approach. In order to avoid the difficulties in the solution of the LQ problems due to the presence of uncontrollable modes with poles on the imaginary axis, the secular and cyclic components of the external disturbances have been modeled with stable models (first order with a large time constant fore the secular part and second order with small damping for the cyclic part). This gives rise to a time periodic filter and a time periodic state feedback.

In (Lovera *et al.*, 2001; De Marchi *et al.*, 1999) the problem is analysed in a similar fashion, by using an extension of the periodic LQ control problem (initially proposed in (Arcara *et al.*, 2000)), by which it is possibile to include marginally stable disturbance models in the plant description. The resulting control design method is applied in a simulation study for the Italian spacecraft MITA.

A similar approach has been also proposed in (Wisniewski, 1996; Wisniewski and Blanke, 1999; Wisniewski and Markley, 1999), where the sole state feedback problem is considered. In addition, in (Wisniewski, 2000) the use of a finite horizon controller is suggested as a means of avoiding the issues related to the implementation of a periodic gain.

Other attempts at avoiding the use of a periodic control law by means of piecewise constant gains computed on board have been reported in (Steyn, 1994; Lagrasta and Bordin, 1996; Curti and Diani, 1999). While such techniques are interesting from an implementation point of view, they have the obvious drawback that no guarantees about closed loop stability of the system can be obtained.

A different approach to the problem of obtaining constant gains via optimal periodic control theory has been explored in (Psiaki, 2000; Psiaki, 2001). The idea is to analyse the asymptotic behaviour of the solutions of the periodic Riccati equation in the limiting case of infinite control weighting, i.e., when in the LQ cost function

$$J = \int_0^{\infty} [x(t)^T Q(t) x(t) + u(t)^T R(t) u(t)] dt$$

one chooses

$$R(t) = \frac{R_0}{\epsilon}$$

and $\epsilon \to 0$. It turns out that the limiting solution is a constant one and it can be computed in an approximate way by means of an appropriate averaging scheme, for a small, but finite, ϵ. This approach has the advantage of providing a constant state feedback gain; unfortunately, the approximations which are introduced in the computation of the limiting solution of the Riccati equation do not guarantee that the computed gain is a stabilising one, so that a posteriori verification via Floquet's Theorem remains necessary.

3.2 *The H_∞ approach*

The H_∞ approach to the problem has been first proposed in (Lovera, 2000) for the design of state feedback attitude controllers and in (Lovera, 2001) for the implementation of momentum management control laws based on magnetic actuators. In the cited contributions, however, no attempt has been made to exploit the H_∞ setting in order to enforce robustness in the design procedure.

3.2.1. *H_∞ periodic state feedback control* Given the T-periodic, linear, continuous-time system:

$$\dot{x}(t) = A(t)x(t) + B_1(t)w(t) + B_2(t)u(t) \quad (11)$$

$$z(t) = C(t)x(t) + D(t)u(t) \quad (12)$$

with $x(t) \in \mathbb{R}^n$ and $u(t) \in \mathbb{R}^m$.

Then, under the assumptions that

(1) $C(t)^T D(t) = 0$, $\forall t$.
(2) $D(t)^T D(t) = I$, $\forall t$.
(3) $(A(\cdot), B_2(\cdot))$ is stabilisable.

the state feedback H_∞ periodic control problem can be stated as follows: find the linear, time periodic state feedback control law, which stabilises the system and guarantees that $||T_{zw}||_\infty < \gamma$ (see, e.g., (Bittanti and Colaneri, 1999; Colaneri, 2000*a*; Colaneri, 2000*b*) for the formal definition of the H_∞ norm for periodic systems).

This leads to the optimal control law $u(t) = -B_2(t)^T P_\infty(t) x(t)$ where $P_\infty(t)$ is the symmetric, positive definite solution of the H_∞ periodic Riccati equation

$$P(t)A(t) + A(t)^T P(t) + P(t)(\gamma^{-2} B_1(t) B_1(t)^T + \\ - B_2(t) B_2(t)^T) P(t) + C(t)^T C(t) = -\dot{P}(t) \quad (13)$$

An H_∞ filter can be designed, either for the sole spacecraft dynamics or for a model augmented with an appropriate description of the external disturbance torques (see (Lovera, 2000) for details).

Clearly, the choice of an H_∞ approach in the formulation of the control problem makes it possible to take into account robustness issues in the tuning of the controller gain. Dealing with parametric and nonparametric uncertainties in the design of magnetic controllers will be the subject of the following Section.

3.2.2. *Robust H_∞ magnetic attitude control* Considering the linearised equations of motion (10) given in the previous Section, one has that the main sources of uncertainty in the model can be reduced to the following:

- Discrepancies between the ideal periodic behaviour of the magnetic field and the actual one (see Figure 1): $b(t) = \bar{b}(t) + \delta b(t)$, where $\bar{b}(t+T) = \bar{b}(t)$ and $||\delta b(t)|| \leq \delta b_{MAX}$. The bound δb_{MAX} can be constructed from high order geomagnetic field models or in-flight measurements.
- Uncertainty in the measurement of the inertial properties of the spacecraft.

Other possible sources of uncertainty are mounting errors in sensors and actuators (which however can be taken into account by means of suitable calibration procedures) and the presence of unmodelled high frequency dynamics such as due to large flexible appendages.

It turns out that the above two main types of parametric uncertainty affect the system matrices in a rational fashion, so that they can be dealt with in an H_∞ framework by writing the attitude equations in LFT form and introducing suitable scalings in order to meet the assumptions of classical H_∞ theory. Simulation results obtained by this approach are however omitted for space limitations.

4. NONLINEAR DESIGN METHODS

Nonlinear control plays a major role in the attitude stabilisation problem, as every spacecraft, in the initial part of its life, will have to deal with an "attitude acquisition" phase in which large rotations and high angular rates must be dealt with. While no specific periodic attitude control laws have been developed so far in the literature, extensive use has been made of periodic systems theory in order to guarantee (at least locally) the stability of a number of attitude acquisition control laws. In particular we mention the recent works (Wang and Shtessel, 1999) and (Wisniewski and Blanke, 1999). It is interesting to point out that the stability proof for the desired equilibrium state of the closed loop system given in (Wisniewski and Blanke, 1999) relies on the 24 hours periodicity of the geomagnetic field rather than on its orbital periodicity, as is commonly done in local linear analyses, in which, as mentioned in the previous Section, variations of b due to the rotation of the Earth are rather treated as uncertain variations of the periodic parameters. Finally, more recent contributions are available, aiming at solving the global magnetic stabilisation problem without resorting to the periodicity assumption. In particular, in (Lovera and Astolfi, 2001) and (Smirnov, 2001) the problem of global stabilisation via magnetic attitude regulation is considered, for an inertially spherical spacecraft; in the former contribution a state feedback control law is considered, with no assumptions on the time evolution of the magnetic field vector; the latter contribution, on the other hand, deals with the case of a control law based on magnetometer measurements feedback only, on the basis of a simplified mathematical model of the magnetic field.

5. FUTURE DEVELOPMENTS

A relevant point in the development of magnetic attitude controllers is the inability of periodic controllers to exploit on-board measurements of the magnetic field. Therefore, future research will aim at developing combinations of periodic and gain scheduling control laws in order to make use of all the available information to the controller.

6. CONCLUDING REMARKS

The attitude control problem for a spacecraft using magnetic actuators has been considered and analysed in the framework of periodic control theory. A set of solutions in terms of classical LQ and H_∞ periodic control and nonlinear control have been discussed and extensions thereof, aiming at achieving robustness of the control law with respect to uncertain parameters have been proposed.

Acknowledgements Paper supported by MURST project "Identification and Control of Industrial systems".

7. REFERENCES

Annoni, G., E. De Marchi, F. Diani, M. Lovera and G.D. Morea (1999). Standardising tools for attitude control system design: the MITA platform experience. In: *Data Systems in Aerospace (DASIA) 1999, Lisbon, Portugal.*

Arcara, P., S. Bittanti and M. Lovera (2000). Active control of vibrations in helicopters by periodic optimal control. *IEEE Transactions on Control Systems Technology* **8**(6), 883–894.

Arduini, C. and P. Baiocco (1997). Active magnetic damping attitude control for gravity gradient stabilised spacecraft. *Journal of Guidance and Control* **20**(1), 117–122.

Bittanti, S. and P. Colaneri (1999). Periodic control. In: *Wiley Encyclopedia of Electrical and Electronic Engineering* (J.G. Webster, Ed.). John Wiley and Sons.

Colaneri, P. (2000*a*). Continuous-time periodic systems in H_2 and H_∞, part I. *Kybernetica* **36**(2), 211–242.

Colaneri, P. (2000*b*). Continuous-time periodic systems in H_2 and H_∞, part II. *Kybernetica* **36**(3), 329–350.

Curti, F. and F. Diani (1999). Study on active magnetic attitude control for the Italian spacecraft bus MITA. In: *Proceedings of the 14th International Conference on Spaceflight Dynamics, Iguassu Falls, Brasil.*

De Marchi, E., L. De Rocco, G.D. Morea and M. Lovera (1999). Optimal magnetic momentum control for inertially pointing spacecraft. In: *4th ESA International Conference on Spacecraft Guidance, Navigation and Control Systems, ESTEC, Noordwijk, The Netherlands.*

Lagrasta, S. and M. Bordin (1996). Normal mode magnetic control of LEO spacecraft with integral action. In: *AIAA Guidance, Navigation and Control Conference, San Diego, USA.*

Lovera, M. (2000). Periodic H_∞ attitude control for satellites with magnetic actuators. In: *3rd IFAC Symposium on Robust Control Design, Prague, Czech Republic.*

Lovera, M. (2001). Optimal magnetic momentum control for inertially pointing spacecraft. *European Journal of Control.*

Lovera, M. and A. Astolfi (2001). Global magnetic attitude regulation. submitted.

Lovera, M., E. De Marchi and S. Bittanti (2001). Periodic attitude control techniques for small satellites with magnetic actuators. *IEEE Transactions on Control Systems Technology.* to appear.

Pittelkau, M. (1993). Optimal periodic control for spacecraft pointing and attitude determination. *Journal of Guidance, Control and Dynamics* **16**(6), 1078–1084.

Psiaki, M. (2000). Magnetic torquer attitude control via asymptotic periodic linear quadratic regulation. In: *AIAA Guidance, Navigation, and Control Conference, Denver, Colorado, USA.*

Psiaki, M. (2001). Magnetic torquer attitude control via asymptotic periodic linear quadratic regulation. *Journal of Guidance, Control and Dynamics* **24**(2), 386–394.

Sidi, M. (1997). *Spacecraft dynamics and control.* Cambridge University Press.

Smirnov, G. (2001). Attitude determination and stabilisation of a spherically symmetric rigid body in a magnetic field. *International Journal of Control* **74**(4), 341–347.

Steyn, W. H. (1994). Comparison of low Earth orbit satellite attitude controllers submitted to controllability constraints. *Journal of Guidance, Control and Dynamics* **17**(4), 795–804.

Stickler, A.C. and K.T. Alfriend (1976). An elementary magnetic attitude control system. *Journal of Spacecraft and Rockets* **13**(5), 282–287.

Wang, P. and Y. Shtessel (1999). Satellite attitude control via magnetorquers using switching control laws. In: *14th IFAC World Congress, Beijing, China.*

Wertz, J. (1978). *Spacecraft attitude determination and control.* D. Reidel Publishing Company.

Wisniewski, R. (1996). Satellite attitude control using only electromagnetic actuation. PhD thesis. Aalborg University, Denmark.

Wisniewski, R. (2000). Linear time-varying approach to satellite attitude control using only electromagnetic actuation. *Journal of Guidance, Control and Dynamics* **23**(4), 640–646.

Wisniewski, R. and L.M. Markley (1999). Optimal magnetic attitude control. In: *14th IFAC World Congress, Beijing, China.*

Wisniewski, R. and M. Blanke (1999). Fully magnetic attitude control for spacecraft subject to gravity gradient. *Automatica* **35**(7), 1201–1214.

Copyright © IFAC Periodic Control Systems,
Cernobbio-Como, Italy, 2001

PERIODIC H_2 SYNTHESIS FOR SPACECRAFT ATTITUDE DETERMINATION AND CONTROL WITH A VECTOR MAGNETOMETER AND MAGNETORQUERS

Rafał Wiśniewski *,[1] **Jakob Stoustrup** *

* *Department of Control Engineering, Aalborg University, Fredrik Bajers Vej 7C, DK-9220 Aalborg Ø, Denmark, E-mail: raf@control.auc.dk, jakob@control.auc.dk*

Abstract: A control synthesis for a spacecraft equipped with a set of mutually perpendicular coils and a vector magnetometer is addressed in this paper. The interaction between the Earth's magnetic field and an artificial magnetic field generated by the coils produces a control torque. Comparison between the expected magnetic field vector and the true magnetometer data is used for the attitude determination. The magnetic attitude control and determination is intrinsically periodic due to periodic nature of the geomagnetic field variation in orbit. The control performance is specified by the generalized H_2 operator norm. The paper proposes an LMI solution to this problem. *Copyright © 2001 IFAC*

Keywords: Attitude control, optimal control, periodic systems, linear matrix inequality

1. INTRODUCTION

A tremendous progress in micro-electronics observed in the last two decade made small, inexpensive spacecraft missions very attractive, and technologically viable. However, due to reduced allocated mission cost, the hardware including the sensors and the actuators is often very simple, furthermore the redundancy is limited or completely avoided. The attitude control system from this perspective has to be sophisticated enough to fully utilize the existent hardware, but at the same time computationally as simple as possible to increase the reliability.

Probably the most typical actuator/sensor configuration currently in use is a combination of magnetorquer coils and a three axis magnetometer. This or similar configuration was used on the British UoSat satellites, Danish Ørsted, South African SunSat, German Champ, Portuguese PoSat. A vital common feature of the magnetometer and the magnetorquer is that they relay on the magnetic field of the Earth. The interaction of the geomagnetic field and artificially generated field in the coils produces a control torque, whereas the comparison of the modeled and the measured magnetic field of the Earth provides attitude information. The attitude estimation and control schemes developed in this paper use an observation that the magnetic field is periodic[2].

The idea in this paper is to consider the spacecraft as a linear periodic system, and to solve the H_2 control synthesis problem. The optimization problem is formulated in this article by certain

[1] This paper was partially supported by the Danish Research Agency under the project Advanced Control Concepts for Precision Pointing of Small Spacecrafts.

[2] The time propagation of the geomagnetic field vector observed from an Earth stabilized spacecraft is a superposition of two periodic motions: orbital and the Earth spin. If a ratio of the two periods is a rational number the geomagnetic field observation is periodic

linear matrix inequalities. There is a great number of publications treating the control synthesis expressed by LMIs, however only very recently periodic systems have been addressed. (Souza and Trofino, 2000) and (Bittanti and Colaneri, 1999) have treated the robust stability problem of a periodic system with the LMI technique. (Dullerud and Lall, 1999) have extended the LMI approach to H_∞ control synthesis for the LTI systems, (Gahinet and Apkarian, 1991), to periodic ones. The attitude control approach presented in this paper uses the method developed in (Wisniewski and Stoustrup, 2001).

The literature on attitude control treats the magnetic estimation and control separately. There is a great number of publications solving the magnetic estimation problem using the Kalman filter with a time dependent Riccati equation, e.g. (Lefferts *et al.*, 1982), (Psiaki *et al.*, 1990), (Natanson, 1993), (Challa *et al.*, 1997), (Psiaki, 1999), (Bak, 1999). A concept for attitude control based on electromagnetic actuation has gained a comparable attention lately. The early work was based on an idea of designing magnetic controller for the system with averaged parameters, rather than time varying. This design strategy was used both for bias momentum satellites (Camillo and Markley, 1980), (Hablani, 1995) , (Hablani, 1997) and three axis control (Martel *et al.*, 1988). In the recent papers more sophisticated control schemes were proposed, where not only the linear, (Cavallo *et al.*, 1993), (Arduini and Baiocco, 1997), (Wisniewski and Markley, 1999), (Wisniewski, 2000) but also nonlinear control methods, (Steyn, 1994), (Wisniewski and Blanke, 1996), (Tabuada *et al.*, 1999) were in focus.

The H_2 attitude control synthesis addressed in this paper is not completely new in (Wisniewski and Markley, 1999) and (Wisniewski, 2000) the equivalent L_2 magnetic control problem was addressed. The solution proposed involved a periodic Riccati equation. This paper tackles the problem using the linear matrix inequalities. General schemes for the control synthesis and its dual problem, estimation are proposed. From the implementation point of view the advantage of the LMI approach is that the computer burden of the control synthesis is in the off-line calculation, whereas the on-board algorithm is simple. This makes the fix-point implementation possible.

2. LMI

The argument for using this paradigm is that the separation principle is valid for a periodic system (Pratò and Ichikawa, 1988).

Consider a system of specifications used for the standard H_2 synthesis

$$\begin{aligned} s_1 &: \begin{bmatrix} w \\ u \end{bmatrix} \mapsto \begin{bmatrix} z \\ y \end{bmatrix}, \\ x(t+1) &= \mathbf{A}(t)x(t) + \mathbf{B}_1(t)w(t) + \mathbf{B}_2(t)u(t) \\ z(t) &= \mathbf{C}_1(t)x(t) + \mathbf{D}_{12}(t)u(t) \\ y(t) &= \mathbf{C}_2(t)x(t) + \mathbf{D}_{21}(t)w(t), \end{aligned} \tag{1}$$

where the system matrices are periodic $\mathbf{B}_1(t+N) = \mathbf{B}_1(t) \in \mathbb{R}^{s\times n}$, $\mathbf{B}_2(t+N) = \mathbf{B}_2(t) \in \mathbb{R}^{m\times n}$, $\mathbf{C}_1(t+N) = \mathbf{C}_1(t) \in \mathbb{R}^{n\times r}$, $\mathbf{C}_2(t+N) = \mathbf{C}_2(t) \in \mathbb{R}^{n\times p}$, $\mathbf{D}_{12}(t+N) = \mathbf{D}_{12}(t) \in \mathbb{R}^{m\times r}$, and $\mathbf{D}_{21}(t+N) = \mathbf{D}_{21}(t) \in \mathbb{R}^{s\times p}$.

We shall first assume full state space information, i.e. $\mathbf{C}_2 = \mathbf{I}$, and $\mathbf{D}_{21} = \mathbf{0}$, and periodic state feedback $u(t) = \mathbf{K}(t)x(t)$, $\mathbf{K}(t+N) = \mathbf{K}(t)$. The objectives of the control design is to compute a gain $\mathbf{K}(t)$ for which the transfer function

$$\begin{aligned} &s_c : w \mapsto z, \\ x(t+1) &= \mathbf{A}_c(t)x(t) + \mathbf{B}_1(t)w(t) \\ z(t) &= \mathbf{C}_c(t)x(t), \end{aligned} \tag{2}$$

where $\mathbf{A}_c(t) = \mathbf{A}(t) + \mathbf{B}_2(t)\mathbf{K}(t)$, $\mathbf{C}_c(t) = \mathbf{C}_1(t) + \mathbf{D}_{12}(t)\mathbf{K}(t)$ satisfies

$$||s_c||_2 < \gamma \tag{3}$$

The main results are summarized in the following theorem

Theorem 1. (Wisniewski and Stoustrup, 2001) Consider a periodic discrete time system s_c, $(\mathbf{A}(t), \mathbf{B}_2(t))$ stabilizable. The suboptimal H_2 problem Eq. (3) is solvable if and only if there exists a symmetric periodic matrix $\mathbf{Q}(t)$ and a periodic $\mathbf{Z}(t)$ such that

$$\begin{aligned} &\left(\mathbf{W}_1(t)^{\mathrm{T}}\mathbf{A}(t) + \mathbf{W}_2(t)^{\mathrm{T}}\mathbf{C}_1(t)\right)\mathbf{Q}(t-1) \\ &\times \left(\mathbf{A}(t)^{\mathrm{T}}\mathbf{W}_1(t) + \mathbf{C}_1(t)^{\mathrm{T}}\mathbf{W}_2(t)\right) \\ &- \mathbf{W}_1(t)^{\mathrm{T}}\mathbf{Q}(t)\mathbf{W}_1(t) - \mathbf{W}_2(t)^{\mathrm{T}}\mathbf{W}_2(t) < \mathbf{0}, \end{aligned} \tag{4}$$

$$\begin{bmatrix} \mathbf{Q}(t) & \mathbf{B}_1(t) \\ \mathbf{B}_1(t)^{\mathrm{T}} & \mathbf{Z}(t) \end{bmatrix} > \mathbf{0}, \tag{5}$$

$$\mathrm{tr}\left(\sum_{t=0}^{N-1} \mathbf{Z}(t)\right) < N\gamma^2, \tag{6}$$

where $\mathrm{im}\begin{bmatrix} \mathbf{W}_1(t) \\ \mathbf{W}_2(t) \end{bmatrix} = \ker\left[\mathbf{B}_2(t)^{\mathrm{T}} \; \mathbf{D}_{12}(t)^{\mathrm{T}}\right]$.

The H_2 control synthesis is decomposed into a feasibility problem of finding symmetric periodic matrices $\mathbf{Q}(t)$ and $\mathbf{Z}(t)$ meeting the inequalities (4) to (6) and a problem of finding a periodic control gain $\mathbf{K}(t)$ satisfying the following LMI (Wisniewski and Stoustrup, 2001)

$$\begin{bmatrix} -\mathbf{Q}(t) & \mathbf{A}(t) & \mathbf{0} \\ \mathbf{A}(t)^{\mathrm{T}} & -\mathbf{Q}^{-1}(t-1) & \mathbf{C}_1(t)^{\mathrm{T}} \\ \mathbf{0} & \mathbf{C}_1(t) & -\mathbf{I} \end{bmatrix} + \left[\mathbf{B}_2(t)^{\mathrm{T}}\ \mathbf{0}\ \mathbf{D}_{12}(t)^{\mathrm{T}}\right]^{\mathrm{T}} \mathbf{K}(t) \left[\mathbf{0}\ \mathbf{I}\ \mathbf{0}\right] + \left[\mathbf{0}\ \mathbf{I}\ \mathbf{0}\right]^{\mathrm{T}} \mathbf{K}(t)^{\mathrm{T}} \left[\mathbf{B}_2(t)^{\mathrm{T}}\ \mathbf{0}\ \mathbf{D}_{12}(t)^{\mathrm{T}}\right] < \mathbf{0}. \quad (7)$$

The following algorithm will be used in Section 3 for the periodic state feedback synthesis

Algorithm 1.

(1) Find using a symmetric matrix $\mathbf{Q}(t)$ and a matrix $\mathbf{Z}(t)$ for $t = 0...N-1$ satisfying LMIs (4) to (6).
(2) For each $t = 0...N-1$ find a matrix $\mathbf{K}(t)$, which satisfies LMI (7).

The observer synthesis reduces to an application of the duality argument. The following system is considered

$$\begin{aligned} s_o &: \boldsymbol{w} \mapsto \boldsymbol{z}, \\ \boldsymbol{x}(t+1) &= \mathbf{A}_o(t)\boldsymbol{x}(t) + \mathbf{B}_o(t)\boldsymbol{w}(t) \\ \boldsymbol{z}(t) &= \mathbf{C}_1(t)\boldsymbol{x}(t), \end{aligned} \quad (8)$$

where $\mathbf{A}_o(t) = \mathbf{A}(t) + \mathbf{L}(t)\mathbf{C}_2(t)$, $\mathbf{B}_o(t) = \mathbf{B}_1(t) + \mathbf{L}(t)\mathbf{D}_{21}(t)$. The observer synthesis is such that the gain $\mathbf{L}(t)$ fulfills

$$||s_o||_2 < \gamma \quad (9)$$

The problem Eq. (9) is solvable if and only if there exists a symmetric periodic matrix $\mathbf{Q}(t)$ and a periodic $\mathbf{Z}(t)$ such that

$$\left(\mathbf{W}_1(t)^{\mathrm{T}}\mathbf{A}(t)^{\mathrm{T}} + \mathbf{W}_2(t)^{\mathrm{T}}\mathbf{B}_1(t)^{\mathrm{T}}\right)\mathbf{Q}(t-1) \times \left(\mathbf{A}(t)\mathbf{W}_1(t) + \mathbf{B}_1(t)\mathbf{W}_2(t)\right) - \mathbf{W}_1(t)^{\mathrm{T}}\mathbf{Q}(t)\mathbf{W}_1(t) - \mathbf{W}_2(t)^{\mathrm{T}}\mathbf{W}_2(t) < \mathbf{0}, \quad (10)$$

$$\begin{bmatrix} \mathbf{Q}(t) & \mathbf{C}_2^{\mathrm{T}}(t) \\ \mathbf{C}_2(t) & \mathbf{Z}(t) \end{bmatrix} > \mathbf{0}, \quad (11)$$

$$\mathrm{tr}\left(\sum_{t=0}^{N-1} \mathbf{Z}(t)\right) < N\gamma^2, \quad (12)$$

where $\mathrm{im}\begin{bmatrix} \mathbf{W}_1(t) \\ \mathbf{W}_2(t) \end{bmatrix} = \ker\left[\mathbf{C}_2(t)\ \mathbf{D}_{21}(t)\right]$. The observer gain is a solution of the following LMI

$$\begin{bmatrix} -\mathbf{Q}(t) & \mathbf{A}(t)^{\mathrm{T}} & \mathbf{0} \\ \mathbf{A}(t) & -\mathbf{Q}^{-1}(t-1) & \mathbf{B}_1(t) \\ \mathbf{0} & \mathbf{B}_1(t)^{\mathrm{T}} & -\mathbf{I} \end{bmatrix} + \left[\mathbf{C}_2(t)\ \mathbf{0}\ \mathbf{D}_{21}(t)\right]^{\mathrm{T}} \mathbf{L}(t)^{\mathrm{T}} \left[\mathbf{0}\ \mathbf{I}\ \mathbf{0}\right] + \left[\mathbf{0}\ \mathbf{I}\ \mathbf{0}\right]^{\mathrm{T}} \mathbf{L}(t) \left[\mathbf{C}_2(t)\ \mathbf{0}\ \mathbf{D}_{21}(t)\right] < \mathbf{0}. \quad (13)$$

The algorithm for periodic observer design is

Algorithm 2.

(1) Find a symmetric matrix $\mathbf{Q}(t)$ and a matrix $\mathbf{Z}(t)$ for $t = 0...N-1$ satisfying LMIs (10) to (12).
(2) For each $t = 0...N-1$ find $\mathbf{L}(t)$ which satisfies (13).

In spacecraft applications a time invariant control/oserver gains are often desirable in simple on-board implementations. In this case the step 1 in the Algorithm 1 and 2 remains unchanged, whereas the periodic $\mathbf{K}(t)$ and $\mathbf{L}(t)$ may be substituted by the time invariant matrix $\mathbf{K}$ and $\mathbf{L}$ in the step 2. The drawback of this approach is that the K and L matrices do not correspond to the optimal constant observer and controller gains.

3. MAGNETICALLY ACTUATED SPACECRAFT

The objectives of this section is to apply Algorithms 1 and 2 for the three-axis attitude control of a spacecraft in a low, highly inclined Earth orbit. The spacecraft is actuated by three mutually perpendicular electromagnetic coils. The interaction between the geomagnetic field and the magnetic field in the coil produces the control torque. The comparison between the expected magnetic field vector and the true magnetometer data is used for the attitude determination.

3.1 *Spacecraft Model*

The satellite considered in this study is modeled as a rigid body in the Earth gravitational field influenced by the aerodynamic drag torque and the control torque generated by the magnetorquers. The attitude is parameterized by the unit quaternion providing a singularity free representation of the kinematics (Goldstein, 1980), (Wertz, 1990).

The control torque, $\boldsymbol{N}_{ctrl}$, of the magnetically actuated satellite always lies perpendicular to the geomagnetic field vector, $\boldsymbol{b}$. Therefore a magnetic moment, $\boldsymbol{m}$, generated in the direction parallel to the local geomagnetic field has no influence on the satellite motion. This can be explained by the following equality

$$\boldsymbol{N}_{ctrl} = (\boldsymbol{m}_{\parallel} + \boldsymbol{m}_{\perp}) \times \boldsymbol{b} = \boldsymbol{m}_{\perp} \times \boldsymbol{b}, \quad (14)$$

where $\boldsymbol{m}_{\parallel}$ is the component of the magnetic moment parallel to $\boldsymbol{b}$, whereas $\boldsymbol{m}_{\perp}$ is perpendicular to the local geomagnetic field.

Concluding, the necessary condition for power optimality of a control law is that the magnetic moment lies on a plane perpendicular to the geomagnetic field vector.

Consider the following mapping

$$\tilde{m} \mapsto m \ : \ m \ = \ \tilde{m} \times b/|b|^2 \qquad (15)$$

where $|\cdot|$ denotes the standard Euclidean norm, and $\tilde{m}$ represents a new control signal for the satellite. Now, the magnetic moment, m, is exactly perpendicular to the local geomagnetic field vector and the control theory for a system with unconstrained input $\tilde{m}$ can be applied. The direction of the signal vector $\tilde{m}$ (contrary to $\mathbf{m}$) can be chosen arbitrarily by the controller.

The model of the sensor, a three axis magnetometer, will be developed in the following. For simplicity of this exposition it is assumed that the magnetometer measurements are provided in a coordinate system spanned on the principal axes, denoted a body coordinate system. The model of the vector magnetometer is then

$$y(t) = \mathbf{R}(q)b(t), \qquad (16)$$

where $\mathbf{R}(q)$ is the rotation matrix corresponding to the attitude quaternion q and describing a rotation from an orbit fixed coordinate system e.g. Local-Vertical-Local-Horizontal Coordinate System (LVLH) to the body coordinate system; for complete definition of the involved coordinate systems the reader is referred to (Wisniewski *et al.*, 2000). The vector $b(t)$ is the local magnetic field seen from the orbit (LVLH) and reproduced in this paper by the 8th-degree IGRF model (Wertz, 1990).

Locally the attitude can be represented by three coordinates. In this work three components of the vector part of the attitude quaternion are used. The continuous time linear model of the satellite motion is given in terms of the angular velocity and the vector part of the attitude quaternion (Wisniewski, 2000)

$$\frac{d}{dt}\begin{bmatrix}\delta\Omega \\ \delta q\end{bmatrix} = \mathbf{A}_s \begin{bmatrix}\delta\Omega \\ \delta q\end{bmatrix} + \mathbf{B}_s(t)\tilde{m} \qquad (17)$$

$$y(t) = \mathbf{C}_s(t)\begin{bmatrix}\delta\Omega \\ \delta q\end{bmatrix},$$

where

$$\mathbf{A}_s = \begin{bmatrix} 0 & 0 & -\sigma_x\omega_o & -6\omega_o^2\sigma_x & 0 & 0 \\ 0 & 0 & 0 & 0 & 6\omega_o^2\sigma_y & 0 \\ -\omega_o\sigma_z & 0 & 0 & 0 & 0 & 0 \\ \frac{1}{2} & 0 & 0 & 0 & 0 & \omega_o \\ 0 & \frac{1}{2} & 0 & 0 & 0 & 0 \\ 0 & 0 & \frac{1}{2} & -\omega_o & 0 & 0 \end{bmatrix},$$

$$\sigma_x = \frac{I_y - I_z}{I_x}, \ \sigma_y = \frac{I_z - I_x}{I_y}, \ \sigma_z = \frac{I_x - I_y}{I_z},$$

$$\mathbf{B}_s(t) = \begin{bmatrix} \frac{\mathbf{I}^{-1}}{|b|^2}\begin{bmatrix} -b_y^2 - b_z^2 & b_x b_y & b_x b_z \\ b_x b_y & -b_x^2 - b_z^2 & b_y b_z \\ b_x b_z & b_y b_z & -b_x^2 - b_y^2 \end{bmatrix} \\ \begin{bmatrix} 0 & 0 & 0 \\ 0 & 0 & 0 \\ 0 & 0 & 0 \end{bmatrix} \end{bmatrix},$$

$$\mathbf{C}_s(t) = 2\begin{bmatrix}\begin{bmatrix} 0 & -b_3 & b_2 \\ b_3 & 0 & -b_1 \\ -b_2 & b_1 & 0 \end{bmatrix} & \mathbf{0}\end{bmatrix},$$

where ω_o is the orbital rate, and I_x, I_y, I_z are components on the diagonal of the inertia tensor $\mathbf{I}$ (the principal moments of inertia). The matrix $\mathbf{C}_s(t)$ corresponds to a vector product $2b(t)\times$, the control matrix $\mathbf{B}_s(t)$ comes from the double cross product operation $-b(t) \times (b(t)\times)$ divided by $\mathbf{I}$. The upper left 3 by 3 submatrix of $\mathbf{A}_s$ is due to Euler coupling, the submatrix in the upper right corner arises from the gravity gradient, and the lower part of the matrix $\mathbf{A}$ is the linearized kinematics. Notice that the matrices $\mathbf{B}_s(t)$ and $\mathbf{C}_s(t)$ are the periodic parts of Eq. (17). This is due to periodicity the magnetic field vector $b(t)$.

The model (17) is used in Algorithms 1 and 2 computing the periodic observer and controller gains $\mathbf{K}(t)$ and $\mathbf{L}(t)$.

4. SIMULATION RESULTS

In the numerical calculations the satellite principal moments of inertia are assigned to $[180\ 150\ 1]^{\mathrm{T}}$, which characterizes a spacecraft with a long gravity gradient boom. The algorithm is implemented in the Matlab LMI toolbox. The orbit is divided into $N = 100$ samples. The normalized Earth magnetic field vector is shown in Figure 1. The components $(5,6)$, $(6,6)$ of $\mathbf{Q}(t)$ and $(2,2)$, $(2,3)$ of the observer gain $\mathbf{L}(t)$ in Algorithm 2 are depicted in Figure 2. The components $(5,6)$, $(6,6)$ of $\mathbf{Q}(t)$ and $(2,2)$, $(2,3)$ of $\mathbf{K}(t)$ in Algorithm 1 are depicted in Figure 3. It is seen that both the observer and the controller gain are periodic as expected. Comparing Figures 1 and 3 it is seen that near the polar zones, where the z-component of the geomagnetic field vector reaches maximum and minimum values, the pitch and roll gains increase, see the gray zones.

An impulse response of the closed loop system gives a reasonable interpretation of a H_2 control performance. The result of the closed-loop simulation for 8 orbits is shown in Figure 4. The initial attitude is $q = [0.82\ 0.02\ 0.05\ 0.57]^{\mathrm{T}}$. The components q_1, q_2, q_3 can be treated for small deviations from the identity quaterion as half values of pitch, roll, and yaw angles respectively. For the spacecraft considered in the simulation study pitch and roll are passively stabilized by the gravity gradient, whereas yaw needs active

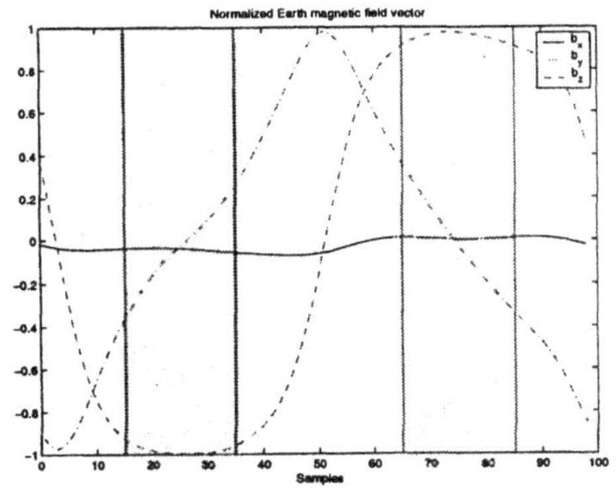

Fig. 1. The normalized magnetic field vector of the Earth computed for one orbit. The grey zones correspond to the polar regions.

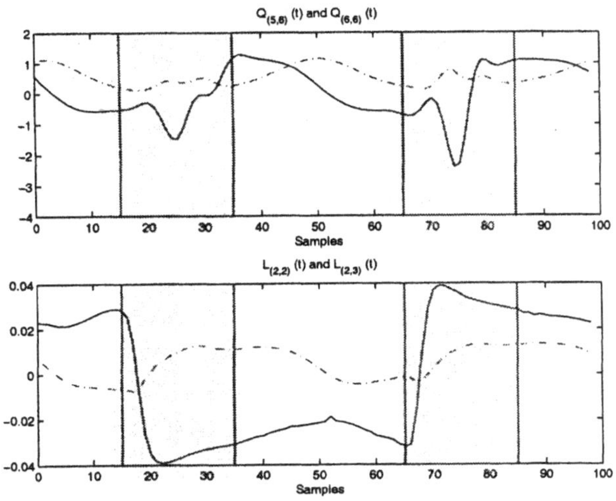

Fig. 2. The components (5,6), (6,6) of the observer matrix $\mathbf{Q}(t)$ and (2,2), (2,3) of $\mathbf{L}(t)$ computed for one orbit in Algorithm 2.

control action. It is seen that after four orbits, yaw is below the specified value 0.1, which corresponds to 10 deg., and the angular velocity is below 10^{-3} rad/sec.

5. CONCLUSIONS

A periodic control scheme for H_2 control synthesis was developed and implemented for the attitude control and estimation. The design algorithm presented showed the potential for onboard implementation on a small spacecraft platform equipped with a vector magnetometer and magnetorquers.

6. REFERENCES

Arduini, C. and P. Baiocco (1997). Active magnetic damping attitude control for gravity gradient stabilized spacecraft. *Journal of Guidance, Control, and Dynamics* **20**(1), 117–122.

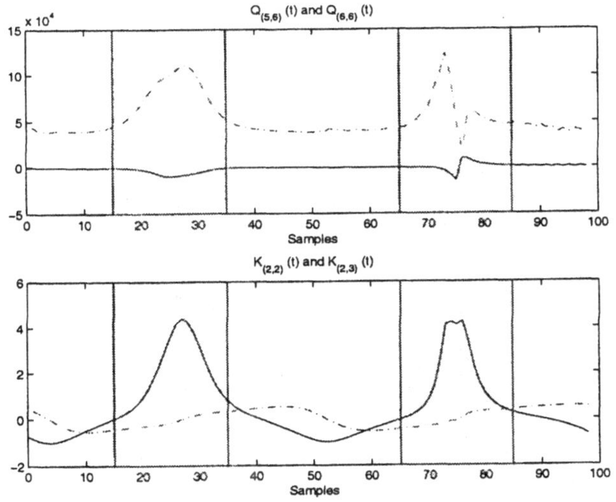

Fig. 3. The components (5,6), (6,6) of the matrix $\mathbf{Q}(t)$ and (2,2), (2,3) of the gain $\mathbf{K}(t)$ computed for one orbit in Algorithms 1. It is seen that the pitch and roll gains increase in the polar regions.

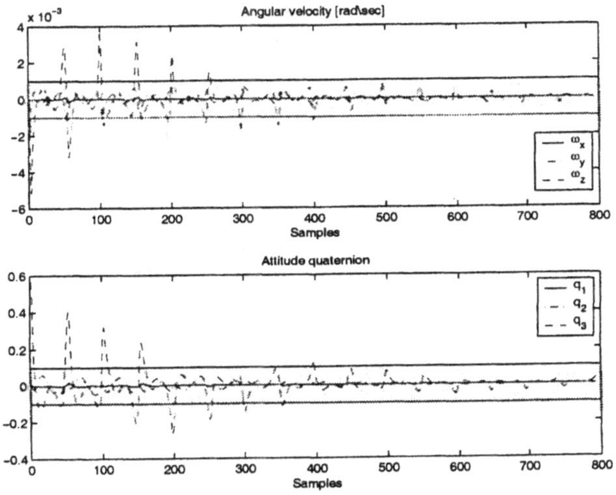

Fig. 4. The result of the closed-loop simulation: the angular velocity with respect to LVLH, and the vector part of the attitude quaternion. It is seen that after four orbits yaw is below the specified value 0.1, which corresponds to 10 deg., and the angular velocity is below 10^{-3} rad/sec.

Bak, T. (1999). Spacecraft Attitude Determination - a Magnetometer Approach. PhD thesis. Aalborg University.

Bittanti, S. and P. Colaneri (1999). An LMI characterization of the class of stabilizing controllers for periodic discrete-time systems. In: *14th IFAC World Congress, Beijing, China.*

Camillo, P.J. and F.L. Markley (1980). Orbit-averaged behavior of magnetic control laws for momentum unloading. *Journal of Guidance and Control* **3**(6), 563–568.

Cavallo, A., G. De Maria, F. Ferrara and P. Nistri (1993). A sliding manifold approach to satellite attitude control. In: *12th World Congress IFAC, Sydney.* pp. 177–184.

Challa, M., S. Kotaru and G. Natanson (1997). Magnetometer-only attitude and rate estimates during the earth radiation budget satellite 1987 control anomaly. In: *AIAA Guidance and Control Conference.*

Dullerud, G.E. and S. Lall (1999). A new approach for analysis and synthesis of time-varying systems. *IEEE Transactions on Automatic Control* **44**, 1486–1497.

Gahinet, P. and P. Apkarian (1991). A linear matrix inequality approach to h_{∞} control. *Int. J. Robust and Nonlinear Control* **4**, 421–448.

Goldstein, H. (1980). *Classical Mechanics.* Addison-Wesley.

Hablani, H. (1997). Pole-placement technique for magnetic momentum removal of earth-pointing spacecraft. *Journal of Guidance, Control, and Dynamics* **20**(2), 268–275.

Hablani, H.B. (1995). Comparative stability analyses and performance of magnetic controllers for momentum bias satellites. *Journal of Guidance, Control and Dynamics* **18**(6), 1313–1320.

Lefferts, E.J., F.L. Markley and M.D. Shuster (1982). Active magnetic damping attitude control for gravity gradient stabilized spacecraft. *Journal of Guidance, Control, and Dynamics* **5**(5), 417–429.

Martel, F., K.P. Parimal and M. Psiaki (1988). Active magnetic control system for gravity gradient stabilized spacecraft. In: *Annual AIAA/Utah State University Conference on Small Satellites.* pp. 1–10.

Natanson, G. (1993). A deterministic method for estimating attitude from magnetometer data. In: *43rd Congress of the International Astronautical Federation.*

Prato, G. Da and A. Ichikawa (1988). Quadratic control for linear periodic systems. *Applied Mathematics and Optimization* **18**, 39–66.

Psiaki, Mark L., Francois Martel and Parimal K. Pal (1990). Three-axis attitude determination via kalman filtering of magnetometer data. *Journal of Guidance, Control and Dynamics* **13**(3), 506–514.

Psiaki, M.L. (1999). Autonomous low-earth-orbit determination from magnetometer and sun sensor data. *Journal of Guidance, Control, and Dynamics* **22**(2), 296–304.

Souza, C. E. De and A. Trofino (2000). An LMI approach to stabilization of linear discrete-time periodic systems. *International Journal of Control* **73**(8), 698–703.

Steyn, W.H. (1994). Comparison of low-earth orbitting satellite attitude controllers submitted to controllability constraints. *Journal of Guidance, Control, and Dynamics* **17**(4), 795–804.

Tabuada, P., P. Alves, P. Tavares and P. Lima (1999). A predictive algorithm for attitude stabilization and spin control of small satellites. In: *European Control Conference - ECC'99, Karlsruhe, Germany.*

Wertz, J.R. (1990). *Spacecraft Attitude Determination and Control.* Kluwer Academic Publishers.

Wisniewski, R. (2000). Linear time-varying approach to satellite attitude control using only electromagnetic actuation. *Journal of Guidance, Control, and Dynamics* **23**(4), 640–647.

Wisniewski, R., A. Astolfi, T. Bak, M. Blanke, P. Lima, K. Spindler, P. Tabuada and P. Tavares (2000). *Satellite Attitude Control in Control of Complex Systems.* Springer Verlag. pp. 127-162.

Wisniewski, R. and F.L. Markley (1999). Optimal magnetic attitude control. In: *14th IFAC World Congress, Beijing, China.*

Wisniewski, R. and J. Stoustrup (2001). Generalized h_2 control synthesis for periodic systems. In: *2001 American Control Conference, Arlington, Virgina.*

Wisniewski, R. and M. Blanke (1996). Three-axis satellite attitude control based on magnetic torquing. In: *13th IFAC World Congress, San Francisco, California.* pp. 291–297.

Copyright © IFAC Periodic Control Systems,
Cernobbio-Como, Italy, 2001

AUTONOMOUS ORBIT CONTROL FOR SPACECRAFT ON ELLIPTICAL ORBITS USING A NON-INERTIAL COORDINATE FRAME

Andreas H. Schubert

ESG Elektroniksystem- und Logistik GmbH
Einsteinstr. 174 , 81675 München
Germany
Tel.: +49-89-9216-2320
Fax: +49-89-9216-2632
e-mail: ASchubert@esg-gmbh.de

Abstract: A method for designing an autonomous orbit controller for spacecraft on elliptical orbits is presented. For the derivation of the linearized differential and difference equations of motion, a non-inertial coordinate frame is utilized which coincides with the one which is applied in the well-known Clohessy-Wiltshire equations for small excentricities. The derivation takes atmospheric drag into account. The resulting time-discrete periodic system is augmented by an integrator state. A control law is obtained by applying linear optimal periodic control. The orbit controller is tested and verified in simulation runs. *Copyright© 2001 IFAC.*

Keywords: Spacecraft autonomy, Satellite control, Periodic motion, Linear optimal control, Riccati equations, Discrete time systems, State space models, Linearization.

1. INTRODUCTION

Autonomous orbit control for spacecraft is of industrial and scientific interest in numerous applications (e.g. autonomous position and constellation keeping), because it facilitates the reduction of ground station costs while maintaining the required orbit strategy. Whereas attitude control of a spacecraft, which relates to controlling the rotational degrees of freedom, is widely used in practical satellite missions, orbit control, which relates to controlling the translational degrees of freedom is usually only performed in an average sense and mostly by manually adjusting (i.e. transmitting actuator commands from a ground control station) the orbit when certain orbit elements exceed a boundary relative to their desired values (e.g. Gurevich *et al.*, 2000; Wertz and Gurevich, 2001). Only recently was the first flight of a satellite with fully autonomous orbit control successfully tested (e.g. Gurevich *et al.*, 2000). In literature, the problem has been often addressed for various applications (e.g. Guinn and Boain, 1996; Ebert *et al*, 2000; Wertz and Gurevich, 2001), but mostly for circular orbits. Furthermore, even though the problem is closely related to satellite rendezvous maneuvers, for which a linear time invariant description is well known, (cf. Clohessy and Wiltshire, 1960), linear control is seldom employed. A control system equipped with a GPS receiver and an ion thruster using linear time invariant control was proposed in (Zeiler *et al.*, 1999) for circular orbits. It was shown that the actual position of the spacecraft can be reliably kept in a narrow box of one to two kilometers around the desired position for an extended period of time of about three years under realistic conditions. In addition to circular orbits, relative motion and control for elliptical orbits are increasingly attracting academic interest (e.g. Wiesel 2001) and in (Schubert, 2001) a method for designing an orbit controller for spacecraft on elliptical orbits using discrete time optimal periodic control was presented. Even though the latter outlined the basic design procedure, it used an inertial coordinate frame and did not allow for atmospheric drag which, in flight path sections close to the earth, causes significant disturbances. This paper, therefore, introduces a novel method for designing an autonomous orbit controller for spacecraft on elliptical orbits taking account of atmospheric drag.

This paper is organized as follows. In the next section, linearized differential and difference equations for spacecraft on elliptical orbits are derived in a non-inertial coordinate frame taking atmospheric drag into consideration. Section three briefly summarizes the optimal periodic controller design procedure which has been utilized. Section four presents the simulation environment, the mission which has been chosen to demonstrate the concept and simulation results. A summary and an perspective for future work conclude this paper.

2. DERIVATION OF STATE SPACE MODEL

For the derivation of linearized equations of motion, the desired state of the spacecraft is compared to a perturbed actual state of the spacecraft. The equations of motion for both states are derived in a non-inertial coordinate frame which has its origin at the center of gravity with the x/y plane in the orbit plane and which rotates with the angular velocity $\underline{\omega}$ of the desired state, cf. figure 1. This coordinate frame offers the advantage that for small eccentricities e it approximates the coordinate frame of the well known Clohessy-Wiltshire equations (Clohessy and Wiltshire, 1960).

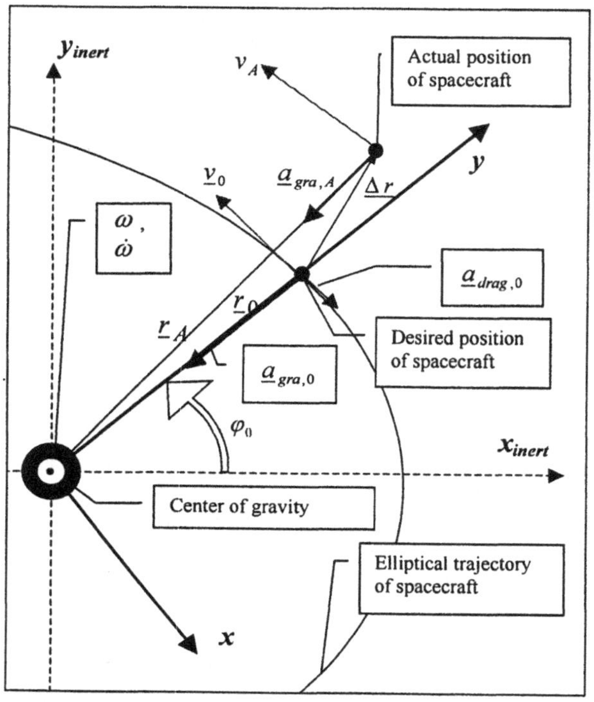

Fig. 1. Actual and desired state of spacecraft.

2.1. *Basic equations*

The use of a rotating coordinate frame necessitates the introduction of the following well-known formulae of a non-inertial coordinate frame, cf. figure 1:

$$\underline{a}_{inert} = \underline{a}_{guide} + \underline{a}_{non-inert} + \underline{a}_{cor} \tag{1}$$

where

$$\underline{a}_{guide} = \dot{\underline{\omega}} \times \underline{r} + \underline{\omega} \times (\underline{\omega} \times \underline{r}) \tag{2}$$

is the guidance acceleration, $\underline{a}_{non-inert}$ is the acceleration in the non-inertial coordinate frame,

$$\underline{a}_{cor} = 2\underline{\omega} \times \dot{\underline{r}} \tag{3}$$

is the Corriolis acceleration and $\underline{a}_{inert}$ is the absolute acceleration, i.e. the acceleration as measured in the inertial coordinate frame.

Three components contribute to the inertial acceleration $\underline{a}_{inert}$

$$\underline{a}_{inert} = \underline{a}_{gra} + \underline{a}_{drag} + \underline{a}_{thr} \,. \tag{4}$$

The gravitational acceleration resulting from a central gravity field (without oblateness effects) is given by

$$\underline{a}_{gra} = -\mu \frac{\underline{r}}{r^3} = -\frac{\mu}{r^2} \frac{\underline{r}}{r} \,, \quad r = |\underline{r}| \tag{5}$$

with the gravitational constant μ. The acceleration resulting from atmospheric drag is given by

$$\underline{a}_{drag} = -\frac{\rho(H)}{2m} v^2 c_D S \frac{\underline{v}}{v} \,, \quad v = |\underline{v}| \tag{6}$$

where $\underline{v}$ is the velocity vector, S is the cross-sectional area and c_D is the drag coefficient of the spacecraft. The density ρ is a function of the elevation of the spacecraft H above the earth surface. Since it is assumed that the actuator is aligned tangentially to the trajectory, the thrust vector is computed as

$$\underline{a}_{thr} = a_{thr} \frac{\underline{v}}{v} \,, \quad a_{thr} = |\underline{a}_{thr}| \,. \tag{7}$$

2.2. *Desired state*

Since the coordinate frame rotates with the angular velocity of the desired state, the position of the spacecraft in the desired state in the non-inertial coordinate frame is given by

$$\begin{bmatrix} x_0 \\ y_0 \\ z_0 \end{bmatrix} = \begin{bmatrix} r_0 \\ 0 \\ 0 \end{bmatrix} \tag{8}$$

and the angular velocity as well as angular acceleration are given by

$$\underline{\omega} = \begin{bmatrix} 0 \\ 0 \\ \omega \end{bmatrix} \text{ and } \dot{\underline{\omega}} = \begin{bmatrix} 0 \\ 0 \\ \dot{\omega} \end{bmatrix} \tag{9}$$

respectively. Following the derivations in (Schubert, 2001) the direction of the velocity vector can be obtained

$$\frac{\underline{v}_0}{v_0}=\frac{1}{\sqrt{e^2+2e\cos(\varphi_0)+1}}\begin{bmatrix}-1-e\cos(\varphi_0)\\ e\sin(\varphi_0)\\ 0\end{bmatrix} := \begin{bmatrix}t_x\\ t_y\\ 0\end{bmatrix} \tag{10}$$

Substituting equations (8) to (10) into (2) to (7) and the latter in (1), the following expression for the equations of motion for the desired state can be derived

$$\frac{-\mu}{r_0^2}\begin{bmatrix}1\\0\\0\end{bmatrix}-\frac{\rho(H)}{2m}v_0^2c_DS\begin{bmatrix}t_x\\t_y\\0\end{bmatrix}+\begin{bmatrix}t_x\\t_y\\0\end{bmatrix}a_{thr,0} = \begin{bmatrix}-\omega^2x_0+\ddot{x}_0\\ \dot{\omega}x_0+2\omega\dot{x}_0+\ddot{y}_0\\ \ddot{z}_0\end{bmatrix} \tag{11}$$

2.3. *Actual state*

The position and velocity of the spacecraft in the perturbed actual state in the non-inertial coordinate frame are given by

$$\underline{r}_A=\begin{bmatrix}x_A\\y_A\\z_A\end{bmatrix} \text{ and } \underline{v}_A=\begin{bmatrix}\dot{x}_A\\ \dot{y}_A\\ \dot{z}_A\end{bmatrix} \tag{12}$$

Observing that the direction of thrust is determined be the corresponding desired state, the equations of motion for the spacecraft in the actual state can be obtained

$$\underline{a}_{gra,A}+\underline{a}_{drag,A}+\begin{bmatrix}t_x\\t_y\\0\end{bmatrix}a_{thr,A} = \begin{bmatrix}-\dot{\omega}y_A-\omega^2x_A-2\omega\dot{y}_A+\ddot{x}_A\\ \dot{\omega}x_A-\omega^2y_A+2\omega\dot{x}_A+\ddot{y}_A\\ \ddot{z}_A\end{bmatrix} \tag{13}$$

Since only small perturbations of the actual state relative to the desired state are considered, linearized approximations can be employed. Expansion of the accelerations resulting from the gravitational forces in the actual state relative to the desired state yields (cf. Schubert, 2001)

$$\underline{a}_{gra,A}\approx-\frac{\mu}{r_0^2}\begin{bmatrix}1\\0\\0\end{bmatrix}+\frac{\mu}{r_0^3}\begin{bmatrix}2(x_A-x_0)\\ -(y_A-y_0)\\ -(z_A-z_0)\end{bmatrix} \tag{14}$$

Concerning the accelerations resulting from atmospheric drag, two effects can be distinguished. Firstly, differences between the actual and the desired velocity result in differences in the corresponding accelerations. Secondly, position differences lead to a change in density and therefore, in drag. Expansion of the accelerations resulting from atmospheric drag of the actual state relative to the desired state yields

$$\underline{a}_{drag,A}=-\frac{\rho(H)}{2m}v_0^2c_DS\begin{bmatrix}t_x\\t_y\\0\end{bmatrix} -\frac{\rho(H)}{m}v_0c_DS\begin{bmatrix}t_x^2(\dot{x}_A-\dot{x}_0)+t_xt_y(\dot{y}_A-\dot{y}_0)\\ t_xt_y(\dot{x}_A-\dot{x}_0)+t_y^2(\dot{y}_A-\dot{y}_0)\\ 0\end{bmatrix} -\partial\rho(H)/\partial H\,\frac{v_0^2c_DS}{2m}\begin{bmatrix}t_x(x_A-x_0)\\ t_y(x_A-x_0)\\ 0\end{bmatrix} \tag{15}$$

Substitution of equations (14) and (15) in (13) results in

$$\frac{-\mu}{r_0^2}\begin{bmatrix}1\\0\\0\end{bmatrix}+\frac{\mu}{r_0^3}\begin{bmatrix}2(x_A-x_0)\\ -(y_A-y_0)\\ -(z_A-z_0)\end{bmatrix}-\frac{\rho}{2m}v_0^2c_DS\begin{bmatrix}t_x\\t_y\\0\end{bmatrix}- \begin{bmatrix}d_{xx}^v(\dot{x}_A-\dot{x}_0)+d_{xy}^v(\dot{y}_A-\dot{y}_0)\\ d_{xy}^v(\dot{x}_A-\dot{x}_0)+d_{yy}^v(\dot{y}_A-\dot{y}_0)\\ 0\end{bmatrix}+\begin{bmatrix}t_x\\t_y\\0\end{bmatrix}a_{thr,A}- \begin{bmatrix}d_x^p(x_A-x_0)\\ d_y^p(x_A-x_0)\\ 0\end{bmatrix}=\begin{bmatrix}-\dot{\omega}y_A-\omega^2x_A-2\omega\dot{y}_A+\ddot{x}_A\\ \dot{\omega}x_A-\omega^2y_A+2\omega\dot{x}_A+\ddot{y}_A\\ \ddot{z}_A\end{bmatrix} \tag{16}$$

where

$$\begin{aligned} d_{xx}^v&=\frac{\rho}{m}v_0c_DSt_x^2\\ d_{xy}^v&=\frac{\rho}{m}v_0c_DSt_xt_y\\ d_{yy}^v&=\frac{\rho}{m}v_0c_DSt_y^2\\ d_x^p&=\partial\rho/\partial H\,\frac{v_0^2c_DS}{2m}t_x\\ d_y^p&=\partial\rho/\partial H\,\frac{v_0^2c_DS}{2m}t_y \end{aligned} \tag{17}$$

2.4. *Derivation of continuous state space model*

Central to control theory are deviations of the actual state from the desired state. Therefore, equation (11) concerning the desired state is subtracted from equation (17) concerning the actual state. By substituting differences between an actual state component and the corresponding desired state component by a difference expression, e.g.

$$\Delta x = x_A - x_0, \ldots, \Delta \dot{y} = \dot{y}_A - \dot{y}_0, \ldots, \text{etc.} \tag{18}$$

the following vector equation can be derived

$$\frac{\mu}{x_D^3}\begin{bmatrix} 2\Delta x \\ -\Delta y \\ -\Delta z \end{bmatrix} - \begin{bmatrix} d_{xx}^v \Delta\dot{x} + d_{xy}^v \Delta\dot{y} \\ d_{xy}^v \Delta\dot{x} + d_{yy}^v \Delta\dot{y} \\ 0 \end{bmatrix} - \begin{bmatrix} d_x^p \Delta x \\ d_y^p \Delta x \\ 0 \end{bmatrix} + \begin{bmatrix} t_x \\ t_y \\ 0 \end{bmatrix} \Delta a_{thr} = \begin{bmatrix} -\dot{\omega}\Delta y - \omega^2 \Delta x - 2\omega\Delta\dot{y} + \Delta\ddot{x} \\ \dot{\omega}\Delta x - \omega^2 \Delta y + 2\omega\Delta\dot{x} + \Delta\ddot{y} \\ \Delta\ddot{z} \end{bmatrix} \tag{19}$$

Equation (19) can be rearranged and transformed into the standard state space representation. The resulting equation can be decoupled into two sets of linear differential equations, one describing the motion in the orbit plane and the other describing the motion perpendicular to it. Since the motion in the orbit plane is more complex and difficult to stabilize, only the first set is considered in the following. This motion is governed by

$$\begin{bmatrix} \Delta\dot{x} \\ \Delta\dot{y} \\ \Delta\ddot{x} \\ \Delta\ddot{y} \end{bmatrix} = \begin{bmatrix} 0 \\ 0 \\ t_x \\ t_y \end{bmatrix} \Delta a_{thr} + \left[\begin{array}{c|c|c|c} 0 & 0 & 1 & 0 \\ \hline 0 & 0 & 0 & 1 \\ \hline \frac{2\mu}{r_0^3} + \omega^2 - d_x^p & \dot{\omega} & -d_{xx}^v & 2\omega - d_{xy}^v \\ \hline -\dot{\omega} - d_y^p & \omega^2 - \frac{\mu}{r_0^3} & -2\omega - d_{xy}^v & -d_{yy}^v \end{array}\right] \begin{bmatrix} \Delta x \\ \Delta y \\ \Delta\dot{x} \\ \Delta\dot{y} \end{bmatrix} \tag{20}$$

$$\hat{=} \underline{\dot{x}} = \underline{B}u + \underline{\underline{A}}\underline{x}$$

With this result, the continuous model of the motion in the orbit plane is complete. The continuous system is periodic with the orbital period *T*. However, it is intractable when used for continuous control, because the functional dependence of the positions *x* and *y* of time *t* cannot be solved analytically. It is therefore transferred to a discrete time system with the constant sample time

$$T_S = T / n_{Step} \tag{21}$$

Keppler's equation is employed to calculate the positions at the n_{Step} discrete time steps on the desired elliptical orbit.

2.5. *Linearized difference equations and integrator state*

Even though an analytical integration of Eq. (20) is impossible, the corresponding difference equations can be derived under the assumption that, for sample periods that are small as compared to the orbital period, the control and system matrix are approximately constant. The time discrete system and control matrices can therefore be calculated as

$$\underline{\underline{A}}_k^D = \exp\left(\underline{\underline{A}}\big|_k \cdot T_S\right)$$
$$\underline{B}_k^D = \left(\exp\left(\underline{\underline{A}}\big|_k \cdot T_S\right) - \underline{\underline{I}}\right)\left(\underline{\underline{A}}\big|_k\right)^{-1} \underline{B}\big|_k \tag{22}$$

constituting the discrete time system

$$\underline{x}_{k+1} = \underline{\underline{A}}_k^D \underline{x}_k + \underline{B}_k^D u_k . \tag{23}$$

Because of the periodicity of the system the relations

$$\underline{\underline{A}}_k^D = \underline{\underline{A}}_{k+n_{Step}}^D \text{ and } \underline{B}_k^D = \underline{B}_{k+n_{Step}}^D \tag{24}$$

hold. The denotation $|_k$ at the continuous system and control matrices is intended to indicate that the respective continuous matrix is appropriately chosen from within the interval that corresponds with the index *k*. Hence, a time discrete periodic system emerges comprising respectively n_{Step} time discrete system and control matrices.

Using linear control, the atmospheric drag would cause an undesirable constant average position error. To eliminate this error, the system state is augmented by a periodic integrator state x^I adding up the position errors in the flight path direction. The resulting state space model for the in-orbit motion can be obtained from (the Δ-symbol with the state variables here and in the following is omitted for clarity)

$$\begin{bmatrix} x \\ y \\ \dot{x} \\ \dot{y} \\ x^I \end{bmatrix}_{k+1} = \left[\begin{array}{cccc|c} & & & & 0 \\ & \underline{\underline{A}}^D & & & 0 \\ & & & & 0 \\ & & & & 0 \\ \hline t_x|_k & t_y|_k & 0 & 0 & 0 \end{array}\right]_k \begin{bmatrix} x \\ y \\ \dot{x} \\ \dot{y} \\ x^I \end{bmatrix}_k + \begin{bmatrix} \underline{B}^D \\ 0 \end{bmatrix}_k u_k \tag{25}$$

$$\hat{=}$$

$$\underline{x}_{k+1}^a = \underline{\underline{A}}_k^{D,a} \underline{x}_k^a + \underline{B}_k^{D,a} u_k$$

It is assumed that appropriate sensors (preferably in combination with a filter algorithm) are available which provide the complete system state with sufficient accuracy. However, to follow the well-established lines of linear optimal periodic control more closely, it is convenient to formally introduce an output

$$\underline{y}_k = \underline{\underline{C}}_k \underline{x}_k^a, \qquad (26)$$

which is of no immediate physical significance, and where $\underline{\underline{C}}_k \in \Re^{5\times5}$ is an appropriately chosen output matrix.

3. OPTIMAL CONTROLLER DESIGN

Analysis of the discrete time periodic system as well as optimal controller design can be performed via the time-invariant reformulation (e.g. Bittanti *et al.*, 1991), which consists of the monodromy matrix $\underline{\underline{\Psi}}_A$, an augmented input matrix $\overline{\underline{\underline{B}}}$, an augmented output matrix $\overline{\underline{\underline{C}}}$, and an augmented direct transmission matrix $\overline{\underline{\underline{D}}}$. These matrices are given by

$$\underline{\underline{\Psi}}_A := \overline{\underline{\underline{A}}} = \prod_{k=0}^{n_{Step}-1} \underline{\underline{A}}_k^{D,a}, \qquad (27)$$

$$\overline{\underline{\underline{B}}} = \left[\underline{\underline{\Phi}}_A (n_{Step},1) \underline{\underline{B}}_0^{D,a} \; \vdots \; \cdots \; \vdots \; \underline{\underline{B}}_{n_{Step}-1}^{D,a} \right], \qquad (28)$$

$$\overline{\underline{\underline{C}}} = \left[\underline{\underline{C}}_0^T \; \vdots \; \cdots \; \vdots \; \underline{\underline{\Phi}}_A (n_{Step}-1,0)^T \underline{\underline{C}}_{n_{Step}-1}^T \right]^T, \qquad (29)$$

$$\overline{\underline{\underline{D}}} = \begin{bmatrix} \overline{\underline{\underline{D}}}_{0,0} & \cdots & \overline{\underline{\underline{D}}}_{0,n_{Step}-1} \\ \vdots & & \vdots \\ \overline{\underline{\underline{D}}}_{n_{Step}-1,0} & \cdots & \overline{\underline{\underline{D}}}_{n_{Step}-1,n_{Step}-1} \end{bmatrix}, \qquad (30)$$

where

$$\underline{\underline{\Phi}}_A (j,i) = \prod_{k=i}^{k<j} \underline{\underline{A}}_k^{D,a} \qquad (31)$$

is the transition matrix and

$$\overline{\underline{\underline{D}}}_{i,j} = \begin{Bmatrix} 0 & \text{if } i \le j \\ \underline{\underline{C}}_{i-1}^T \underline{\underline{\Phi}}_A (i-1,j) \underline{\underline{B}}_{j-1}^{D,a} & \text{if } i > j \end{Bmatrix}. \qquad (32)$$

The matrices defined above constitute the time-invariant system

$$\begin{aligned} \overline{\underline{x}}_{l+1} &= \overline{\underline{\underline{A}}\underline{x}}_l + \overline{\underline{\underline{B}}\underline{u}}_l \\ \overline{\underline{y}}_l &= \overline{\underline{\underline{C}}\underline{x}}_l + \overline{\underline{\underline{D}}\underline{u}}_l \end{aligned}. \qquad (33)$$

A stabilizing feedback for the spacecraft motion can be obtained by application of linear optimal periodic control minimizing the quadratic performance index

$$J = \sum_{k=0}^{\infty} \left\{ \underline{y}_k^T \underline{y}_k + \underline{u}_k^T \underline{u}_k \right\} \qquad (34)$$

The computational procedure for the determination of the optimal control law involves solving the well-known Discrete time Algebraic Riccati Equation (DARE) (cf. Bittanti *et al.*, 1991) to which only one positive definite symmetric solution $\overline{\underline{\underline{P}}}$ exists if $(\overline{\underline{\underline{A}}}, \overline{\underline{\underline{B}}})$ is controllable and $(\overline{\underline{\underline{C}}}, \overline{\underline{\underline{A}}})$ observable (e.g. Anderson and Moore, 1971). The optimal feedback gains can be obtained from

$$\begin{aligned} \tilde{\underline{\underline{K}}} &= \left(\underline{\underline{I}} + \overline{\underline{\underline{D}}}^T \overline{\underline{\underline{D}}} + \overline{\underline{\underline{B}}}^T \overline{\underline{\underline{P}}\underline{\underline{B}}} \right)^{-1} \overline{\underline{\underline{B}}}^T \overline{\underline{\underline{P}}} \tilde{\underline{\underline{A}}} \\ &= \left[\underline{K}_0^T \; \vdots \; \underline{K}_1^T \; \vdots \; \cdots \; \vdots \; \underline{K}_i^T \; \vdots \; \cdots \; \vdots \; \underline{K}_{n_{Step}-1}^T \right]^T \end{aligned}. \qquad (35)$$

The closed loop system is given by

$$\begin{aligned} \underline{x}_{k+1}^a &= \underline{\underline{A}}_k^{D,a} \underline{x}_k^a + \underline{B}_k^{D,a} u_k \\ &= \left(\underline{\underline{A}}_k^{D,a} - \underline{B}_k^{D,a} \underline{K}_k \right) \underline{x}_k^a \end{aligned}. \qquad (36)$$

4. SIMULATION RESULTS

For verification of the mathematical derivation, the controller design has been tested in a nonlinear simulation comprising gravitational forces, atmospheric drag and, as control input, thrust. The gravity field is modeled as an inverse square gravity field as indicated by Eq. (5). For simulation of the atmospheric drag, the international reference atmosphere (CIRA, 1972) has been utilized. In addition to deterministic drag forces, stochastic disturbances have been included. It is assumed that the spacecraft is equipped with ion thrusters that provide a thrust magnitude which is linear within bounds and that the thrust is acting tangential to the trajectory (e.g. Zeiler *et al.*, 1999).

The controller consists of two components, the orbit generator and the implementation of the control algorithm. The orbit generator provides the desired spacecraft position and velocity at the n_{Step} discrete time steps. It is assumed that appropriate sensors are available (e.g. GPS or GALILEO receivers, accelerometers, star trackers, preferably in combination with a filter algorithm) that provide the actual system state with sufficient accuracy. The control algorithm determines the difference between the actual and the desired state and transforms it into the non-inertial coordinate frame. The integrator state is evaluated by recursively computing

$$x^I_{k+1} = x^I_k + t_x\big|_k x_k + t_y\big|_k y_k \tag{37}$$

Selection of the appropriate gain matrix is performed according to the position on the elliptical orbit and the thrust magnitude is given by

$$u_k = \underline{K}_k \underline{x}^a_k \tag{38}$$

A telecommunications mission on a highly elliptical orbit with orbit elements according to table 1 (cf. Potti *et al.*, 1995) has been selected to demonstrate the feasibility and validity of the concept.

Table 1 Orbit elements of spacecraft mission

Orbital period	8 h
Semi-major axis	20.278 km
Eccentricity	0.636
Perigee height	1000 km
Apogee height	26.800 km
Inclination	63.435 deg

The augmented system defined by these data is controllable and observable. Figure 2 displays a time plot of the position errors during three orbital periods.

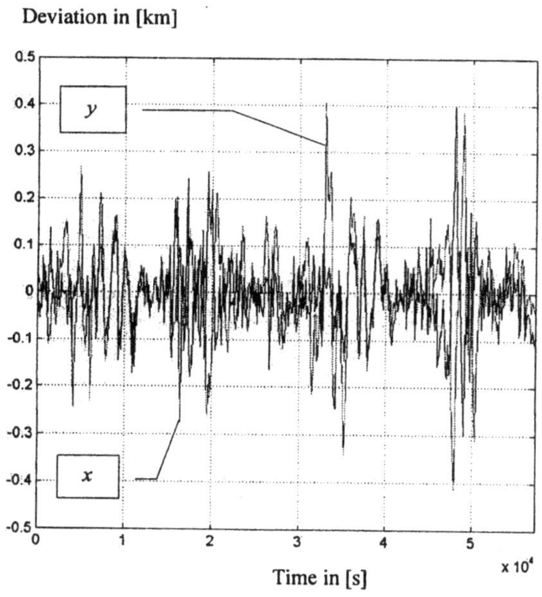

Fig. 2. Position errors of spacecraft with linear optimal control

The linear quadratic optimal periodic controller reliably stabilizes the spacecraft system. The position error in the orbit plane is limited to less than 500 m in both coordinate axes. The controller is also capable of leveling out initial errors. Figure 3 exhibits a time plot over one orbital period of the controlled spacecraft system after an initial error of 10 km at perigee.

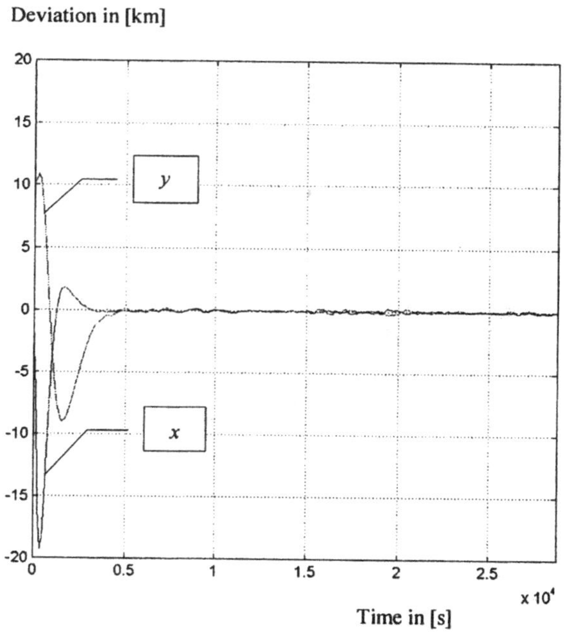

Fig. 3. Response of controlled system to initial error.

5. CONCLUSION

Linearized differential and difference equations and a controller design procedure for spacecraft on elliptical orbits have been presented using a non-inertial coordinate frame which facilitates results that approximate the well known Clohessy-Wiltshire equations for small eccentricities. In addition to gravitational forces, atmospheric drag has been taken into account, firstly in the derivation of the state space model and secondly by introducing a periodic integrator. In simulation runs it has been proven that the closed loop system is reliably stabilized, verifying the feasibility of the concept. Calculation of the gain matrices is complex; however, the complete design procedure can be performed offline and the results stored on a mission computer.

In view of growing academic and industrial interest in autonomous orbit control, the definition and realization of a spacecraft mission on an elliptical orbit equipped with an onboard autonomous orbit controller implementing a linear quadratic optimal control law would be the subsequent step. Constellation keeping for several spacecraft on an elliptical orbit is an interesting field of application of the presented controller design procedure. Furthermore, because of the duality theorem, the design procedure could also be easily transferred to the design of periodic estimators to determine the state of spacecraft on elliptical orbits.

6. ACKNOWLEDGEMENT

I would like to thank Prof. Dickmanns from the University of the German Armed Forces in Munich, Germany, for his support of this research and for many valuable ideas.

7. REFERENCES

Anderson, B.D.O. and J.B. Moore (1971): Linear optimal control. Prentice Hall, Englewood Cliffs, N.J..

Bittanti, S., P. Colaneri, G. De Nicolao (1991): The Periodic Riccati Equation. In: *The Riccati Equation* (S. Bittanti, A.J. Laub and J.C. Willems (Eds.)), pp. 127-162. Springer-Verlag; Berlin, Heidelberg.

CIRA (1972). Cospar International Reference Atmosphere (A. C. Stickland (Executive Ed.)). Akademie-Verlag GmbH, Berlin.

Clohessy, W. H. and R. S. Wiltshire (1960): Terminal Guidance System for Satellite Rendevous. *Journal of the Aerospace Sciences*, **27**, pp. 653-658.

Ebert, K., W. Fichter, E. Gottzein, O. Juckenhöfel, M. Mittnacht, C. Mueller, W. Oesterlin (2000): Autonomous GPS-based On-Board System for Geostationary Station Keeping. *Deutscher Luft- und Raumfahrtkongreß 2000 / DGLR-Jahrestagung 2000, Leipzig, September 18th-21st 2000*, pp. 925-932.

Guinn, J. R. and R. J. Boain (1996): Spacecraft Autonomous Navigation for Formation Flying Earth Orbiters Using GPS. *AIAA/AAS Astrodynamics Conference, Reston, Virgina, USA, July 29-31, 1996*, pp. 722-732.

Gurevich, G., R. Bell and J. Wertz (2000): Autonomous Orbit Control: Flight Results and Applications. *AIAA 2000 Conference and Exposition, Long Beach, California, September 19-21, 2000*. Paper AIAA 2000-5226.

Potti, J., P. Bernedo and A. Pasetti (1995): Applicability of GPS-based Orbit Determination Systems to a wide Range of HEO Missions. *Proceedings of the 8th International Meeting of the Satellite Division of the Institute of Navigation; Palm Springs, Ca.; Sept. 12-15 1995*, pp. 589-598.

Schubert, A. (2001): Linear Optimal Periodic Position Control for Elliptical Orbits. *AAS/AIAA Space Flight Mechanics Meeting, Santa Barbara, California, February 11-15, 2001*. Paper AAS 01-237.

Wiesel, W. E. (2001): The Dynamics of Relative Satellite Motion. *AAS/AIAA Space Flight Mechanics Meeting, Santa Barbara, California, February 11-15, 2001*. Paper AAS 01-163.

Zeiler, O., A. Schubert, B. Häusler and J. Puls (1999): Autonomous Orbit Control for Earth-Observation Satellites. *Proceedings 4th ESA International Conference on Spacecraft Guidance, Navigation and Control Systems; ESTEC, Noordwijk, The Netherlands; 18-21 October 1999* (**ESA SP-425**, February 2000).

Copyright © IFAC Periodic Control Systems,
Cernobbio-Como, Italy, 2001

Periodic Control of Systems with Delayed Observation Sharing Patterns [1]

Petros G. Voulgaris
Dept. of Aeronautical and Astronautical Engineering and Coordinated Science Laboratory
University of Illinois at Urbana-Champaign
1308 W. Main St.
Urbana IL 61801
petros@ktisivios.csl.uiuc.edu

Abstract

In this paper we present an input-output point of view of certain optimal control problems with constraints on the processing of the measurement data. In particular, we consider norm minimization optimal control problems under the so-called one-step delay observation sharing pattern. We present a Youla parametrization approach that leads to their solution by converting them to nonstandard, yet convex, model matching problems. This conversion is always possible whenever the part of the plant that relates controls to measurements possesses the same structure in its feedthrough term with the one imposed by the observation pattern on the feedthrough term of the controller, i.e., (block-)diagonal. When that is not the case, it amounts to the so-called non-classical information pattern problems. For the $\mathcal{H}^\infty$ case, using loop-shifting ideas, a simple sufficient condition is given under which the problem can be still converted to a convex, model matching problem. We also demonstrate that there are several nontrivial classes of problems satisfying this condition. Finally, we extend these ideas to the case of a N-step delay observation sharing pattern. *Copyright ©2001 IFAC*

Keywords: optimal, decentralized, input-output, discrete-time, lifting

1 Introduction

Optimal control under decentralized information structures is a topic that, although it has been studied extensively over the last forty years or so, still remains a challenge to the control community. The early encounters with the problem date back in the fifties and early sixties under the framework of team theory (e.g., [7, 8].) Soon it was realized that, in general, optimal decision making is very difficult to obtain when decision makers have access to private information, but do not exchange their information [17]. Nonetheless, under particular decentralized information schemes such as the partially nested information structures [6] certain optimal control problems admit trackable solutions. Several results exist by now when exchange of information is allowed with a one-step time delay (which is a special case of the partially nested information structure.) To mention only a few we refer to [3, 9] where LQG criteria are of interest, [10, 11, 12, 13] where linear exponential-quadratic Gaussian (LEQG) problems are considered and certain connections to minimax quadratic problems are furnished. The interested reader may further refer to [1] for further bibliographical information

In this paper, in contrast to the state-space viewpoint of the works previously cited, we undertake an input-output approach to optimal control under the information scheme known as the one-step delay observation sharing pattern (e.g., [3]). Under this pattern, measurement information can be exchanged between the decision makers with a delay of one time step. In the paper we provide an approach for solving the ℓ^1, $\mathcal{H}^\infty$ and $\mathcal{H}^2$ (or LQG) optimal disturbance rejection in the case of quasi-classical information exchange [2]. This is the case whenever the part of the plant that relates controls to measurements possesses the same structure in its feedthrough term with the one imposed by the observation pattern on the feedthrough term of the controller, i.e., (block-)diagonal. The key ingredient in this approach is the transformation of

[1] This work is supported by National Science Foundation Grant CCR 00-85917 ITR

the decentralization constraints on the controller to *linear* constraints on the Youla parameter used to characterize all controllers. Hence, the resulting problems in the input-output setting are, although nonstandard, convex. These problems are of the same form as the ones appearing in optimal control of periodic systems when lifting techniques are employed [4, 14], and can be solved analogously by employing Duality, Nehari and Projection theorems respectively.

When the part of the plant that relates controls to measurements *does not* possess the same structure in its feedthrough term with the one imposed by the observation pattern on the feedthrough term of the controller, the information exchange is nonclassical (e.g., [3]). The previous approach leads in general to nonconvex problems since the constraints on the Youla parameter are not linear any more. For the $\mathcal{H}^\infty$ case however, using loop-shifting ideas, a simple sufficient condition is given under which the problem can be still converted to a convex, model matching problem. We also demonstrate that there are nontrivial classes of problems satisfying this condition. These ideas can be extended in the case of a N-step delay observation sharing pattern using lifting techniques to obtain similar simple conditions for convexity [15].

2 Problem Definition

The standard block diagram for the disturbance rejection problem is depicted in Figure 1. In this figure, P denotes some fixed linear time invariant (LTI) causal plant, C denotes the compensator, and the signals w, z, y, and u are defined as follows: w, exogenous disturbance; z, signals to be regulated; y, measured plant output; and u, control inputs to the plant. In what follows we will assume that both P and C are LTI systems; we comment on this restriction on C later. Furthermore, we assume that there is a predefined information structure that the controller C has to respect when operating on the measurement signal y. The particular information structure is precisely defined in the sequel.

The one-step delay observation sharing pattern

To simplify our analysis we will consider the case where the control input u and plant output y are partitioned into two (possibly vector) components u_1, u_2 and y_1, y_2 respectively, i.e.,

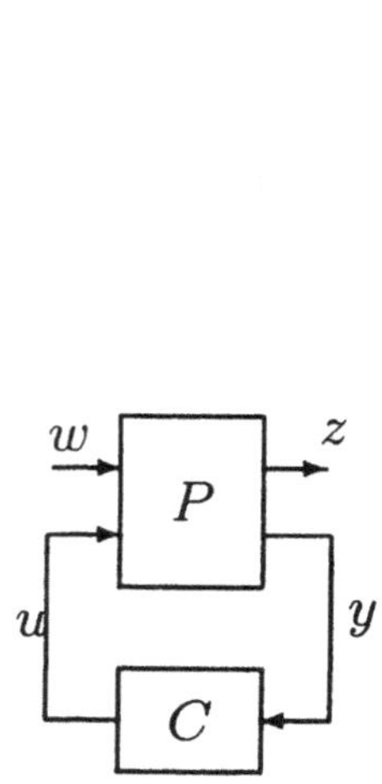

Figure 1: Block Diagram for Disturbance Rejection.

$u = (u_1^T\ u_2^T)^T$ and $y = (y_1^T\ y_2^T)^T$. Let $Y_k := \{y_1(0), y_2(0), \ldots, y_1(k), y_2(k)\}$ represent the measurement set at time k. The controllers that we are considering (henceforth, admissible controllers) are such that $u_1(k)$ is a function of the data $\{Y_{k-1}, y_1(k)\}$ and $u_2(k)$ is a function of the data $\{Y_{k-1}, y_2(k)\}$. I.e., there is one-step delay in passing the currently received information $y_2(k)$ to the channel that computes $u_1(k)$ and similarly for $y_1(k)$ and $u_2(k)$. We refer to this particular information processing structure imposed on the controller as the *one-step delay observation sharing pattern.*

Alternatively, partitioning the controller C accordingly as $C = \begin{pmatrix} C_{11} & C_{12} \\ C_{12} & C_{22} \end{pmatrix}$, the one-step delay observation sharing pattern requires that both C_{12} and C_{21} be strictly causal operators. Let now

$$\mathcal{S} := \{C \text{ stabilizing and LTI} : C_{12},\ C_{21} \text{ strictly causal}\}$$

and let T_{zw} represent the resulting map from w to z for a given compensator $C \in \mathcal{S}$ which we also denote as the linear fractional map $\mathcal{F}(P, C)$. The problem of interest is

$$\mu := \inf_{C \in \mathcal{S}} \|T_{zw}\| = \inf_{C \in \mathcal{S}} \|\mathcal{F}(P, C)\|$$

where $\|\bullet\|$ represents a system norm such as $\mathcal{H}^2$, $\mathcal{H}^\infty$ or ℓ^1.

To solve the above problems the following standard assumption is introduced. Let $P = \begin{pmatrix} P_{11} & P_{12} \\ P_{21} & P_{22} \end{pmatrix}$ then,

Assumption 2.1 *P is finite dimensional and stabilizable and the closed loop in Figure 1 is well-posed.*

Hence, if P has a state space description $P \sim (A, (B_1\ B_2), \begin{pmatrix} C_1 \\ C_2 \end{pmatrix}, \begin{pmatrix} D_{11} & D_{12} \\ D_{12} & D_{22} \end{pmatrix})$ then the pairs

(A, B_2) and (A, C_2) are stabilizable and detectable respectively.

3 Problem Solution

We consider separately two cases. In the first case it is assumed that the feedthrough term D_{22} is block diagonal which corresponds to a quasi-classical information pattern. The second case is when D_{22} is *not* block diagonal which corresponds to a non-classical information pattern. Using the Youla parametrization, the problem converts to a convex minimization in the first case, while in the second, sufficient conditions for which convexification is possible are provided.

3.1 Case (i): D_{22} block diagonal

The problem defined in the previous section can be related to problems in periodic systems where additional constraints that ensure causality appear in the so-called lifted system [4, 14]. These constraints are of similar nature as with the problem at hand. A basic step in the solution is the convenient characterization of all controllers that are in $\mathcal{S}$. This is done in the sequel.

Since we have assumed that P is finite dimensional with a stabilizable and detectable state space description we can obtain a doubly coprime factorization (dcf) of P_{22} using standard formulas (e.g., [5, 16]) i.e., having P_{22} associated with the state space description $P_{22} \sim (A, B_2, C_2, D_{22})$ a coprime factorization is $P_{22} = N_r D_r^{-1} = D_l^{-1} N_l$ with

$$\begin{pmatrix} X_l & -Y_l \\ -N_l & D_l \end{pmatrix} \begin{pmatrix} D_r & Y_r \\ N_r & X_r \end{pmatrix} = I$$

where $N_r \sim (A_K, B_2, C_K, D_{22})$, $D_r \sim (A_K, B_2, K, I)$, $N_l \sim (A_M, B_M, C_2, D_{22})$, $D_l \sim (A_M, M, C_2, I)$, $X_r \sim (A_K, -M, C_K, I)$, $Y_r \sim (A_K, B_2, -M, K, 0)$, $X_l \sim (A_M, -B_M, K, I)$, $Y_l \sim (A_M, -M, K, 0)$ with K, M selected such that $A_K = A + B_2 K$, $A_M = A + MC_2$ are stable (eigenvalues in the open unit disk) and $B_M = B_2 + MD_{22}$, $C_K = C_2 + D_{22}K$. Note that the above formulas indicate that the coprime factors of P_{22} have as feedforward terms the matrices D_{22} or I or 0 which are all block diagonal. The following is a well-known result (e.g.,[5, 16]):

Fact 3.1 *All ℓ^p-stabilizing LTI controllers C (possibly not in $\mathcal{S}$) of P are given by*

$$C = (Y_r - D_r Q)(X_r - N_r Q)^{-1} = (X_l - QN_l)^{-1}(Y_l - QD_l).$$

where Q is an ℓ^p bounded and causal LTI map.

The above fact characterizes the set of all stabilizing controllers in terms of the so-called Youla parameter Q and it amounts to the observer based parametrization in Figure ?? The set $\mathcal{S}$ of interest is clearly a subset of the set implied by Fact 3.1. The constraints on C amount to the constraint that the feedforward term of C should be block diagonal i.e.,

$$C(0) = \begin{pmatrix} C_{11}(0) & 0 \\ 0 & C_{22}(0) \end{pmatrix}.$$

A simple characterization of such a constraint is possible as the following lemma indicates

Lemma 3.1 *All ℓ^p-stabilizing controllers C in $\mathcal{S}$ of P are given by*

$$C = (Y_r - D_r Q)(X_r - N_r Q)^{-1} = (X_l - QN_l)^{-1}(Y_l - QD_l).$$

where Q is an ℓ^p bounded and causal LTI map with $Q(0)$ block diagonal.

Proof: It follows from the particular structure of the doubly coprime factors of P_{22} since $C(0) = -Q(0)(I - D_{22}Q(0))^{-1}$ with D_{22} block diagonal and hence $C(0)$ is block diagonal if and only if $Q(0)$ is block diagonal. ∎

Using the above lemma it follows that all feasible closed-loop maps are given as $T_{zw} = H - UQV$ where H, U, V stable depending only on P, and, Q is ℓ^p-stable with $Q(0)$ block diagonal. Hence the problem transforms to the minimization

$$\mu = \inf_{Q,\ Q(0)\ \text{block diagonal}} \|H - UQV\|$$

The problem above is an infinite dimensional, minimization of a convex functional over a convex domain (subspace). It represents a non-standard model matching problem due to the constraint on $Q(0)$. However, this constraint is of the same nature as in the optimal control problems of periodic systems considered and solved in [4] for ℓ^1 via duality theory, and, in [14] for $\mathcal{H}^\infty$ and $\mathcal{H}^2$ via Nehari and Projection Theorems respectively. The solution procedures in [4, 14] carry through exactly for the problem at hand and thus we refer the reader to these references for further details. We would also like to mention that considering smooth nonlinear controllers does not improve performance [15]

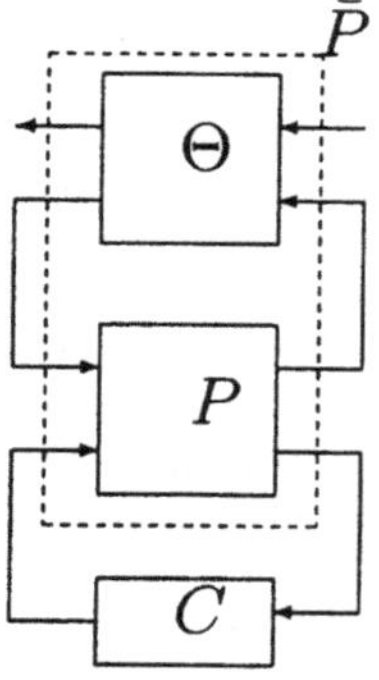

Figure 2: Loop shifting

3.2 Case (ii): D_{22} not block diagonal
When D_{22} is not block diagonal the constraints on C do not transform to convex constraints on the Youla parameter Q as in Lemma 3.1. Hence the resulting problem becomes a nonlinear and non-convex optimization which is in general difficult to handle. The idea in this section is to transform, if possible, the original problem to a problem of Case (i) of the previous section i.e., obtain an equivalent system for which the new D_{22} has the block diagonal structure that ensures convexity of the (new) problem. This is done herein *only for $\mathcal{H}^\infty$ optimal control* by using the so-called loop-shifting property (e.g., [18] chap. 16).

In particular, consider the loop-shifted system $\bar{P}$ of Figure 2 where $\Theta = \begin{pmatrix} \Theta_{11} & \Theta_{12} \\ \Theta_{21} & \Theta_{22} \end{pmatrix}$ is an inner matrix i.e., $\Theta^*\Theta = I$ with Θ_{21} invertible. Then the following equivalence holds

$$\|\mathcal{F}(P,C)\| < 1 \iff \|\mathcal{F}(\bar{P},C)\| < 1.$$

If $\bar{D}_{22} := \bar{P}_{22}(0)$ i.e., the new D_{22}-term in $\bar{P}$, then it holds that

$$\bar{D}_{22} = D_{22} + D_{21}\Theta_{22}(I - D_{11}\Theta_{22})^{-1}D_{12}.$$

The goal here is to select Θ_{22} so that $\bar{D}_{22}$ is block diagonal and thus apply the methods of the previous subsection to solve a convex yet nonstandard model matching problem. Splitting

$$D_{22} = D_{22,d} + D_{22,u}$$

where $D_{22,d}$ is the block diagonal part and $D_{22,u}$ is the remaining (unwanted) part of D_{22}, we would like to have a Θ_{22} such that

$$D_{22,u} = -D_{21}\Theta_{22}(I - D_{11}\Theta_{22})^{-1}D_{12}.$$

If we assume further that D_{21} and D_{12} have full row and column rank respectively then a matrix Θ_{22} that satisfies the above equation generically exists. However, such a Θ_{22} may not in general lead to an inner Θ: note that for Θ inner it is necessary $\|\Theta_{22}\| \leq 1$. To investigate when an appropriate Θ exists let

$$\rho := \inf\{\|\Theta_{22}\| : D_{22,u} = -D_{21}\Theta_{22}(I - D_{11}\Theta_{22})^{-1}D_{12}\}$$

and let Θ_{22}^o denote a minimizer [1] of the above definition (i.e., $\|\Theta_{22}^o\| = \rho$).

If we *assume* that $\mu\rho < 1$ then letting $\gamma > 0$ be such that $\mu < \gamma < 1/\rho$ where $\mu = \inf_{C \in S} \|\mathcal{F}(\bar{P},C)\|$ we have that for

$$\Theta := \begin{pmatrix} -\gamma\Theta_{22}^o & (I - \gamma^2\Theta_{22}^o\Theta_{22}^{o*})^{1/2} \\ (I - \gamma^2\Theta_{22}^{o*}\Theta_{22}^o)^{1/2} & \gamma\Theta_{22}^{o*} \end{pmatrix}$$

it holds that $\Theta^*\Theta = I$ and Θ_{21} invertible. This leads to

$$\|\mathcal{F}(P,C)\| < \gamma \iff \|\mathcal{F}(P_\gamma,C)\| < 1$$

in Figure 3. Therefore, the above equivalence allows as to solve the original problem by solving for

$$\inf\{\gamma : \|\mathcal{F}(P_\gamma,C)\| < 1\}$$

for each fixed γ which is a Case (i) problem of the previous section since P_γ has a block triangular D_{22} term. Hence, the solution procedure outlined in Case (i) can readily be applied. The following summarizes the developments so far.

Theorem 3.1 *If $\mu\rho < 1$, the $\mathcal{H}^\infty$ optimal disturbance rejection under the one step delay observation sharing pattern is a convex problem.*

To check whether the sufficient condition for convexity holds in the above theorem, one needs to know μ. This would not typically be known. However, one can check using computable upper bounds $\bar{\mu} \geq \mu$, i.e., check if $\bar{\mu}\rho < 1$. For example, if P is stable, a readily obtained upper bound on μ is obviously $\bar{\mu} = \|P_{11}\|$. A more general way could be to solve a *standard* $\mathcal{H}^\infty$ problem restricting C to be strictly causal. That amounts to placing a delay in the measured output (or control input) in both

[1] in situations where a minimizer does not exist Θ_{22}^o is taken to mean arbitrarily close to ρ in size, i.e., $\|\Theta_{22}^o\| = \rho + \delta$ with $\delta > 0$ arbitrarily small

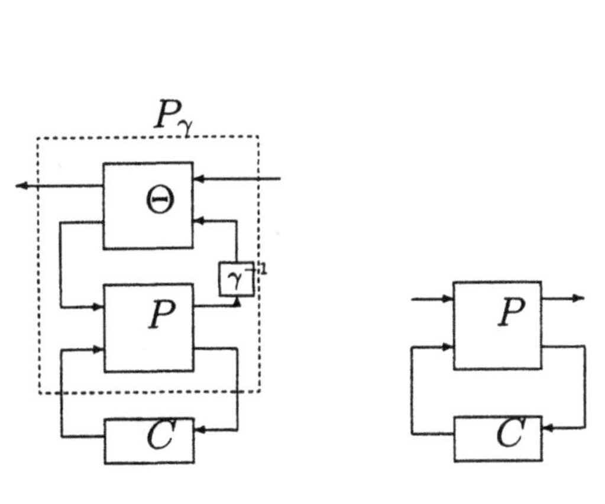

Figure 3: $\|\mathcal{F}(P, C)\| < \gamma$ iff $\|\mathcal{F}(P_\gamma, C)\| < 1$.

channels in the original system P i.e., augmenting P_{22} to ΛP_{22} where Λ is the delay operator, and design for the $\mathcal{H}^\infty$ optimal cost; it is simple to show that the augmented system is stabilizable given Assumption 2.1 and that the so obtained $\mathcal{H}^\infty$ optimal cost is an upper bound on μ. Obviously, the aforementioned bounds may be very far from μ and better ones can be sought for.

Irrespective of what upper bounds may be available for a-priori checks the following algorithm provides the solution to the problem whenever convexification via the loop shifting procedure described above is possible.

- **step 1**: Find
$$\rho := \inf\{\|\Theta_{22}\| : D_{22,u} = -D_{21}\Theta_{22}(I - D_{11}\Theta_{22})^{-1}D_{12}\}$$
and a minimizer Θ_{22}^o.
- **step 2**: Let $\gamma_0 = \frac{1}{\rho+\epsilon}$ where $\epsilon > 0$ is arbitrarily small
- **step 3**: Solve for $C \in \mathcal{S}$ such that
$$\|\mathcal{F}(P_{\gamma_0}, C)\| < 1$$
using the procedures indicated in Case (i)
- **step 4**: If step 3 feasible then $\mu\rho < 1$ and $\inf_{C \in \mathcal{S}} \|\mathcal{F}(P, C)\|$ is convex: solve equivalent
$$\mu = \inf\{\gamma < \gamma_0 : \|\mathcal{F}(P_\gamma, C)\| < 1\}$$
by the methods indicated in Case (i)
- **step 5**: If step 3 not feasible, then $\mu > 1/\rho$ but convexification fails

An interesting interpretation of Theorem 3.1 can be given in the special case (yet quite common) where $D_{11} = \begin{pmatrix} 0 & 0 \\ 0 & 0 \end{pmatrix}$, $D_{12} = \begin{pmatrix} 0 \\ I \end{pmatrix}$ and $D_{21} = (0 \;\; I)$. Letting
$$D_{22,u} = \begin{pmatrix} 0 & D_{2212} \\ D_{2221} & 0 \end{pmatrix}$$
we get
$$\Theta_{22}^o = \begin{pmatrix} 0 & 0 \\ 0 & -D_{22,u} \end{pmatrix}$$
and $\rho = \|D_{22,u}\| = \max(\|D_{2212}\|, \|D_{2221}\|)$. The sufficient condition for convexity of Theorem 3.1 gives
$$\mu \max(\|D_{2212}\|, \|D_{2221}\|) < 1.$$
This condition implies that if the size of the unwanted terms in D_{22} i.e., the off-block diagonal, are strictly smaller than the inverse of the optimal $\mathcal{H}^\infty$ cost, the problem is convex. If P is stable, $\mu \leq \|P_{11}\|$. Based on that, several problems with non-classical information pattern can be constructed that lead to convex problems. Consider as an example the norm minimization of weighted sensitivity and complementary sensitivity given as
$$\mathcal{F}(P, C) = \begin{pmatrix} W_{z_1}(I - P_{22}C)^{-1}W_{w_1} & W_{z_1}P_{22}C(I - P_{22}C)^{-1}W_{w_2} \\ W_{z_2}C(I - P_{22}C)^{-1}W_{w_1} & W_{z_2}C(I - P_{22}C)^{-1}W_{w_2} \end{pmatrix}$$
with W_{z_i}, W_{w_i} stable weights for $i = 1, 2$ satisfying $W_{w_1}(0) = W_{z_1}(0) = 0$ and $W_{z_2}(0) = W_{w_2}(0) = I$ and where P_{22} is stable. In this case $P_{11} = \begin{pmatrix} W_{z_1}W_{w_1} & 0 \\ 0 & 0 \end{pmatrix}$, $P_{12} = \begin{pmatrix} W_{z_1}P_{22} \\ W_{z_2} \end{pmatrix}$, $P_{21} = (W_{w_1} \;\; W_{w_2})$. If the weights are selected to satisfy $\|W_{z_1}W_{w_1}\| \max(\|D_{2212}\|, \|D_{2221}\|) < 1$ then $\mu\rho < 1$ and the problem is therefore convex. This shows that there are several nontrivial classes of problems for which the condition of Theorem 3.1 is satisfied.

4 Conclusions

We presented an input-output point of view for norm minimization optimal control problems under the one-step delay observation sharing pattern. In the case where the part of the plant that relates controls to measurements possesses the same structure in its feedthrough term with the one imposed by the observation pattern on the feedthrough term of the controller, i.e., (block-)diagonal, the problem is convex and a procedure that leads to its solution was given. It was also documented in this case of such a quasi-classical information pattern, that smooth nonlinear time-varying controllers do not outperform linear time invariant.

The case where the part of the plant that relates controls to measurements does not posses the same structure in its feedthrough term with the one imposed by the observation pattern on the feedthrough term of the controller, a case of non-classical information pattern, was also considered. For the $\mathcal{H}^\infty$ case, using loop-shifting ideas a simple sufficient condition was given under which the problem can be still converted to a convex, model matching problem. In this condition, the size of the unwanted terms for convexity in the part of the plant that relates controls to measurements is related to the optimal cost. It was also demonstrated that there are several nontrivial classes of problems satisfying this condition and a general high level algorithm was given to solve the problem if convexification with loop-shifting is possible.

References

[1] T. Başar and R. Bansal. "The theory of teams: a selective annotated bibliography," *Differential Games and Applications*, Lecture Notes in Control and Information Sciences, vol. 119, Springer-Verlag, pp. 186-201, 1989.

[2] T. Başar and J.B. Cruz, Jr. "Concepts and methods in multiperson coordination and control," *Optimization and Control of Dynamic Operational Research Models*, North Holland, 1982.

[3] T. Başar. "Two-criteria LQG decision problems with one-step delay observation sharing pattern," *Information and Control*, vol. 38, pp. 21-50, 1978.

[4] M. A. Dahleh, P.G. Voulgaris, and L. Valavani, "Optimal and robust controllers for periodic and multirate systems," *IEEE Trans. Automat. Control*, vol. AC-37, pp. 90–99, January 1992.

[5] B.A. Francis. *A Course in H_∞ Control Theory*, Springer-Verlag, 1987.

[6] Y.C. Ho and K.C. Chu. "Team decision theory and information structures in optimal control problems-parts I and II," *IEEE Trans. A-C*, Vol AC-17, 15-22, 22-28, 1972.

[7] J. Marschak and R. Rander. "The firm as a team," *Econometrica*, 22, 1954

[8] R. Rander. "Team decision problems," *Ann. Math. Statist.*, vol 33, pp. 857-881, 1962

[9] N. Sandell and M. Athans. "Solution of some nonclassical LQG stochastic decision problems," *IEEE Trans. A-C*, Vol AC-19, pp. 108-116, 1974.

[10] J.L. Speyer, S.I. Marcus and J.C. Krainak. "A decentralized team decision problem with an exponential cost criterion," *IEEE Trans. A-C*, Vol AC-25,pp. 919-924, 1980.

[11] J.C. Krainak, J.L. Speyer and S.I. Marcus. "Static team problems-Part I: Sufficient conditions and the exponential cost criterion," *IEEE Trans. A-C*, Vol AC-27,pp. 839-848, 1982.

[12] C. Fan, J.L. Speyer and C. Jaensch. "Centralized and decentralized solutions to the linear exponential-Gaussian problem," *IEEE Trans. A-C*, Vol AC-39, pp. 1986-2003, 1994.

[13] R. Srikant, "Relationships between decentralized controllers design using $\mathcal{H}^\infty$ and stochastic risk-averse criteria," *IEEE Trans. A-C*, Vol AC-39,pp. 861-864, 1994.

[14] P.G. Voulgaris, M.A. Dahleh and L.S. Valavani, "$\mathcal{H}^\infty$ and $\mathcal{H}^2$ optimal controllers for periodic and multirate systems," *Automatica*, vol. 30, no. 2, pp. 252-263, 1994.

[15] P.G. Voulgaris, "Convex problems in systems with delayed observation sharing patterns," Tech. Report AAE 00-02, UILU ENG 00-0502, U. of Illinois at U-C, Sept. 2000.

[16] M. Vidyasagar. *Control Systems Synthesis: A Factorization Approach*, MIT press, 1985.

[17] H.S. Witsenhausen, "A counterexample in stochastic optimal control," *SIAM J. Contr.*, vol 6, pp. 131-147, 1968.

[18] K. Zhou, J. C. Doyle and K. Glover, *Robust and Optimal Control*. New Jersey: Prentice Hall, 1996

Copyright © IFAC Periodic Control Systems,
Cernobbio-Como, Italy, 2001

LINEAR PERIODIC CONTROL FOR ρ–STABILIZATION AND ASYMPTOTIC TRACKING UNDER UNBOUNDED OUTPUT MULTIPLICATIVE PERTURBATIONS [2]

Sergio Galeani, Osvaldo Maria Grasselli and Laura Menini [*,1]

* *Dip. Informatica, Sistemi e Produzione*
Università di Roma Tor Vergata
Via di Tor Vergata, 110 -00133 Roma, Italy

Abstract: In this paper the problem of the asymptotic tracking and stabilization with infinite gain margin, is addressed for linear time-invariant discrete-time multivariable systems in the case when unknown different scalar gains act on the outputs. Necessary and sufficient conditions for the solvability of the problem by means of a linear periodic discrete-time error feedback dynamic controller are derived. A procedure is given for designing the proposed periodic controller (containing an internal model of the reference signals). *Copyright ©2001 IFAC*

Keywords: Robustness, tracking, gain margins, MIMO, discrete-time system

1. INTRODUCTION

The problem of the robust asymptotic tracking and stabilization of a linear time-invariant multivariable system was studied by several authors both for continuous-time and for discrete-time plants (see, e.g., (Davison, 1976),(Francis and Wonham, 1976), (Grasselli and Longhi, 1991*b*), (Grasselli *et al.*, 1993), (Kimura and Tanaka, 1981) and the references therein). In such papers, both independent perturbations of the matrices characterizing the state-space description of the plant, and perturbations of some physical parameters, were considered. Here, the same kind of problem is dealt with for a system whose outputs are subject to unknown multiplicative perturbations of scalar gains whose values may be unbounded.

This kind of problem, for the mere asymptotic stability, is called the gain margin problem. For SISO systems, and for the case when the unknown scalar gain may vary in a finite interval and a linear time-invariant controller is used, the problem was studied and solved in (Tannenbaum, 1980; Tannenbaum, 1982; Khargonekar and Tannenbaum, 1985). If the unknown scalar gain is allowed to vary in an infinite interval the problem is usually called the infinite gain margin (IGM) problem. Periodic controllers to obtain an arbitrarily large gain margin were described in (Khargonekar *et al.*, 1985) for discrete-time plants, in (Lee *et al.*, 1987) for continuous-time plants, since periodic control is strongly helpful to face some structured uncertainties (Khargonekar *et al.*, 1985).

For MIMO systems, the same problem can be extended in two different ways: by considering a scalar unknown constant gain acting on every output (or input) of the system, or by considering different independent unknown constant gains acting on each scalar output (or each scalar input) of the system. Note that, in the second case, in general a controller designed to solve the problem when the unknown gains act on the input is not able to solve the problem when the unknown gains act on the output.

An arbitrarily large gain margin for the first kind of extension by linear periodic controllers was obtained in (Francis and Georgiou, 1988; Yan *et al.*, 1994) both for discrete-time systems and for continuous-time systems by digital control schemes (in (Francis and Georgiou, 1988) by means of a conventional periodic digital controller, in (Yan *et al.*, 1994) by the use of Generalized Sampled-data Hold Functions (GSHF)). For the second kind of extension, the authors are aware of two kinds of results for continuous-time systems: the well known LQR, state feedback results and some sufficient conditions (which become necessary if decoupling is required too) in (Maeda and Vidyasagar, 1985) for dynamic output feedback; for the same kind of extension, the IGM problem is solved in (Galeani *et al.*, 2000) for discrete-time systems, in the case when the unknown scalar gains act on the inputs of the plant, making use of linear periodic controllers.

[1] E-mail: {galeani,grasselli,menini}@disp.uniroma2.it
[2] This work was supported by ASI, CNR and MURST

Here the same kind of problem, i.e., the IGM problem for discrete-time linear time invariant (LTI) MIMO systems, will be addressed in the case of different independent unknown scalar gains acting on *each scalar output* of the system, with the additional requirement of the asymptotic tracking of reference signals belonging to a general known class of polynomial-exponential kind. Necessary and sufficient conditions, and a design procedure, will be given for the solution of the problem via a linear periodic error feedback controller, in the case of a prescribed rate of convergence of the free responses of the closed-loop system. The proposed compensator is the series connection of a LTI subcompensator, which guarantees the internal model needed to asymptotically track the exogenous signals, and of a linear periodically time-varying (LPTV) subcompensator whose role is to guarantee the prescribed robust asymptotic stability of the control system.

Notations For a linear discrete-time system $\widehat{S}$, its input, state and output will be denoted by $u^{\widehat{S}}(\cdot)$, $x^{\widehat{S}}(\cdot)$ and $y^{\widehat{S}}(\cdot)$, respectively, unless otherwise specified, and the matrices giving its state-space description will be denoted by $A^{\widehat{S}}(\cdot)$, $B^{\widehat{S}}(\cdot)$, $C^{\widehat{S}}(\cdot)$, and $D^{\widehat{S}}(\cdot)$, so that $\widehat{S} = \left(A^{\widehat{S}}(k), B^{\widehat{S}}(k), C^{\widehat{S}}(k), D^{\widehat{S}}(k)\right)$ will be characterized by:

$$x^{\widehat{S}}(k+1) = A^{\widehat{S}}(k)x^{\widehat{S}}(k) + B^{\widehat{S}}(k)u^{\widehat{S}}(k), \quad \text{(1a)}$$
$$y^{\widehat{S}}(k) = C^{\widehat{S}}(k)x^{\widehat{S}}(k) + D^{\widehat{S}}(k)u^{\widehat{S}}(k). \quad \text{(1b)}$$

Throughout the paper, $\mathbf{Z}^+$ will denote the set of non negative integers, and, for non-dynamic systems, either LTI or LPTV, the same symbol will be used for the matrix of the gains which characterizes its input-output behaviour and for the system itself.

The identity matrix of dimension ν will be denoted by I_ν or, when confusion cannot arise, simply by I. Zero vectors, of dimension s, and matrices, of dimensions $s \times r$, will be denoted by 0_s and $0_{s\times r}$, respectively, whenever specifying such dimensions will be useful. Further, given a vector $v \in \mathbb{R}^a$ and an ordered set of positive integers $s = \{s_1, \dots, s_e\}$ of cardinality $e \le a$ and such that $1 \le s_1 < s_e \le a$, two new vectors $\langle v\rangle_{(s)} \in \mathbb{R}^e$ and $\langle v\rangle^{(s)} \in \mathbb{R}^a$ are defined, where the i-th component of $\langle v\rangle_{(s)}$ is equal to the s_i-th component of v for each $i \in \{1, \dots, e\}$, whereas $\langle v\rangle^{(s)}$ is equal to v with the exception of the components in positions $s_1, \dots, s_e$ which are equal to zero. For singletons like $\{s_1\}$, the shorthand $\langle v\rangle_{(s_1)}$ will be used instead of $\langle v\rangle_{(\{s_1\})}$. Given a matrix $M \in \mathbb{R}^{a\times b}$ and two ordered sets of positive integers $s = \{s_1, \dots, s_e\}$ of cardinality $e \le a$ and $t = \{t_1, \dots, t_f\}$ of cardinality $f \le b$, and such that $1 \le s_1 < s_e \le a$, $1 \le t_1 < t_f \le b$, two new matrices $\langle M\rangle_{(s,t)} \in \mathbb{R}^{e\times f}$ and $\langle M\rangle^{(s,t)} \in \mathbb{R}^{a\times b}$ are defined, where the (i,j) element of $\langle M\rangle_{(s,t)}$ is equal to the element (s_i, t_j) of M for each $i \in \{1, \dots, e\}, j \in \{1, \dots, f\}$, whereas all the elements of $\langle M\rangle^{(s,t)}$ are equal to the corresponding elements of M with the exception of the elements in positions (s_i, t_j) for $i \in \{1, \dots, e\}, j \in \{1, \dots, f\}$ which are equal to zero. For singletons like $\{s_1\}$, the shorthand $\langle M\rangle_{(s_1,t)}$ will be used instead of $\langle M\rangle_{(\{s_1\},t)}$ and a similar notation will be used if t or, possibly, both s and t are singletons.

For any given square matrix $M \in \mathbb{R}^{\nu\times\nu}$, the spectral radius of M (i.e. the maximum of the moduli of the eigenvalues of M) will be denoted by $r(M)$. Note that the matrix norm $\|M\|_\infty$, defined by $\|M\|_\infty := \max_{i\in\{1,\dots,\nu\}} \sum_{j=1}^{\nu} \left|\langle M\rangle_{(i,j)}\right|$, satisfies the relation $r(M) \le \|M\|_\infty$.

For any matrix $N \in \mathbb{R}^{\mu\times\nu}$ of rank μ, $N^\sharp$ will denote its right pseudoinverse, expressed by $N^\sharp = N'(NN')^{-1}$. For $j, k \in \mathbb{Z}$, $k < j$, the symbol $\sum_{i=j}^{k} s(i)$, for any kind of argument $s(i)$, will have the meaning of zero, even when the argument $s(i)$ cannot be computed for $i < j$ or for $i > k$.

2. PRELIMINARIES

In order to formally state the problem studied in this paper, some definitions are needed, related with LTI systems and LPTV systems. Then, let $S = (A, B, C, D)$ be a LTI discrete-time system, whose state space description is given by:

$$x(k+1) = Ax(k) + Bu(k), \quad \text{(2a)}$$
$$y(k) = Cx(k) + Du(k), \quad \text{(2b)}$$

where $k \in \mathbb{Z}$, $x(k) \in \mathbb{R}^n$, $u(k) \in \mathbb{R}^p$, $y(k) \in \mathbb{R}^q$, and A, B, C and D are constant. Later on, the definitions and properties stated for S will be referred to any LTI system, e.g., the plant to be controlled, which will be introduced in Section 3.

In addition, since a periodic compensator will be used, consider a LPTV discrete-time system $\tilde{\Theta} = (\tilde{A}(k), \tilde{B}(k), \tilde{C}(k), \tilde{D}(k))$, whose state space description is characterized by the equations:

$$\tilde{x}(k+1) = \tilde{A}(k)\tilde{x}(k) + \tilde{B}(k)\tilde{u}(k), \quad \text{(3a)}$$
$$\tilde{y}(k) = \tilde{C}(k)\tilde{x}(k) + \tilde{D}(k)\tilde{u}(k), \quad \text{(3b)}$$

where $k \in \mathbb{Z}$, $\tilde{x}(k) \in \mathbb{R}^{\tilde{n}}$, $\tilde{u}(k) \in \mathbb{R}^{\tilde{p}}$, $\tilde{y}(k) \in \mathbb{R}^{\tilde{q}}$, and $\tilde{A}(\cdot)$, $\tilde{B}(\cdot)$, $\tilde{C}(\cdot)$, $\tilde{D}(\cdot)$, are ω-periodic matriceswith entries in $\mathbb{R}$. Later on, the definitions and notations given for $\tilde{\Theta}$ will be referred to any LPTV system, e.g., the compensator to be synthesized. Specifically, from now on, denote by $\tilde{\Phi}(i,j)$ the state transition matrix of $\tilde{\Theta}$ from time j to time $i \ge j$ (where $\tilde{\Phi}(i,j) = \tilde{A}(i-1)\tilde{A}(i-2)\cdots\tilde{A}(j)$ if $i > j$, and $\tilde{\Phi}(j,j) = I_{\tilde{n}}$). It is well known that, for any initial time $k_0 \in \mathbb{Z}$, and for all initial states

$\tilde{x}(k_0) \in \mathbb{R}^{\tilde{n}}$ and all input functions $\tilde{u}(\cdot)$, the output response $\tilde{y}(\cdot)$ of the system $\tilde{\Theta}$ for $k \geq k_0$ can be obtained from the output response of a lifted representation of $\tilde{\Theta}$, and specifically from that introduced in (Meyer and Burrus, 1975), that is a LTI system $\tilde{\Theta}_{k_0}$ whose input and output sequences are the lifted representations of the input and output sequences of $\tilde{\Theta}$. The state space description of system $\tilde{\Theta}_{k_0}$ is characterized by the equations:

$$\tilde{x}_{k_0}(h+1) = \tilde{A}_{k_0}\tilde{x}_{k_0}(h) + \tilde{B}_{k_0}\tilde{u}_{k_0}(h), \quad \text{(4a)}$$
$$\tilde{y}_{k_0}(h) = \tilde{C}_{k_0}\tilde{x}_{k_0}(h) + \tilde{D}_{k_0}\tilde{u}_{k_0}(h), \quad \text{(4b)}$$

where $\tilde{x}_{k_0}(h) \in \mathbb{R}^{\tilde{n}}$, $\tilde{u}_{k_0}(h) \in \mathbb{R}^{\omega\tilde{p}}$, $\tilde{y}_{k_0}(h) \in \mathbb{R}^{\omega\tilde{q}}$, $\tilde{A}_{k_0} = \tilde{\Phi}(k_0 + \omega, k_0)$, $\tilde{B}_{k_0} = [\tilde{\Phi}(k_0 + \omega, k_0 + 1)\tilde{B}(k_0) \ldots \tilde{\Phi}(k_0 + \omega, k_0 + \omega)\tilde{B}(k_0 + \omega - 1)]$, $\tilde{C}_{k_0} = [(\tilde{C}(k_0)\tilde{\Phi}(k_0, k_0))' \ldots (\tilde{C}(k_0 + \omega - 1)\tilde{\Phi}(k_0 + \omega - 1, k_0))']'$, and $\tilde{D}_{k_0}$ is a $(\omega \times \omega)$ block matrix whose (i,j) block is zero if $i < j$, is equal to $\tilde{D}(k_0 + j - 1, k_0 + j - 1)$ if $i = j$ and is equal to $\tilde{C}(k_0 + i - 1)\tilde{\Phi}(k_0 + i - 1, k_0 + j)\tilde{B}(k_0 + j - 1)$ otherwise. In fact, if $\tilde{x}_{k_0}(0) = \tilde{x}(k_0)$ and $\tilde{u}_{k_0}(h) = [\tilde{u}'(k_0 + h\omega) \ldots \tilde{u}'(k_0 + \omega - 1 + h\omega)]'$ for all $h \in \mathbb{Z}^+$, then $\tilde{x}_{k_0}(h) = \tilde{x}(k_0 + h\omega)$ and $\tilde{y}_{k_0}(h) = [\tilde{y}'(k_0 + h\omega) \ldots \tilde{y}'(k_0 + \omega - 1 + h\omega)]'$ for all $h \in \mathbb{Z}^+$. From now on, $\tilde{\Theta}_{k_0}$ will be called *the associated system of $\tilde{\Theta}$ at (the initial) time k_0*.

Now consider the special case when system $\tilde{\Theta}$ is time invariant, i.e., consider $S = (A, B, C, D)$. In such a case, one can always look at system S as a LPTV system of arbitrary period $\omega \geq 1$; moreover, for every choice of the period ω, the associated systems of S at all initial times k_0 have the same state space description of $S_0 = (A_0, B_0, C_0, D_0)$ where $A_0 = A^{\omega}$, $B_0 = [A^{\omega-1}B \ldots B]$, $C_0 = [C' \ldots (CA^{\omega-1})']'$, and D_0 a $(\omega\times\omega)$–block matrix whose (i,j) block is 0 if $i < j$, is equal to D if $i = j$ and is equal to $CA^{i-j-1}B$ if $i > j$.

In the following, periodic systems will be denoted by a tilded capital letter, whereas LTI systems will be denoted by a capital letter with no tilde. However, the associated systems of LPTV systems like $\tilde{\Theta}$ at a given time k_0, despite the fact of being time-invariant, will be denoted by the tilded capital letter used to denote the LPTV system, with the subscript k_0, like in the case of $\tilde{\Theta}_{k_0}$. Notice also that the general type of notation introduced in Section 1 through equations (1) could apply, in principle, also to systems S and S_0, so that symbols such as $x^S(k)$, $u^S(k)$, $y^S(k)$, $x^{S_0}(h)$, $u^{S_0}(h)$ and $y^{S_0}(h)$ could actually be used instead of the symbols $x(k)$, $u(k)$, $y(k)$, $x_0(h)$, $u_0(h)$ and $y_0(h)$, respectively; although this might seem confusing, sometimes it can be helpful.

In addition, recall that the characteristic multipliers of $\tilde{A}(\cdot)$, i.e., the eigenvalues of $\tilde{A}_{k_0}$, are independent of k_0, together with their algebraic multiplicities in the characteristic polynomial of $\tilde{A}_{k_0}$; therefore they characterize the asymptotic stability of $\tilde{\Theta}$, and also the rate of convergence of the state free motions of $\tilde{\Theta}$, according to the following definition.

Definition 1. *For a given positive constant $\rho \leq 1$, the LPTV system $\tilde{\Theta}$, of period ω, is said to be ρ-stable if there exist positive $\overline{\rho} < \rho$ and $\overline{c} > 1$ such that for all initial times $k_0 \in \mathbb{Z}$, and for all initial states $\tilde{x}(k_0) \in \mathbb{R}^{\tilde{n}}$, the state free motions of $\tilde{\Theta}$ satisfy the following relation:*

$$\|\tilde{x}(k)\| < \overline{c}\,\overline{\rho}^{(k-k_0)}\|\tilde{x}(k_0)\|, \quad \forall k \geq k_0,\, k \in \mathbb{Z}.$$

The form of Definition 1 is such that for $\rho = 1$ it implies the mere exponential stability of $\tilde{\Theta}$, characterized by some $\overline{\rho} < 1$. By applying Definition 1 to the LTI system $\tilde{\Theta}_{k_0}$ described by (4), whose period is 1 and whose time variable is h, it is easy to derive the following proposition.

Proposition 1. *The ω-periodic system $\tilde{\Theta}$ is ρ-stable if and only if all the characteristic multipliers of $\tilde{A}(\cdot)$ are smaller than ρ^{ω}, in modulus, or, equivalently, if and only if, for an arbitrary $k_0 \in \mathbb{Z}$, $\tilde{\Theta}_{k_0}$ is ρ^{ω}-stable.*

In the following, the characteristic multipliers of $\tilde{A}(\cdot)$ will be called also the characteristic multipliers of system $\tilde{\Theta}$. In addition, the ω-periodic system $\tilde{\Theta}$ will be said to be ρ-stabilizable if there exists an ω-periodic linear map $\tilde{Q}(k) \in \mathbb{R}^{\tilde{p}\times\tilde{n}}$ such that all the characteristic multipliers of $\tilde{A}(\cdot) + \tilde{B}(\cdot)\tilde{Q}(\cdot)$ are smaller than ρ^{ω}, in modulus; it will be said to be ρ-detectable if a dual property holds.

Proposition 2. *(Grasselli and Longhi, 1991a) System $\tilde{\Theta}$ is ρ-stabilizable [or ρ-detectable] if and only if, for an arbitrary $k_0 \in \mathbb{Z}$,*

$$\operatorname{rank}\begin{bmatrix} z I_{\tilde{n}} - \tilde{A}_{k_0} & \tilde{B}_{k_0} \end{bmatrix} = \tilde{n},\ \forall z \in \mathbb{C},\ |z| \geq \rho^{\omega}, \quad (5)$$

[or

$$\operatorname{rank}\begin{bmatrix} z I_{\tilde{n}} - \tilde{A}_{k_0} \\ \tilde{C}_{k_0} \end{bmatrix} = \tilde{n},\ \forall z \in \mathbb{C},\ |z| \geq \rho^{\omega}]. \quad (6)$$

Since the purpose of this paper is to make use of a LPTV dynamic compensator in order to control a LTI system, all the previous definitions, notations and discussion concerning the LPTV system $\tilde{\Theta}$ can be applied to the LPTV connection of a LPTV (sub)compensator and a LTI subsystem. In particular, when referring to the LTI system S by itself (whose period ω is 1), its characteristic multipliers are the eigenvalues of A (whose modulus is to be smaller than ρ for the ρ-stability of S) and conditions (5) and (6) reduce, respectively, to

$$\operatorname{rank}[z I_n - A \quad B] = n, \quad \forall z \in \mathbb{C},\ |z| \geq \rho, \quad (7)$$

and

$$\operatorname{rank}\begin{bmatrix} z I_n - A \\ C \end{bmatrix} = n, \quad \forall z \in \mathbb{C},\ |z| \geq \rho. \quad (8)$$

Therefore conditions (5) and (6) are equivalent, respectively, to the ρ^{ω}-stabilizability and the ρ^{ω}-detectability of the LTI system $\tilde{\Theta}_{k_0}$. It is also recalled that the controllability and the reconstructibility of S can be tested by checking the rank of the matrices in (7) and (8), respectively, for all nonzero $z \in \mathbb{C}$ (Grasselli, 1980; Grasselli and Longhi, 1991*a*).

3. PROBLEM STATEMENT AND SOLUTION

The problem here studied of the asymptotic tracking with ρ-stability and infinite gain margin (to be formally defined) will be dealt with for the family $\mathcal{P}$ of LTI systems to be controlled defined in the subsequent Definition 3, on the basis of the nominal description of the plant P to be controlled – which will be assumed to be subject to output multiplicative perturbations. The state space description of P is given by:

$$x^P(k+1) = A^P x^P(k) + B^P u^P(k), \quad (9a)$$

$$y^P(k) = C^P x^P(k) + D^P u^P(k), \quad (9b)$$

where $x^P(k) \in \mathbb{R}^{n^P}$, $u^P(k) \in \mathbb{R}^p$ is the control input, $y^P(k) \in \mathbb{R}^q$ is the output, and A^P, B^P, C^P, D^P, are real matrices of suitable dimensions.

Definition 2. *The family $\mathcal{X}_q$ of admissible output multiplicative perturbations for P is defined as $\mathcal{X}_q = \{\Xi \in \mathbb{R}^{q \times q}, \Xi = diag(\xi_1, \ldots, \xi_q) : \xi_i \in [1, +\infty), \forall i \in \{1, \ldots, q\}\}$.*

Definition 3. *For the nominal plant $P = (A^P, B^P, C^P, D^P)$, the family $\mathcal{P}$ of output perturbed LTI plants P_Ξ is defined as the set of the LTI systems obtained as the series connection of P and an admissible output multiplicative perturbation $\Xi \in \mathcal{X}_q$, that is $\mathcal{P} = \{P_\Xi = (A^{P_\Xi}, B^{P_\Xi}, C^{P_\Xi}, D^{P_\Xi}) : A^{P_\Xi} = A^P, B^{P_\Xi} = B^P, C^{P_\Xi} = \Xi C^P, D^{P_\Xi} = \Xi D^P, \Xi \in \mathcal{X}_q\}$.*

Remark 1. Throughout the paper the output y^{P_Ξ} of the actual plant to be controlled P_Ξ will be assumed to be its controlled and measured output and will be denoted simply by y. □

Remark 2. Notice that the choice of the interval $[1, +\infty)$ for the admissible scalar gains ξ_i in Definition 2 does not imply any loss of generality. As a matter of fact, if more general intervals of the kind $[\xi_{i,0}, +\infty)$, with $\xi_{i,0} \neq 1$ (and $\xi_{i,0} > 0$), were to be considered, it would be sufficient to modify the nominal plant P by multiplying each row of C^P by the corresponding value of $\xi_{i,0}$. □

The class $\mathbf{R}$ of reference signals $r(\cdot)$ to be asymptotically tracked is assumed to be:

$$\mathbf{R} := \mathbf{R}_1 \oplus \mathbf{R}_2 \oplus \ldots . \oplus \mathbf{R}_\mu, \quad (10)$$

$$\mathbf{R}_i := \left\{ r(\cdot) : r(k) = \sum_{j=0}^{\nu_i - 1} [\delta_j \vartheta_i^{k-j} + \delta_j^* (\vartheta_i^*)^{k-j}] \binom{k}{j}, \right.$$

$$\left. \forall k \in \mathbb{Z}, \delta_j \in \mathbb{C}^q \right\}, \; i = 1, 2, \ldots, \mu, \quad (11)$$

for some positive integers μ, ν_i, $i = 1, 2, \ldots, \mu$, and some $\vartheta_i \in \mathbb{C}, i = 1, 2, \ldots, \mu$, which are assumed to be all distinct and to satisfy $|\vartheta_i| \geq 1$, $\mathrm{im}[\vartheta_i] \geq 0, i = 1, 2, \ldots, \mu$, where * means complex conjugate, and, for $i, j \in \mathbb{Z}^+$,

$$\binom{i}{j} := 0 \text{ if } i < j, \quad \binom{i}{j} := \frac{i!}{j!(i-j)!} \text{ if } i \geq j.$$

Assume that, for each $i = 1, 2, \ldots, \mu_1$, $\mathrm{im}[\vartheta_i] > 0$ and, for each $i = \mu_1 + 1, \ldots, \mu$, $\mathrm{im}[\vartheta_i] = 0$, for some $\mu_1 \in \mathbb{Z}^+$, $\mu_1 \leq \mu$.

Now, the problem, here considered, of the *asymptotic tracking with ρ–stability and infinite gain margin* can be formally stated as follows.

Problem 1. *For the given nominal plant P, and for a prescribed positive $\rho \leq 1$, find (if any) a period $\omega \geq 1$ and a LPTV dynamic output feedback compensator $\tilde{K}$ of period ω, having $y(k)$ and $r(k)$ as inputs and $u^P(k)$ as output, such that the closed-loop LPTV control system $\tilde{\Sigma}$ obtained by connecting $\tilde{K}$ and P_Ξ satisfies the following three requirements for all $P_\Xi \in \mathcal{P}$, i.e., for all $\Xi \in \mathcal{X}_q$:*

(a) $\tilde{\Sigma}$ is well-posed;
(b) $\tilde{\Sigma}$ is ρ-stable;
(c) its error response $e(k) := y(k) - r(k)$ asymptotically goes to zero for all reference signals $r(\cdot) \in \mathbf{R}$, for all the initial states of $\tilde{\Sigma}$ and for all initial times $k_0 \in \mathbb{Z}$.

The following theorem gives a solution to Problem 1. For the sake of brevity its proof is omitted. However, a procedure (i.e., the subsequent Procedure 1) for the design of a compensator that, under the conditions of Theorem 1, constitutes a solution to Problem 1, will be explicitly reported.

Theorem 1. *A LPTV dynamic error feedback compensator, having $e(k)$ as input and $u^P(k)$ as output, that solves Problem 1 exists if and only if*

(i) system P is ρ–stabilizable and ρ–detectable;
(ii) rank $\begin{bmatrix} A^P - \vartheta_i I & B^P \\ C^P & D^P \end{bmatrix} = n^P + q, \; \forall i \in \{1, \ldots, \mu\}$;
(iii) each row of D^P is nonzero.

Remark 3. For the second kind of extension of the IGM problem to MIMO discrete-time systems, (Galeani *et al.*, 2000) is the only contribution of which the authors are aware, although there a stable controller is required and the unknown scalar gains act on each scalar input of the plant (instead of each scalar output as here); namely, if requirement (c) is substituted with the asymptotic stability of the required compensator $\tilde{K}$ in the statement of Problem 1, a problem dual to the problem solved in (Galeani *et al.*, 2000) is obtained. Theorem 1 of (Galeani *et al.*, 2000) gives the necessary and sufficient conditions for the solvability of the latter, under the assumption that all the inputs are needed for the ρ-stabilizability of the plant; such conditions are the dual ones of conditions (i) and (iii) of the above stated Theorem 1. However, the part of the compensator

$\tilde{K}$ here proposed that is devoted to the mere ρ-stabilization of the plant is much simpler than the dual of the compensator proposed in (Galeani *et al.*, 2000). For a comparison of our results with the other existing ones, mentioned in the introduction, see Remark 4 in (Galeani *et al.*, 2000). □

Remark 4. Condition (iii) of Theorem 1 can be seen as a severe restriction, since it makes unsolvable Problem 1 for strictly causal plants (i.e., for when $D^P = 0$). The necessity of condition (iii) of Theorem 1 when P is SISO (and then condition (iii) is equivalent to the condition $D^P \neq 0$) was conjectured in (Khargonekar *et al.*, 1985). □

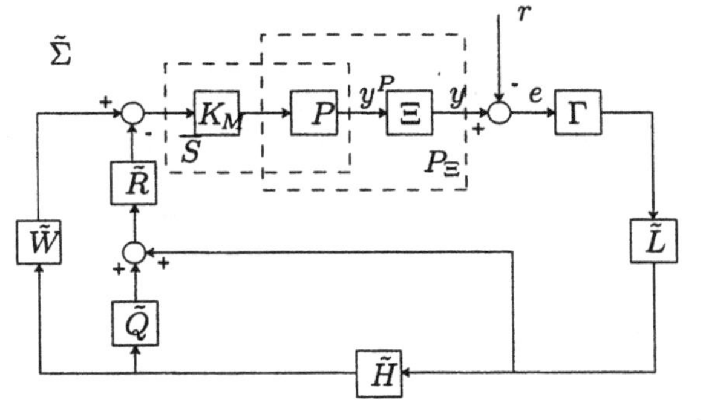

Fig. 1. The structure of the control system proposed to solve Problem 1.

Remark 5. As opposed to the (obvious) necessity of conditions (i) (for output feedback ρ–stabilization) and (ii) (for asymptotic tracking) of Theorem 1, the necessity of condition (iii) is linked to the error feedback structure chosen in Theorem 1 for the compensator, so that one could think that a weaker condition is necessary if a more general structure is chosen. However, it is well known that a tracking controller (based on the internal model principle) won't work in the presence of small uncertainties, unless an error feedback structure is chosen; on the other hand, if such a structure is used, the tracking capabilities given by the internal model will be preserved as long as the closed loop system is asymptotically stable, even for large uncertainties. □

The following design procedure of a solution to Problem 1 will refer to fig. 1, where the structure of the proposed control system is depicted.

Procedure 1. *(valid under the hypotheses that system P satisfies conditions i-iii of Theorem 1): Design of a solution to Problem 1.*

Step 1 *(Design of subcompensator K_M (internal model of reference signals)) Set $\psi_1(z) = \dots = \psi_p(z) = 1$. Fix μ ordered subsets $f_1 = \{f_{11}, \dots, f_{1q}\}, \dots, f_\mu = \{f_{\mu 1}, \dots, f_{\mu q}\}$ of $\{1, \dots, p\}$ such that for all $i \in \{1, \dots, \mu\}$*

$$rank\left(\begin{bmatrix} A^P - \vartheta_i I & \left\langle \begin{matrix} B^P \\ D^P \end{matrix} \right\rangle_{(\{1,\dots,n^P+q\},f_i)} \\ C^P & \end{bmatrix}\right) = n^P + q. \quad (12)$$

For each $i \in \{1, \dots, \mu_1\}$, set $\psi_{f_{ij}}(z) = \psi_{f_{ij}}(z)(z-\vartheta_i)^{\nu_i}(z - \vartheta_i^*)^{\nu_i}$, $j = 1, \dots, q$. For each $i \in \{\mu_1 + 1, \dots, \mu\}$, set $\psi_{f_{ij}}(z) = \psi_{f_{ij}}(z)(z - \vartheta_i)^{\nu_i}$, $j = 1, \dots, q$. For all $j \in \{1, \dots, p\}$, set $\varepsilon_j(z) := z^{\deg(\psi_j(z))}$. Compute system $K_M = (A^{K_M}, B^{K_M}, C^{K_M}, D^{K_M})$ as a minimal realization of the transfer matrix

$$diag\left(\frac{\varepsilon_1(z)}{\psi_1(z)}, \dots, \frac{\varepsilon_p(z)}{\psi_p(z)}\right).$$

Step 2 *(Computation of S) Compute system $\overline{S} = (\overline{A}, \overline{B}, \overline{C}, \overline{D})$ as the cascade connection of K_M and P, by putting $u^P(k) = y^{K_M}(k)$, for all $k \in \mathbb{Z}$. If $\overline{S}$ is reachable and observable, set $S = (A, B, C, D)$ equal to $\overline{S}$, else compute a minimal realization of $\overline{S}$, and set $S = (A, B, C, D)$ equal to it. Denote by n the dimension of the state space of system S (which has the same number of inputs and outputs than system P has).*

Step 3 *(Fix the integer r and the sets of integers r_i and c_i for $i = 0, \dots, r-1$) Define $\pi = \{1, \dots, p\}$ and $\chi = \{1, \dots, q\}$. Select $r \in \mathbb{N}, 1 \leq r \leq q$, such that it is possible to partition χ into r ordered subsets $r_0 = \{r_{0,1}, \dots, r_{0,\alpha_0}\}, \dots, r_{r-1} = \{r_{r-1,1}, \dots, r_{r-1,\alpha_{r-1}}\}$, $\sum_{i=0}^{r-1} \alpha_i = q$, such that, for every set r_i, $i = 0, \dots, r-1$, it is possible to find an ordered subset $c_i = \{c_{i,1}, \dots, c_{i,\alpha_i}\}$ of π such that matrix $\langle D \rangle_{(r_i,c_i)}$ is nonsingular.*

Step 4 *(Fix matrices L_i for $i = 0, \dots, r-1$) For each $i = 0, \dots, r-1$, define $L_i \in \mathbb{R}^{q \times q}$ as follows:*

$$\langle L_i \rangle_{(r_i,r_i)} = I_{\alpha_i}, \quad \langle L_i \rangle^{(r_i,r_i)} = 0.$$

Step 5 *(Fix matrices R_i for $i = 0, \dots, r-1$) For each $i = 0, \dots, r-1$, define $R_i \in \mathbb{R}^{p \times q}$ as follows:*

$$\langle R_i \rangle_{(c_i,r_i)} = \left(\langle D \rangle_{(r_i,c_i)}\right)^{-1}, \quad \langle R_i \rangle^{(c_i,r_i)} = 0.$$

Step 6 *(Fix matrices Q_i for $i = 0, \dots, r-1$) For each $i = 0, \dots, r-1$, define $Q_i \in \mathbb{R}^{q \times q}$ as:*

$$Q_0 := 0, \quad (13a)$$

$$Q_i := -\sum_{j=0}^{i-1} \left[CA^{i-1-j}BR_j + \sum_{k=j+1}^{i-1} CA^{i-1-k}BR_kQ_k\right] L_j. \quad (13b)$$

Step 7 *(Choice of the period ω) Call ν the reachability index of system S and set $F = \sum_{j=0}^{r-1} L_jCA^j$. Set $\omega \in \mathbb{Z}$ equal to the smallest integer satisfying $\omega \geq \nu + r$ such that the pair (A^ω, F) is reconstructible.*

Step 8 *(Fix matrices W_i for $i = r, \dots, \omega - 1$) Compute $V \in \mathbb{R}^{n \times q}$ such that all the eigenvalues of $A^\omega + VF$ are equal to zero. Then, define* $U_1 := \begin{bmatrix} -R_0 (I_q + Q_0) \\ \vdots \\ -R_{r-1} (I_q + Q_{r-1}) \end{bmatrix}$, $B_1 = [\,A^{\omega-1}B \,\dots\, A^{\omega-r}B\,]$ *and* $B_2 = [\,A^{\omega-r-1}B \,\dots\, AB \; B]$ *and compute $U_2 := B_2^\sharp (V - B_1U_1)$. Partition*

U_2 *as* $U_2 = \begin{bmatrix} W_r \\ \vdots \\ W_{\omega-1} \end{bmatrix}$, *with* $W_i \in \mathbb{R}^{p\times q}$, $i = r, \ldots, \omega - 1$.

Step 9 *(Choice of matrix* Γ*) Set* $\eta = \{1, \ldots, n\}$. *Let* $T \in \mathbb{R}^{n\times n}$ *such that* $T(A^\omega + VF)T^{-1}$ *is in Jordan form, and let* τ *be the degree of the minimal polynomial of* $(A^\omega + VF)$. *Fix* $\gamma_1, \ldots, \gamma_q \in \mathbb{R}^+$ *such that, for each* $i \in \{1, \ldots, q\}$, $\gamma_i > 1$ *and*

$$\frac{1}{1+\gamma_i} \left\| \langle TV \rangle_{(\eta,i)} \langle FT^{-1} \rangle_{(i,\eta)} \right\|_\infty < \frac{(\rho^{\omega\tau}+1)^{1/\tau} - 1}{q}.$$

Set $\Gamma = diag(\gamma_1, \ldots, \gamma_q)$. (14)

Step 10 *(Fix subcompensators* $\tilde{L}$, $\tilde{R}$, $\tilde{Q}$, $\tilde{W}$ *and* $\tilde{H}$*) For each* $i = r, \ldots, \omega - 1$, *set*

$$L_i := 0_{q\times q}, \qquad R_i := 0_{p\times q} \qquad Q_i := 0_{q\times q}.$$

For each $i = 0, \ldots, r-1$, *set* $W_i := 0_{p\times q}$. *Then, define* $\tilde{L}$, $\tilde{R}$, $\tilde{Q}$ *and* $\tilde{W}$ *as periodic non dynamic gain matrices* $\tilde{L}(k)$, $\tilde{R}(k)$, $\tilde{Q}(k)$ *and* $\tilde{W}(k)$ *such that* $\tilde{L}(h\omega + i) = L_i$, $\tilde{R}(h\omega + i) = R_i$, $\tilde{Q}(h\omega + i) = Q_i$ *and* $\tilde{W}(h\omega + i) = W_i$, *for all* $h \in \mathbb{Z}$ *and for each* $i = 0, \ldots, \omega - 1$. *Furthermore, set* $\tilde{H}$ *as the LPTV dynamic system whose state space description is characterized by the matrices:*

$$\left.\begin{array}{l} A^{\tilde{H}}(k) = I_q,\ B^{\tilde{H}}(k) = I_q \\ C^{\tilde{H}}(k) = I_q,\ D^{\tilde{H}}(k) = 0_{q\times q} \end{array}\right\}$$

$$\text{for } k = h\omega + i,\ h \in \mathbb{Z},\ i = 0, \ldots, \omega - 2, \quad (15a)$$

$$\left.\begin{array}{l} A^{\tilde{H}}(k) = 0_{q\times q},\ B^{\tilde{H}}(k) = 0_{q\times q} \\ C^{\tilde{H}}(k) = I_q, \quad D^{\tilde{H}}(k) = 0_{q\times q} \end{array}\right\}$$

$$\text{for } k = h\omega + \omega - 1,\ h \in \mathbb{Z}. \quad (15b)$$

Step 11 *Define the overall compensator* $\tilde{K}$ *as the connection of* Γ, $\tilde{L}$, $\tilde{H}$, $\tilde{Q}$, $\tilde{R}$, $\tilde{W}$ *and* K_M *according to the block diagram depicted in fig. 1.* ◇

Remark 6. If, at step 9 of Procedure 1, $\gamma_1, \ldots, \gamma_q \in \mathbb{R}^+$ are fixed such that, $\forall i \in \{1, \ldots, q\}$, $\gamma_i > 1$ and the following relation is satisfied:

$$\frac{1}{\gamma_i - 1} \left\| \langle TV \rangle_{(\eta,i)} \langle FT^{-1} \rangle_{(i,\eta)} \right\|_\infty < \frac{(\rho^{\omega\tau}+1)^{1/\tau} - 1}{q}$$

(where $\eta = \{1, \ldots, n\}$), instead of (14), then the control system in fig. 1 satisfies all the requirements of Problem 1 for the whole family of output perturbations defined by $\overline{\mathcal{X}_q} := \{\Xi \in \mathbb{R}^{q\times q}, \Xi = diag(\xi_1, \ldots, \xi_q) : \xi_i \in (-\infty, -1] \cup [+1, +\infty), \forall i \in \{1, \ldots, q\}\}$. □

4. REFERENCES

Davison, E.J. (1976). The robust control of a servomechanism problem for linear time-invariant multivariable systems. *IEEE Trans. Aut. Control* **AC-21**, 25–34.

Francis, B.A. and T.T. Georgiou (1988). Stability theory for linear time-invariant plants with periodic digital controllers. *IEEE Trans. Automatic Control* **33**(9), 820–832.

Francis, B.A. and W.M. Wonham (1976). The internal model principle of control theory. *Automatica* **12**, 457–465.

Galeani, S., O. M. Grasselli and L. Menini (2000). Linear periodic control for strong ρ–stabilization with infinite gain margin. In: *Proc. of the American Control Conference.* Chicago, IL, USA. pp. 1164–1168.

Grasselli, O.M. (1980). Conditions for controllability and reconstructibility of discrete-time linear composite systems. *Int. J. of Control* **31**(3), 433–441.

Grasselli, O.M. and S. Longhi (1991*a*). Finite zero structure of linear periodic discrete-time systems. *Int.J.Syst.Science* **22**(10),1785-1806.

Grasselli, O.M. and S. Longhi (1991*b*). Robust output regulation under uncertainties of physical parameters. *Syst. Contr. Lett.* **16**, 33–40.

Grasselli, O.M., S. Longhi and A. Tornambè (1993). Robust tracking and performance for multivariable systems under physical parameter uncertainties. *Automatica* **29**, 169–179.

Khargonekar, P.P. and A. Tannenbaum (1985). Non-euclidian metrics and the robust stabilization of systems with parameter uncertainty. *IEEE Trans. Automatic Control* **30**(10), 1005–1013.

Khargonekar, P.P., K. Poolla and A. Tannenbaum (1985). Robust control of linear time-invariant plants using periodic compensation. *IEEE Trans. Automatic Control* **30**(11), 1088–1096.

Kimura, H. and Y. Tanaka (1981). Minimal-time minimal order deadbeat regulators with internal stability. *IEEE Trans. Automatic Control* **AC-26**, 1276–1282.

Lee, S., S.M. Meerkov and T. Runolfsson (1987). Vibrational feedback control: Zeros placement capabilities. *IEEE Trans. Automatic Control* **32**(7), 604–610.

Maeda, H. and M. Vidyasagar (1985). Design of multivariable feedback systems with infinite gain margin and decoupling. *Systems&Control Letters* **6**, 127–130.

Meyer, R.A. and C.S. Burrus (1975). A unified analysis of multirate and periodically time-varying digital filters. *IEEE Trans. Circuit Systems* **22**, 162–168.

Tannenbaum, A. (1980). Feedback stabilization of linear dynamical plants with uncertainty in the gain factor. *Int. J. of Control* **32**(1), 1–16.

Tannenbaum, A. (1982). Modified Nevanlinna-Pick interpolation and feedback stabilization of linear plants with uncertainty in the gain factor. *Int. J. of Control* **36**(2), 331–6.

Yan, W.-Y., B.D.O. Anderson and R.R. Bitmead (1994). On the gain margin improvement using dynamic compensation based on generalized sampled-data hold functions. *IEEE Trans. Automatic Control* **39**(11), 2347–2354.

Copyright © IFAC Periodic Control Systems,
Cernobbio-Como, Italy, 2001

THE PERIODIC OPTIMALITY OF LQ CONTROLLERS SATISFYING STRONG STABILIZATION [1]

Jonathan D. Wolfe * Jason L. Speyer *

* UCLA Mechanical and Aerospace Engineering Department
Los Angeles, California 90095

Abstract: A Π test is presented for determining when a controller with periodic gains is superior to a LTI compensator for a class of LQ strong stabilization problems. For systems with strictly proper transfer functions, it is proven that stable high frequency periodic controllers based on weak variations from the LTI case cannot give better performance than stable LTI compensators. In the development, a means to evaluate the second partials of functions with respect to matrix valued parameters is introduced. Two examples detailing the application of the Π test are provided. *Copyright ©2001 IFAC*

Keywords: optimal control, chattering, stability properties, LQG control, periodic

1. INTRODUCTION

Often it is desired that an output feedback controller not only stabilize a plant, but be stable itself. The process of designing such a controller is referred to as the strong stabilization problem. Savkin and Petersen (Savkin and Petersen, 1998) have recently proposed a scheme for constructing periodic controllers to strongly stabilize linear systems. In their procedure, a full state controller operates using an old state estimate that is propagated without measurements forward in time. At periodic intervals, this state estimate is updated by a Luenberger estimator. This method will strongly stabilize any detectable and stabilizable linear system.

[1] This research was supported in part by the U.S. Air Force Office of Scientific Research under Grant F49620-97-0272.

Nevertheless, this method has some drawbacks. There is a minimum period length T_0 required to ensure strong stabilization, which depends on the gain of the controller between the periodic updates. Large T_0 implies poor performance in the presence of disturbances, and reducing T_0 requires high controller gains. Also, it is worrisome from a robustness standpoint that the controller runs open-loop over each period. A continuous feedback controller should reject disturbances better.

Instead of designing a strongly stabilizing controller, we investigate the following related question: If we restrict ourselves to considering only observer-structure controllers, and require the controller to be stable, when can a control with periodic gains potentially reduce a quadratic cost function compared to one with fixed gains? In answering this question, the Π test (Bittanti *et al.*, 1973; Bernstein and Gilbert, 1980) is a useful tool for determining when periodic control may improve on time invariant results. One contribution of this paper is a cost formulation that induces strong stability and the application of the Π test to show that periodic strongly stabilizing controllers produce a lower cost. A second contribution is a means of converting problems where the optimization parameters are gain matrices into a form amenable to application of the Π test. Since a large number of fixed structure problems (including the static output feedback problem and several decentralized control problems) involve optimizing over gain matrices, the method derived here appears to have many extensions.

This paper is organized as follows: Section 2 formulates a new cost function that penalizes unstable controllers. In section 3, we state some results on second derivatives of traces of matrix functions, which are interesting in themselves for numerical optimization of fixed structure controllers. Section 4 is a review of the Kronecker product, while section 5 discusses a new product that is similar to the Kronecker product. In section 6 the techniques of the preceding sections are used to create a compact expression of the second variation for the optimization problem. Section 7 applies the products discussed in sections 4 and 5 to linearizing the dependence of the state covariance on the controller gains. We present the Π test for strongly stabilizing controllers in section 8, and show that when a plant transfer function is strictly proper, high frequency gains based on weak variations from the static gains cannot reduce the cost function below the cost with an LTI controller. The Π test is applied to example systems in sections 9 and 10. Section 11 concludes the paper.

2. OPTIMAL CONTROL PROBLEM WITH STRONG STABILIZATION CONSTRAINT

Consider the linear time-invariant system

$$dx = (Ax + Bu)\,dt + [\Gamma_1\ 0]\,d\tilde{w}, \qquad (1)$$

$$dy = (Cx + Du)\,dt + [0\ \Gamma_2]\,d\tilde{w}, \qquad (2)$$

where $\tilde{w}$ is a Brownian motion process whose independent increment processes $d\tilde{w}$ have the statistics

$$E[d\tilde{w}\,d\tilde{w}^T] = I\,dt, \qquad E[d\tilde{w}] = 0. \qquad (3)$$

Without loss of generality, Γ_2 is assumed to have full row rank. Our cost function is the expectation of the cost function in (Bittanti *et al.*, 1973):

$$J[u,\tau] = \lim_{\tau\to\infty} \frac{1}{\tau} E[\int_0^{\tau} (x^T Q x + u^T R u)\,dt], \qquad (4)$$

where Q is symmetric positive semidefinite and R is symmetric positive definite. The answer to this optimization problem is the well known linear quadratic Gaussian controller:

$$d\hat{x} = A\hat{x}\,dt + Bu\,dt + L(dy - C\hat{x} - Du)\,dt, \qquad (5)$$

$$u = -K\hat{x}, \qquad (6)$$

where K is the linear-quadratic regulator gain and L is the gain of the Kalman-Bucy filter.

Note that the dynamics of the controller are described by $A_c \triangleq A - BK - LC + LDK$, and that the eigenvalues of this matrix need not be stable. Suppose we were to add a cost term that would penalize an unstable controller. If we constrained the controller to have the same observer structure as before, the dynamics and cost would look like:

$$dx_{cl} = A_{cl}x_{cl}dt + B_{cl}dw \qquad (7)$$

$$z = C_{cl}x_{cl} \qquad (8)$$

$$J[K,L,\tau] = \lim_{\tau\to\infty} \frac{1}{\tau} E[\int_0^{\tau} (z^T z)\,dt], \qquad (9)$$

with

$$E[dw\,dw^T] = I\,dt, \quad E[dw] = 0,$$
$$x_{cl}^T = [x^T e^T x_f^T], \quad e = x - \hat{x},$$
$$A_{cl} = \begin{bmatrix} A - BK & BK & 0 \\ 0 & A - LC & 0 \\ 0 & 0 & A_c \end{bmatrix},$$
$$B_{cl} = \begin{bmatrix} \Gamma_1 & 0 & 0 \\ \Gamma_1 & -L\Gamma_2 & 0 \\ 0 & 0 & \Gamma_f \end{bmatrix}, \qquad (10)$$
$$C_{cl} = \begin{bmatrix} Q^{\frac{1}{2}} & 0 & 0 \\ -R^{\frac{1}{2}}K & R^{\frac{1}{2}}K & 0 \\ 0 & 0 & Q_f^{\frac{1}{2}} \end{bmatrix},$$

where the "fake" controller state x_f is forced by noise and included in the cost function via the weight Q_f. To insure that all controller states are penalized in the cost, it is required that that Q_f be positive definite and that Γ_f have full row rank.

The cost expression above can be written in terms of the covariance of x_{cl}, $P \triangleq [x_{cl}x_{cl}^T]$:

$$J = \lim_{\tau\to\infty} \mathbf{tr}\frac{1}{\tau}\int_0^{\tau} \{PC_{cl}(t)^T C_{cl}(t)\}\,dt, \qquad (11)$$

subject to

$$\dot{P} = A_{cl}(t)P + PA_{cl}(t)^T + B_{cl}(t)B_{cl}(t)^T, \qquad (12)$$

$$P > 0, \qquad (13)$$

where **tr** denotes the trace operation. Let us partition P into $n \times n$ pieces as follows:

$$P = \begin{bmatrix} P_1 & P_{12} & P_{13} \\ P_{12}^T & P_2 & P_{23} \\ P_{13}^T & P_{23}^T & P_3 \end{bmatrix}. \quad (14)$$

An equivalent expression of the cost is then

$$J = \lim_{\tau \to \infty} \mathbf{tr} \frac{1}{\tau} \int_0^{\tau} \{P_1 Q + (P_1 - P_{12} - P_{12}^T + P_2) K^T R K + P_3 Q_f\}\, dt. \quad (15)$$

The Hamiltonian for this optimization problem is

$$\mathcal{H} = \mathbf{tr}\{P_1 Q + (P_1 - P_{12} - P_{12}^T + P_2) K^T R K + P_3 Q_f + \Lambda(A_{cl} P + P A_{cl}^T + B_{cl} B_{cl}^T)\}, \quad (16)$$

which is almost identical to the Hamiltonian used in (Denham and Speyer, 1964) and is similar to one (for a case with no process noise) used in (Athans, 1968). Following standard derivations, we can find the necessary conditions for minimizing J:

Theorem 1. (Pontryagin's necessary conditions). The following are necessary for minimizing J:

(1) $\mathcal{H}$ is minimized
(2) $\mathcal{H}_X = -d\Lambda/\mathrm{dt}$

If $\mathcal{H}$ is continuously differentiable, a necessary condition for minimizing $\mathcal{H}$ is that $\mathcal{H}_K = \mathcal{H}_L = 0$. If we further assume that there is a steady state stationary solution to the optimization problem, then $d\Lambda/dt = 0$ and $dP/dt = 0$. These conditions may be used to search for steady state optimal gains, as in (Geromel and Bernussou, 1979).

2.1 *When is there a steady state stationary solution to the optimization problem?*

The conditions for determining when a LTI system may be stabilized by a stable controller were found by (Youla *et al.*, 1974). Later, Vidyasagar (Vidyasagar, 1985) found several useful alternative forms of the conditions for strong stabilizability. For brevity, we will state only one of the results for MIMO systems from (Vidyasagar, 1985):

Theorem 2. (Parity interlacing property). Let $\mathbb{C}_{+e}$ denote the extended right half of the complex plane ($\{s \in \mathbb{C} : \mathbf{Re}(s) \geq 0\}$ together with positive infinity). A plant P is strongly stabilizable if and only if the number of poles of P (counted according to their McMillan degrees) between any pair of real $\mathbb{C}_{+e}$-blocking zeros of P is even.

Note that the stable compensator that stabilizes the system in the above theorem is a proper matrix transfer function of arbitrary order – *i.e.* , a strictly proper stabilizing compensator with the same order as the plant may not exist. However, there are constructive sufficient conditions for stable, strictly proper full-order compensation (Wang and Bernstein, 1994). If such a compensator can be found, it can be used as a starting point for an iterative scheme to find a stationary point of our optimization problem (Geromel and Bernussou, 1979; Toivonen and Mäkilä, 1985).

3. SECOND DERIVATIVES OF TRACES

The sequel will require some results on second order derivatives of traces of matrix functions. As the proofs are trivial but lengthy, we excluded them from this paper.

Proposition 3. Let X, Y, A, B be complex matrices of appropriate dimension. Denote the (i, j)th component of a matrix by $()_{ij}$. Then

- $\frac{\partial^2}{\partial y_{kl} \partial x_{ij}} \mathbf{tr}(XAYB) = b_{li} a_{jk}$.
- $\frac{\partial^2}{\partial y_{kl} \partial x_{ij}} \mathbf{tr}(XAY^T B) = b_{ki} a_{jl}$.
- $\frac{\partial^2}{\partial y_{kl} \partial x_{ij}} \mathbf{tr}(XAY^T BYC) = c_{li}[B^T Y A^T]_{kj} + [BYC]_{ki} a_{jl}$.
- $\frac{\partial^2}{\partial x_{kl} \partial x_{ij}} \mathbf{tr}(XAX^T B) = b_{ki} a_{jl} + b_{ik} a_{lj}$.
- $\frac{\partial^2}{\partial y_{kl} \partial x_{ij}} \mathbf{tr}(X^T AYB) = a_{ik} b_{lj}$.
- $\frac{\partial^2}{\partial y_{kl} \partial x_{ij}} \mathbf{tr}(X^T AY^T B) = a_{il} b_{kj}$.
- $\frac{\partial^2}{\partial y_{kl} \partial x_{ij}} \mathbf{tr}(X^T AY^T BYC) = a_{il}[BYC]_{kj} + [B^T Y A^T]_{ki} c_{lj}$.

4. REVIEW OF THE KRONECKER PRODUCT

Definition 4. (Kronecker Operator). Let $\mathcal{F}$ denote a field. If $A \in \mathcal{F}_{m \times n}$ and $B \in \mathcal{F}_{o \times p}$ then the Kronecker operation on A and B, written $A \otimes B$, is defined to be the partitioned matrix

$$A \otimes B = \begin{bmatrix} a_{11}B & a_{12}B & \ldots & a_{1n}B \\ a_{21}B & a_{22}B & \ldots & a_{2n}B \\ \vdots & \vdots & & \vdots \\ a_{m1}B & a_{m2}B & \ldots & a_{mn}B \end{bmatrix} \in \mathcal{F}_{mo \times np}$$

Note that if we write

$$k = (r-1)o + s, \qquad l = (i-1)p + j$$

then $[A \otimes B]_{kl} = a_{ri}b_{sj}$.

Proposition 5. (Kronecker Product). If $A \in \mathcal{F}_{m\times n}$ and $B \in \mathcal{F}_{o\times p}$, then the Kronecker operation $A \otimes B$ is a well defined product.

Proposition 6. (Kronecker Product and Linear Matrix Equations). Consider the following matrix linear equation for the unknown matrix $X \in \mathcal{F}_{n\times n}$: $AXB = C$ where $A, B, C \in \mathcal{F}_{n\times n}$. We can consider this equation as an abbreviation for n^2 scalar equations for the n^2 elements of X. Let us define the "vectorized" versions of X and C in $\mathcal{F}_{n^2}$ by

$$\mathbf{x} = \left[X_{1*}\ X_{2*}\ \dots\ X_{n*}\right]^T,$$
$$\mathbf{c} = \left[C_{1*}\ C_{2*}\ \dots\ C_{n*}\right]^T$$

where X_{i*}, C_{j*} denote the ith row of X and the jth row of C, respectively. Then the equation $AXB = C$ is equivalent to $G\mathbf{x} = \mathbf{c}$ for $G = A \otimes B^T$.

Proposition 7. (Kronecker Product of Positive Definite Matrices). If A and B are two positive definite matrices, then $A \otimes B$ is also positive definite.

These are well known results. For further information on Kronecker products, refer to elementary texts on matrices (e.g. (Lancaster, 1969)).

5. ANOTHER PRODUCT

In the previous section, we showed that the matrix equation $AXB = C$ can be transformed to the form $(A \otimes B^T)\mathbf{x} = \mathbf{c}$, where $\mathbf{x}$ and $\mathbf{c}$ "vectorize" X and C by rows. Now, suppose we wished to express the matrix equation $AX^TB = C$ as $G\mathbf{x} = \mathbf{c}$, where $\mathbf{x}$ and $\mathbf{c}$ are the same as before. Motivated by this problem, we define a new operator:

Definition 8. (KT-operator). Let $\mathcal{F}$ denote a field. If $A \in \mathcal{F}_{m\times n}$ and $B \in \mathcal{F}_{o\times p}$ then the KT-operation on A and B, written $A \overset{T}{\otimes} B$, is defined as follows:

$$[A \overset{T}{\otimes} B]_{kl} = a_{rj}b_{si},$$
$$\text{where} \quad k = (r-1)o + s, \quad l = (i-1)n + j.$$

Proposition 9. (KT-product). If $A \in \mathcal{F}_{m\times n}$ and $B \in \mathcal{F}_{o\times p}$, then the KT-operation $A \overset{T}{\otimes} B$ is a well defined product.

Proposition 10. Let $C \in \mathcal{F}_{m\times n}$, $A \in \mathcal{F}_{m\times o}$, $X \in \mathcal{F}_{p\times o}$, $B \in \mathcal{F}_{p\times n}$. Let $AX^TB = C$ be a linear matrix equation in X. "Vectorize" X and C as follows:

$$\mathbf{x} = \left[X_{1*}\ X_{2*}\ \dots\ X_{n*}\right]^T,$$
$$\mathbf{c} = \left[C_{1*}\ C_{2*}\ \dots\ C_{n*}\right]^T.$$

Then $AX^TB = C$ is equivalent to the equation $(A \overset{T}{\otimes} B^T)\mathbf{x} = \mathbf{c}$.

The proofs of the above propositions are trivial modifications of the corresponding proofs for the Kronecker product case.

6. PARTIAL DERIVATIVES OF $\mathcal{H}$

The second partials of $\mathcal{H}$ (16) can be found using the results on second partials of traces with respect to matrices developed in Section 3. These results can be expressed in a more compact notation if we "vectorize" the parameter matrices and write our results in terms of Kronecker products and KT-products. As an illustrative example, we will derive the second partial of $\mathcal{H}$ with respect to K.

Using the formulas in Section 3, one finds that

$$\frac{\partial^2}{\partial k_{gh}\partial k_{ef}}\mathcal{H}(P, K, L, \Lambda) = R_{ge}[P_1 - P_{12} - P_{12}^T + P_2]_{fh} + R_{eg}[P_1 - P_{12} - P_{12}^T + P_2]_{hf}. \quad (17)$$

Let δK be a small variation in K. Vectorize δK by rows, *i.e.* $\delta\mathbf{k}_{(q-1)n+r} = \delta K_{qr}$. Define $\mathcal{H}(X, K, L, \Lambda)_{\mathbf{kk}}$ as follows

$$\delta\mathbf{k}^T\mathcal{H}(P, K, L, \Lambda)_{\mathbf{kk}}\delta\mathbf{k} = \sum_{g=1}^{m}\sum_{h=1}^{n}\sum_{e=1}^{m}\sum_{f=1}^{n}\delta K_{gh}\frac{\partial^2\mathcal{H}(P, K, L, \Lambda)}{\partial k_{gh}\partial k_{ef}}\delta K_{ef} \quad (18)$$

Then, using what we know about Kronecker products and equations (18), (17), we have:

$$\mathcal{H}(P, K, L, \Lambda)_{\mathbf{kk}} = 2R \otimes [P_1 - P_{12} - P_{12}^T + P_2]. \quad (19)$$

In an analogous way, we can define $\mathbf{l}, \mathbf{p_1}, \mathbf{p_{12}}, \mathbf{p_{13}}, \mathbf{p_2}, \mathbf{p_{23}}, \mathbf{p_3}$ with respect to L and P. Then $\mathcal{H}_{\mathbf{kl}}, \mathcal{H}_{\mathbf{kp_1}}, \mathcal{H}_{\mathbf{kp_{12}}}, \mathcal{H}_{\mathbf{kp_{13}}}, \mathcal{H}_{\mathbf{kp_2}}, \mathcal{H}_{\mathbf{kp_{23}}}, \mathcal{H}_{\mathbf{kp_3}}, \mathcal{H}_{\mathbf{ll}}, \mathcal{H}_{\mathbf{lp_1}}, \mathcal{H}_{\mathbf{lp_{12}}}, \mathcal{H}_{\mathbf{lp_{13}}}, \mathcal{H}_{\mathbf{lp_2}}, \mathcal{H}_{\mathbf{lp_{23}}}, \mathcal{H}_{\mathbf{lp_3}}$ can be determined in terms of Kronecker and KT-products of the system matrices. For brevity, these expressions have not been included. Note that since P appears linearly in $\mathcal{H}$, $\frac{\partial^2\mathcal{H}}{\partial p_{kl}\partial p_{ij}} = 0 \quad \forall i, j, k, l$.

7. LINEARIZATION OF EQUATIONS OF MOTION

The equation of motion of the covariance P is

$$\dot{P}(t) = A_{cl}(t)P(t) + P(t)A_{cl}(t)^T + B_{cl}(t)B_{cl}(t)^T. \quad (20)$$

To linearize this bilinear form, suppose that P^o, K^o, L^o are nominal solutions that satisfy the equations of motion. Then take small variations so that $P = P^o + \delta P, K = K^o + \delta K, L = L^o + \delta L$. We can eliminate the higher order terms and express the result in terms of "vectorized" quantities. This is easily accomplished using the rules for "vectorizing" matrix equations given in the sections discussing the Kronecker and KT products. For instance,

$$\begin{aligned}\delta\dot{\mathbf{p}}_1 = {} & [(A - BK^o) \otimes I + I \otimes (A - BK^o)]\delta\mathbf{p}_1 + \\ & [(BK^o) \overset{T}{\otimes} I + I \otimes (BK^o)]\delta\mathbf{p}_{12} + \\ & [B \otimes P_{12}^o - B \otimes P_1^o - P_1^o \overset{T}{\otimes} B + P_{12}^o \overset{T}{\otimes} B]\delta\mathbf{k}. \quad (21)\end{aligned}$$

The state space equations for all the $\delta\mathbf{p}$'s can be put together into a large linear system:

$$\delta\dot{\mathbf{p}} = \bar{F}\delta\mathbf{p} + \bar{G}\mathbf{u}, \qquad \mathbf{u} = \left[\delta\mathbf{k}^T \ \delta\mathbf{l}^T\right]^T. \quad (22)$$

A transfer function $H(s)$ from the parameter variations $\delta\mathbf{k}$ and $\delta\mathbf{l}$ to the states $\delta\mathbf{p}$ can then be computed in the standard way:

$$H(s) = (sI - \bar{F})^{-1}\bar{G}, \qquad \delta\mathbf{p}(s) = H(s)\mathbf{u}(s). \quad (23)$$

8. CONSTRUCTING A Π TEST

We will now create a Π test for the fixed structure strong stabilization problem, following the same general strategy used in the state feedback case (Bittanti *et al.*, 1973; Bernstein and Gilbert, 1980). Suppose that K_o and L_o meet the first order necessary conditions for optimality for the system (7) - (10) with cost (11).

Because the term "proper" has historically been used to describe the optimality of periodic optimal control problems, as well as describing transfer functions that have more poles than zeros, we will distinguish the two definitions by always referring to the subject of the properness.

Definition 11. An optimal periodic control problem is said to be *proper* if there exists a period $\hat{\tau}$ and a admissible control gains $\hat{K}(t), \hat{L}(t)$ such that

$$J[\hat{K}(t), \hat{L}(t), \hat{\tau}] < \bar{J}^o, \quad (24)$$

where $\bar{J}^o$ is the cost corresponding to the optimal steady state solution of the problem, using the static gains $\bar{K}^o, \bar{L}^o$. Hence, a strong variation from the steady state solution costs less than the static cost.

Definition 12. An optimal periodic control problem is said to be *locally proper* if there exists a period $\hat{\tau}$ and admissible weak variations $\hat{\delta}K(t), \hat{\delta}L(t)$ in the controller gains such that

$$J[\bar{K}^o + \hat{\delta}K(t), \bar{L}^o + \hat{\delta}L(t), \hat{\tau}] < \bar{J}^o, \quad (25)$$

where $\bar{J}^o$ is the cost corresponding to the optimal steady state solution of the problem, using the static gains $\bar{K}^o, \bar{L}^o$. Here, a weak variation from the optimal steady state solution yields a lower cost.

Definition 13. Let $(\bar{P}, \bar{K}, \bar{L})$ be a steady-state admissible triple. The optimal periodic control problem is *normal* at $(\bar{P}, \bar{K}, \bar{L})$ if the following condition is satisfied for some τ:

$$\operatorname{rank}\left[(e^{\bar{F}\tau} - I_n) \ \bar{G} \ \bar{F}\bar{G} \ \ldots \ \bar{F}^{n-1}\bar{G}\right] = n. \quad (26)$$

For convenience, we will drop the use of functional notation for $\mathcal{H}$ and its derivatives – any usage is assumed to occur at the stationary point. Using the techniques of the previous sections, we can construct $\mathcal{H}_{\mathbf{up}}, \mathcal{H}_{\mathbf{uu}}$. We also know that $\mathcal{H}_{\mathbf{pp}} = 0, \mathcal{H}_{\mathbf{pu}} = \mathcal{H}_{\mathbf{up}}^T$.

Theorem 14. If the local minimum of the optimal steady state problem is normal and the $(m \times m)$ Hermitian matrix

$$\Pi(\omega) = \mathcal{H}_{\mathbf{pu}}^T H(j\omega) + H(-j\omega)^T \mathcal{H}_{\mathbf{pu}} + \mathcal{H}_{\mathbf{uu}} \quad (27)$$

is partially negative for some $\omega > 0$, then the optimal periodic control problem is locally proper (and hence proper). Conversely, if the optimal periodic control problem is locally proper, then there exists $\omega > 0$ such that $\Pi(\omega)$ is not positive definite.

PROOF. The proof for this theorem is the same as that given in (Bittanti *et al.*, 1973; Bernstein and Gilbert, 1980), where the control input is the vectorized parameters $\mathbf{u}$. □

Corollary 15. (Implications for strictly proper plants). If the plant transfer function is strictly proper (*i.e.* $D = 0$), then there is a frequency Ω such that the optimal periodic control problem cannot be locally proper for frequencies greater than Ω.

PROOF. The magnitude of $H(j\omega)$ must attenuate at high frequencies due to the asymptotic stability of the stationary solution. Hence, $\Pi(j\omega) \to \mathcal{H}_{\mathbf{uu}}$ as $\omega \to \infty$. Now, the elements of $\mathcal{H}_{\mathbf{uu}}$ are given by

$$\mathcal{H}_{\mathbf{kk}} = 2R \otimes [P_1 - P_{12} - P_{12}^T + P_2], \quad (28)$$

$$\mathcal{H}_{\mathbf{ll}} = 2\Lambda_2 \otimes [\Gamma_2 \Gamma_2^T], \quad (29)$$

$$\mathcal{H}_{\mathbf{kl}} = 2D^T \overset{T}{\otimes} [P_{13}^T \Lambda_{13} + P_3 \Lambda_3 + P_{23}^T \Lambda_{23}], \quad (30)$$

so since $D = 0$, $\mathcal{H}_{\mathbf{uu}}$ is positive definite if and only if both $\mathcal{H}_{\mathbf{kk}}$ and $\mathcal{H}_{\mathbf{ll}}$ are positive definite. We know that R, Λ_2, and $\Gamma_2 \Gamma_2^T$ are all positive definite. Also

$$[P_1 - P_{12} - P_{12}^T + P_2] = [I \ -I \ 0] \, P \, [I \ -I \ 0]^T$$

must be positive definite, since P is. Hence $\mathcal{H}_{\mathbf{kk}}$ and $\mathcal{H}_{\mathbf{ll}}$ are the Kronecker products of positive definite matrices, which means they are positive definite themselves. So $\Pi(j\omega)$ converges to a positive definite matrix as $\omega \to \infty$, which implies that there is a frequency Ω such that $\Pi(j\omega)$ is positive definite for all $\omega > \Omega$. By the results of the previous theorem, the optimal periodic control problem cannot be locally proper for frequencies $\omega > \Omega$. □

Thus, a chattering control that is a weak variation from the static optimum can never produce a better cost than the static optimum for any plant with a strictly proper transfer function.

9. EXAMPLE: PLANT WITH A DC TERM

Consider the linear system and cost given by

$$A = 1, \quad B = 1, \quad C = 1.5, \quad D = 1, \quad \Gamma_1 = 1,$$
$$\Gamma_2 = 1, \quad Q = 1, \quad R = 1, \quad Q_f = 0.01, \quad \Gamma_f = 1.$$

The open loop transfer function here is $\frac{s+0.5}{s-1}$, which satisfies the parity interlacing property (Youla *et al.*, 1974). Hence the plant may be stabilized by a stable linear time invariant controller. However, the gains given by a conventional LQG design lead to an unstable controller.

A static solution for the modified cost given by (9) was found using the methods in (Toivonen and Mäkilä, 1985). The results of the local optimization were $K^o = 3.9112$, $L^o = 1.1774$, resulting in the pole -0.0724 for the static optimal controller.

We also calculated the static optimal gains for several other values of R, and performed the Π test on each case. For each case, the minimum eigenvalue of Π is plotted vs. frequency in Figure 1. Note that when $R = 1$, the minimum eigenvalue of Π is never negative. For this value of R, there is no instance at

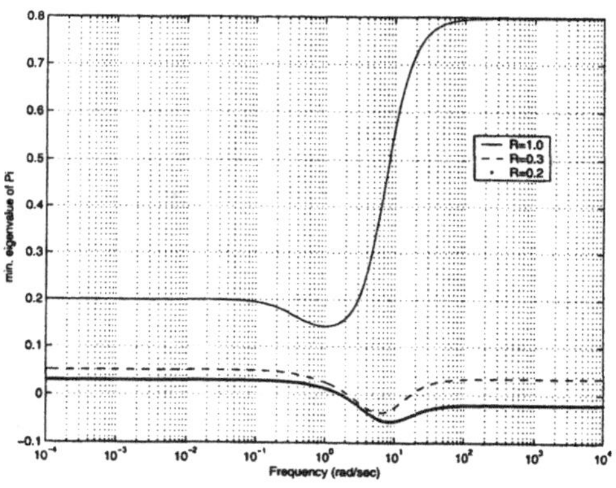

Fig. 1. Minimum eigenvalue of $\Pi(j\omega)$ vs. ω

which a lower cost can be realized via periodic gains. When $R = 0.3$, the minimum eigenvalue of Π falls below 0 for frequencies between 2 and 10.5 rad/sec. If R is reduced to 0.2, the minimum eigenvalue of Π is negative for all frequencies greater than 2 rad/sec, which means a chattering solution may reduce the cost below that of the static optimum. Note that this does not violate Corollary 15, as the plant's transfer function is not strictly proper.

10. FLEXIBLE STRUCTURE EXAMPLE

Consider a system consisting of two masses connected by a spring. Suppose that we sense the position of the first mass and that our control input is a force applied to the first mass. Suppose also that the exogenous input is a force applied to the second mass and that only the position of the second mass is weighted in the state cost term Q. This problem could be viewed as one of positioning a flexible robot arm using only sensors and actuators at the base of the arm. One such problem is described by the following linear system and cost:

$$A = \begin{bmatrix} 0 & 1 & 0 & 0 \\ -1 & 0 & 1 & 0 \\ 0 & 0 & 0 & 1 \\ 1 & 0 & -1 & 0 \end{bmatrix}, \quad B = \begin{bmatrix} 0 \\ 1 \\ 0 \\ 0 \end{bmatrix}, \quad C = \begin{bmatrix} 1 & 0 & 0 & 0 \end{bmatrix},$$

$$D = 0, \quad \Gamma_1 = \begin{bmatrix} 0 \\ 0 \\ 0 \\ 1 \end{bmatrix}, \quad \Gamma_2 = 1, \quad Q = \begin{bmatrix} 0 & 0 & 0 & 0 \\ 0 & 0 & 0 & 0 \\ 0 & 0 & 1 & 0 \\ 0 & 0 & 0 & 0 \end{bmatrix},$$

$$R = 10^{-3}, \quad Q_f = 10^{-2} \cdot I_4, \quad \Gamma_f = I_4.$$

where I_4 denotes the 4×4 identity matrix. The open loop transfer function for this system is $\frac{(s+j)(s-j)}{s^2(s+1.4142j)(s-1.4142j)}$, which meets the parity interlacing property (Youla *et al.*, 1974), so this plant may be stabilized by a stable LTI controller. However, the LQG gains generate an unstable controller.

A static solution for the modified cost given by (9) was found using the methods in (Toivonen and Mäkilä, 1985). The resulting gains were

$$K = \begin{bmatrix} 8.1188 & 2.0586 & -3.7766 & 4.9878 \end{bmatrix}$$
$$L = \begin{bmatrix} 7.8756 & 6.0895 & 5.0344 & -1.7341 \end{bmatrix}^T.$$

The poles of the static optimal controller were then -5.0269, -3.3677, $-0.7698 \pm 1.4055j$.

The Π test was then performed; Figure 2 plots the lowest eigenvalue of Π vs. frequency. Note that high frequency behavior is as predicted by Corollary 15.

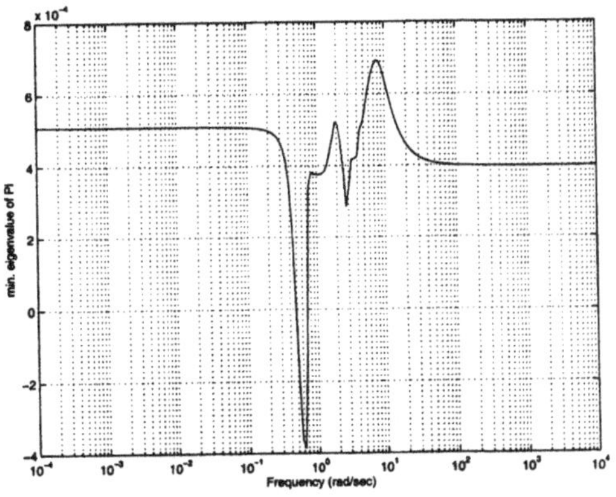

Fig. 2. Minimum eigenvalue of $\Pi(j\omega)$ vs. ω

11. CONCLUSION

A Π test applicable to a linear quadratic Gaussian strong stabilization problem has been developed, determining when periodic coefficients in the gain matrices can potentially reduce the cost.

The method proposed here is restricted in several ways. First, adding a term in the cost to ensure strong stabilization is more restrictive than adding inequality constraints on A_c. Unfortunately, a convenient means of determining the solution to the resulting static problem has not been found. If such a solution were known, the Π test can be modified to include provisions for inequality constraints (Bernstein and Gilbert, 1980). The other restriction is that the open loop system must satisfy the parity interlacing condition described in (Youla *et al.*, 1974) to ensure that the plant can be stabilized by a stable LTI controller. Furthermore, a stable, strictly proper controller of plant order must be found to ensure the existence of a strongly stabilizing static solution. If no such solution exists, one must use time varying control.

Methods used to derive the Π test in this paper can be applied to other control problems. In particular, the techniques used here can be trivially modified to deal with problems involving optimizing decentralized controllers of systems with fixed modes (Wang and Davison, 1973).

12. REFERENCES

Athans, M. (1968). The matrix maximum principle. *Information and Control* (11), 592–606.

Bernstein, D. S. and E. G. Gilbert (1980). Optimal periodic control: The π test revisited. *IEEE Transactions on Automatic Control* **AC-25**(4), 673–684.

Bittanti, S., G. Fronza and G. Guardabassi (1973). Periodic control: A frequency domain approach. *IEEE Transactions on Automatic Control* **AC-18**(1), 33–38.

Denham, W. F. and J. L. Speyer (1964). Optimal measurement and velocity correction programs for midcourse guidance. *AIAA Journal* **2**(5), 896–907.

Geromel, J. C. and J. Bernussou (1979). An algorithm for optimal decentralized regulation of linear quadratic interconnected systems. *Automatica* **15**, 489–491.

Lancaster, P. (1969). *Theory of Matrices*. Chap. 8. Academic Press. New York.

Savkin, A. V. and I. R. Petersen (1998). Almost optimal LQ-control using stable periodic controllers. *Automatica* **34**(10), 1251–54.

Toivonen, H. T. and P. M. Mäkilä (1985). A descent Anderson-Moore algorithm for optimal decentralized control. *Automatica* **21**(6), 743–744.

Vidyasagar, M. (1985). *Control System Synthesis*. number 7 In: *MIT Press series in signal processing, optimization, and control*. The MIT Press. Cambridge, MA.

Wang, S. H. and E. D. Davison (1973). On the stabilization of decentralized control systems. *IEEE Transactions on Automatic Control* **AC-18**(5), 473–8.

Wang, Y. W. and D. S. Bernstein (1994). H_2-suboptimal stable stabilization. *Automatica* **30**(11), 1797–800.

Youla, D. C., J. J. Bongiorno and C. N. Lu (1974). Single-loop feedback-stabilization of linear multivariable dynamical plants. *Automatica* **10**, 159–173.

Copyright © IFAC Periodic Control Systems,
Cernobbio-Como, Italy, 2001

STABILIZATION OF PERIODIC SYSTEMS: OVERVIEW AND ADVANCES

S. Bittanti, P. Colaneri

Politecnico di Milano
Dipartimento di Elettronica e Informazione
Piazza Leonardo da Vinci 32, 20133 Milano (Italy)
bittanti@elet.polimi.it colaneri@elet.polimi.it

Abstract: In this paper we consider the problem of stability and the problem of stabilization via state-feedback of periodic linear systems in discrete-time. First, the case of known parameter systems is considered. Then, both norm-bounded and polytopic uncertain systems are dealt with. The basic tools for the analysis are the robust Lyapunov paradigm and the periodic Linear Matrix Inequality technique.
Copyright © 2001 IFAC

Keywords: Periodic Control, Discrete Time Linear Systems, Stability, Quadratic Stability, Linear Matrix Inequality

1. INTRODUCTION AND PROBLEM POSITION

Since long time, the problem of stability for systems with periodically varying coefficients attracted the attention of many scientists. In the second half of nineteenth century, the focus was on systems free of external influences, with milestone contributions by Floquet and Lyapunov. One century later, the rise of systems theory led to the consideration of an enlarged class of systems, namely dynamical systems driven by exogenous signals. This fact required the treatment of the stability problem in a new framework. A further evolution took place recently, motivated by the developments of robust control of uncertain systems. The issue became then that of finding a robust stabilizing feedback, namely a feedback law guaranteeing closed loop stability for all possible plants in a given family.

In this paper, we treat the problem of stabilization via state-feedback of linear periodic systems in discrete time. For an overview on periodic control see the recent chapter in [1]. To this purpose, we first study the stability of the system

$$x(t+1) = A(t)x(t) \tag{1}$$

where $A(\cdot)$is a periodic matrix of period T and the state $x(t) \in R^n$.

The associate transition matrix is

$$\Psi_A(t,\tau) = \begin{cases} A(t)A(t-1)\cdots A(\tau) & , t > \tau \\ I_n & , t = \tau \end{cases}$$

When $\tau = t - T$, then the transition matrix takes the name of *monodromy matrix*, denoted as $\Phi_A(t)$:

$$\Phi_A(t) = \Psi_A(t+T,t)$$

Notice in passing that, although $\Phi_A(t)$ changes with time, the characteristics multipliers are time-invariant, [2]. System (1), or, equivalently, $A(\cdot)$, is stable if and only if the multipliers are all located inside the unit disk in the complex plane.

We will first analyze system (1) when matrix $A(\cdot)$ is completely known. Then, we will turn to study systems subject to parametric uncertainties. Two classes of uncertainties are considered. The first class is that of norm bounded uncertainties where the perturbed system (1) is described by

$$A(t) = A_n(t) + L(t)\Delta(t)N(t) \quad (2)$$

The T-periodic matrix $A_n(\cdot)$ is the nominal dynamical matrix whereas $L(\cdot)$, $N(\cdot)$ and $M(\cdot)$ are T-periodic matrices, used to describe the structure of the uncertainty. The only unknown matrix is $\Delta(\cdot)$ which characterizes the model uncertainty and obeys the the norm bounded condition

$$||\Delta(t)|| \leq 1, \quad \forall t \quad (3)$$

Another class of uncertaint systems is the so-called class of polytopic uncertainty, where the system matrix $A(\cdot)$ is described by

$$A(t) = \sum_{i=1}^{N} \alpha_i A_i(t), \quad \alpha_i > 0, \quad \sum_{i=1}^{N} \alpha_i = 1 \quad (4)$$

In this case, matrices $A_i(\cdot)$ are known T-periodic matrices and the class of uncertain systems is obtained by letting the coefficients α_i free over the range defined in (4). In this way, $A(\cdot)$ is a linear combination of the matrices $A_i(\cdot)$ which are known as *vertices* of the polytope.

In the known parameter case, we will characterize the stability of matrix $A(\cdot)$ in terms of suitable equivalent periodic linear matrix inequalities. This equivalence between different formulations is then useful to work out the stability conditions for the classes of uncertain systems.

We address the stabilization problem via state-feedback. Precisely, we will make reference to the system

$$x(t+1) = A(t)x(t) + B(t)u(t) \quad (5)$$

with the state-feedback

$$u(t) = K(t)x(t) \quad (6)$$

Here, both the input matrix $B(\cdot)$ and the feedback gain $K(\cdot)$ are periodic of period T.

Also for the stabilization problem, we first consider the case where the matrices $A(\cdot)$ and $B(\cdot)$ are known. For such a case, a parametrization of all stabilizing gains is provided. Then we move to the perturbed case. Again we make reference to the two classes of perturbations introduced above. Precisely, the norm bounded perturbations amounts to assuming

$$A(t) = A_n(t) + L(t)\Delta(t)N(t) \quad (7)$$
$$B(t) = B_n(t) + L(t)\Delta(t)M(t) \quad (8)$$

where $A_n(\cdot)$, $B_n(\cdot)$, $L(\cdot)$, $N(\cdot)$ and $M(\cdot)$ are T-periodic matrices and $\Delta(\cdot)$ satisfies (3). As for the polytopic family, the structure of the system is as follows

$$A(\cdot) = \sum_{i=1}^{N} \alpha_i A_i(t), \; B(\cdot) = \sum_{i=1}^{M} \beta_i B_i(t) \quad (9)$$

where

$$\alpha_i > 0, \; \sum_{i=1}^{N} \alpha_i = 1, \; \beta_i > 0, \; \sum_{i=1}^{M} \beta_i = 1$$

In the perturbed case, the problem is the so-called robust stabilization problem, in which one has to find and characterize the class of T-periodic gains $K(\cdot)$ capable of stabilize the system for any pair constituted by the dynamic matrix and the input matrix in the corresponding uncertain sets.

We close the introduction by mentioning that, in the time-invariant case, these issues have been extensively treated, see e.g. [3]. For periodic systems, [4] and [5] have to be mentioned. The present paper is to provide missing results and overview the overall material into a general unifying scenario.

2. THE LYAPUNOV PARADIGM

The purpose of this section is to provide a basic result in Lyapunov stability, while further important issues on the subject are postponed to Sections 3 and 4.

By *difference Lyapunov inequalities* we mean precisely the following matrix inequalities

$$A(t)P(t)A(t)' - P(t+1) < 0, \; \forall t \quad (10)$$
$$A(t)'Q(t+1)A(t) - Q(t) < 0, \forall t \quad (11)$$

The two inequalities above can be equivalently written in two different matrix block forms, which are useful in analysis and design. For their derivation, a main role is played the celebrated Schur Lemma, which is not reported here since it is a standard tool in matrix theory.

Lemma 1. The following statements are equivalent each other.

(i) $A(\cdot)$ is stable
(ii) There exists a T-periodic positive definite solution $Q(\cdot)$ of the Lyapunov inequality

$$A(t)'Q(t+1)A(t) - Q(t) < 0$$

(iii) There exists a T-periodic solution $Q(\cdot)$ of the inequality

$$\begin{bmatrix} -Q(t) & A(t)'Q(t+1) \\ Q(t+1)A(t) & -Q(t+1) \end{bmatrix} < 0 \quad (12)$$

(iv) There exist two T-periodic matrices $Q(\cdot)$ and $Z(\cdot)$ solving the following inequality

$$\begin{bmatrix} -Q(t) & A(t)'Z(t) \\ Z(t)'A(t) & Q(t+1) - Z(t) - Z(t)' \end{bmatrix} < 0 \quad (13)$$

Proof. We first show the equivalence between (i) and (ii). ¿From (ii), by consecutive iterations one obtains

$$\begin{aligned} Q(t) &> A(t)'Q(t+1)A(t) \\ &\leq A(t)'[A(t+1)'Q(t+2)A(t+1)]A(t) \\ &\leq \cdots \leq \\ &\leq \Phi_A(t)'Q(t+T)\Phi_A(t) \end{aligned}$$

Then, setting $Q(t+T) = Q(t)$ it follows

$$Q(t) > \Phi_A(t)'Q(t)\Phi_A(t)$$

The positive semidefiteness of $Q(\cdot)$ implies that $\Phi_A(t)$ must be contractive, i.e. $A(\cdot)$ is stable.

Now, assume that $A(\cdot)$ is a T-periodic stable matrix (point (i)) and let

$$Q(t) = \sum_{i=0}^{\infty} \Psi_A(t+i,t)'\Psi_A(t+i,t)$$

This matrix series is convergent. Indeed, for any positive integer r,

$$\sum_{i=0}^{rT-1} \Psi_A(t+i,t)'\Psi_A(t+i,t) = \sum_{i=0}^{r-1} \Phi_A(t)'^k M \Phi_A(t)^k$$

with

$$M = \sum_{j=0}^{T-1} \Psi_A(t+j,t)'\Psi_A(t+j,t)$$

The stability assumption implies that $||\Phi_A(t)^k|| \leq K\lambda^k$ for some $K > 0$ and $\lambda < 1$. Hence

$$|| \sum_{i=0}^{rT-1} \Psi_A(t+i,t)'\Psi_A(t+i,t)|| \leq K^2||M|| \sum_{k=0}^{r-1} \lambda^{2k}$$

Therefore the quantity

$$\sum_{i=0}^{\tau} \Psi_A(t+i,t)'\Psi_A(t+i,t)$$

is bounded for any τ; moreover it is obviously monotonically nondecreasing in τ. In conclusion

$$Q(t) = \sum_{i=0}^{\infty} \Psi_A(t+i,t)'\Psi_A(t+i,t)$$

exists. Moreover, $Q(t)$ is T-periodic and positive semidefinite. Finally, a simple computation shows that

$$Q(t) = A(t)'Q(t+1)A(t) + I$$

Hence (i) is satisfied.

Equivalence between (ii) and (iii) are indeed equivalent thanks to the Schur Lemma. As for the equivalence between (iii) and (iv), assume first that (iii) holds. Then, taking $Z(t) = Q(t+1)$ one obtains (12). Conversely, assume that condition (iv) is met with. Then, $Z(t)+Z(t)' > Q(t+1) > 0$ so that $Z(t)$ is invertible (for each t). Consider now $M(t) = (Z(t)' - Q(t+1))Q(t+1)^{-1}(Z(t) - Q(t+1))$. Obviously, $M(t) \geq 0$. This entails that $Q(t+1) - Z(t) - Z(t)' \geq -Z(t)'Q(t+1)^{-1}Z(t)$. Therefore

$$\begin{aligned} 0 &> \begin{bmatrix} -Q(t) & A(t)'Z(t) \\ Z(t)'A(t) & Q(t+1) - Z(t) - Z(t)' \end{bmatrix} \\ &\geq \begin{bmatrix} -Q(t) & A(t)'Z(t) \\ Z(t)'A(t) & -Z(t)'Q(t+1)^{-1}Z(t) \end{bmatrix} \\ &= \begin{bmatrix} I & 0 \\ 0 & Z(t)' \end{bmatrix} \begin{bmatrix} -Q(t) & A(t)' \\ A(t) & -Q(t+1)^{-1} \end{bmatrix} \begin{bmatrix} I & 0 \\ 0 & Z(t) \end{bmatrix} \end{aligned}$$

Since $Z(t)$ is invertible, condition (iii) follows. ∎

Lemma 1 above can be "dualized" in the following result, the proof of which is omitted since it can be immediately obtained by duality.

Lemma 2. The following statements are equivalent each other.

(i) $A(\cdot)$ is stable

(ii) There exists A T-periodic positive definite solution $P(\cdot)$ of the Lyapunov inequality

$$A(t)P(t)A(t)' - P(t+1) < 0$$

(iii) There exists a T-periodic solution $P(\cdot)$ of the inequality

$$\begin{bmatrix} -P(t+1) & A(t)P(t) \\ P(t)A(t)' & -P(t) \end{bmatrix} < 0 \quad (14)$$

(iv) There exist two T-periodic matrices $P(\cdot)$ and $Z(\cdot)$ solving the following inequality

$$\begin{bmatrix} -P(t+1) & A(t)Z(t) \\ Z(t)'A(t)' & -P(t) + Z(t) + Z(t)' \end{bmatrix} < 0 \quad (15)$$

Lemma 1 is the key to deal with the robust prediction problem, whereas Lemma 2 is the key for the treatment of the robust state-feedback stabilization problem. Obviously, inequality (10) admits a periodic positive semidefinite solution $P(\cdot)$ iff inequality (11) admits a positive semidefinite solution $Q(\cdot)$.

We are now in a position to introduce the concept of Lyapunov function for the system (1). The Lyapunov function is any function $V(x,t)$ of the state coordinate $x \in R^n$ and time t such that

(i) $V(x,t)$ is continuous in x, $x \neq 0$
(ii) $V(x,t) > 0, \ \forall t$
(iii) $V(0,t) = 0, \ \forall t$
(iv) $V(x(t+1), t+1) - V(x(t), t) < 0, \forall x(\cdot)$ solving (1).

The existence of a Lyapunov function is equivalent to the stability of (1). If one has a T-periodic positive definite solution $Q(\cdot)$ of inequality (11), then it is straightforward to verify that

$$V(x,t) = x'Q(t)x$$

is a Lyapunov function.

Remark 2.1. The periodic Lyapunov Lemma can be easily generalized to cope with the property of exponential stability of $A(\cdot)$. Precisely, given a real T-periodic function $r(\cdot)$, with $0 < r(t) < 1$, $\forall t$, matrix $A(\cdot)$ is said to be r-exponentially stable if $r(\cdot)^{-1}A(\cdot)$ is stable. The necessary and sufficient condition for $A(\cdot)$ to be r-exponentially stable is that one of the following Lyapunov inequalities

$$A(t)P(t)A(t)' - r(t)^2P(t+1) < 0, \ \forall t$$
$$A(t)'Q(t+1)A(t) - r(t)^2Q(t) < 0, \forall t$$

admits a T-periodic positive semidefinite solution. Of course, the existence of a T-periodic positive semidefite solution $P(\cdot)$ to the first equation is equivalent to the existence of a T-periodic positive semidefite solution $Q(\cdot)$ to the second one. Notice that r-exponential stability entails that the characteristic multipliers of $A(\cdot)$ are inside the disk of radius $r(T-1)r(T-2)\cdots r(0)$. Such a radius becomes r^T if $r(t) = r = constant$.

3. ROBUST STABILITY

In recent years the issue of robust stability came to the stage [6], [7]. Basically the problem is to assess the stability of any matrix $A(\cdot)$ in a given set $\mathcal{A}$ of admissble matrices. In this regard, the following definition is in order [7].

Definition 3.1. The system (1) is *$\mathcal{A}$-robustly stable* if $A(\cdot)$ is stable $\forall A(\cdot) \in \mathcal{A}$.

An useful tool to assess robust stability is provided by the notion of *quadratic stability*.

Definition 3.2. The system (1) is said to be *quadratically stable* if there exists a function $V(x,t) = x'Q(t)x$, $Q(t+T) = Q(t)$, $\forall t$ which can serve as a Lyapunov function for each $A(\cdot) \in \mathcal{A}$.

Obviously, quadratic stability implies robust stability, whilst the converse statement is false in general.

The problem of robust or quadratic stability can take different facets depending upon the definition of $\mathcal{A}$.

3.1 *Robust stability for norm bounded perturbations*

If one focuses on the norm bounded perturbations, then it is possible to give a simple necessary and sufficient condition for quadratic stability based on a periodic matrix inequality.

First of all consider the norm bounded perturbation described in (2), i.e.

$$\mathcal{A} = \{A(\cdot) = A_n(\cdot) + L(\cdot)\Delta(\cdot)N(\cdot)\}$$

where $||\Delta(t)|| \leq 1, \forall t$. Moreover, consider the Periodic Linear Matrix Inequality (PLMI)

$$0 > \begin{bmatrix} -P(t+1) + L(t)L(t)' & 0 \\ 0 & -I \end{bmatrix} + \begin{bmatrix} A_n(t) \\ N(t) \end{bmatrix} P(t) \begin{bmatrix} A_n(t)' & N(t)' \end{bmatrix} \quad (16)$$

As a straightforward consequence of the Schur lemma, it is easy to see that this periodic matrix inequality is equivalent to the following inequalities:

$$\begin{aligned} P(t+1) > {} & A_n(t)P(t)A_n(t)' + L(t)L(t)' + \\ & + A_n(t)P(t)N(t)'[I - N(t)P(t)N(t)']^{-1} \\ & \times N(t)P(t)A_n(t)' \quad (17) \\ 0 < {} & I - N(t)P(t)N(t)' \quad (18) \end{aligned}$$

These inequalities are of lower dimension but no longer linear in $P(\cdot)$. The equivalence is a straightforward consequence of the Schur Lemma. Now, recalling that $||\Delta(t)|| \leq 1, \forall t$, from (17), (18)

$$\begin{aligned} P(t+1) > {} & A_n(t)P(t)A_n(t) + L(t)\Delta(t)\Delta(t)'L(t)' \\ & + A_n(t)P(t)N(t)'[I - N(t)P(t)N(t)']^{-1} \\ & \times N(t)P(t)A_n(t)' \\ & - [A_n(t)P(t)N(t)' - L(t)\Delta(t)S(t)]S(t)^{-1} \\ & \times [A_n(t)P(t)N(t)' - L(t)\Delta(t)S(t)]' \\ & = A(t)P(t)A(t)' \end{aligned}$$

Hence $P(\cdot)$ is a positive semidefinite solution of the Lyapunov inequality (10). Since $P(\cdot)$ is independent of $\Delta(\cdot)$ (see (17)), quadratic stability is ensured.

As a matter of fact, it can also be shown that quadratic stability implies the existence of a periodic positive semidefinite matrix $P(\cdot)$ satisfying both (17) and (18), or, equivalently, the PLMI (16). In conclusion, for the quadratic stability of periodic systems the basic result can be stated as follows.

Proposition 3. The T-periodic linear system (1), with dynamic matrix $A(t)$ subject to a norm bounded perturbation (2) with $\|\Delta(t)\| \leq 1$, $\forall t$, is quadratically stable if and only if there exists a T-periodic positive semidefinite matrix $P(\cdot)$ to the PLMI (16).

Remark 3.1. Note that the equation

$$x(t+1) = (A_n(t) + L(t)\Delta(t)N(t))x(t)$$

can be equivalently rewritten in the form

$$x(t+1) = A_n(t)x(t) + \beta L(t)w(t) \qquad (19)$$

$$y(t) = \frac{N(t)}{\beta}x(t) \qquad (20)$$

with the auxiliary signal $w(\cdot)$ obtained by the feedback law

$$w(t) = \Delta(t)y(t)$$

and β any positive scalar. This interpretation of a system subject to a norm bounded perturbation is very useful to establish a bridge with H_∞ considerations. Precisely, the quadratic stability of $A(\cdot)$ is equivalent to the fact that the H_∞ norm of the system $(A_n(\cdot), \beta L(\cdot), N(\cdot)\beta^{-1})$ is less than one. In turn this is equivalent to the existence of a periodic positive semidefinite solution $P(\cdot)$ of

$$\begin{aligned} P(t+1) &> A_n(t)P(t)A_n(t)' - \beta^2 L(t)L(t)' \\ &\quad + A_n(t)P(t)N(t)'[\beta^2 I - N(t)P(t)N(t)']^{-1} \\ &\quad \times N(t)P(t)A_n(t)' \\ 0 &< \beta^2 I - N(t)P(t)N(t)' \end{aligned}$$

Notice that in this framework the role of parameter β is immaterial. Indeed, without any loss of generality, it is always possible to scale $P(t)$ by β^2 in the last inequalities. As a matter of fact, the role of parameter β can be significant in the optimization of the upper bound of the covariance matrix associated with the cyclostationary process defined by (19), (20). See [8] for the time-invariant case. ■

The previous result can be given an useful dual version by referring to the dual PLMI

$$0 \geq \begin{bmatrix} -Q(t) + N(t)'N(t) & 0 \\ 0 & -I \end{bmatrix} + \begin{bmatrix} A_n(t)' \\ L(t)' \end{bmatrix} Q(t+1) \begin{bmatrix} A_n(t) & L(t) \end{bmatrix} \qquad (21)$$

The proof of the following proposition is omitted since follows the same rationale of Proposition 3.

Proposition 4. The T-periodic linear system (1), with dynamic matrix $A(t)$ subject to a norm bounded perturbation (2) with $\|\Delta(t)\| \leq 1$, $\forall t$, is quadratically stable if and only if there exists a T-periodic positive semidefinite matrix $Q(\cdot)$ to the PLMI (21).

3.2 *Robust stability for polytopic perturbations*

We now turn to the polytopic class of perturbations introduced in (4),

$$\mathcal{A} = \{A(\cdot) = \sum_{i=1}^{N} \alpha_i A_i(t)\}$$

where $\alpha_i > 0$ and $\sum_{i=1}^{N} \alpha_i = 1$.

As already said, this set $\mathcal{A}$ is obtained as the convex linear combination of matrices $A_i(\cdot)$ which are known as *vertices* of the polytope. For such an uncertainty set, the quadratic stability is easily checkable. Indeed, consider the Lyapunov inequalities

$$A_i(t)'Q(t+1)A_i(t) - Q(t) < 0, \quad i = 1, 2, \cdots, N$$

and assume that for each i there exists the same T-periodic positive definite solution $Q(\cdot)$. Now, recalling (12), it is apparent that $Q(\cdot)$ satisfies

$$\begin{bmatrix} -Q(t) & A_i(t)'Q(t+1) \\ Q(t+1)A_i(t) & -Q(t+1) \end{bmatrix} < 0$$

for each $i = 1, 2, \cdots, N$. Multiplying each of these matrices by α_i and summing up, one obtains (12), which is equivalent to

$$A(t)'Q(t+1)A(t) - Q(t) < 0$$

This means that stability is ensured for each $A(\cdot) \in \mathcal{A}$. Conversely, if the system is quadratically stable, then there exists a periodic positive definite matrix $Q(\cdot)$ satisfying the Lyapunov inequality for all matrices $A(\cdot) \in \mathcal{A}$. In particular, this matrix must satisfy the Lyapunov inequality for all vertices. In conclusion the following proposition can be stated.

Proposition 5. The T-periodic linear system (1), with dynamic matrix $A(t)$ belonging to the polytopic family (4) is quadratically stable if and only if there exists a T-periodic positive definite matrix $Q(\cdot)$ satisfying

$$A_i(t)'Q(t+1)A_i(t) - Q(t) < 0, \quad i = 1, 2, \cdots, N \blacksquare$$

The property of quadratic stability has the disadvantage of requiring one Lyapunov function for all members of the uncertainty set. This may be too conservative since there might exist sets of matrices which are stable in a robust sense without being quadratically stable. Below, a sufficient condition of robust stability for polytopic uncertainty which does not necessarily entail quadratic stability is provided.

Proposition 6. The T-periodic linear system (1), with dynamic matrix $A(t)$ belonging to the polytopic family (4) is robustly stable if there exist a T-periodic matrix $Z(\cdot)$ and T-periodic matrices $Q_i(\cdot)$, $i = 1, 2, \cdots, N$ such that

$$\begin{bmatrix} -Q_i(t) & A_i(t)'Z(t) \\ Z(t)'A_i(t) & Q_i(t+1) - Z(t) - Z(t)' \end{bmatrix} < 0 \quad (22)$$

Proof. Multiply (22) by α_i and sum with respect to i, $i = 1, 2, \cdots, N$. Then, recalling that $\sum_{i=0}^{N} \alpha_i = 1$ it follows

$$\begin{bmatrix} -\bar{Q}(t) & A(t)'Z(t) \\ Z(t)'A(t) & \bar{Q}(t+1) - Z(t) - Z(t)' \end{bmatrix} < 0$$

where $\bar{Q}(\cdot) = \sum_{i=1}^{N} \alpha_i Q_i(\cdot)$. The stability of any $A(\cdot) \in \mathcal{A}$ follows from Lemma 1.

Remark 3.2. A possible extension of the above analysis for polytopic uncertainty set, amounts to replacing the constant coefficients α_i in the uncertainty description (4) with T-periodically varying coefficients $\alpha_i(t)$. As can be easily seen, this does not affect the validity of Proposition 5 for quadratic stability. On the opposite, the fact that the coefficients are constant cannot be relaxed in Proposition 6.

4. THE CLASS OF STABILIZING CONTROLLERS

The basic stabilizability problem is that of finding a feedback law (6) such as to stabilize system (5) for $A(\cdot)$ and $B(\cdot)$ known. A general parametrization of all periodic stabilizing gains $K(\cdot)$ can be worked out by means of a suitable matrix inequality. Precisely, the *Lyapunov inequality condition* already worked out enables one to conclude that the closed-loop system associated with a periodic gain $K(\cdot)$ is stable, if and only if there exists a positive definite periodic matrix $P(\cdot)$ satisfying the inequality:

$$P(t+1) > (A(t) + B(t)K(t))P(t) \times \\ \times (A(t) + B(t)K(t))'$$

A sensible way of approaching this problem is to write $K(\cdot)$ in the following factorized form

$$K(t) = W(t)'P(t)^{-1} \quad (23)$$

where $W(\cdot)$ is a periodic matrix to be determined. It is important to notice that this position does not imply any lack of generality. Now, it is possible to use the PLMI (12) so that the following conclusion can be directly drawn.

Proposition 7. All the stabilizing gains $K(\cdot)$ for system (5) are given by (23), where $P(\cdot)$ and $W(\cdot)$ are T-periodic solution of the PLMI

$$\begin{bmatrix} -P(t+1) & A(t)P(t) + B(t)W(t)' \\ * & -P(t) \end{bmatrix} < 0$$

The symbol $*$ in the $(2,1)$ position denotes the transpose of the $(1,2)$ entry.

The class of solutions of a PLMI constitutes reportedly a convex set. So, the family of generating pairs $(W(\cdot), P(\cdot))$ with $P(t) > 0$, $\forall t$, is convex. However, by no means, this entails that the set of $K(\cdot) = W(\cdot)'P(\cdot)^{-1}$ is convex. Note that the idea of working out a convex set generator for the class of stabilizing controllers originates from the time-invariant case where it was explored in a number of papers, such as [3].

5. ROBUST STABILIZATION

Consider now the stabilization problem in the case in which both $A(\cdot)$ and $B(\cdot)$ are uncertain matrices belonging to the given classes (7), (8) or (9), respectively.

Then, the objective is to design a T-periodic state-feedback controller

$$u(t) = K(t)x(t)$$

which provides a stable closed-loop system for any possible choice of $A(\cdot)$ and $B(\cdot)$ in the considered uncertain sets.

First of all we consider the class of norm bounded perturbations defined in (7), (8).

Proposition 8. The T-periodic linear system with matrices $A(t)$ and $B(t)$ belonging to the norm bounded sets (7), (8) is quadratically stabilizable by a T-periodic state-feedback control law if and only if there exist T-periodic matrices $W(\cdot)$ and $P(t)$ such that the matrix

$$\begin{bmatrix} -P(t+1)+L(t)L(t)' & 0 & A_n(t)P(t)+B_n(t)W(t)' \\ * & -I & N(t)P(t)+M(t)W(t)' \\ * & * & -P(t) \end{bmatrix}$$

is negative definite. Moreover, all the quadratically stabilizing T-periodic gains are given by (23).

Proof Again the symbols $*$ denote the transpose elements of the entries $(1,2)$, $(1,3)$ and $(2,3)$. This result is easily obtained from (16) by taking $K(t)P(t) = W(t)'$ and recognizing that the closed-loop system can be written as

$$\begin{aligned} x(t+1) &= (A_n(t)+B_n(t)K(t))x(t) \\ &\quad + L(t)\Delta(t)(N(t)+B(t)M(t))x(t) \end{aligned}$$

Indeed (16) transforms into

$$\begin{aligned} 0 > &\begin{bmatrix} -P(t+1)+L(t)L(t)' & 0 \\ 0 & -I \end{bmatrix} \\ &+ \begin{bmatrix} A_n(t)P(t)+B_n(t)W(t)' \\ N(t)P(t)+B_n(t)W(t)' \end{bmatrix} P(t)^{-1} \times \\ &\times \begin{bmatrix} A_n(t)P(t)+B_n(t)W(t)' \\ N(t)P(t)+B_n(t)W(t)' \end{bmatrix}' \end{aligned}$$

In view of the Schur Lemma, the above condition is equivalent to the one introduced in the statement. ■

We end the paper by considering the second class of uncertainties, namely the polytopic family defined in (9). Two results are provided. The first one concerns the quadratic stabilization problem, which aims at finding a T-periodic gain $K(\cdot)$ for which there exists a *unique* Lyapunov function associated with $A(t)+B(t)K(t)$, for any $A(\cdot)$ and any $B(\cdot)$ in the uncertainty sets. The second result provides a sufficient condition of robust stabilization based on parameter dependent Lyapunov function. Obviously, the second approach leads to a less conservative condition with respect to that achieved with the quadratic stabilizability viewpoint.

Proposition 9. The T-periodic linear system with matrices $A(t)$ and $B(t)$ belonging to the polytopic families (9) is quadratically stabilizable by a T-periodic state-feedback control law if and only if there exist T-periodic matrices $W(\cdot)$ and $P(\cdot)$ such that, for each $i = 1,2,\cdots,N$, $j = 1,2,\cdots,M$, the matrix

$$\begin{bmatrix} -P(t+1) & B_j(t)W(t)'+A_i(t)P(t) \\ W(t)B_j(t)'+P(t)A(t) & -P(t) \end{bmatrix}$$

is negative definite. Moreover, all the quadratically stabilizing T-periodic gains are given by (23).

Proof If the system is quadratically stabilizable then there exists a T-periodic gain $K(\cdot)$ such that $A(\cdot)+B(\cdot)K(\cdot)$ is quadratically stable. This obviously implies the stability for any pair $(A_i(\cdot), B_j(\cdot))$, so that there exists a positive definite T-periodic matrix $P(\cdot)$ and $W(\cdot) = P(\cdot)K(\cdot)'$ such that the condition is satisfied. Conversely, if the condition is met with for some T-periodic $W(\cdot)$ and for some positive definite T-periodic matrix $P(\cdot)$, then the thesis follows from Lemma 2 by multiplying the matrix by $\alpha_i\beta_j$ and summing up with respect to i, $i = 1,2,\cdots,N$ and $j = 1,2,\cdots,M$. Indeed, by doing so, it follows that

$$\begin{bmatrix} -P(t+1) & (A(t)+B(t)K(t))P(t) \\ P(t)(A(t)+B(t)K(t))' & -P(t) \end{bmatrix}$$

is negative with $K(t) = W(t)'P(t)^{-1}$. ■

The next result provides a less conservative sufficient robust stabilizability condition based on a parameter dependent Lyapunov function.

Proposition 10. The T-periodic linear system with matrices $A(t)$ and $B(t)$ belonging to the polytopic families (9) is robustly stabilizable with a T-periodic state-feedback if there exist T-periodic matrices $Z(\cdot)$, $W(\cdot)$ and T-periodic matrices $P_{ij}(\cdot)$, such that, for each $i = 1,2,\cdots,N$, $j = 1,2,\cdots,M$, the matrix

$$\begin{bmatrix} -P_{ij}(t+1) & A_i(t)Z(t)+B_j(t)W(t)' \\ Z(t)'A_i(t)'+W(t)B_j(t)' & P_{ij}(t)-Z(t)-Z(t)' \end{bmatrix}$$

is negative definite. If this condition is satisfied, then a robustly stabilizing T-periodic control law is given by $K(t) = W(t)'Z(t)^{-1}$.

Proof. Multiply the matrix by $\alpha_i\beta_j$ and sum with respect to i, $i = 1,2,\cdots,N$ and $j = 1,2,\cdots,M$. It follows that

$$\begin{bmatrix} -P(t+1) & (A(t)+B(t)K(t))Z(t) \\ Z(t)'(A(t)+B(t)K(t))' & P(t)-Z(t)-Z(t)' \end{bmatrix}$$

is negative definite where

$$P(t) = \sum_{i=0}^{N} \alpha_i \sum_{j=0}^{M} \beta_j P_{ij}(t), \quad K(t) = W(t)'Z(t)^{-1}$$

The result finally follows from Lemma 2.

6. CONCLUSIONS

In the paper the stability problem and state-feedback stabilization problem of siscrete-time periodic systems have been addressed. A number of results have been provided for systems with known parameters and for uncertain systems. In the latter case, two types of uncertainty structure have been dealt with, namely the class of norm bounded uncertainty and the family of matrices in a polytopic set. In this way, a wide range of cases is treated and a comprehensive overview of the area is provided. Future work will concern the stability and stabilization problems via canonical forms.

ACNOWLEDGEMENTS

The paper was partially supported by CESTIA - Centro di Teoria dei Sistemi of the Italian National Research Council (CNR), by the Italian Ministry of Education (MURST), and by the European Network Europoly.

7. REFERENCES

[1] S. Bittanti, P. Colaneri, Periodic Control. John Wiley Encyclopedia of Electrical and Electronic Engineering, Vol. 16, pp. 240 - 253, John Wiley and sons, J. Webster ed., 1999.

[2] S. Bittanti, Deterministic and stochastic periodic systems, in "Time series and linear systems" (S. Bittanti ed.). Springler-Verlag,1986.

[3] J.C. Geromel, P.L.D. Peres and J. Bernussou, On a convex parameter space method for linear control design of uncertain systems, *SIAM J. Control and Optimization*, Vol. 29, n. 2, pp. 381-402, 1991.

[4] C.E. de Souza, A.Trofino, Stabilization of linear discrete-time periodic systems, *IFAC Conference on System Structure and Control*, Nantes, France, June 1998.

[5] S. Bittanti, P. Colaneri, An LMI characterization of the class of stabilizing controllers for periodic discrete-time systems, *World IFAC Conference*, Beijing, 1999.

[6] J. Ackermann, Robust control, Springer Verlag, 1993.

[7] B.R. Barmish, New tools for robustness of linear systems, Mc Millan Publishing Company, New York, 1994.

[8] P.Bolzern, P.Colaneri, G.De Nicolao, Covariance bounds discrete-time linear systems with parameter uncertainty, Int. J. of Control, Vol. 60, N. 6, pp. 1307-1317, 1994.

Copyright © IFAC Periodic Control Systems,
Cernobbio-Como, Italy, 2001

OUTPUT-FEEDBACK RECEDING HORIZON CONTROLLER DESIGN FOR LTV SYSTEMS

Seung Cheol Jeong [1] **PooGyeon Park** [2]

Electrical and Computer Engineering Division, Pohang University of Science and Technology, Pohang, Kyungbuk, 790-784, Korea

Abstract: This paper presents a dynamic output-feedback receding horizon controller (RHC) for LTV systems including periodic systems. The existing output-feedback RHC in the literature is composed of a state observer and a static controller associated with the observer states (similar to LQG control), where the fundamental assumption is that the state observer will supply the exact states as time goes up. When the system suffers from disturbances and uncertainties or is time-varying, however, the performance of those controllers may be much degraded and even the closed-loop stability may not be guaranteed. The proposed controller, which is not necessary to have the state-observer, overcomes such difficulties. Using matrix inequality conditions on the terminal weighting matrix, the closed-loop system stability is guaranteed. *Copyright © 2001 IFAC*

Keywords: Output feedback, receding horizon control, linear time varying, matrix inequality condition, stability.

1. INTRODUCTION

Receding horizon control (RHC), also known as model predictive control, has received much attention in control societies because of its good tracking performance and many applications to industrial processing systems such as distillation, paper processing, and etc (Kwon and Peason, 1977; Garcia *et al.*, 1991; Cutler and Hawkins, 1988). The basic concept of the receding horizon control is to solve an optimization problem over a fixed number of future time instants at the current time and to implement the first optimal control law as the current control law. The procedure is then repeated at each subsequent instant. The advantage of the RHC is that constraints may be directly incorporated into the on-line optimization. In particular, it is a suitable control strategy for time-varying systems (Rawlings and Muske, 1993; Lee *et al.*, 1998).

Recently, some results are presented on the output feedback RHC for the time invariant system (Lee, 2000; Lee and Kouvaritakis, 2000). In those control schemes, the authors used state observers. However, using state observers made the analysis and synthesis problem complicated because of estimation errors. Furthermore, when the system suffers from disturbances and uncertainties, such controllers may not yield good performance and moreover closed-loop stability.

In this paper, a dynamic output-feedback RHC for LTV systems including periodic systems is suggested. The proposed controller, which is not necessary to have the state observer, overcomes such

[1] abraham@postech.ac.kr
[2] Corresponding author. Tel.: +82-54-279-2238, fax: +82-54-279-2903, ppg@postech.ac.kr

difficulties. Using matrix inequality conditions on the terminal weighting matrix, the closed-loop system stability is guaranteed. Numerical example is suggested in order to show how the proposed controller improves performance compared with the observer-based RHC.

2. MAIN RESULTS

Let us consider a linear time-varying system

$$x_{k+1} = A_k x_k + B_k u_k, \quad (1)$$
$$y_k = C_k x_k \quad (2)$$

where $x_k \in \mathcal{R}^n$ and $u_k \in \mathcal{R}^m$ are state and input vectors, respectively, and all matrices are known. Assume that one can only measure the output y_k of the given system so that the state-feedback control law cannot be used. Furthermore, let us assume that the system described by equations (1) and (2) is stabilizable and detectable. For simple notations, time indexes of all matrices shall be henceforth omitted . Therefore, A_k, B_k and C_k are replaced with A, B, and C, respectively.

Let us design a dynamic output-feedback controller such as; for $j = N-1, ..., 0$,

$$z_{k+1+j} = A^c_{j,N} z_{k+j} + B^c_{j,N} y_{k+j}, \quad (3)$$
$$u_{k+j} = C^c_{j,N} z_{k+j} + D^c_{j,N} y_{k+j}, \quad (4)$$

to minimize the following linear quadratic cost

$$J^T_{k,k+N} \triangleq \sum_{j=0}^{N-1} \{x^T_{k+j} Q x_{k+j} + u^T_{k+j} R u_{k+j}\} + x^T_{k+N} \Phi x_{k+N}. \quad (5)$$

In (3) and (4), $z_k \in \mathcal{R}^l$ and the order of the controller l is not fixed *a priori* and possibly includes a reduced-order controller ($l < n$) or static controller ($l = 0$) as its special case.

Combining (1) and (3) provides a closed-loop system

$$\eta_{k+1+j} = (A_0 + B_0 \Sigma_{j,N} C_0)\eta_{k+j}, \quad (6)$$

where

$$\eta_{k+j} = \begin{bmatrix} x_{k+j} \\ z_{k+j} \end{bmatrix}, \quad \Sigma_{j,N} = \begin{bmatrix} A^c_{j,N} & B^c_{j,N} \\ C^c_{j,N} & D^c_{j,N} \end{bmatrix}$$
$$A_0 = \begin{bmatrix} A & 0 \\ 0 & 0 \end{bmatrix}, \quad B_0 = \begin{bmatrix} 0 & B \\ I & 0 \end{bmatrix}, \quad C_0 = \begin{bmatrix} 0 & I \\ C & 0 \end{bmatrix}$$
$$E_1 = [I \ 0], \quad E_2 = [0 \ I]$$

Since it holds that

$$\begin{bmatrix} x_{k+j} \\ u_{k+j} \end{bmatrix} = \begin{bmatrix} E_1 \\ E_2 \Sigma_{j,N} C_0 \end{bmatrix} \eta_{k+j}$$

(5) can be represented as

$$J^T_{k,k+N} \triangleq \sum_{j=0}^{N-1} \{\eta^T_{k+j}[E_1^T Q E_1 + (E_2 \Sigma_{j,N} C_0)^T R(E_2 \Sigma_{j,N} C_0)]\eta_{k+j}\} + \eta^T_{k+N} E_1^T \Phi E_1 \eta_{k+N} \quad (7)$$

where

$$E_1^T \Phi E_1 = \begin{bmatrix} \Phi & 0 \\ 0 & 0 \end{bmatrix}. \quad (8)$$

Note that in the state-feedback case, Φ plays a very important role in guaranteeing the stability of the closed-loop system. However, there has been no result on the dynamic output-feedback receding horizon control due to the singularity of the final weighting matrix (8), we guess. In this paper this problem shall be handled by relaxation of the final cost as follows. Let us introduce a full-rank matrix P_f such that

$$E_1^T \Phi E_1 \le P_f. \quad (9)$$

which shall be also found to guarantee the stability of the closed-loop system in this paper. The following inequality is then satisfied

$$J^T_{k,k+N} \le \sum_{j=0}^{N-1} \{\eta^T_{k+j}[E_1^T Q E_1 + (E_2 \Sigma_{j,N} C_0)^T R(E_2 \Sigma_{j,N} C_0)]\eta_{k+j}\} + \eta^T_{k+N} P_f \eta_{k+N} \triangleq J_{k,k+N} \quad (10)$$

From now on, the cost itself, $J^T_{k,k+N}$ is not dealt with , but the upper bound of the cost, $J_{k,k+N}$. To begin with, let us consider the following equality; for $i = N-1, ..., 0$,

$$J_{k+i,\ k+i+1} = \eta^T_{k+i}\{E_1^T Q E_1 + (E_2 \Sigma_{i,N} C_0)^T R(E_2 \Sigma_{i,N} C_0)\}\eta_{k+i} + \eta^T_{k+i+1} P_{i+1,N} \eta_{k+i+1} \quad (11)$$

The optimal controller $\Sigma^*_{i,N}$ that minimizes the cost $J_{k+i,\ k+i+1}$ can be obtained by the following linear programming; minimize the trace of $P_{i,N}$, for the given $P_{i+1,N}$, subject to

$$P_{i,N} \ge E_1^T Q E_1 + (E_2 \Sigma_{i,N} C_0)^T R(E_2 \Sigma_{i,N} C_0) + + \mathcal{F}^T P_{i+1,N} \mathcal{F}. \quad (12)$$

where $\mathcal{F} = A_0 + B_0 \Sigma_{i,N} C_0$. With some efforts, (12) can be transformed into the following inequality,

$$\begin{bmatrix} -R^{-1} & E_2\Sigma_{i,N}C_0 & 0 \\ (E_2\Sigma_{i,N}C_0)^T & E_1^TQE_1 - P_{i,N} & \mathcal{F}^T \\ 0 & \mathcal{F} & -P_{i+1,N}^{-1} \end{bmatrix} \leq 0 \quad (13)$$

with the boundary condition $P_{N,N} = P_f$. The optimal cost of (10) with the optimal controller $\Sigma_{i,N}^*$ for $i = N-1, ..., 0$ is given by

$$J_{k+i,k+N}^* = \eta_{k+i}^T P_{i,N} \eta_{k+i}. \quad (14)$$

Remark 1: For the given $P_{i+1,N}$, although there exist a lot of $\Sigma_{i,N}$ satisfying (13), the minimal $P_{i,N}$ is unique. ∎

Remark 2: Let us think about the trajectories of $P_{i,N}^1$ and $P_{i,N}^2$ (for $i = N, ..., 0$) subject to (13). If the initial values of the two trajectories, i.e., $P_{N,N}^1$ and $P_{N,N}^2$ have the relation of $P_{N,N}^1 \leq P_{N,N}^2$, then it holds that $P_{i,N}^1 \leq P_{i,N}^2$ for all the subsequent times, $i = N-1, ..., 0$. ∎

The receding horizon control strategy is to solve the finite optimal problem for (10), $i = N-1, ..., 0$ at the current time k and implement its first solution, i.e. $\Sigma_{0,N}^*$, then to repeat the procedure at the next time $k+1$.

In the following theorem, the terminal inequality condition on P_f is suggested, which play a crucial role in guaranteeing the stability of the closed-loop system.

Theorem 1. **(Terminal inequality condition)** Assume that P_f in (10) satisfies the following inequality

$$P_f \geq E_1^TQE_1 + (E_2\Sigma_{N,N+1}C_0)^T R(E_2\Sigma_{N,N+1}C_0) + (A_0 + B_0\Sigma_{N,N+1}C_0)^T \; P_f \; (A_0 + B_0\Sigma_{N,N+1}C_0). \quad (15)$$

The optimal cost $J_{k+i,k+N}^*$ in (14) then satisfies the following relation

$$J_{k+i,k+N+1}^* \leq J_{k+i,k+N}^*. \quad (16)$$

Proof:

$$\begin{aligned} & J_{k+i,k+N+1}^* - J_{k+i,k+N}^* \\ &= \sum_{j=i}^{N-1} \{\eta_{k+j}^{1T}[E_1^TQE_1 + (E_2\Sigma_{j,N}^1C_0)^T \\ &\quad R(E_2\Sigma_{j,N}^1C_0)]\eta_{k+j}^1\} + J_{k+N,k+N+1}^{1*} \\ &\quad - \sum_{j=i}^{N-1} \{\eta_{k+j}^{2T}[E_1^TQE_1 + (E_2\Sigma_{j,N}^2C_0)^T \\ &\quad R(E_2\Sigma_{j,N}^2C_0)]\eta_{k+j}^2\} - \eta_{k+N}^{2T}P_f\eta_{k+N}^2 \end{aligned}$$

where $\Sigma_{j,N}^1$, $\Sigma_{j,N}^2$ and η_{k+j}^1, η_{k+j}^2 are optimal controls to minimize and states corresponding to $J_{k+i,k+N+1}^*$ and $J_{k+i,k+N}^*$, respectively. If we replace $\Sigma_{j,N}^1$ by $\Sigma_{j,N}^2$ for $j = i$ through $N-1$, then we have

$$\begin{aligned} & (J_{k+i,k+N+1}^* - J_{k+i,k+N}^*) \\ &\leq (J_{k+N,k+N+1} - \eta_{k+N}^{2T}P_f\eta_{k+N}^2) \end{aligned}$$

From (11) and (15), we can rewrite this

$$\begin{aligned} & J_{k+N,k+N+1} - \eta_{k+N}^{2T}P_f\eta_{k+N}^2 \\ &= \eta_{k+N}^{2T}\{E_1^TQE_1 + (E_2\Sigma_{N,N+1}C_0)^T R(E_2 \\ &\quad \Sigma_{N,N+1}C_0) + (A_0 + B_0\Sigma_{N,N+1}C_0)^T \; P_f \\ &\quad (A_0 + B_0\Sigma_{N,N+1}C_0) - P_f\}\eta_{k+N}^2 \leq 0 \end{aligned}$$

The proof is completed. ∎

Remark 3: The Theorem 1 is the extension of (Lee *et al.*, 1998) for the output-feedback case. The result of (Lee *et al.*, 1998) is originally done for the system whose states can be measured. However, the terminal inequality condition is constructed to ensure the stability of the closed-loop system. If one concerns about the stability of the resulting closed-loop system, one can still use the same terminal inequality condition for the case of output feedback. ∎

Remark 4: The Theorem 1 says that if it holds that $J_{k+N,k+N+1}^* \leq J_{k+N,k+N}^*$, then $J_{k+i,k+N+1}^* \leq J_{k+i,k+N}^*$. One can generalize this result as follows. Assume that

$$J_{k+k',k+k_f+m}^* \leq J_{k+k',k+k_f}^*, \quad m \geq 1, \quad (17)$$

for some k'. Then, for $0 \leq k'' \leq k'$,

$$\begin{aligned} & J_{k+k'',k+k_f+m}^* - J_{k+k'',k+k_f}^* \\ &= \sum_{j=k''}^{k'-1} \{\eta_{k+j}^{1T}[E_1^TQE_1 + (E_2\Sigma_{j,N}^1C_0)^T \\ &\quad R(E_2\Sigma_{j,N}^1C_0)]\eta_{k+j}^1\} + J_{k+k',k+k_f+m}^{1*} \\ &\quad - \sum_{j=k''}^{k'-1} \{\eta_{k+j}^{2T}[E_1^TQE_1 + (E_2\Sigma_{j,N}^2C_0)^T \\ &\quad R(E_2\Sigma_{j,N}^2C_0)]\eta_{k+j}^2\} - J_{k+k',k+k_f}^{2*} \end{aligned}$$

where $\Sigma_{j,N}^1$ and $\Sigma_{j,N}^2$ are optimal controls to minimize $J_{k+k'',k+k_f+m}$ and $J_{k+k'',k+k_f}$, respectively. If we replace $\Sigma_{j,N}^1$ by $\Sigma_{j,N}^2$ for $j = k''$ to $k'-1$,

$$\left\{J^*_{k+k'',k+k_f+m} - J^*_{k+k'',k+k_f}\right\} \leq \left\{J^*_{k+k',k+k_f+m} - J^*_{k+k',k+k_f}\right\} \leq 0.$$

This result says that when the monotonicity of the optimal cost holds once, it holds for all subsequent times. Using this property, one can develop the block-shift output-feedback receding horizon control. ∎

If there exists such a P_f satisfying (15), it can be obtained by the linear programming. With some efforts, we can transform (15) into the following; for any N',

$$0 > W + U\Sigma_{N',N'+1}V^T + V\Sigma^T_{N',N'+1}U^T \quad (18)$$

where

$$W = \begin{bmatrix} -R^{-1} & 0 & 0 \\ 0 & E_1^T Q E_1 - P_f & A_0^T P_f \\ 0 & P_f A_0 & -P_f \end{bmatrix},$$

$$U = \begin{bmatrix} E_2 \\ 0 \\ P_f B_0 \end{bmatrix}, \quad V = \begin{bmatrix} 0 \\ C_0^T \\ 0 \end{bmatrix}$$

The variable elimination procedure (Boyd *et al.*, 1994) states that (18) is solvable for some $\Sigma_{N',N'+1}$ if and only if

$$0 > U_\perp^T W U_\perp \quad \text{and} \quad 0 > V_\perp^T W V_\perp$$

where $U_\perp$ and $V_\perp$ is orthogonal complements of U and V, respectively.

The main result of this paper is summarized in the following theorem.

Theorem 2. **(Dynamic output-feedback receding horizon controller)** Suppose that there exist positive definite matrices X and $\bar{X}$ subject to

$$\begin{bmatrix} -\bar{X} - BR^{-1}B^T + A\bar{X}A^T & A\bar{X} \\ \bar{X}A^T & -Q^{-1} + \bar{X} \end{bmatrix} < 0 \quad (19)$$

$$C^\perp \left[Q - X + A^T X A\right] (C^T)^\perp < 0 \quad (20)$$

$$\begin{bmatrix} X & I \\ I & \bar{X} \end{bmatrix} \geq 0 \quad (21)$$

where $(C^T)^\perp$ denotes an orthogonal complement of (C^T) and $C^\perp = ((C^T)^\perp)^T$. After forming P_f from X and $\bar{X}$ such as

$$P_f = \begin{bmatrix} X & Y \\ Y^T & Z \end{bmatrix}, \quad (YZ^{-1}Y^T) = \text{SVD}(X - \bar{X}^{-1}),$$

one can find the optimal controller $\Sigma^*_{i,N}$ for $i = N-1, \ldots, 0$ by the following linear programming; minimize the trace of $P_{i,N}$, for the given $P_{i+1,N}$, subject to (13). Then the output-feedback receding horizon control implementing only $\Sigma^*_{0,N}$ at time k asymptotically stabilizes the system (1) and (2). ∎

Proof: To prove this theorem, one need to show the monotonic decreasing property of the upper bound of the cost. Therefore, it shall be claimed that

$$J^*_{k+1,k+N+1} \leq J^*_{k,k+N}. \quad (22)$$

If there exist P_f satisfying (19)-(21), $J^*_{k,k+N} \geq J^*_{k,k+N+1}$ by Theorem 1. Therefore

$$\begin{aligned} J^*_{k,k+N} &= \eta_k^T [E_1^T Q E_1 + (E_2 \Sigma^*_{k,N} C_0)^T R \\ &\quad (E_2 \Sigma^*_{k,N} C_0)]\eta_k + J^*_{k+1,k+N} \quad (23) \\ &\geq \eta_k^T [E_1^T Q E_1 + (E_2 \Sigma^*_{k,N} C_0)^T R \\ &\quad (E_2 \Sigma^*_{k,N} C_0)]\eta_k + J^*_{k+1,k+N+1} \quad (24) \\ &\geq J^*_{k+1,k+N+1} \quad (25) \end{aligned}$$

The proof ends. ∎

Remark 5: The dimension of controller l is equal to the rank of $(X - \bar{X}^{-1})$ and the rank minimization problem is NP-hard. ∎

3. SIMULATION RESULTS

Consider a periodic system (which is originally a radar system, but modified as a periodic one).

$$\begin{aligned} x_{k+1} &= \begin{bmatrix} 1 + 2\sin(2\pi * 0.1k) & 0.1 \\ 0 & 0.495 \end{bmatrix} x_k \\ &\quad + \begin{bmatrix} 0 \\ 0.0787 \end{bmatrix} u_k, \\ y_k &= \begin{bmatrix} 1 & 0 \end{bmatrix} x_k. \end{aligned}$$

The weighting matrices $Q = \text{diag}\left[100\ 100\right]$, $R = 10$ and the control horizon $N = 3$. It is assumed that $u_0 = 0$, the order of dynamic controller $k = 2$ and $z_0 = \left[0\ 0\right]^T$.

The performance of the proposed controller is compared with that of observer-based output-feedback RHC. Simulation results show that the proposed controller yield better or at least the same regulation performance. However, if there exist disturbances and uncertainties in the periodic system, the proposed controller may yield better performance than the observer-based output-feedback RHC.

4. CONCLUDING REMARKS

In this paper, a dynamic output-feedback RHC for LTV systems including periodic systems is suggested. The proposed controller yields good performance and overcomes some difficulties which the existing observer-based output-feedback RHC copes with. Using matrix inequality conditions on the terminal weighting matrix, the closed-loop system stability is guaranteed. Numerical example shows how the proposed controller improves performance compared with the observer-based RHC.

5. REFERENCES

Boyd, S., L. El Ghaoui, E. Feron and V. Balakrishanan (1994). *Linear Martix Inequalities in System and Control Theory.* SIAM.

Cutler, C. R. and R. B. Hawkins (1988). Applications of a large predictive multivariable controller to a hydrocracker second stage reactor. *Proceedings of 1988 American Control conference* pp. 284–291.

Garcia, C. E., D. M. Prett and M. Morari (1991). Model predictive control: Theory and practice-a survey. *Automatica* **25**, 335–348.

Kwon, W. H. and A. E. Peason (1977). A modified quadratic cost problem and feedback stabilization of a linear system. *IEEE Trans. Automatic Control* **22**, 838–842.

Lee, J. W. (2000). Exponential stability of constrained receding horizon control with terminal ellipsoid constraints. *IEEE Trans. Automatic Control* **45**, 83–88.

Lee, J. W., W. H. Kwon and J. H. Choi (1998). On stability of constrained receding horizon control with finite terminal weighting matrix. *Automatica* **34**, 1607–1612.

Lee, Y. I. and B. Kouvaritakis (2000). Receding horizon output feedback control for linear systems with input saturation. *Proceedings of 2000 conference on Decision and Control* pp. Australia(pp.656–661).

Rawlings, J. B. and K. R. Muske (1993). The stability of constrained receding horizon control. *IEEE Trans. Automatic Control* **38**, 1512–1516.

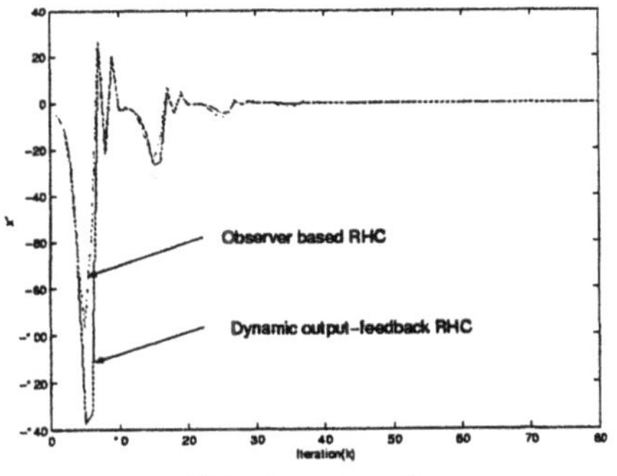

Fig.1 state x1

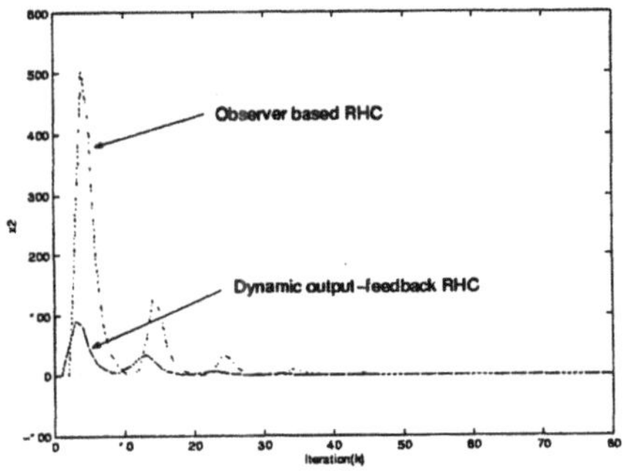

Fig.2 state x2

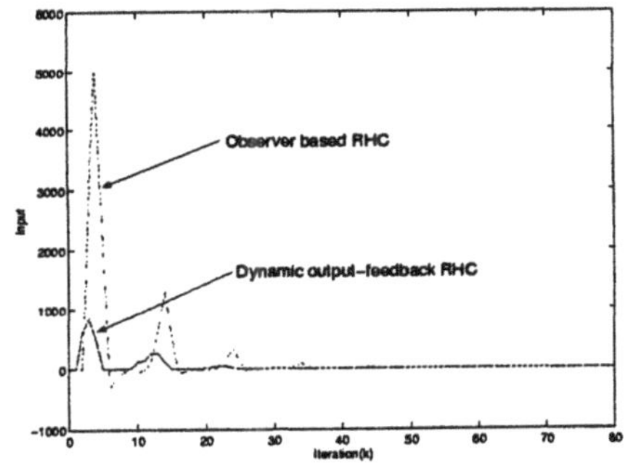

Fig.3 Control input

Copyright © IFAC Periodic Control Systems,
Cernobbio-Como, Italy, 2001

COMPUTATIONAL METHODS FOR PERIODIC SYSTEMS - AN OVERVIEW

A. Varga *, **P. Van Dooren** **

* *German Aerospace Center, DLR - Oberpfaffenhofen*
Institute of Robotics and Mechatronics
D-82234 Wessling, Germany
Andras.Varga@dlr.de

** *Centre for Engineering Systems and Applied Mechanics*
Université catholique de Louvain
B-1348 Louvain-la-Neuve, Belgium
VDooren@csam.ucl.ac.be

Abstract: We present an up-to-date survey of numerical methods for the analysis and design of linear discrete-time periodic systems. The basic tool is the periodic Schur form and its variants, for which a certain form of numerical stability can be ensured.
Copyright ©2001 IFAC

Keywords: Periodic systems, time-varying systems, discrete-time systems, numerical methods

1. INTRODUCTION

The theory of linear discrete-time periodic systems has received a lot of attention in the last 25 years (Bittanti and Colaneri, 1996). Almost all results for standard discrete-time systems have been extended to periodic systems of the form

$$\begin{aligned} x_{k+1} &= A_k x_k + B_k u_k \\ y_k &= C_k x_k + D_k u_k \end{aligned} \tag{1}$$

where the matrices $A_k \in \mathbb{R}^{n_{k+1}\times n_k}$, $B_k \in \mathbb{R}^{n_{k+1}\times m}$, $C_k \in \mathbb{R}^{p\times n_k}$, $D_k \in \mathbb{R}^{p\times m}$ are periodic with period $K \geq 1$. Most theoretical results are based on two lifting techniques which reduce the problem for the periodic system (1) to an equivalent problem for a time-invariant system of increased dimensions. The first lifting approach, proposed by Meyer and Burrus (1975), involves forming products of up to K matrices, while the second lifting approach, proposed by Flamm (1991), leads to a large order standard system representation with sparse and highly structured matrices. Although these lifting techniques are useful for their theoretical insight, their sparsity and structure may not be suited for numerical computations. This is why, in parallel to the theoretical developments, numerical methods have been developed that try to exploit this structure. For most analysis and design problems of standard state space systems, there are good numerical algorithms available that meet the standard requirements of speed and accuracy. The purpose of this paper is to present a short overview of recently developed numerical methods for the analysis and design of *periodic systems*. We also mention some open areas where there is still a need for new algorithmic developments.

Notation. To simplify the presentation we introduce first some notation. For a K-periodic matrix X_k we use alternatively the *script* notation

$$\mathcal{X} := \operatorname{diag}(X_0, X_1, \ldots, X_{K-1}),$$

which associates the block-diagonal matrix $\mathcal{X}$ to the cyclic matrix sequence X_k, $k = 0, \ldots, K-1$. This notation is consistent with the standard matrix operations as for instance addition, multiplication, inversion as well as with several standard matrix decompositions (Cholesky, SVD). We denote with $\sigma\mathcal{X}$ the K-cyclic shift

$$\sigma\mathcal{X} = \text{diag}\,(X_1, \ldots, X_{K-1}, X_0)$$

of the cyclic sequence X_k, $k = 0, \ldots, K-1$.

By using the script notation, the periodic system (1) will be alternatively denoted by the quadruple $(\mathcal{A}, \mathcal{B}, \mathcal{C}, \mathcal{D})$. The transition matrix of the system (1) is defined by the $n_j \times n_i$ matrix $\Phi_A(j,i) = A_{j-1}A_{j-2}\cdots A_i$, where $\Phi_A(i,i) := I_{n_i}$. The state transition matrix over one period $\Phi_A(j+K,j) \in \mathrm{R}^{n_j \times n_j}$ is called the *monodromy matrix* of system (1) at time j and its eigenvalues are called *characteristic multipliers* at time j.

2. DESCRIPTOR PERIODIC SYSTEMS

Descriptor periodic systems of the form

$$\begin{aligned} E_k x_{k+1} &= A_k x_k + B_k u_k \\ y_k &= C_k x_k + D_k u_k \end{aligned} \qquad (2)$$

where the matrices $A_k, E_k \in \mathrm{R}^{n\times n}$, $B_k \in \mathrm{R}^{n\times m}$, $C_k \in \mathrm{R}^{p\times n}$, $D_k \in \mathrm{R}^{p\times m}$ are periodic with period $K \geq 1$, have been considered in (Sreedhar and Van Dooren, 1997; Sreedhar *et al.*, 1998). These systems may also arise in the context of ordinary periodic systems, when for instance forming the inverse or conjugate periodic system. Provided the matrices E_k are invertible we can divide the first equation from the left by E_k which then reduces to a standard periodic model with system quadruple $(\mathcal{E}^{-1}\mathcal{A}, \mathcal{E}^{-1}\mathcal{B}, \mathcal{C}, \mathcal{D})$. The monodromy matrix in this case becomes the $n \times n$ matrix $\Phi_{E^{-1}A}(j,i) := E_{j-1}^{-1}A_{j-1}E_{j-2}^{-1}A_{j-2}\cdots E_i^{-1}A_i$. It should be pointed out that analysis and design algorithms for such systems should nevertheless work even when the matrices E_k are singular, provided these problems are well defined.

3. SATISFACTORY ALGORITHMS

We first briefly discuss three key requirements for a satisfactory numerical algorithm for periodic system: generality, numerical stability, and efficiency. A *general* algorithm is one which has no limitations for its applicability of any technical nature. For the periodic system (1) it should be able to handle the most general class of periodic systems. For example, a pole assignment algorithm for a periodic system able to assign *only* distinct poles should not be considered satisfactory. Since the minimal realization of a periodic system has in general a time-varying state dimension, it is highly desirable to develop algorithms for the analysis and design of periodic systems which are able to handle systems with time-varying dimensions.

Numerical stability (more precisely, *backward stability*) of an algorithm means that the results computed by that algorithm are exact for slightly perturbed original data. As a consequence, a numerically stable algorithm applied to a well conditioned problem will produce guaranteed accurate results. This is why numerical stability is a key feature for a satisfactory algorithm. A basic ingredient to achieve numerical stability is the use of orthogonal transformations wherever possible. The use of these transformations often leads to bounds for perturbations of the initial data which are equivalent to the cumulative effect of round-off errors occurring during the computations. This is a way to prove the numerical stability of such an algorithm. This immediately suggests that one should avoid forming matrix products as those appearing in the lifted formulation of Meyer and Burrus (1975), since these amount to non-orthogonal transformations of the data matrices. The main idea when developing numerically stable algorithms for periodic systems is to exploit the problem structure by applying only orthogonal transformations on the original problem data, and thereby trying to reduce the original problem to an equivalent one which is easier to solve.

Because of the intrinsic complexity of several computational problems in systems theory, it is not always possible to develop numerically stable algorithms for them. Therefore one often imposes this requirement only on the substeps of the algorithm. Although this is not enough to guarantee numerical stability of the global algorithm, one can still expect that it will perform accurately on well-conditioned problems.

The *efficiency* of an algorithm involves two main aspects: avoiding extensive storage use and keeping the computational complexity as low as possible. For periodic systems the first requirement implies that the storage should be proportional to the amount of data defining the system, i.e. it should be $O(K\overline{n}^2) + O(K\overline{n}m) + O(K\overline{n}p)$, where $\overline{n} = \max\{n_i\}$. Explicitly forming the lifted representation of Flamm (1991) should thus be avoided. Concerning the second requirement applied to a periodic system of period K, one would hope for a complexity of at most $O(K\overline{n}^3)$ since the complexity for standard state-space algorithms is typically $O(n^3)$. This implies again that one should not use large dimensional lifted representations.

4. BASIC NUMERICAL INGREDIENTS

The use of condensed forms of the system matrices, obtained under orthogonal transformations, is a basic ingredient for solving many computational problems (Van Dooren and Verhaegen, 1985). The system matrices are transformed to a particular coordinate system in which they are condensed, such that the solution of the computational problem is straightforward. For periodic systems with constant dimensions, the *periodic real Schur form* (PRSF) plays an important role in solving many computational problems. According to Bojanczyk *et al.* (1992), given the matrices $A_k \in \mathbb{R}^{n\times n}$, $k = 0, 1, \ldots, K{-}1$, there exist orthogonal matrices Z_k, $k = 0, 1, \ldots, K{-}1$, $Z_K := Z_0$, such that

$$\widetilde{A}_k := Z_{k+1}^T A_k Z_k \qquad (3)$$

where $\widetilde{A}_{K-1}$ is in *real Schur form* (RSF) and the matrices $\widetilde{A}_k$ for $k = 0, \ldots, K-2$ are upper triangular. Numerically stable algorithms to compute the PRSF have been proposed in (Bojanczyk *et al.*, 1992; Hench and Laub, 1994). By using these algorithms, we can determine the orthogonal matrices Z_k, $k = 0, \ldots, K-1$ to reduce the cyclic product $A_{K-1} \cdots A_1 A_0$ to the RSF without forming explicitly this product. An intermediate condensed form with potential applications in computational algorithms is the *periodic Hessenberg form* (PHF), where $\widetilde{A}_{K-1}$ is in a Hessenberg form, while $\widetilde{A}_k$ for $k = 0, \ldots, K{-}2$ are upper triangular.

For systems with time-varying dimensions, the *extended periodic real Schur form* (EPRSF) represents a generalization of the PRSF which allows to address many problems with varying dimensions. According to Varga (1999), given the matrices $A_k \in \mathbb{R}^{n_{k+1}\times n_k}$, $k = 0, 1, \ldots, K{-}1$, with $n_K = n_0$ there exist orthogonal matrices $Z_k \in \mathbb{R}^{n_k \times n_k}$, $k = 0, 1, \ldots, K-1$, $Z_K := Z_0$, such that the matrices

$$\widetilde{A}_k := Z_{k+1}^T A_k Z_k = \begin{bmatrix} \widetilde{A}_{k,11} & \widetilde{A}_{k,12} \\ 0 & \widetilde{A}_{k,22} \end{bmatrix}, \qquad (4)$$

are block upper triangular, where $\widetilde{A}_{k,11} \in \mathbb{R}^{\underline{n}\times\underline{n}}$, $\widetilde{A}_{k,22} \in \mathbb{R}^{(n_{k+1}-\underline{n})\times(n_k-\underline{n})}$ for $k = 0, 1, \ldots, K-1$ and $\underline{n} = \min_k\{n_k\}$. Moreover, $\widetilde{A}_{K-1,11}$ is in RSF, $\widetilde{A}_{k,11}$ for $k = 0, \ldots, K{-}2$ are upper triangular and $\widetilde{A}_{k,22}$ for $k = 0, \ldots, K-1$ are upper trapezoidal.

For descriptor systems with fixed dimensions, the *generalized periodic real Schur form* (GPRSF) extends the PRSF to so-called regular periodic systems (see e.g. Bojanczyk *et al.* (1992)). Given the matrices $A_k, E_k \in \mathbb{R}^{n\times n}$, $k = 0, 1, \ldots, K-1$, there exist orthogonal matrices $Z_k, Q_k \in \mathbb{R}^{n\times n}$, $k = 0, 1, \ldots, K-1$, $Z_K := Z_0$, such that the matrices

$$\widetilde{A}_k := Q_k^T A_k Z_k, \quad \widetilde{E}_k := Q_k^T E_k Z_{k+1}, \qquad (5)$$

are all upper triangular, except for $\widetilde{A}_{K-1}$, which is in RSF.

5. PERIODIC MATRIX EQUATIONS

The reduction of a periodic matrix A_k to PRSF and EPRSF is the principal ingredient in solving important linear equations for periodic systems as the periodic Lyapunov and Sylvester equations. Periodic Lyapunov equations appear in solving periodic state-feedback stabilization problems or in computing gradients for optimal periodic output feedback problems (Varga and Pieters, 1998). Consider for example the *discrete-time periodic Lyapunov equations* (DPLE)

$$\mathcal{X} = \mathcal{A}^T \sigma \mathcal{X} \mathcal{A} + \mathcal{V} \qquad (6)$$

$$\sigma \mathcal{X} = \mathcal{A} \mathcal{X} \mathcal{A}^T + \sigma \mathcal{W} \qquad (7)$$

where V_k, W_k are symmetric K-periodic matrices of appropriate dimensions. To ensure the existence of a unique solution of these equations we assume that the monodromy matrix $\Phi_A(K, 0)$ has no reciprocal eigenvalues.

The orthogonal Lyapunov transformation in (3) or (4) can be expressed as $\widetilde{\mathcal{A}} = \sigma \mathcal{Z}^T \mathcal{A} \mathcal{Z}$. By multiplying equation (6) with $\mathcal{Z}^T$ from left and with $\mathcal{Z}$ from right, and multiplying equation (7) with $\sigma\mathcal{Z}^T$ from left and with $\sigma\mathcal{Z}$ from right, one obtains

$$\widetilde{\mathcal{X}} = \widetilde{\mathcal{A}}^T \sigma \widetilde{\mathcal{X}} \widetilde{\mathcal{A}} + \widetilde{\mathcal{V}}, \qquad (8)$$

$$\sigma \widetilde{\mathcal{X}} = \widetilde{\mathcal{A}} \widetilde{\mathcal{X}} \widetilde{\mathcal{A}}^T + \sigma \widetilde{\mathcal{W}}, \qquad (9)$$

where $\widetilde{\mathcal{X}} = \mathcal{Z}^T \mathcal{X} \mathcal{Z}$, $\widetilde{\mathcal{V}} = \mathcal{Z}^T \mathcal{V} \mathcal{Z}$ and $\widetilde{\mathcal{W}} = \mathcal{Z}^T \mathcal{W} \mathcal{Z}$. By this transformation the resulted transformed equations (8) and (9) have exactly the same form as the original ones in (6) and (7), but this time the periodic matrix $\widetilde{A}_k$ is in PRSF or EPRSF. After solving these equations for $\widetilde{\mathcal{X}}$, the solution results as $\mathcal{X} = \mathcal{Z}\widetilde{\mathcal{X}}\mathcal{Z}^T$. The reduced DPLEs (8) and (9) can be solved by using special substitution algorithms (Varga, 1997). Important computational subproblems are in this context the efficient and numerically stable solution of low order DPLEs and periodic Sylvester equations. Computational approaches for these subproblems are also described in detail in (Varga, 1997).

The computation of periodic reachability and observability grammians for periodic systems involves the solution of periodic Lyapunov equations having nonnegative solutions. For example, assuming that $\Phi_A(K, 0)$ has only eigenvalues in the interior of the unit circle, the DPLE

$$\mathcal{X} = \mathcal{A}^T \sigma \mathcal{X} \mathcal{A} + \mathcal{R}^T \mathcal{R} \qquad (10)$$

can be solved directly for the Cholesky factor $\mathcal{U}$ of the nonnegative definite solution $\mathcal{X} = \mathcal{U}^T\mathcal{U}$. When solving minimal realization or model reduction problems, the periodic matrices A_k and R_k in (10) have often time-varying dimensions. Efficient algorithms to solve nonnegative DPLE are based on transformation techniques involving the PRSF for constant dimensions (Varga, 1997) or the EPRSF for time-varying dimensions (Varga, 1999). Similar comments hold for the dual equation

$$\sigma\mathcal{X} = \mathcal{A}\mathcal{X}\mathcal{A}^T + \sigma\mathcal{R}\sigma\mathcal{R}^T. \quad (11)$$

A class of periodic robust state-feedback pole assignment problems can be reduced to the solution of a *periodic Sylvester equation* (PSE) of the form (Varga, 2000*b*)

$$\mathcal{A}\mathcal{X} + \sigma\mathcal{X}\mathcal{B} = \mathcal{C} \quad (12)$$

where A_k, B_k and C_k are K-periodic matrices with constant dimensions. By reducing A_k and B_k to the PRSF, the periodic solution X_k can be computed using a transformation method which generalizes the well-known Bartels-Stewart method (Byers and Rhee, 1995; Varga, 2000*b*).

Periodic Riccati equations appear when solving periodic LQ-design problems (Sreedhar and Van Dooren, 1994*b*). Using our notation it can be written as follows

$$\mathcal{P} = \mathcal{A}^T[\sigma\mathcal{P} - \sigma\mathcal{P}\mathcal{B}(\mathcal{R} + \mathcal{B}^T\sigma\mathcal{P}\mathcal{B})^{-1}\mathcal{B}^T\mathcal{P}]\mathcal{A} + \mathcal{S}.$$

The solution of this periodic Riccati equation can be obtained from the generalized periodic Schur decomposition of $(\hat{E}_k, \hat{A}_k)$, where

$$\hat{E}_k \doteq \begin{bmatrix} I_n & B_k R_k^{-1} B_k^T \\ 0 & A_k^T \end{bmatrix}, \; \hat{A}_k \doteq \begin{bmatrix} A_k & 0 \\ -S_k & I_n \end{bmatrix} \quad (13)$$

Upon partitioning the $2n \times 2n$ matrices Z_k of the GPRSF of (13) as

$$\hat{Z}_k \doteq \begin{bmatrix} X_k & V_k \\ Y_k & W_k \end{bmatrix}, \quad (14)$$

one obtains the matrices P_k that constitute $\mathcal{P}$ as $P_k = Y_k X_k^{-1}$, provided the GPRSF has ordered eigenvalues, i.e. the stable eigenvalues appear first in the Schur form. Algorithms for this have been proposed in (Bojanczyk *et al.*, 1992; Hench and Laub, 1994; Sreedhar and Van Dooren, 1994*b*). Similar results hold for the dual periodic Riccati equation occurring in the filtering problem :

$$\sigma\mathcal{P} = \mathcal{A}[\mathcal{P} - \mathcal{P}\mathcal{C}^T(\mathcal{R} + \mathcal{C}\mathcal{P}\mathcal{C}^T)^{-1}\mathcal{C}\mathcal{P}]\mathcal{A}^T + \sigma\mathcal{S}.$$

6. ALGORITHMS FOR THE ANALYSIS OF PERIODIC SYSTEMS

Structural properties of stable periodic systems such as reachability, observability, minimality can be analyzed by computing the reachability and observability grammians. With a straightforward scaling, the same techniques can be used to study unstable systems as well. The computation of minimal (i.e., completely reachable and completely observable) realizations can be done using efficient and numerically reliable algorithms proposed by Varga (1999). Order reduction of periodic systems using balancing techniques can be performed by using accuracy enhancing *square-root* and *balancing-free* algorithms developed in (Varga, 2000*a*). The main computation in these algorithms is the solution of two nonnegative definite periodic Lyapunov equations to determine directly the Cholesky factors of the controllability and observability grammians. The *square-root* term signifies that all subsequent computations for determining the projection matrices are based exclusively on square-root information (i.e., Cholesky factors of the grammians). The accuracy can be further enhanced by avoiding the computation of the possibly ill-conditioned balancing transformation. Instead, well-conditioned projection matrices are constructed which leads to so-called *balancing-free* order reduction.

The computation of poles is important in many applications. To compute the poles of a periodic system, the eigenvalues of the *monodromy* matrix product $\Phi_A(K, 0)$ must be determined. The computation can be done without forming this matrix product explicitly, but by reducing the K-periodic matrices A_k to the PRSF in the case of constant dimensions, or to the EPRSF in the case of time-varying dimensions. If only poles are requested one can also use the more economical approach suggested by Van Dooren (1999).

The evaluation of system norms can be done using reliable algorithms. The Hankel-norm of a periodic system can be computed as the maximal singular value of the products of periodic grammians. This computation is part of the algorithm to determine minimal realizations of periodic systems using balancing techniques (Varga, 1999). The computation of the H_2-norm involves the solution of a periodic Lyapunov equation to determine the controllability or observability grammians (Bittanti and Colaneri, 1996). This can be done using algorithms proposed in (Varga, 1997; Varga, 1999). For the computation of the H_∞-norm an algorithm with quadratic convergence is given in (Sreedhar *et al.*, 1997). The main step there is the computation of generalized eigenvalues of a periodic symplectic pencil using the periodic QZ-decomposition (Bojanczyk *et al.*, 1992).

7. ALGORITHMS FOR THE DESIGN OF PERIODIC SYSTEMS

Basic design algorithms for periodic systems with constant state dimension have been proposed by several authors. A Schur method for pole assignment has been proposed by Sreedhar and Van Dooren (1993) and a stabilization algorithm has been proposed by Sreedhar and Van Dooren (1994*a*). A robust pole assignment method relying on Sylvester equations has been proposed in (Varga, 2000*b*).

A computational approach for the periodic LQG methods involves the solution of periodic Riccati equations. Efficient methods have been proposed using the ordered PRSF or *generalized periodic real Schur form* (GPRSF) (Bojanczyk *et al.*, 1992; Hench and Laub, 1994). A computational approach for the solution of the optimal periodic output feedback problem has been developed by Varga and Pieters (1998).

8. ALGORITHMS FOR DESCRIPTOR PERIODIC SYSTEMS

All of the equations for standard periodic systems $(\mathcal{A}, \mathcal{B}, \mathcal{C}, \mathcal{D})$ can be extended to descriptor periodic systems $(\mathcal{E}, \mathcal{A}, \mathcal{B}, \mathcal{C}, \mathcal{D})$ provided $\mathcal{E}$ is invertible. When $\mathcal{E}$ is singular, one has to show that the underlying control problem makes sense and is solved by the corresponding equations. For regular periodic systems this is typically the case (see e.g., Sreedhar and Van Dooren (1999)).

The generalized Lyapunov and Riccati equations are

$$\mathcal{E}\sigma\mathcal{X}\mathcal{E}^T = \mathcal{A}\mathcal{X}\mathcal{A}^T + \sigma\mathcal{W}, \qquad (15)$$

and

$$\mathcal{E}\sigma\mathcal{P}\mathcal{E}^T = \mathcal{A}[\mathcal{P} - \mathcal{P}\mathcal{C}^T(\mathcal{R}+\mathcal{C}\mathcal{P}\mathcal{C}^T)^{-1}\mathcal{C}\mathcal{P}]\mathcal{A}^T + \sigma\mathcal{S},$$

respectively. Using the GPRSF (5), which can be written as follows :

$$\widetilde{\mathcal{E}} = \mathcal{Q}^T\mathcal{E}\sigma\mathcal{Z}, \quad \widetilde{\mathcal{A}} = \mathcal{Q}^T\mathcal{A}\mathcal{Z}, \qquad (16)$$

these can easily be transformed to the same equations where now $\mathcal{E}$ and $\mathcal{A}$ are in GPRSF form. From this, the required solution then easily follows in much the same way as for the standard periodic equations. It is shown by Sreedhar and Van Dooren (1994*b*) how these equations can been used to solve stabilization and optimal control problems of periodic descriptor systems.

The generalized Sylvester equation has the form

$$\mathcal{E}\sigma\mathcal{X}\mathcal{B} + \mathcal{A}\mathcal{X}\mathcal{F} = \mathcal{C}, \qquad (17)$$

where the matrix pairs $\mathcal{E}, \mathcal{A}$ and $\mathcal{B}, \mathcal{F}$ can again be put in GPRSF. Using the latter, a lifting technique for periodic descriptor systems has been introduced by Sreedhar *et al.* (1998) which, in conjunction with the backward/forward decomposition technique of Sreedhar and Van Dooren (1997), allows to determine minimal order representations of descriptor periodic systems.

Further, poles of descriptor periodic systems can be computed using the periodic QZ decomposition proposed by Bojanczyk *et al.* (1992) or the more economical variant described by Van Dooren (1999).

9. SOME EXTENSIONS AND OPEN PROBLEMS

There are several computational problems for periodic systems for which it is in principle straightforward to develop reliable computational methods by extending algorithms for standard systems. Frequency-domain methods for the analysis of periodic systems rely on the $Kp \times Km$ *transfer-function matrix* (TFM) of the associated lifted systems (Meyer and Burrus, 1975; Flamm, 1991). The computation of frequency responses can be done by computing first the corresponding TFM and then evaluating the frequency response using the resulting rational matrix. A method to compute the TFM can be devised along the lines of the poles/zeros approach proposed by Varga and Sima (1981) for standard systems and in by Varga (1989) for descriptor systems. Alternatively, the frequency response can be computed by exploiting the sparse structure of the lifted representation of the periodic system. Here the periodic Hessenberg form can play potentially the same role as the Hessenberg form in the case of standard systems (Laub, 1981).

Recursive Schur techniques to compute coprime factorizations (Varga, 1998) can be extended to the periodic case along the line of the periodic Schur form method for pole assignment proposed by Sreedhar and Van Dooren (1993). This has been done for the coprime factorization with inner denominator by Varga (2001) in the context of balancing and model reduction of unstable periodic systems. A computational approach to determine normalized coprime factorizations can probably be developed based on results for standard systems of Bongers and Heuberger (1990).

There are several open problems for which still efficient algorithms are to be developed, as for example, the computation of controllability and observability canonical forms, zeros of the associated TFM, Kronecker-structure of associated system pencil, inner-outer and spectral factorization, and so on. Recent developments also look at the solution of periodic Linear Matrix Inequalities for solving various design problems (Bittanti and Cuzzola, 2000). The use of periodic matrix decompositions could be useful here as well.

10. REFERENCES

Bittanti, S. and F. Cuzzola (2000). An LMI approach to periodic unbiased filtering. Technical report. Dipt. Elet. e Inf., Politecnico di Milano.

Bittanti, S. and P. Colaneri (1996). Analysis of discrete-time linear periodic systems. In: *Digital Control and Signal Processing Systems and Techniques* (C. T. Leondes, Ed.). Vol. 78 of *Control and Dynamics Systems*. pp. 313–339. Academic Press.

Bojanczyk, A. W., G. Golub and P. Van Dooren (1992). The periodic Schur decomposition. Algorithms and applications. In: *Proceedings SPIE Conference* (F. T. Luk, Ed.). Vol. 1770. pp. 31–42.

Bongers, P. M. M. and P. S. C. Heuberger (1990). Discrete normalized coprime factorization. In: *Proc. 9th INRIA Conf. Analysis and Optimization of Systems* (A. Bensoussan and J. L. Lions, Eds.). Vol. 144 of *Lect. Notes Control and Inf. Scie.*. Springer-Verlag, Berlin. pp. 307–313.

Byers, R. and N. Rhee (1995). Cyclic Schur and Hessenberg-Schur numerical methods for solving periodic Lyapunov and Sylvester equations. Technical report. Dept. of Mathematics, Univ. of Missouri at Kansas City.

Flamm, D. S. (1991). A new shift-invariant representation of periodic linear systems. *Systems & Control Lett.* **17**, 9–14.

Hench, J. J. and A. J. Laub (1994). Numerical solution of the discrete-time periodic Riccati equation. *IEEE Trans. Autom. Control* **39**, 1197–1210.

Laub, A. J. (1981). Efficient multivariable frequency response computations. *IEEE Trans. Autom. Control* **26**, 407–408.

Meyer, R. A. and C. S. Burrus (1975). A unified analysis of multirate and periodically time-varying digital filters. *IEEE Trans. Circuits and Systems* **22**, 162–168.

Sreedhar, J. and P. Van Dooren (1993). Pole placement via the periodic Schur decomposition. In: *Proc. 1993 American Control Conference, San Francisco, CA*. pp. 1563–1567.

Sreedhar, J. and P. Van Dooren (1994*a*). On finding stabilizing state feedback gains for a discrete-time periodic system. In: *Proc. 1994 American Control Conference, Baltimore, MD*. pp. 1167–1168.

Sreedhar, J. and P. Van Dooren (1994*b*). A Schur approach for solving some periodic matrix equations. In: *Systems and Networks: Mathematical Theory and Applications* (U. Helmke, R. Mennicken and J. Saurer, Eds.). Vol. 77 of *Mathematical Research*. pp. 339–362.

Sreedhar, J. and P. Van Dooren (1996). Forward/backward decomposition of periodic descriptor systems. In: *Proc. 1997 ECC, Brussels, Belgium*, paper FR-A-L7.

Sreedhar, J., P. Van Dooren and B. Bamieh (1995). Computing H_∞-norm of discrete-time descriptor systems–a quadratically convergent algorithm. In: *Proc. 1997 ECC, Brussels, Belgium*, paper FR-A-L8.

Sreedhar, J., P. Van Dooren and P. Misra (1998). Minimal order time invariant representation of periodic descriptor systems (submitted for publication)

Sreedhar, J. and P. Van Dooren (1999). Periodic descriptor systems : solvability and conditionability, *IEEE Transactions on Automatic Control* **AC-44**, 310-313.

Van Dooren, P. (1999). Two point boundary value and periodic eigenvalue problems. In: *Proc. CACSD'99 Symposium, Kohala Coast, Hawaii.*

Van Dooren, P. and M. Verhaegen (1985). *On the use of unitary state-space transformations*. Vol. 47 of *Special Issue of Contemporary Mathematics in Linear Algebra and Its Role in Systems Theory*. Amer. Math. Soc.. Providence, R.I.

Varga, A. (1989). Computation of transfer function matrices of generalized state-space models. *Int. J. Control* **50**, 2543–2561.

Varga, A. (1997). Periodic Lyapunov equations: some applications and new algorithms. *Int. J. Control* **67**, 69–87.

Varga, A. (1998). Computation of coprime factorizations of rational matrices. *Lin. Alg. & Appl.* **271**, 83–115.

Varga, A. (1999). Balancing related methods for minimal realization of periodic systems. *Systems & Control Lett.* **36**, 339–349.

Varga, A. (2000*a*). Balanced truncation model reduction of periodic systems. In: *Proc. CDC'2000, Sydney, Australia.*

Varga, A. (2000*b*). Robust and minimum norm pole assignment with periodic state feedback. *IEEE Trans. Autom. Control* **45**, 1017–1022.

Varga, A. (2001). On balancing and order reduction of unstable periodic systems. In: *Proc. of IFAC Workshop on Periodic Control Systems, Como, Italy.*

Varga, A. and S. Pieters (1998). Gradient-based approach to solve optimal periodic output feedback control problems. *Automatica* **34**, 477–481.

Varga, A. and V. Sima (1981). A numerically stable algorithm for transfer-function matrix evaluation. *Int. J. Control* **33**, 1123–1133.

Copyright © IFAC Periodic Control Systems,
Cernobbio-Como, Italy, 2001

ON BALANCING AND ORDER REDUCTION OF UNSTABLE PERIODIC SYSTEMS

A. Varga *

* *German Aerospace Center, DLR - Oberpfaffenhofen*
Institute of Robotics and Mechatronics
D-82234 Wessling, Germany
Andras.Varga@dlr.de

Abstract: We consider the direct application of balancing techniques to unstable periodic systems by extending the balancing concepts to arbitrary periodic systems. We extend first the balancing concepts to unstable discrete-time systems by defining the reachability and observability grammians from appropriate right and left coprime factorizations with inner denominators. Further, we extend this new balancing method to unstable linear time-varying discrete-time systems with periodically varying coefficient matrices. The new balancing approach serves as basis to develop balancing related order reduction methods for unstable periodic systems using accuracy enhancing square-root and balancing-free algorithms. *Copyright ©2001 IFAC*

Keywords: Periodic systems, time-varying systems, discrete-time systems, balanced truncation, model reduction, numerical methods

1. INTRODUCTION

For a stable periodic system, numerical procedures for balancing (Varga, 1999) and model reduction (Longhi and Orlando, 1999; Varga, 2000) have been developed recently. These methods extend the well-known balancing and model reduction techniques for standard systems to the periodic case. To reduce unstable periodic systems, two embedding approaches extending similar techniques for standard systems, can be used in conjunction with balancing techniques.

In the first approach, the unstable periodic system is additively decomposed as the sum of its stable and an unstable parts. Then the order reduction is performed only on the stable part using appropriate balancing related methods. The reduced model is formed as the sum of the reduced stable part and the unstable part. This approach has the disadvantage that the unstable part is copied unmodified back into the reduced model, although sometimes a lower order approximation would be possible if this part is also reduced. In particular, if the unstable part is non-minimal, the reduced model results non-minimal too.

The second approach relies on coprime factorization techniques and therefore implicitly involves the reduction of both stable and unstable parts. The unstable periodic system can be expressed in a coprime factorized representation, where the factors are stable periodic systems. Using balancing related techniques, the compound system formed by appending the two factors is reduced and the reduced factors are recovered. Finally, the reduced periodic system is constructed from the coprime factorization of the reduced factors.

In this paper we consider a third approach which addresses the direct application of balancing techniques to unstable periodic systems by extending the balancing concepts to arbitrary periodic systems. Such an approach has been proposed recently for standard continuous-time systems by Zhou *et al.* (1999). We extend first this approach to unstable standard discrete-time systems. Further, we extend this new balancing method to unstable time-varying linear discrete-time systems with periodically varying coefficient matrices.

The new methods rely essentially on computing the controllability and observability grammians from appropriate right and left coprime factorizations with inner denominators. Transformation techniques allow to reduce the computational burden for computing these factorizations by solving reduced order Lyapunov equations instead full order Riccati equations. By using recursive factorization techniques, the factorizations can be determined directly with the state matrices in quasi-upper triangular forms which allows an efficient computation of the grammians.

The new balancing approach can serve as basis to develop balancing related order reduction methods for unstable periodic systems using accuracy enhancing square-root and balancing-free algorithms. The grammians can be computed directly in Cholesky factorized forms which can be employed to determine appropriate truncation matrices to perform model reduction of unstable periodic systems, analogously to methods developed for stable periodic systems in (Varga, 1999; Varga, 2000).

2. THE STANDARD CASE

Let $G(z)$ be a given discrete-time *transfer-function matrix* (TFM) without poles on the unit circle and let (A, B, C, D) be a stabilizable and detectable state-space representation satisfying

$$G(z) = C(zI - A)^{-1}B + D$$

In analogy to the continuous-time case (Zhou *et al.*, 1999) we define the controllability grammian P and the observability grammian Q as

$$P = \frac{1}{2\pi}\int_0^{2\pi}(e^{j\theta}I - A)^{-1}BB^T(e^{-j\theta}I - A^T)^{-1}d\theta \quad (1)$$

$$Q = \frac{1}{2\pi}\int_0^{2\pi}(e^{-j\theta}I - A^T)^{-1}C^TC(e^{j\theta}I - A)^{-1}d\theta \quad (2)$$

In the case of a stable system, P and Q are the usual positive semidefinite grammians satisfying the discrete-time matrix Lyapunov equations

$$\begin{aligned} APA^T + BB^T &= P \\ A^TQA + C^TC &= Q \end{aligned} \quad (3)$$

In the case of an unstable system consider the right coprime factorization

$$(zI - A)^{-1}B = N(z)M^{-1}(z)$$

with $M(z)$ an inner TFM. The factors can be computed according to (Zhou *et al.*, 1996) in the form

$$\begin{bmatrix} N(z) \\ M(z) \end{bmatrix} = \left[\begin{array}{c|c} A+BF & BW \\ \hline I & 0 \\ F & W \end{array}\right] := \left[\begin{array}{c|c} A_r & B_r \\ \hline I & 0 \\ F & W \end{array}\right]$$

where

$$\begin{aligned} F &= -WW^TB^TXA \\ W^T(I + B^TXB)W &= I \end{aligned} \quad (4)$$

and X is the stabilizing symmetric positive semidefinite solution of the Riccati equation

$$A^TX(I + BB^TX)^{-1}A - X = 0 \quad (5)$$

It follows from (1) that

$$\begin{aligned} P &= \frac{1}{2\pi}\int_0^{2\pi} N(e^{j\theta})N^T(e^{-j\theta})d\theta \\ &= \frac{1}{2\pi}\int_0^{2\pi}(e^{j\theta}I - A_r)^{-1}B_rB_r^T(e^{-j\theta}I - A_r^T)^{-1}d\theta \end{aligned}$$

and thus, for an unstable system, P fulfills the Lyapunov equation

$$(A + BF)P(A + BF)^T + BWW^TB^T = P \quad (6)$$

Similarly we compute the observability grammian Q by first determining the left coprime factorization

$$C(zI - A)^{-1} = \widetilde{M}^{-1}(z)\widetilde{N}(z)$$

where $\widetilde{M}(z)$ is an inner TFM. The factors can be computed in the form (Zhou *et al.*, 1996)

$$[\,N(z)\ M(z)\,] = \left[\begin{array}{c|cc} A+LC & I & L \\ \hline VC & 0 & V \end{array}\right]$$

where

$$\begin{aligned} L &= -A\widetilde{X}C^TV^TV \\ V(I + C\widetilde{X}C^T)V^T &= I \end{aligned} \quad (7)$$

and $\widetilde{X}$ is the stabilizing symmetric positive semidefinite solution of the Riccati equation

$$A(I + \widetilde{X}C^TC)^{-1}\widetilde{X}A^T - \widetilde{X} = 0 \quad (8)$$

The observability grammian Q thus satisfies

$$(A + LC)^TQ(A + LC) + C^TV^TVC = Q \quad (9)$$

For a minimal system, the grammians P and Q are positive definite (i.e., nonsingular), and they can be used to perform a system balancing by determining a coordinate transformation such that both grammians in the new coordinate system are equal and diagonal. Alternatively, P and Q can be employed to determine left and right truncation matrices L and T, respectively, to obtain a minimal or a reduced order system G_r with state space representation (LAT, LB, CT, D). Note that for a stable system $X = 0$, $\widetilde{X} = 0$, $W = I$ and $V = I$, thus P and Q are the standard grammians for a stable discrete-time system.

The emphasis on improving the accuracy of computations has led to so-called model reduction algorithms with *enhanced accuracy*. The grammians can be always determined directly in Cholesky factorized forms $P = SS^T$ and $Q = R^T R$, where S and R are upper-triangular matrices (Hammarling, 1982). The computation of L and T can be done from the singular value decomposition

$$RS = \begin{bmatrix} U_1 \ U_2 \end{bmatrix} \mathrm{diag}(\Sigma_1, \Sigma_2) \begin{bmatrix} V_1 \ V_2 \end{bmatrix}^T,$$

where

$$\Sigma_1 = \mathrm{diag}(\sigma_1, \ldots, \sigma_r), \quad \Sigma_2 = \mathrm{diag}(\sigma_{r+1}, \ldots, \sigma_n)$$

and $\sigma_1 \geq \ldots \geq \sigma_r > \sigma_{r+1} \geq \ldots \geq \sigma_n \geq 0$.

The so-called *square-root* (**SR**) methods determine L and T as (Tombs and Postlethwaite, 1987)

$$L = \Sigma_1^{-1/2} U_1^T R, \qquad T = S V_1 \Sigma_1^{-1/2}.$$

This approach is usually numerically very accurate for well-equilibrated systems. However if the original system is highly unbalanced, potential accuracy losses can be induced in the reduced model if either of the truncation matrices L or T is ill-conditioned (i.e., nearly rank deficient).

A *balancing-free square-root* (**BFSR**) algorithm proposed in (Varga, 1991) combines the advantages of a *balancing-free* (**BF**) approach (Safonov and Chiang, 1989) and of the **SR** approach. L and T are determined as

$$L = (Y^T X)^{-1} Y^T, \qquad T = X,$$

where X and Y are $n \times r$ matrices with orthogonal columns computed from two QR decompositions

$$SV_1 = XW, \quad R^T U_1 = YZ$$

with W and Z non-singular and upper-triangular. The accuracy of the **BFSR** algorithm is usually better than either of **SR** or **BF** approaches.

We have the following analogous result to Theorem 4 of Zhou *et al.* (1999):

Theorem 1. Suppose $G(z)$ has no poles on the unit circle and let $G_r(z)$ be the TFM of the reduced order model with state-space realization (LAT, LB, CT, D), where L and T are the truncation matrices computed above. Then $G_r(z)$ has no poles on the unit circle and

$$\|G(z) - G_r(z)\|_\infty \leq 2 \sum_{r+1}^{n} \sigma_i$$

3. FURTHER NUMERICAL ASPECTS

The computation of grammians involves apparently the solution of two Riccati equations (5) and (8) of particular types. This can be however avoided easily using the technique developed in (Varga, 1993) to compute coprime factorizations with inner denominators. Let U_1 be an orthogonal transformation matrix which reduces A to an ordered *real Schur form* (RSF)

$$U_1^T A U_1 = \begin{bmatrix} A_{11} & A_{12} \\ 0 & A_{22} \end{bmatrix}, \quad U_1^T B = \begin{bmatrix} B_1 \\ B_2 \end{bmatrix}$$

where A_{11} contains the stable eigenvalues of A (i.e., those lying inside the unit circle), A_{22} contains the unstable eigenvalues of A, and the transformed $U_1^T B$ is partitioned in accordance with $U_1^T A U_1$. Using the transformed forms of the system matrices, the stabilizing feedback in (4) can be determined as

$$F = \begin{bmatrix} 0 \ F_2 \end{bmatrix} U_1^T$$

where F_2 and W_2 are computed as

$$F_2 = -W_2 W_2^T B_2^T X_2 A_{22}$$

$$W_2^T (I + B_2^T X_2 B_2) W_2 = I$$

with X_2 being the stabilizing solution of the reduced order Riccati equation

$$A_{22}^T X_2 (I + B_2 B_2^T X_2)^{-1} A_{22} - X_2 = 0$$

Because A_{22} is anti-stable, X_2 is positive definite. Thus, since both A_{22} and X_2 are invertible, we can rewrite the above Riccati equation as a Lyapunov equation in the variable $Y = X_2^{-1}$

$$A_{22}^T Y A_{22} - B_2 B_2^T = Y$$

and F_2 can be alternatively computed as

$$F_2 = -B_2^T (Y + B_2 B_2^T)^{-1} A_{22}$$

To reduce the overall computational costs, it is possible to determine F_2 using the recursive approach of (Varga, 1993; Varga, 1998). With this method a second orthogonal transformation matrix U_2 is determined such that $U_2^T (A_{22} + B_2 F_2) U_2$ is further in a RSF. Thus, with

$$U = U_1 \begin{bmatrix} I & 0 \\ 0 & U_2 \end{bmatrix}$$

we have

$$\begin{bmatrix} N(z) \\ M(z) \end{bmatrix} = \left[\begin{array}{c|c} U^T(A+BF)U & U^TBW_2 \\ \hline U & 0 \\ FU & W_2 \end{array}\right]$$

and $U^T(A+BF)U$ is in RSF. The controllability grammian $\widehat{P}$ corresponding to the transformed matrices $\widehat{A} = U^T(A+BF)U$, $\widehat{B} = U^TB$ satisfies the Lyapunov equation

$$\widehat{A}\widehat{P}\widehat{A}^T + \widehat{B}W_2W_2^T\widehat{B}^T = \widehat{P}$$

and the controllability grammian in the original coordinates is given by $P = U^T\widehat{P}U$. If $\widehat{P}$ is determined in a Cholesky-factorized form $\widehat{P} = \widehat{S}\widehat{S}^T$ (e.g., by using the algorithm of Hammarling (1982)), then P can be easily computed in a similar form $P = SS^T$, where S is the upper triangular factor in the QR-decomposition of $U^T\widehat{S}$.

An entirely similar computational approach can be devised to determine the observability grammian Q in a Cholesky factorized form $Q = R^TR$. As before, we can avoid the solution of a Riccati equation by solving instead a reduced order Lyapunov equation. All computational details follow by duality formulas.

4. THE PERIODIC CASE

In this section we extend the previous results for standard discrete-time systems to periodic systems of the form

$$\begin{aligned} x_{k+1} &= A_kx_k + B_ku_k \\ y_k &= C_kx_k + D_ku_k \end{aligned} \qquad (10)$$

where the matrices $A_k \in R^{n_{k+1}\times n_k}$, $B_k \in R^{n_{k+1}\times m}$, $C_k \in R^{p\times n_k}$, $D_k \in R^{p\times m}$ are periodic with period $K \geq 1$.

To simplify the presentation we introduce first some notation. For a K-periodic matrix X_k we use alternatively the *script* notation

$$\mathcal{X} := \text{diag}\,(X_0, X_1, \ldots, X_{K-1}),$$

which associates the block-diagonal matrix $\mathcal{X}$ to the cyclic matrix sequence X_k, $k = 0, \ldots, K-1$. This notation is consistent with the standard matrix operations as for instance addition, multiplication, inversion as well as with several standard matrix decompositions (Cholesky, SVD). We denote with $\sigma\mathcal{X}$ the K-cyclic shift

$$\sigma\mathcal{X} = \text{diag}\,(X_1, \ldots, X_{K-1}, X_0)$$

of the cyclic sequence X_k, $k = 0, \ldots, K-1$. By using the script notation, the periodic system (10) will be alternatively denoted by the quadruple $(\mathcal{A}, \mathcal{B}, \mathcal{C}, \mathcal{D})_{\mathbf{n}}$, where the time-varying state vector dimensions are denoted compactly by $\mathbf{n} = (n_0, \ldots, n_{K-1})$. We denote with $\mathcal{I}_{\mathbf{n}}$ the K-periodic identity matrix I_{n_k} with time-varying dimensions. The transition matrix of the system (10) is defined by the $n_j \times n_i$ matrix $\Phi_A(j,i) = A_{j-1}A_{j-2}\cdots A_i$, where $\Phi_A(i,i) := I_{n_i}$. The state transition matrix over one period $\Phi_A(j{+}K, j) \in R^{n_j\times n_j}$ is called the *monodromy matrix* of system (10) at time j and its eigenvalues are called *characteristic multipliers* at time j.

In what follows, we assume that the monodromy matrix $\Phi_A(\tau + K, \tau)$ has no eigenvalues on the unit circle. Using a similar approach as for the standard case, we define the controllability grammian of a possibly unstable periodic system $(\mathcal{A}, \mathcal{B}, \mathcal{C}, \mathcal{D})_{\mathbf{n}}$ as the periodic semipositive definite matrix $\mathcal{P}$ satisfying the periodic Lyapunov equation analogous to (6)

$$\sigma\mathcal{P} = (\mathcal{A} + \mathcal{BF})\mathcal{P}(\mathcal{A} + \mathcal{BF})^T + \mathcal{BWW}^T\mathcal{B}^T \quad (11)$$

where

$$\begin{aligned} \mathcal{F} &= -\mathcal{WW}^T\mathcal{B}^T\mathcal{A}\sigma\mathcal{X} \\ \mathcal{W}^T(\mathcal{I} &+ \mathcal{B}^T\sigma\mathcal{XB})\mathcal{W} = \mathcal{I} \end{aligned} \qquad (12)$$

and $\mathcal{X}$ is the stabilizing symmetric positive semidefinite solution of the periodic Riccati equation

$$\mathcal{A}^T\sigma\mathcal{X}(\sigma\mathcal{I}_{\mathbf{n}} + \mathcal{BB}^T\sigma\mathcal{X})^{-1}\mathcal{A} - \mathcal{X} = 0 \quad (13)$$

These equations can be deduced from the conditions characterizing an all-pass periodic system established by Xie *et al.* (1996, Theorem 4.2).

Similarly, we define the periodic observability grammian $\mathcal{Q}$ satisfying the periodic Lyapunov equation analogous to (9)

$$\mathcal{Q} = (\mathcal{A} + \mathcal{LC})^T\sigma\mathcal{Q}(\mathcal{A} + \mathcal{LC}) + \mathcal{C}^T\mathcal{V}^T\mathcal{VC} \quad (14)$$

where

$$\begin{aligned} \mathcal{L} &= -\mathcal{A}\widetilde{\mathcal{X}}\mathcal{C}^T\mathcal{V}^T\mathcal{V} \\ \mathcal{V}(\mathcal{I} &+ \mathcal{C}\widetilde{\mathcal{X}}\mathcal{C}^T)\mathcal{V}^T = \mathcal{I} \end{aligned} \qquad (15)$$

and $\widetilde{\mathcal{X}}$ is the stabilizing symmetric semipositive definite solution of the periodic Riccati equation

$$\mathcal{A}(\mathcal{I}_{\mathbf{n}} + \widetilde{\mathcal{X}}\mathcal{C}^T\mathcal{C})^{-1}\widetilde{\mathcal{X}}\mathcal{A}^T - \sigma\widetilde{\mathcal{X}} = 0 \quad (16)$$

Let $\mathcal{P} = \mathcal{S}^T\mathcal{S}$ and $\mathcal{Q} = \mathcal{R}^T\mathcal{R}$ be the Cholesky factorizations of grammians. For a minimal system, in analogy with the standard stable case (Tombs and Postlethwaite, 1987), we can use the singular value decomposition

$$\mathcal{RS} = \mathcal{U}\Sigma\mathcal{V}^T, \quad (17)$$

to compute the balancing transformation matrix $\mathcal{T}$ and its inverse $\mathcal{T}^{-1}$ as

$$\mathcal{T} = \mathcal{S}\mathcal{V}\Sigma^{-1/2}, \qquad \mathcal{T}^{-1} = \Sigma^{-1/2}\mathcal{U}^T\mathcal{R}.$$

It is easy to show that the Lyapunov-transformed system $(\sigma\mathcal{T}^{-1}\mathcal{A}\mathcal{T}, \sigma\mathcal{T}^{-1}\mathcal{B}, \mathcal{C}\mathcal{T}, \mathcal{D})_{\mathbf{n}}$ has the controllability and observability grammians equal and diagonal. We call such a realization of the possibly unstable periodic system (10) a *balanced* realization.

Algorithms with enhanced accuracy for periodic model reduction have been developed by Varga (2000). The truncation formulas to determine directly the matrices of the reduced system $(\mathcal{A}_r, \mathcal{B}_r, \mathcal{C}_r, \mathcal{D}_r)$ generalize those in the standard case. Let us write the singular value decomposition (17) at each time instant k in the partitioned form

$$R_kS_k = [\,U_{k,1}\; U_{k,2}\,]\begin{bmatrix}\Sigma_{k,1} & 0\\ 0 & \Sigma_{k,2}\end{bmatrix}[\,V_{k,1}\; V_{k,2}\,]^T \quad (18)$$

where $\Sigma_{k,1} \in \mathbf{R}^{r_k\times r_k}$, $U_{k,1} \in \mathbf{R}^{n_k\times r_k}$, $V_{k,1} \in \mathbf{R}^{n_k\times r_k}$ and $\Sigma_{k,1} > 0$. From the above decomposition define, with $\widetilde{\Sigma}_1 = \mathrm{diag}\,(\Sigma_{0,1},\ldots,\Sigma_{K-1,1})$, the *truncation* matrices

$$\mathcal{L} = \widetilde{\Sigma}_1^{-\frac{1}{2}}\mathcal{U}_1^T\mathcal{R}, \qquad \mathcal{T} = \mathcal{S}\mathcal{V}_1\widetilde{\Sigma}_1^{-\frac{1}{2}}. \quad (19)$$

Then the reduced system can be computed as

$$(\mathcal{A}_r, \mathcal{B}_r, \mathcal{C}_r, \mathcal{D}_r)_{\mathbf{r}} = (\sigma\mathcal{L}\mathcal{A}\mathcal{T}, \sigma\mathcal{L}\mathcal{B}, \mathcal{C}\mathcal{T}, \mathcal{D})_{\mathbf{r}} \quad (20)$$

Since the computation of the reduced model relies exclusively on square-root information (the Cholesky factors of grammians), this model reduction method is called the *square-root* approach. This approach leads to a guaranteed enhancement of the overall numerical accuracy of computations. The key computation in determining the truncation matrices $\mathcal{L}$ and $\mathcal{T}$ is the solution of the two periodic Lyapunov equations (11) and (14) with time-varying dimensions directly for the Cholesky factors of the grammians. Numerically stable algorithms for these computations have been developed recently by Varga (1999).

The *square-root* method is essentially a balancing based truncation approach. To avoid accuracy losses potentially induced by balancing, an alternative is to use a *balancing-free* approach to determine the truncation matrices. A *square-root balancing-free* approach for the periodic case, which combines both these desirable features, has been proposed recently by Varga (2000). Consider the QR-decompositions

$$\mathcal{S}\mathcal{V}_1 = \widetilde{\mathcal{T}}\mathcal{X}, \quad \mathcal{R}^T\mathcal{U}_1 = \widetilde{\mathcal{Z}}\mathcal{Y}, \quad (21)$$

where $\mathcal{X}$ and $\mathcal{Y}$ are nonsingular matrices and $\widetilde{\mathcal{T}}$ and $\widetilde{\mathcal{Z}}$ are matrices with orthonormal columns. With the already computed $\widetilde{\mathcal{T}}$ we define the corresponding $\widetilde{\mathcal{L}}$ as

$$\widetilde{\mathcal{L}} = (\widetilde{\mathcal{Z}}^T\widetilde{\mathcal{T}})^{-1}\widetilde{\mathcal{Z}}^T. \quad (22)$$

It is easy to show that the periodic system $(\sigma\widetilde{\mathcal{L}}\mathcal{A}\widetilde{\mathcal{T}}, \sigma\widetilde{\mathcal{L}}\mathcal{B}, \mathcal{C}\widetilde{\mathcal{T}}, \mathcal{D})_{\mathbf{r}}$ with $\widetilde{\mathcal{L}}$ and $\widetilde{\mathcal{T}}$ defined in (21) and (22) is Lyapunov-similar to the reduced system $(\mathcal{A}_r, \mathcal{B}_r, \mathcal{C}_r, \mathcal{D}_r)_{\mathbf{r}}$ obtained with the *square-root* approach.

Similarly to the standard case we can avoid the solution of periodic Riccati equations by solving instead periodic Lyapunov equations of lower order. Consider the periodic orthogonal matrix $\mathcal{U}$ such that $\mathcal{A}$ is in an ordered EPRSF (Bojanczyk *et al.*, 1992; Varga, 1999)

$$\begin{aligned} U_{k+1,1}^T A_k U_{k,1} &= \begin{bmatrix} A_{k,11} & A_{k,12}\\ 0 & A_{k,22}\end{bmatrix}\\ U_{k+1,1}^T B_k &= \begin{bmatrix} B_{k,1}\\ B_{k,2}\end{bmatrix}\end{aligned} \quad (23)$$

where $\mathcal{A}_{11}$ contains the stable characteristic values of $\mathcal{A}$ (i.e., those lying in the unit circle), $\mathcal{A}_{22}$ contains the unstable characteristic values of $\mathcal{A}$, and the transformed $\sigma\mathcal{U}_1^T\mathcal{B}$ is partitioned in accordance with $\sigma\mathcal{U}_1^T A U_1$.

Using the transformed forms of the system matrices, the stabilizing periodic state feedback can be determined as

$$F_k = [\,0\;\; F_{k,2}\,]\,U_{k,1}^T$$

where $\mathcal{F}_2$ and $\mathcal{W}_2$ can be computed as

$$\mathcal{F}_2 = -\mathcal{W}_2\mathcal{W}_2^T\mathcal{B}_2^T\sigma\mathcal{X}_2\mathcal{A}_{22}$$

$$\mathcal{W}_2^T(\mathcal{I} + \mathcal{B}_2^T\sigma\mathcal{X}_2\mathcal{B}_2)\mathcal{W}_2 = \mathcal{I}$$

with $\mathcal{X}_2$ being the stabilizing solution of the reduced order periodic Riccati equation

$$\mathcal{A}_{22}^T\sigma\mathcal{X}_2(\mathcal{I} + \mathcal{B}_2\mathcal{B}_2^T\sigma\mathcal{X}_2)^{-1}\mathcal{A}_{22} - \mathcal{X}_2 = 0$$

Because $\mathcal{A}_{22}$ is anti-stable, $\mathcal{X}_2$ is positive definite. Thus, since both $\mathcal{A}_{22}$ and $\mathcal{X}_2$ are invertible, we can rewrite the above periodic Riccati equation as a periodic Lyapunov equation in the variable $\mathcal{Y} = \mathcal{X}_2^{-1}$

$$\mathcal{A}_{22}^T\mathcal{Y}\mathcal{A}_{22} - \mathcal{B}_2\mathcal{B}_2^T = \sigma\mathcal{Y} \quad (24)$$

and $\mathcal{F}_2$ can be computed as

$$\mathcal{F}_2 = -\mathcal{B}_2^T(\sigma\mathcal{Y} + \mathcal{B}_2\mathcal{B}_2^T)^{-1}\mathcal{A}_{22}$$

Since the periodic submatrix $A_{k,22}$ in the EPRSF (23) has constant dimension, the algorithm of Varga (1997) can be employed to solve (24).

It is also possible in the periodic case to reduce the overall computational costs, by determining $\mathcal{F}_2$ using a recursive approach similar to the approach for standard system presented in (Varga, 1993). For this purpose, the periodic Schur method for pole assignment (Sreedhar and Van Dooren, 1993) can be extended to compute recursively coprime factorizations for periodic systems. We will not enter into the details of such a method, but discuss only the outcome of it. The periodic state feedback $\mathcal{F}_2$ is determined simultaneously with a second orthogonal periodic transformation matrix $\mathcal{U}_2$ such that $\sigma\mathcal{U}_2^T(\mathcal{A}_{22} + \mathcal{B}_2\mathcal{F}_2)\mathcal{U}_2$ is in a PRSF. Thus, with

$$U_k = U_{k,1}\begin{bmatrix} I & 0 \\ 0 & U_{k,2} \end{bmatrix}$$

we have that the controllability grammian $\widehat{\mathcal{P}}$ corresponding to the transformed system matrices $\widehat{\mathcal{A}} = \sigma\mathcal{U}^T(\mathcal{A} + \mathcal{B}\mathcal{F})\mathcal{U}$ and $\widehat{\mathcal{B}} = \sigma\mathcal{U}^T\mathcal{B}$ satisfies the periodic Lyapunov equation

$$\widehat{\mathcal{A}}\widehat{\mathcal{P}}\widehat{\mathcal{A}}^T + \widehat{\mathcal{B}}\mathcal{W}_2\mathcal{W}_2^T\widehat{\mathcal{B}}^T = \sigma\widehat{\mathcal{P}}$$

The controllability grammian in the original coordinates is given by $\mathcal{P} = \mathcal{U}^T\widehat{\mathcal{P}}\mathcal{U}$. If $\widehat{\mathcal{P}}$ is determined in a Cholesky-factorized form $\widehat{\mathcal{P}} = \widehat{\mathcal{S}}\widehat{\mathcal{S}}^T$ using the algorithm developed by Varga (1999) for time-varying dimensions, then $\mathcal{P}$ can be easily computed in a similar form $\mathcal{P} = \mathcal{S}\mathcal{S}^T$, where $\mathcal{S}$ is the upper triangular factor in the QR-decomposition of $\mathcal{U}^T\widehat{\mathcal{S}}$.

An entirely similar computational approach can be devised to determine the periodic observability grammian $\mathcal{Q}$. As before, we can avoid the solution of a periodic Riccati equation by solving instead a reduced order periodic Lyapunov equation. All computational details follow by dual formulas.

5. CONCLUSIONS

We extended the balancing concepts for stable systems to unstable standard and periodic discrete-time systems. This allows the application of balancing related accuracy enhancing order reduction methods to unstable periodic systems. The main computational problems are the computation of coprime factorizations with inner denominators, the solution of sign definite periodic Lyapunov equations with constant and time-varying dimensions, and the computation of truncation matrices for model reduction using square-root and balancing-free techniques. The main computational ingredient to solve these problems is the computation of extended periodic real Schur form.

6. REFERENCES

Bojanczyk, A. W., G. Golub and P. Van Dooren (1992). The periodic Schur decomposition. Algorithms and applications. In: *Proceedings SPIE Conference* (F. T. Luk, Ed.). Vol. 1770. pp. 31–42.

Hammarling, S. J. (1982). Numerical solution of the stable, non-negative definite Lyapunov equation. *IMA J. Numer. Anal.* **2**, 303–323.

Longhi, S. and G. Orlando (1999). Balanced reduction of linear periodic systems. *Kybernetika* **35**, 737–751.

Safonov, M. G. and R. Y. Chiang (1989). A Schur method for balanced-truncation model reduction. *IEEE Trans. Autom. Control* **34**, 729–733.

Sreedhar, J. and P. Van Dooren (1993). Pole placement via the periodic Schur decomposition. In: *Proc. 1993 American Control Conference, San Francisco, CA*. pp. 1563–1567.

Tombs, M. S. and I. Postlethwaite (1987). Truncated balanced realization of a stable non-minimal state-space system. *Int. J. Control* **46**, 1319–1330.

Varga, A. (1991). Efficient minimal realization procedure based on balancing. In: *Prepr. of IMACS Symp. on Modelling and Control of Technological Systems* (A. El Moudni, P. Borne and S. G. Tzafestas, Eds.). Vol. 2. pp. 42–47.

Varga, A. (1993). A Schur method for computing coprime factorizations with inner denominators and applications in model reduction. In: *Proc. 1993 American Control Conference, San Francisco, CA*. pp. 2130–2131.

Varga, A. (1997). Periodic Lyapunov equations: some applications and new algorithms. *Int. J. Control* **67**, 69–87.

Varga, A. (1998). Computation of coprime factorizations of rational matrices. *Lin. Alg. & Appl.* **271**, 83–115.

Varga, A. (1999). Balancing related methods for minimal realization of periodic systems. *Systems & Control Lett.* **36**, 339–349.

Varga, A. (2000). Balanced truncation model reduction of periodic systems. In: *Proc. CDC'2000, Sydney, Australia.*

Xie, B., R. K. A. V. Aripirala and V. Syrmos (1996). Model reduction of linear discrete-time periodic systems using hankel-norm approximations. In: *Proc. 13th IFAC Congress, San Francisco, USA*. pp. 245–250.

Zhou, K., G. Salomon and E. Wu (1999). Balanced realization and model reduction for unstable systems. *Int. J. Robust and Nonlinear Control* **9**, 183–198.

Zhou, K., J. C. Doyle and K. Glover (1996). *Robust and Optimal Control.* Prentice Hall.

Copyright © IFAC Periodic Control Systems,
Cernobbio-Como, Italy, 2001

STABILITY RADIUS AND OPTIMAL SCALING OF DISCRETE-TIME PERIODIC SYSTEMS

Y. Genin * I. Ipsen ** R. Ştefan * P. Van Dooren *

* *CESAME, Université catholique de Louvain,*
Av. G. Lemaître 4,
B-1348 Louvain-la-Neuve, Belgium
** *Center for Research in Scientific Computation,*
Department of Mathematics,
North Carolina State University,
Raleigh, NC 27695-8205, USA.

Abstract: Robust stability properties of periodic discrete time systems are investigated. Analytic expressions are derived for the stability radius in the scalar case.
Copyright ©2001 IFAC

Keywords: stability radii, periodic systems, optimal scaling

1. INTRODUCTION

Let us consider a discrete periodic system of the form

$$E_k x_{k+1} = A_k x_k \quad k = 0, 1, 2, \dots \qquad (1)$$

where x_0 is the given initial state and where the matrices E_k, $A_k \in \mathbb{C}^{n\times n}$ vary periodically over a period of length K, i.e.

$$E_{k+K} = E_k, \quad A_{k+K} = A_k, \text{for all } k \geq 0.$$

When x_0 is given, one can solve this initial value problem provided the E_k matrices are invertible, which we will assume throughout this paper. We define the monodromy matrix

$$\Phi := E_{K-1}^{-1} A_{K-1} \dots E_1^{-1} A_1 E_0^{-1} A_0 \qquad (2)$$

and point out that the behaviour over K steps is easily found from (1) to be time invariant :

$$x_{(i+1)K} = \Phi x_{iK} \quad i = 0, 1, \dots \qquad (3)$$

The system (1) is said to be stable, if all the eigenvalues of Φ are in the open unit disc, *i.e* $\Lambda(\Phi) \subset \mathbb{D} = \{z \in \mathbb{C} : |z| < 1\}$. The eigenvalues of Φ can also be obtained from the bicyclic eigenvalue problem

$$\lambda\mathcal{E} - \mathcal{A}\mathcal{Z} := \begin{bmatrix} \lambda E_0 & & & -A_0 \\ -A_1 & \ddots & & \\ & \ddots & \ddots & \\ & & -A_{K-1} & \lambda E_{K-1} \end{bmatrix} \qquad (4)$$

where

$$\mathcal{E} = \text{diag}\{E_0, \cdots, E_{K-1}\},$$
$$\mathcal{A} = \text{diag}\{A_0, \cdots, A_{K-1}\},$$
$$\mathcal{Z} = \begin{bmatrix} 0_n & & & I_n \\ I_n & \ddots & & \\ & \ddots & \ddots & \\ & & I_n & 0_n \end{bmatrix}.$$

Indeed, one easily finds that for $N := nK$

$$\det(\lambda\mathcal{E} - \mathcal{A}\mathcal{Z}) = \det\mathcal{E}\ \det(\lambda I_N - \mathcal{E}^{-1}\mathcal{A}\mathcal{Z})$$
$$= \det\mathcal{E}\ \det(\lambda^K I_n - \Phi)$$

and hence that the generalized eigenvalues of $(\lambda\mathcal{E} - \mathcal{A}\mathcal{Z})$ are the K-th roots of the eigenvalues of Φ. But the system (1) is not a unique representation of the difference equation. Scaling the equations with a scalar $\alpha_k \neq 0$ will not alter the solution x_k, and substituting $\hat{x}_{k+1} = \beta_k x_{k+1}$ with scalars $\beta_k \neq 0$ always allows to retrieve x_{k+1} from $\hat{x}_{k+1}$.

If we choose a K-periodic scaling $\alpha_k = \alpha_{k+K}$ and $\beta_k = \beta_{k+K}$, for all $k \geq 0$, then we obtain a new periodic system characterized by

$$\lambda\hat{\mathcal{E}} - \hat{\mathcal{A}}\mathcal{Z} = \begin{bmatrix} \alpha_0 I_n & & \\ & \ddots & \\ & & \alpha_{k-1} I_n \end{bmatrix} (\lambda\mathcal{E} - \mathcal{A}\mathcal{Z}) \begin{bmatrix} \beta_0 I_n & & \\ & \ddots & \\ & & \beta_{k-1} I_n \end{bmatrix}^{-1} = D_\alpha(\lambda\mathcal{E} - \mathcal{A}\mathcal{Z})D_\beta^{-1} \tag{5}$$

corresponding to the scaled difference equations

$$\underbrace{(\alpha_k \; E_k \; \beta_k^{-1})}_{\hat{E}_k} \; \hat{x}_{k+1} \;=\; \underbrace{(\alpha_k \; A_k \; \beta_{k-1}^{-1})}_{\hat{A}_k} \; \hat{x}_k. \tag{6}$$

2. STABILITY RADIUS OF DISCRETE-TIME PERIODIC SYSTEMS

The robustness issue is a crucial problem for the application of control theory; for example, one of the basic goals of feedback control is to enhance system robustness. Robust stability is also an important topic in linear algebra as well as in numerical analysis. A fundamental problem in robustness analysis is to determine the ability of a system matrix to maintain its stability under a certain class of perturbations. A natural robustness measure is the *distance* of a stable system to the set of unstable systems, defined by Hinrichsen and Pritchard (Hinrichsen and Pritchard, 1986) as the ***stability radius*** of the system.

Assuming the system (1) to be stable implies that $(\lambda\mathcal{E} - \mathcal{A}\mathcal{Z})$ has only generalized eigenvalues inside the unit circle. Therefore $\mathcal{E}$ is invertible, Φ is well defined and Φ has its eigenvalues inside the unit circle. One is interested in determining the smallest perturbations of the coefficients E_k, A_k that will make the system unstable. Equivalently, one analyzes the sensitivity of the generalized eigenvalues of $\lambda\mathcal{E} - \mathcal{A}\mathcal{Z}$ to structured perturbations in this pencil. To be more specific, one has to find the smallest perturbations

$$\Delta\,\mathcal{E} = \operatorname{diag}\{\delta E_0, \cdots, \delta E_{K-1}\},$$
$$\Delta\,\mathcal{A} = \operatorname{diag}\{\delta A_0, \cdots, \delta A_{K-1}\}$$

such that the system matrix $[\lambda\,(\mathcal{E} + \Delta\mathcal{E}) - (\mathcal{A} + \Delta\mathcal{A})\,\mathcal{Z}]$ has at least one generalized eigenvalue in the ***unstable*** part of the complex plane, *i.e.* outside the open unit disc. Measured in term of the 2-norm, the minimality of the perturbations in question is characterized by the **stability radius** defined as follows:

$$r_{\mathcal{E},\mathcal{A}} = \inf_{\Delta\mathcal{E},\Delta\mathcal{A}}\{\max(\|\Delta\mathcal{E}\|_2, \|\Delta\mathcal{A}\|_2) \;:\; \exists\lambda \notin \mathbb{D} \text{ s.t.} \det[\lambda(\mathcal{E} + \Delta\mathcal{E}) - (\mathcal{A} + \Delta\mathcal{A})\,\mathcal{Z}] = 0\}. \tag{7}$$

Two subproblems of special interest can derived from this general setting by imposing either the constraint $\Delta\,\mathcal{E} = 0$ or the constraint $\Delta\,\mathcal{A} = 0$. The corresponding stability radii then take respectively the form

$$r_{\mathcal{E}} = \inf_{\Delta\mathcal{E}}\{\|\Delta\mathcal{E}\|_2 \;:\; \exists\lambda \notin \mathbb{D} \text{ s.t.} \det[\lambda(\mathcal{E} + \Delta\mathcal{E}) - \mathcal{A}\,\mathcal{Z}] = 0\} \tag{8}$$

or

$$r_{\mathcal{A}} = \inf_{\Delta\mathcal{A}}\{\|\Delta\mathcal{A}\|_2) \;:\; \exists\lambda \notin \mathbb{D} \text{ s.t.} \det[\lambda\,\mathcal{E} - (\mathcal{A} + \Delta\mathcal{A})\,\mathcal{Z}] = 0\}. \tag{9}$$

Because eigenvalues move continuously with $\Delta\mathcal{E}$, $\Delta\mathcal{A}$, equalities (7), (8) and (9) can be rewritten into the form

$$r_{\mathcal{E},\mathcal{A}} = \inf_{|\lambda|=1}\;\inf_{\Delta\mathcal{E},\Delta\mathcal{A}}\{\max(\|\Delta\mathcal{E}\|_2, \|\Delta\mathcal{A}\|_2) \;:\; \det[\lambda(\mathcal{E} + \Delta\mathcal{E}) - (\mathcal{A} + \Delta\mathcal{A})\,\mathcal{Z}] = 0\}. \tag{10}$$

Note that one has the relations

$$\|\Delta\,\mathcal{E}\|_2 = \max_k \|\delta E_k\|_2, \quad \|\Delta\mathcal{A}\|_2 = \max_k \|\delta A_k\|_2$$

by definition, since the matrices $\Delta\,\mathcal{E}$ and $\Delta\,\mathcal{A}$ are diagonal.

Next, we focus our attention on the *scalar* case, when $n = 1$ and E_k, A_k are real numbers, which can be simply denoted by e_k, a_k. Accordingly, for the perturbations matrices δE_k, δA_k we write δe_k, δa_k.

To deal with the problem, let us introduce from the data the two polynomials of the x variable:

$$P_e(x) = \prod_{k=0}^{K-1}(1 - x/|e_k|), \; P_a(x) = \prod_{k=0}^{K-1}(1 + x/|a_k|). \tag{11}$$

Define also the constant $a \in \mathbb{C}$ by

$$a^K = \frac{a_0\,a_1, \ldots, a_{K-1}}{e_0\,e_1, \ldots, e_{K-1}} \tag{12}$$

which is just another way to write (2) in the scalar case. Note that, since the unperturbed system (1) is assumed to be stable, one has the property $|a| < 1$.

A closed formula for the stability radius (10) is given by the next theorem and it is expressed in terms of a polynomial equation involving both P_e and P_a introduced above.

Theorem 1. Let ζ_0 be the *smallest positive real zero* of the polynomial equation

$$P_e(x) - |a|^K\,P_a(x) = 0. \tag{13}$$

Then

$$r_{\mathcal{E},\mathcal{A}} = \zeta_0 \quad (14)$$

i.e the stability radius (10) is the smallest positive real root of the polynomial equation (13).
Moreover, a minimal perturbation is obtained by setting for $k = 0, 1, \ldots, K-1$:

$$\delta e_k = -\, e_k\, \zeta_0/|e_k|,\ \delta a_k = +\, a_k\, \zeta_0/|a_k|,\ \hat{\lambda} = a/|a|. \quad (15)$$

As an immediate consequence, one has

Corollary 2. The spectral radii $r_{\mathcal{E}}$ and $r_{\mathcal{A}}$ will be determined as the *smallest positive real zero* of the polynomial equations $P_e(x) - |a|^K = 0$ and $1 - |a|^K P_a(x) = 0$, respectively, while minimal coefficient perturbations can be still defined by the first or second relation (15).

The case $a = 0$ is treated separately.

3. OPTIMAL SCALING OF BICYCLIC MATRICES

The second part of the paper is dedicated to an alternative derivation of Theorem 1 and Corollary 2. Recall the definitions of $r_{\mathcal{E},\mathcal{A}}$, $r_{\mathcal{E}}$ and $r_{\mathcal{A}}$, respectively. Introduce the following matrices, parametrized by a variable λ:

$$M_\lambda := \lambda\mathcal{E} - \mathcal{A}\mathcal{Z}, \quad (16)$$
$$N_\lambda := (\lambda\mathcal{E} - \mathcal{A}\mathcal{Z})\mathcal{Z}^{-1}, \quad (17)$$
$$L_\lambda := \begin{bmatrix} I \\ \mathcal{Z} \end{bmatrix} M_\lambda^{-1} \left[\lambda I \ -I\right]. \quad (18)$$

Using structured perturbation results (Van Dooren and Vermaut, 1997), one can show that appropriate *lower bounds* for the stability radii (7), (8), (9) are given by the following optimization problems

$$r_{\mathcal{E}} \geq \left\{ \sup_{|\lambda|=1} \inf_{D} \sigma_{\max}\left(D M_\lambda^{-1} D^{-1}\right)\right\}^{-1} \quad (19)$$
$$r_{\mathcal{A}} \geq \left\{ \sup_{|\lambda|=1} \inf_{\tilde{D}} \sigma_{\max}\left(\tilde{D} N_\lambda^{-1} \tilde{D}^{-1}\right)\right\}^{-1} \quad (20)$$
$$r_{\mathcal{E},\mathcal{A}} \geq \left\{ \sup_{|\lambda|=1} \inf_{\hat{D}} \sigma_{\max}\left(\hat{D} L_\lambda \hat{D}^{-1}\right)\right\}^{-1} \quad (21)$$

where $\hat{D} = \mathrm{diag}\{D_1, D_2\}$. For more details, see (Van Dooren and Vermaut, 1997).

We show in this paper that these lower bounds are actually **equalities** in the scalar case ($n = 1$) and that the optimal scaling in (5) can be computed relatively easily. Furthermore, one can also construct the "optimal" perturbations $\Delta\mathcal{E}$, $\Delta\mathcal{A}$ which actually attain the lower bounds in (19)–(21).

The key point in determining the stability radii $r_{\mathcal{E},\mathcal{A}}$, $r_{\mathcal{E}}$ and $r_{\mathcal{A}}$ is to find an optimal scaling of the matrix M_λ introduced by (16), such that the smallest singular value of $D_\alpha\ M_\lambda\ D_\beta^{-1}$ is maximized, and whereby we have special relations between D_α and D_β. We point out that M_λ has bicyclic structure, i.e. it is a linear combination of a block diagonal and a block cyclic matrix. This property will be crucial in proving our extremal properties.

We introduce two classes of diagonal and unitary block scaling matrices:

$$\mathcal{D} = \{D = \mathrm{diag}\{d_1 I_n, \ldots, d_K I_n\}\ d_i \in \mathbb{R}_+\} \quad (22)$$
$$\mathcal{U} = \{U = \mathrm{diag}\{U_1, \ldots, U_K\}\ U_i^* U_i = I_n\} \quad (23)$$

which play an important role in this analysis. It is clear that any $V \in \mathcal{U}$ and $D \in \mathcal{D}$ commute with each other. Therefore if $D_\alpha, D_\beta \in \mathcal{D}$ we have

$$\sigma_{\max}\{D_\alpha\ M_\lambda\ D_\beta^{-1}\} = \sigma_{\max}\{D_\alpha\ U M_\lambda\ V^*\ D_\beta^{-1}\}$$

for any $U, V \in \mathcal{U}$. So instead of optimizing the scaling of M_λ one can as well optimize the scaling of $U M_\lambda V^*$. This is exploited in the sequel.
Let us also point out that for any matrix M and scaling $D \in \mathcal{D}$ and $U, V \in \mathcal{U}$, since $\sigma_{\min} \leq |\lambda_{\min}|$ and $\sigma_{\max} \geq |\lambda_{\max}|$, we have the inequalities

$$\sigma_{\min}(DMD^{-1}) \leq |\lambda_{\min}(UMV^*)| \iff \sigma_{\max}(DM^{-1}D^{-1}) \geq |\lambda_{\max}(VM^{-1}U^*)|. \quad (24)$$

3.1 *The scalar case* $n = 1$

We consider this case separately since it has a closed form solution. Recall the notation introduced in Section 2 for the special case when $n = 1$.

By using the well-known Floquet transform (see (Sreedhar. and Van Dooren, 1997)) and some elementary algebraic manipulations, one can prove the following important result.

Lemma 3. For the λ-family of matrices M_λ in (16), there exist unitary matrices $U, V \in \mathcal{U}$ such that, for $\hat{\lambda} := a/|a|$, one has

$$V^* M_{\hat{\lambda}}^{-1} U = |M_{\hat{\lambda}}^{-1}|.$$

We are now ready to state the key result of this section. The proof appeals to Perron-Frobenius theory and uses the inequalities (24), as well as Lemma 3.

Lemma 4. Let M_λ be given by (16). Then there exists a unit modulus value $\hat{\lambda}$ such that

$$\sup_{|\lambda|=1} \sup_{U,V\in\mathcal{U}} |\lambda_{\max}(V^* M_\lambda^{-1} U)| = \lambda_{\max}(|M_{\hat{\lambda}}^{-1}|)$$
$$= \sup_{|\lambda|=1} \inf_{D\in\mathcal{D}} \sigma_{\max}(D M_\lambda^{-1} D^{-1}). \quad (25)$$

Moreover, there exist unitary matrices $U, V \in \mathcal{U}$ and a diagonal matrix $D \in \mathcal{D}$ such that actually attain both equalities in (25).

As an immediate consequence, one can find easily an optimal scaling for the λ-matrix of interest showing up in (21), that is, L_λ defined by (18). If we now choose $\hat{V} := \mathrm{diag}\{V, \mathcal{Z}V\mathcal{Z}^T\}$, $\hat{U} := \mathrm{diag}\{\hat{\lambda}^{-1}U, -U\}$, then we have indeed

$$\hat{V}^* L_{\hat{\lambda}} \hat{U} = \begin{bmatrix} I \\ \mathcal{Z} \end{bmatrix} V^* M_{\hat{\lambda}}^{-1} U \,[\, I \;\, I \,]$$
$$= \begin{bmatrix} I \\ \mathcal{Z} \end{bmatrix} |M_{\hat{\lambda}}^{-1}| \,[\, I \;\, I \,] = |L_{\hat{\lambda}}|$$

which is the required scaling for $L_{\hat{\lambda}}$. From this, the next result trivially follows.

Corollary 5. Let L_λ be given by (18). Then there exists a unit modulus value $\hat{\lambda}$ such that

$$\sup_{|\lambda|=1} \sup_{\hat{U},\hat{V}\in\mathcal{U}} |\lambda_{\max}(\hat{V}^* L_\lambda \hat{U})| = \lambda_{\max}(|L_{\hat{\lambda}}|)$$
$$= \sup_{|\lambda|=1} \inf_{\hat{D}\in\mathcal{D}} \sigma_{\max}(\hat{D} L_\lambda \hat{D}^{-1}). \quad (26)$$

Similarly, one can look at the optimal scaling for $r_{\mathcal{A}}$, which is again a similarity scaling but on the matrix $N_\lambda^{-1} = \mathcal{Z} M_\lambda^{-1}$. Take $W := \mathcal{Z}V\mathcal{Z}^T$ (which is diagonal and unitary) and check that

$$W^* N_{\hat{\lambda}}^{-1} U = \mathcal{Z} V^* M_{\hat{\lambda}}^{-1} U = \mathcal{Z} |M_{\hat{\lambda}}^{-1}| = |N_{\hat{\lambda}}^{-1}|,$$

which is now the optimal scaling for $N_{\hat{\lambda}}^{-1}$.

Let us finally note that the case $a = 0$ is treated separately.

We may now state the main result of the paper.

Theorem 6. Let $D_{\mathcal{E},\mathcal{A}}$, $D_{\mathcal{E}}$ and $D_{\mathcal{A}}$ be the *optimal scalings* for the λ-matrices L_λ, M_λ^{-1} and N_λ^{-1}, respectively; these scalings are *directly* obtained from the Perron vectors of $|L_\lambda|$, $|M_\lambda^{-1}|$ and $|N_\lambda^{-1}|$. Then

$$r_{\mathcal{E},\mathcal{A}} = \left(\lambda_{\max}(|L_{\hat{\lambda}}|)\right)^{-1} \quad (27)$$
$$r_{\mathcal{E}} = \left(\lambda_{\max}(|M_{\hat{\lambda}}^{-1}|)\right)^{-1} \quad (28)$$
$$r_{\mathcal{A}} = \left(\lambda_{\max}(|N_{\hat{\lambda}}^{-1}|)\right)^{-1} \quad (29)$$

Remark 7. One can now check without difficulty that equalities (27) and (14) coincide.

3.2 *The case $n > 1$*

Here we use an algorithm that tries to narrow down the gap between the upper and lower bounds

$$\inf_D \sigma_{\max}(DVM_\lambda^{-1}U^*D^{-1})$$
$$\geq \sup_{U,V} |\lambda_{\max}(VDM_\lambda^{-1}D^{-1}U^*)|. \quad (30)$$

and we try to reach a scaled matrix that the gap is zero (and hence satisfies the equal modulus property). Notice that the structure of M is such that every eigenvector has non-zero subvectors (corresponding to the blocks). So the iterative procedure works as follows : compute the dominant left and right eigenvectors of the current matrix M. Scale with D so that subblocks satisfy the equal modulus property. These left and right vectors are not equal but if we "rotate" the subvectors then they become equal. Then iterate on this new M matrix. This appeared to converge quadratically on matrices with real elements.

One should show that at each step the singular value can only decrease and the eigenvalue only increase.

ACKNOWLEDGMENTS

This paper presents research results of the Belgian Programme on Interuniversity Poles of Attraction, initiated by the Belgian State, Prime Minister's Office for Science, Technology and Culture. The scientific responsibility rests with its authors. This research was supported in part by grant DMS-9714811 from the National Science Foundation, USA. The third author has been partially supported by NATO.

4. REFERENCES

Genin, Y., P. Van Dooren and V. Vermaut (1998). Convergence of the calculation of $\mathcal{H}_\infty$-norms and related questions. In: *Proceedings MTNS-98*. Padua, Italy. pp. 429–432.

Hinrichsen, D. and A.J. Pritchard (1986). Stability radii of linear systems. *Systems & Control Letters* **7**, 1–10.

Horn, R. and C. Johnson (1985). *Matrix Analysis*. Cambridge University Press. Cambridge, UK.

Sreedhar., J. and P. Van Dooren (1997). Forward/backward decomposition of periodic descriptor systems and two point boundary value problems. In: *Proceedings European Control Conference*. Bruxelles, Belgium. paper FR-A-L7.

Van Dooren, P. and V. Vermaut (1997). On stability radii of generalized eigenvalue problems. In: *Proceedings European Control Conference*. Bruxelles, Belgium. paper FR-M-H6.

Copyright © IFAC Periodic Control Systems,
Cernobbio-Como, Italy, 2001

AN EFFICIENT AND RELIABLE IMPLEMENTATION OF THE PERIODIC QZ ALGORITHM

Daniel Kressner *

* *Department of Mathematics, University of Chemnitz, Chemnitz, Germany*

Abstract: We discuss performance and accuracy aspects of the periodic QZ algorithm. Blocked formulations of the involved orthogonal transformations increase the data locality and thus address the first task. For the sake of reliability the proposed implementation includes balancing, implicit methods for computing shifts and carefully chosen deflation strategies. Algorithms for pole placement and other tasks arising from periodic discrete-time systems could benefit from these improvements.
Copyright ©2001 IFAC

Keywords: Accuracy, Discrete-time systems, Efficient algorithms, Eigenvalue problems, Exponentially stable, Factorization methods, Implementation

1. INTRODUCTION

For matrices $E_i, F_i \in \mathbb{R}^{n,n}$ and $G_i \in \mathbb{R}^{n,m}$ consider the linear discrete-time system

$$E_i x_{i+1} = F_i x_i + G_i u_i, \; i \in \mathbb{N}, \tag{1}$$

where x_i, u_i are vectors of states and inputs, respectively. The coefficient matrices shall satisfy $E_{i+k} = E_i$, $F_{i+k} = F_i$ and $G_{i+k} = G_i$ for some fixed $k \in \mathbb{N}$. Such periodic systems naturally arise when performing multirate sampling of continuous-time systems.

The corresponding monodromy matrix is associated with the formal product

$$E_k^{-1} F_k E_{k-1}^{-1} F_{k-1} \dots E_1^{-1} F_1. \tag{2}$$

Some of the equations in (1) might have no algebraic constraints, that is, E_i is the identity matrix for some $i \in [1, k]$. In this context it is more appropriate to replace (2) by the general product

$$S = A_1 A_2^{s_2} A_3^{s_3} \dots A_k^{s_k}, \tag{3}$$

where $A_i \in \mathbb{R}^{n,n}$ and $s_i \in \{-1, 1\}$. The implicit assumption $s_1 = 1$ can always be achieved by a suitable reordering or formal inversion of the coefficient matrices. Note that the invertibility of factors A_i with $s_i = -1$ is not assumed. Even when this condition is satisfied it is not favorable to form (3) explicitly.

The periodic Schur decomposition is often the first and most expensive step in numerically reliable methods for pole placement and several other tasks related to linear periodic systems (Sreedhar and Van Dooren, 1993). In this decomposition k orthogonal matrices Q_i are constructed so that $Q_i^T A_i Q_{i+1}$, for $s_i = 1$, and $Q_{i+1}^T A_i Q_i$, for $s_i = -1$, are upper triangular, where $i = 2, \dots, k$ and $Q_{k+1} = Q_1$. The first transformed factor $Q_1^T A_1 Q_2$ is upper quasi-triangular. Illustrated:

$$\underbrace{Q_1^T A_1 Q_2} \; \underbrace{Q_2^T A_2^{s_2} Q_3} \; \dots \; \underbrace{Q_k^T A_k^{s_k} Q_1} \tag{4}$$

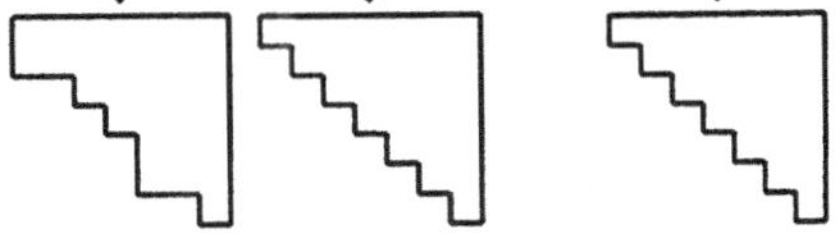

A numerically stable method to compute (4) is the so called Periodic QZ algorithm established by (Van Loan, 1975; Bojanczyk *et al.*, 1992; Hench and Laub, 1994). Sections 2 and 3 of this work are concerned with variations of this algorithm which significantly decrease its execution time. Reliability, a crucial aspect of any competitive implementation, is treated in Sections 4, 5 and 6.

2. BLOCKED REDUCTION TO PERIODIC HESSENBERG FORM

As usual for algorithm which compute a variant of the Schur decomposition, the first step consists of the reduction to a Hessenberg like form. In the context of general products, this form is almost identical to (4) besides the first factor which stays upper Hessenberg.

An efficient implementation should, in a first attempt, reduce A_1 only to block Hessenberg form, that is, $n_b \geq 1$ subdiagonals are nonzero. This concept was successfully applied to the QZ algorithm (Dackland and Kågström, 1999). Since the technique easily generalizes to the periodic case, only a brief outline of the method for the special product $A_1 A_2^{-1} A_3 A_4$ with $n = 9$ and $n_b = 3$ is presented.

The first stage starts with an RQ decomposition of A_2 which alters the matrix A_1. Next, $A_4(:, 1:3)$ is triangularized by three Householder reflectors from the left. Their WY representation is applied to the remaining part of A_4 as well as the matrix A_3 (Golub and Loan, 1996, Section 5.1.7).

A QR decomposition triangularizes $A_3(4:9, 1:3)$ introducing nonzeros in the lower triangular part of $A_2(4:9, 4:9)$.

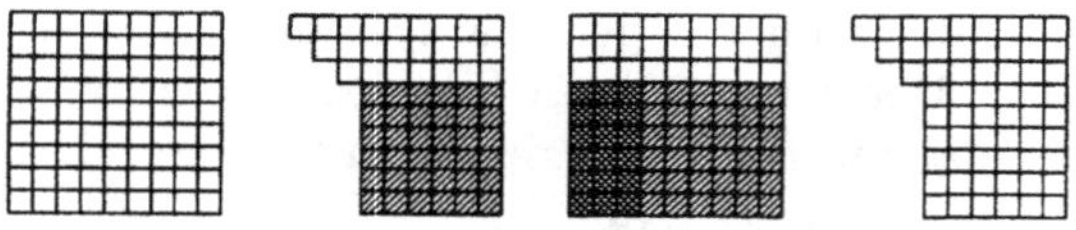

This part is immediately annihilated by an appropriate RQ decomposition.

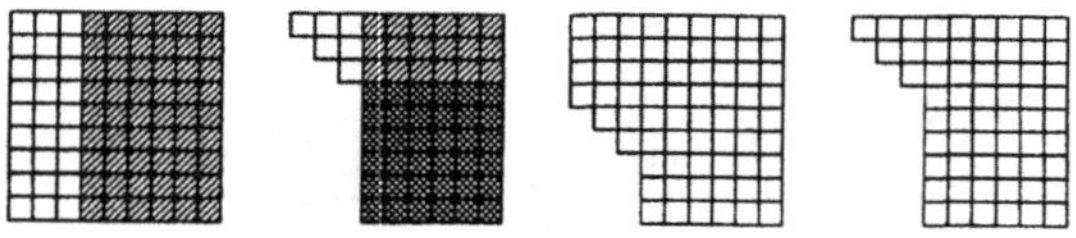

Repeating the procedure for $A_3(1:6, 1:3)$ yields the following pattern.

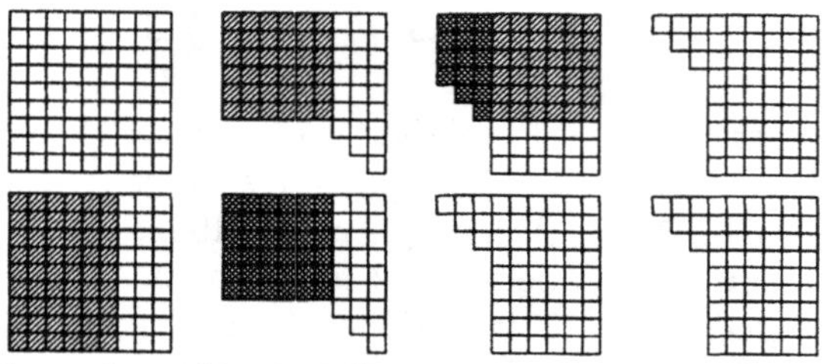

Now, three blocked Householder reflectors triangularize $A_1(:, 1:3)$ from the left.

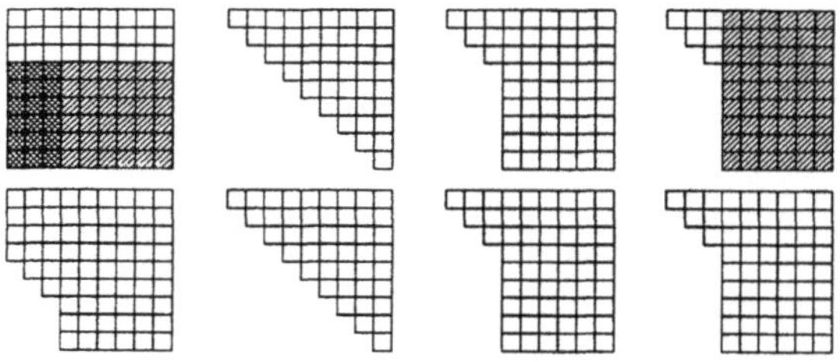

Repeating the procedure for the remaining south-east 6-by-6 subproduct finally leads to the following block Hessenberg form.

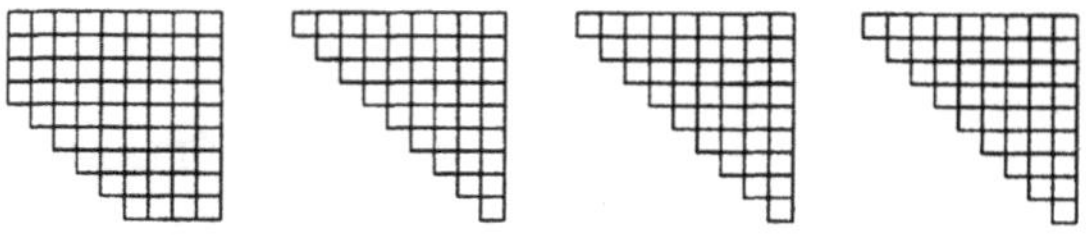

It should be noted that the fill-ins in A_2 overlap for consecutive iterations. Hence, in an actual implementation RQ decompositions of smaller sized r-by-$2r$ matrices are applied in order to annihilate these fill-ins.

The second stage is to annihilate the remaining $n_b - 1$ subdiagonals of A_1, to get an upper Hessenberg matrices while keeping the other matrices upper triangular. This is accomplished via a super-sweep routine as for the general product AB^{-1} described in (Dackland and Kågström, 1999). There is virtually no difference when going to larger products, only that the description of algorithm becomes rather awkward.

Several benchmarks were run to compare the above described blocked with the original version (Bojanczyk *et al.*, 1992) of the Hessenberg reduction algorithm. The FORTRAN 77 implementations were compiled and serially executed on an Origin 2000 computer equipped with 400 MHz IP27 R12000 processors. The programs call optimized BLAS and LAPACK (Anderson *et al.*, 1994) subroutines from the SGI/Cray Scientific Library version 1.2.0.0. The block size n_b was chosen to be 64.

s	n	Original	Blocked
$(1, -1)$	1024	432	195
$(1, -1)$	2048	4740	2481
$(1, -1, 1, -1, 1, -1)$	512	110	60
$(1, -1, 1, -1, 1, -1)$	1024	1295	595
$(1, 1, 1, 1, 1, 1)$	512	62	39
$(1, 1, 1, 1, 1, 1)$	1024	1047	308

Table 1. Execution times in seconds for the reduction to periodic Hessenberg form.

In Table 1 an extract of the observed execution times is presented. Overall, significantly lower times for the blocked algorithms were noted as soon as $n \geq 128$.

3. BLOCKED PERIODIC QZ ITERATION

A similiar algorithm as for the supersweep algorithm can be used to speed up the generalized QZ iterations as well (Dackland and Kågström, 1999). Again, there is a straightforward way to adapt this technique to the periodic case. From the results in Section 2 it can be extrapolated that the execution times of such a blocked periodic QZ iteration will be significantly lower than the original formulation.

4. BALANCING

The periodic QZ algorithm is backward stable (Bojanczyk *et al.*, 1992). That is, the computed Schur decomposition corresponds to a slightly perturbed product $\prod_{i=1}^{k} \hat{A}_i^{s_i}$, where the backward error $\|\hat{A}_i - A_i\|$ is of the order unit round off times $\|A_i\|$. However, even such small perturbations can be harmful when dealing with ill-conditioned problems.

For example, consider the formal product

$$\begin{bmatrix} 5^{-26} & 3^{-14} & 6^{-16} \\ 6^{-06} & 2^{+06} & 3^{+04} \\ 4^{-16} & 2^{-04} & 5^{-06} \end{bmatrix} \begin{bmatrix} 6^{-28} & 3^{-16} & 5^{-18} \\ 7^{-09} & 3^{+03} & 7^{+01} \\ 6^{-23} & 3^{-11} & 3^{-13} \end{bmatrix}^{-1}$$

$$\cdot \begin{bmatrix} 8^{-02} & 6^{-24} & 6^{-11} \\ 5^{+17} & 5^{-05} & 6^{+08} \\ 3^{+03} & 4^{-19} & 7^{-06} \end{bmatrix} \begin{bmatrix} 9^{+00} & 4^{-22} & 3^{-09} \\ 7^{+20} & 2^{-02} & 9^{+11} \\ 4^{+10} & 6^{-12} & 7^{+01} \end{bmatrix}^{-1}, \quad (5)$$

where the signed integer superscript at the end of a number represents its exponential exponent. The eigenvalues, given by the general product of the diagonal elements of the periodic Schur decomposition, and the corresponding condition numbers (Benner *et al.*, 2000) are tabulated below.

Eigenvalue	Condition number
2.88728	4.32×10^{21}
0.39941	1.77×10^{21}
0.07459	2.59×10^{21}

Not surprisingly the periodic QZ algorithm completely fails to compute eigenvalues with acceptable accuracy.

Such effects, caused from matrix entries of widely varying magnitudes, can be removed by a preceding balancing step. For positive definite diagonal matrices D_α, D_β, D_γ, D_ξ the eigenvalues of the formal product $AB^{-1}CE^{-1}$ and

$$(D_\alpha A D_\beta)(D_\gamma B D_\beta)^{-1}(D_\gamma C D_\xi)(D_\alpha E D_\xi)^{-1} \quad (6)$$

are equivalent. Different sign patterns do not pose a problem; if for example $s_B = 1$, then in the following discussion B can virtually be replaced by the matrix

$$\tilde{B} = \left[\tilde{b}_{ij}\right]_{i,j=1}^{n} := \left[\delta(b_{ji} \neq 0) \cdot \frac{1}{b_{ji}}\right]_{i,j=1}^{n}.$$

The diagonal transformations shall reduce the condition numbers and thus improve the accuracy of the computed eigenvalues. However, minimizing the conditioning of the periodic eigenvalue problem is certainly an unrealistic goal. On the other hand, reducing the magnitude ranges of the elements in the factors seems to be reasonable.

Analogously to the generalized eigenvalue problem (Ward, 1981), the balancing step can be formulated as the solution of an optimization problem. Let α_i, β_i, γ_i and ξ_i denote the binary logarithm of the i-th diagonal entry in the corresponding diagonal matrix. Then one wants to minimize the expression

$$\begin{aligned} S(\alpha,\beta,\gamma,\xi) = \sum_{i,j=1}^{n} & (\alpha_i + \beta_j + \log_2 |a_{ij}|)^2 \\ & +(\gamma_i + \beta_j + \log_2 |b_{ij}|)^2 \quad (7) \\ & +(\gamma_i + \xi_j + \log_2 |c_{ij}|)^2 \\ & +(\alpha_i + \xi_j + \log_2 |e_{ij}|)^2. \end{aligned}$$

By differentiation a minimal point $(\alpha,\beta,\gamma,\xi)$ satisfies the linear system of equations with system matrix

$$\begin{bmatrix} F(E,A) & H(A) & 0 & H(E) \\ H^T(A) & G(A,B) & H^T(B) & 0 \\ 0 & H(B) & F(B,C) & H(C) \\ H^T(E) & 0 & H^T(C) & G(C,E) \end{bmatrix}$$

and right hand side

$$-\begin{bmatrix} \text{row}(A) + \text{row}(E) \\ \text{col}(B) + \text{col}(A) \\ \text{row}(C) + \text{row}(B) \\ \text{col}(E) + \text{col}(C) \end{bmatrix},$$

where the notation is as follows:

(1) $F(X,Y)$ / $G(X,Y)$ is a diagonal matrix whose elements are given by the number of nonzero entries in the rows / columns of X and Y,
(2) $H(X)$ is the incidence matrix of X,
(3) row(X) / col(X) is the vector of row / column sums of the matrix

$$\left[\delta(x_{ij} \neq 0) \cdot \log_2 |x_{ij}|\right]_{i,j=1}^{n}.$$

It can be shown that this linear system is symmetric, positive semidefinite and consistent. For its solution a generalized conjugate gradient iteration is used as described in (Ward, 1981).

To reduce the computational it is desirable to construct a suitable preconditioner. Under the assumption of completely dense factors the system matrix is for even k given by a kn-kn block circulant $M_{k,n}$ with first n rows

$$\left[\, 2nI \;\; ee^T \;\; 0 \; \ldots \; 0 \;\; ee^T \,\right],$$

and for odd k by a kn-by-kn block skew circulant $N_{k,n}$ with first n rows

$$\left[\, 2nI \;\; ee^T \;\; 0 \; \dots \; 0 \;\; -ee^T \,\right],$$

where e is the n-vector containing a one in each element.

To be useful for preconditioning the application of the Moore-Penrose generalized inverses $M_{k,n}^{\dagger}$ and $N_{k,n}^{\dagger}$ should be cheap. Indeed, for $x \in \mathbb{R}^k$ the products $M_{k,1}^{\dagger}x$ and $N_{k,1}^{\dagger}x$ can be formed within $O(k)$ operations by using an incomplete Cholesky factorization.

One can show that for general $n \geq 1$, now $x \in \mathbb{R}^{kn}$,

$$M_{k,n}^{\dagger}x = \frac{1}{n^2}\left[\frac{n}{2}I_{kn} + \left(M_{k,1}^{\dagger} - \frac{1}{2}I_k\right) \otimes ee^T\right]x.$$

An analogous result holds for $N_{k,n}^{bot}x$.

Hence, per iteration of the conjugate gradient method $O(kn^2)$ operations are required. If the factors are reasonably dense, then the iterative scheme usually converges within 3 iterations, which was already observed in the context of the generalized eigenvalue problem (Ward, 1981).

For Example (5) the binary logarithms of the optimal scaling parameters are given by

$$\alpha = \left[36.3 \; -30.0 \; 3.14\right] \quad \beta = \left[47.6 \; 8.85 \; 14.7\right]$$
$$\gamma = \left[42.7 \; -20.4 \; 26.5\right] \quad \xi = \left[-38.8 \; 34.7 \; -8.96\right]$$

The eigenvalues of the balanced product are substantially less sensitive as shown below.

Eigenvalue	Condition number
2.88728	2.49
0.39941	4.40
0.07459	3.44

As expected, the periodic QZ algorithm now reveals eigenvalues nearly to machine precision.

5. SHIFT COMPUTATION

At the start of each periodic QZ iteration an initial orthogonal matrix Q_0 is applied to both sides of the product. Given m shifts σ_i the matrix Q_0 is required to satisfy the condition that its first column is parallel to the first column of the shift polynomial

$$P_\sigma = \left(\prod_{i=1}^{k} A_i^{s_i} - \sigma_1\right) \dots \left(\prod_{i=1}^{k} A_i^{s_i} - \sigma_m\right).$$

An usual choice of shifts is to take the two eigenvalues of the southeast two-by-two part of the product,

$$\begin{bmatrix} a_{mm}^{(1)} & a_{mn}^{(1)} \\ a_{nm}^{(1)} & a_{nn}^{(1)} \end{bmatrix} \begin{bmatrix} a_{mm}^{(2)} & a_{mn}^{(2)} \\ 0 & a_{nn}^{(2)} \end{bmatrix}^{s_2} \dots \begin{bmatrix} a_{mm}^{(k)} & a_{mn}^{(k)} \\ 0 & a_{nn}^{(k)} \end{bmatrix}^{s_k},$$

where $m = n-1$ and $a_{jl}^{(i)}$ denotes the (j,l)-th entry of A_i.

Especially for long products, computing the shifts and the shift polynomial desires for great care to avoid unnecessary over-/underflow and disastrous cancellations. From this point of view it is more favorable to construct Q_0 directly from the given data. For example, if the shift polynomial can be rewritten as a product of matrices, then Q_0 can be computed by a partial product QR factorization (De Moor and Van Dooren, 1992). For the double shift strategy described above a suitable product embedding is given by

$$P_\sigma = \left[\, A_1 \;\; I_n \,\right] \cdot \prod_{i=2}^{k} \begin{bmatrix} A_i & 0 \\ 0 & a_{mm}^{(i)} I_n \end{bmatrix}^{s_i}$$
$$\cdot \begin{bmatrix} -I_n & 0 \\ a_{mm}^{(1)} I_n & -a_{nm}^{(1)} I_n \end{bmatrix} \begin{bmatrix} -A_1 & a_{nm}^{(1)} I_n & a_{nn}^{(1)} I_n \\ 0 & a_{mm}^{(1)} I_n & a_{mn}^{(1)} I_n \end{bmatrix}$$
$$\cdot \prod_{i=2}^{k} \begin{bmatrix} A_i & 0 & 0 \\ 0 & a_{mm}^{(i)} I_n & a_{mn}^{(i)} I_n \\ 0 & 0 & a_{nn}^{(i)} I_n \end{bmatrix}^{s_i} \cdot \begin{bmatrix} I_n \\ 0 \\ I_n \end{bmatrix}.$$

By carefully exploiting the underlying structure the recursive computation of Q_0 from this embedding requires approximately $37k$ operations.

6. DEFLATION AND EXPONENTIAL SPLITTINGS

Convergence is certainly the most important aspect of an iterative algorithm.

Consider the product with factors

$$A_1 = \begin{bmatrix} 9.0 & 4.0 & 1.0 & 4.0 & 3.0 & 4.0 \\ 6.0 & 8.0 & 2.0 & 4.0 & 0.0 & 2.0 \\ 0.0 & 7.0 & 4.0 & 4.0 & 6.0 & 6.0 \\ 0.0 & 0.0 & 8.0 & 4.0 & 6.0 & 7.0 \\ 0.0 & 0.0 & 0.0 & 8.0 & 9.0 & 3.0 \\ 0.0 & 0.0 & 0.0 & 0.0 & 5.0 & 0.0 \end{bmatrix}, \qquad (8)$$

$$A_2 = \dots = A_k = \mathrm{diag}(10^{-1}, 10^{-2}, 10^{-3}, 1, 1, 1),$$

and $s_2 = \dots = s_k = 1$. For $k = 5$ the periodic QZ algorithm requires 29 iterations to converge, 62 for $k = 10$, 271 for $k = 40$ and for $k \geq 50$ it does not converge at all. Even worse, the breakdown can not be cured by using standard ad hoc shifts.

The reason is basically that the leading diagonal entries in the triangular factors diverge exponentially, that is, the relation

$$\prod_{i=1}^{k} \left(\frac{a_{j+1,j+1}^{(i)}}{a_{jj}^{(i)}}\right)^{s_i} = O(\alpha^k), \quad 0 \leq \alpha < 1, \qquad (9)$$

is satisfied for $j = 1, 2$. A Givens rotator acting on such a $(j, j+1)$ plane is likely to converge to the 2-by-2 identity matrix when propagated over $A_k, A_{k-1}, \dots, A_2$ back to A_1.

It is important to note that (9) is *not* an exceptional situation. Exponentially splitted products in the sense of (Oliveira and Stewart, 2000) have the pleasant property that even for extremely large k the eigenvalues can be computed to high relative accuracy. Moreover, such products hardly ever fail to satisfy (9). One of the prominent examples is the infinite product where all factors have random entries chosen from a uniform distribution on the interval $(0, 1)$. It can be shown that the sequence of periodic Hessenberg forms related to finite truncations of this product satisfies (9) for all $j = 1, \ldots, n-1$.

In the original algorithm (Bojanczyk *et al.*, 1992), a direct deflation is only performed when a small subdiagonal element in A_1 or a small diagonal element in $A_2, \ldots, A_k$ is encountered. For the purpose that exponentially diverging diagonal entries do not represent a convergence barrier the following additional deflation strategy is proposed.

A QR decomposition is applied to the Hessenberg matrix A_0. If $s_k = 1$, the resulting $n-1$ Givens rotators (c_j, s_j) are successively applied to the columns of A_k,

$$\begin{bmatrix} a_{j,j}^{(k)} & a_{j+1,j}^{(k)} \\ 0 & a_{j+1,j+1}^{(k)} \end{bmatrix} \begin{bmatrix} c_j & s_j \\ -s_j & c_j \end{bmatrix} = \begin{bmatrix} c_j a_{j,j}^{(k)} - s_j a_{j+1,j}^{(k)} & s_j a_{j,j}^{(k)} + c_j a_{j+1,j}^{(k)} \\ -s_j a_{j+1,j+1}^{(k)} & c_j a_{j+1,j+1}^{(k)} \end{bmatrix}.$$

Whenever it happens that $|s_j a_{j+1,j+1}^{(k)}|$ is small compared to

$$\max \left(|c_j a_{j,j}^{(k)} - s_j a_{j+1,j}^{(k)}|, |c_j a_{j+1,j+1}^{(k)}| \right),$$

or, being more generous, compared to $\|A_k\|_F$, then in the following steps (c_j, s_j) can be safely set to $(1, 0)$. Otherwise, the $(j+1, j)$-th element of A_k is annihilated by a Givens rotator acting on rows $(j, j+1)$. (c_j, s_j) is overwritten with the parameters of this rotator.

The process, being similiar when $s_k = 1$, is recursively applied to $A_{k-1}, \ldots, A_2$. At the end, the rotator sequence is applied to the columns of A_1 and each pair $(c_j, s_j) = (1, 0)$ results in a zero element at position $(j+1, j)$ in A_1.

Since the above procedure is as expensive as a single shift periodic QZ iteration it should only occasionally be applied.

For Example (8) with $k = 40$ two applications of the proposed deflation strategy result in zeros at positions $(2, 1)$, $(3, 2)$ and $(7, 6)$ in A_1. Barely 7 periodic QZ iterations are required to reduce the remaining 3-by-3 product to quasi upper triangular form.

7. SOFTWARE IMPLEMENTATION

A FORTRAN 77 software package based on the described algorithms is being developed. The routines conform to the SLICOT implementation and documentation standards (Benner *et al.*, 1999) and are readily available from

`http://www.math.tu-berlin.de/~kressner/`

8. ACKNOWLEDGEMENT

The author sincerly thanks Ralph Byers and Volker Mehrmann for useful discussions and help throughout this work.

9. REFERENCES

Anderson, E., Z. Bai, C. Bischof, J. Demmel, J. Dongarra, J. Du Croz, A. Greenbaum, S. Hammarling, A. McKenney, S. Ostrouchov and D. Sorensen (1994). *LAPACK Users' Guide*. second ed.. SIAM. Philadelphia, PA.

Benner, P., V. Mehrmann and H. Xu (2000). Perturbation analysis for the eigenvalue problem of a formal product of matrices. Berichte aus der Technomathematik 00-01. Fachbereich Mathematik und Informatik, Univ. Bremen.

Benner, P., V. Mehrmann, V. Sima, S. Van Huffel and A. Varga (1999). SLICOT—a subroutine library in systems and control theory. In: *Applied and computational control, signals, and circuits, Vol. 1*. pp. 499–539. Birkhäuser Boston. Boston, MA.

Bojanczyk, A., G. H. Golub and P. Van Dooren (1992). The periodic Schur decomposition; algorithm and applications. In: *Proc. SPIE Conference*. Vol. 1770. pp. 31–42.

Dackland, K. and B. Kågström (1999). Blocked algorithms and software for reduction of a regular matrix pair to generalized Schur form. *ACM Trans. Math. Software* **25**(4), 425–454.

De Moor, B. and P. Van Dooren (1992). Generalizations of the singular value and *QR*-decompositions. *SIAM J. Matrix Anal. Appl.* **13**(4), 993–1014.

Golub, G. H. and C. F. Van Loan (1996). *Matrix Computations*. third ed.. Johns Hopkins University Press. Baltimore.

Hench, J. J. and A. J. Laub (1994). Numerical solution of the discrete-time periodic Riccati equation. *IEEE Trans. Automat. Control* **39**(6), 1197–1210.

Oliveira, S. and D. E. Stewart (2000). Exponential splittings of products of matrices and accurately computing singular values of long products. In: *Proceedings of the International Workshop on Accurate Solution of Eigenvalue*

Problems (University Park, PA, 1998). Vol. 309:1-3. pp. 175–190.

Sreedhar, J. and P. Van Dooren (1993). Pole placement via the periodic Schur decomposition. In: *Proceedings Amer. Contr. Conf.*. pp. 1563–1567.

Van Loan, C. F. (1975). A general matrix eigenvalue algorithm. *SIAM J. Numer. Anal.* **12**(6), 819–834.

Ward, Robert C. (1981). Balancing the generalized eigenvalue problem. *SIAM J. Sci. Statist. Comput.* **2**(2), 141–152.

Copyright © IFAC Periodic Control Systems,
Cernobbio-Como, Italy, 2001

CAD TOOLS FOR CONTROL DESIGN IN LINEAR PERIODIC DISCRETE-TIME SYSTEMS SUBJECT TO INPUT CONSTRAINTS

R. Ciferri *[1] **P. Colaneri** ** **S. Longhi** *[2]

* *Università di Ancona*
Dipartimento di Elettronica e Automatica
Via Brecce Bianche, 60131 Ancona (Italy)
[1] *e-mail: R.Ciferri@ee.unian.it*
[2] *e-mail: S.Longhi@ee.unian.it*
** *Politecnico di Milano*
Dipartimento di Elettronica e Informazione
Piazza Leonardo da Vinci 32, 20133 Milano (Italy)
e-mail: colaneri@elet.polimi.it

Abstract: A polynomial approach is recently investigated for the stabilization of linear periodic discrete-time systems subject to input constraints. This approach requires the use of a wide set of CAD tools to manipulate periodic polynomials and complex algorithms. The paper presents a toolbox for manipulating and solving periodic diophantine equations, necessary for the design of stabilizing controllers of periodic systems subject to input constraints. *Copyright © 2001 IFAC*

Keywords: Control system design, Coprime factorization, Periodic systems

1. INTRODUCTION

Modern control theory provides efficient computational methods to design control laws able to assure stability and performance requirements in the closed-loop systems. Generally, control design does not consider constraints on control and state variables introduced by the control technologies and actuators. For example, actuators provide a limited energy to the plant and this induces significant input constraints. Especially, in multivariable control systems risks of degradation in performance or failures increase as a consequence of unmodelled phenomenons due to constraints and plant complexity.

During last decades this fact led researches to process analysis and controller design using new techniques, implicit or explicit, in constrained situations. Among these techniques we consider an implicit one proposed by (Henrion *et al.*, 1999) in stationary conditions and extended by (Colaneri *et al.*, 2000) to the general case of periodic systems. This method is base on a polynomial approach that considers the case of constrained inputs and combines Youla-Kucera parameterization of all stabilizing controllers of a plant, extended Farkas Lemma and linear programming formulation of the control problem. Linear periodic discrete-time (LPDT) SISO systems are considered, but this is not a limit of the proposed method that can be developed even for MIMO ones. The main aim of this technique is the achievement of the stabilization of LPDT systems subject to input constraints using periodic polynomial equations. This polynomial approach requires the use of a wide set of CAD tools to manipulate periodic polynomial and complex algorithms. In this paper the problems and solutions in developing such instruments will be presented and discussed. In particular, the PERiodic POLynomial toolbox (PERPOL) for MATLAB is here introduced. The paper is organized in the following way. In Section 2 the preliminaries on periodic polynomial representation and manipulation are recalled. In Section 3 the control problem formulation is recalled. In Section 4 the PERPOL toolbox is introduced. The paper ends with a numerical example and concluding remarks.

2. PRELIMINARIES

The aim of this section is to recall some elements of periodic polynomial algebra, that are the main instruments, the basic concepts, the properties and the operations associated to periodic polynomial. Such instruments are developed as an extension to periodic case of time-invariant polynomial algebra, developed in (Kucera, 1979), (Kailath, 1980) and (Blomberg and Ylinen, 1983). Consider the T-periodic polynomial

$$a(d,t) = a_0(t) + a_1(t)d + \ldots + a_n(t)d^n, \quad (1)$$

whose order $deg\ a = n$ is assumed time-invariant and where $a^{(T)}(d,t) := a(d,t-T) = a(d,t)$, for $T \in \mathbb{Z}$ and for all $t \in \mathbb{Z}$, where $\mathbb{Z}$ is the set of integer, i.e. any real coefficient $a_i^{(T)}(t) := a_i(t-T) = a_i(t) \in \mathbb{R}$, for $0 \le i \le n$. The commutation rule between the indefinite operator d and each polynomial coefficient is

$$d^k a_i(t) = a_i^{(k)}(t)d^k = a_i(t-k)d^k, \quad \forall k \in \mathbb{Z}^+, \forall t \in \mathbb{Z}, \quad (2)$$

where $\mathbb{Z}^+$ is the set of not negative integers.

The asymmetric result introduced by the operator d on the periodic polynomial (1) calls attention to polynomial representation. There are two different representations for polynomial (1), the right and left ones, having the forms

$$a(d,t) = \bar{a}_0(t) + \bar{a}_1(t)d + \ldots + \bar{a}_n(t)d^n, \quad (3)$$
$$a(d,t) = \tilde{a}_0(t) + d\,\tilde{a}_1(t) + \ldots + d^n\tilde{a}_n(t), \quad (4)$$

respectively. In the following all the polynomials are in the right form.

Definition 2.1 *Denoting with $S[d]$ the ring of polynomials, define the set*

$$\Re = \{a = (\sum_{i=0}^{n}(a_i\ d^i)) \in S[d], a_i \in \mathbb{R}, i \in \mathbb{Z}^+ : a_i^{(T)}(t) = a_i(t), \forall t \in \mathbb{Z}\}$$

as the ring of T-periodic polynomials. $\Re$ is a non commutative ring.

In the periodic case there are significant differences in multiplication and division operations respect to the time-invariant ones. Consider two T-periodic polynomials $a(d,t)$ and $b(d,t)$:

$$a(d,t) = a_0(t) + a_1(t)d + \cdots + a_n(t)d^n, \quad (5)$$
$$b(d,t) = b_0(t) + b_1(t)d + \cdots + b_m(t)d^m. \quad (6)$$

Definition 2.2 *The T-periodic polynomial $p(d,t)$ product of the multiplication of $a(d,t)$ with $b(d,t)$ has the form*

$$p(d,t) = a(d,t)b(d,t) \quad (7)$$
$$= p_0(t) + p_1(t)d + \ldots + p_l(t)d^l \quad (8)$$

with

$$p_i(t) = \sum_{j=0}^{i}[a_j(t)b_{i-j}(t)], i = 0,1,\ldots,l, \forall t \in \mathbb{Z}, (9)$$

where $l = n + m$. The multiplication is non commutative.

Definition 2.3 *The right (left) division between $a(d,t)$ and $b(d,t)$ belonging to $\Re$, with $n \ge m$ and $b(d,t) \neq 0$, has the polynomial form*

$$a(d,t) = \tilde{q}(d,t)b(d,t) + \tilde{r}(d,t), \forall t \in \mathbb{Z} \quad (10)$$
$$(a(d,t) = b(d,t)\bar{q}(d,t) + \bar{r}(d,t), \forall t \in \mathbb{Z}), \quad (11)$$

where $\tilde{q}(d,t)$ ($\bar{q}(d,t)$) is the quotient of order $n-m$ and $\tilde{r}(d,t)$ ($\bar{r}(d,t)$) is the rest of order $l \le m-1$.

The asymmetric commutation rule of the delay operator d produces in the periodic case a more complex algorithm than the time-invariant one. Quotients and rests of the two kinds of division are not the same, except for polynomials whose orders are zero.

Given two T-periodic polynomials $a(d,t)$ and $b(d,t)$, the common right (left) divisor is a polynomial $\tilde{s}(d,t)$ ($\bar{s}(d,t)$) $\in \Re$ such that $a(d,t) = \tilde{s}(d,t)\hat{a}(d,t)$, $b(d,t) = \tilde{s}(d,t)\hat{b}(d,t)$ $(a(d,t) = \bar{s}(d,t)\hat{a}(d,t), b(d,t) = \bar{s}(d,t)\hat{b}(d,t))$.

Definition 2.4 (El Mrabet and Bourles, 1998) *The T-periodic polynomials $a(d,t)$ and $b(d,t)$ are weakly right (left) coprime in $\Re$ if and only if their common right (left) divisors are unimodular in $\Re$ (i.e. its inverse is in $\Re$).*

Definition 2.5 (El Mrabet and Bourles, 1998) *The T-periodic polynomials $a(d,t)$ and $b(d,t)$ are strongly right (left) coprime in $\Re$ if and only if there are $\tilde{x}(d,t)$ and $\tilde{w}(d,t)$ ($\bar{x}(d,t)$ and $\bar{w}(d,t)$) belonging to $\Re$, where $deg\ \tilde{x} < m$ and $deg\ \tilde{w} < n$ ($deg\ \bar{x} < m$ and $deg\ \bar{w} < n$), such that*

$$\tilde{x}(d,t)a(d,t) + \tilde{w}(d,t)b(d,t) = 1_T \quad (12)$$
$$(a(d,t)\bar{x}(d,t) + b(d,t)\bar{w}(d,t) = 1_T), \quad (13)$$

where 1_T is the T-periodic unit polynomial.

Definition 2.6 *Given the T-periodic polynomials $a(d,t)$ and $b(d,t)$, where $b(d,t) \neq 0$, for all $t \in \mathbb{Z}$, the right (left) factorization of $f(d,t) := a(d,t)b(d,t)^{-1}$ is the reduction to the form*

$$f(d,t) = \tilde{a}(d,t)\tilde{b}(d,t)^{-1} \quad (14)$$
$$(f(d,t) = \bar{b}(d,t)^{-1}\bar{a}(d,t)), \quad (15)$$

where $\tilde{a}(d,t)$ and $\tilde{b}(d,t)$ ($\bar{a}(d,t)$ and $\bar{b}(d,t)$) belonging to $\Re$ are strongly left (right) coprime.

The factorization of a pair of periodic polynomials is obtained by the extended Euclidean algorithm, i.e. by the extraction of the greatest common divisor (*GCD*) and the minimum common multiple (*mcm*) of the pair of polynomials, using a particular polynomial equations system with strongly coprime elements, as the following preposition states.

Proposition 2.1 *Given the T-periodic no zero divisors polynomials $a(d,t)$ and $b(d,t)$, whose orders are time-invariant, there are two pairs of T-periodic polynomials ($\tilde{p}(d,t)$,$\tilde{q}(d,t)$) and ($\tilde{r}(d,t)$, $\tilde{s}(d,t)$) (($\bar{p}(d,t)$,$\bar{q}(d,t)$) and ($\bar{r}(d,t)$,$\bar{s}(d,t)$)), strongly left (right) coprime, such that*

$$\tilde{p}(d,t)a(d,t) + \tilde{q}(d,t)b(d,t) = \tilde{g}(d,t) \quad (16)$$
$$(a(d,t)\bar{p}(d,t) + b(d,t)\bar{q}(d,t) = \bar{g}(d,t)), \quad (17)$$
$$\tilde{r}(d,t)a(d,t) + \tilde{s}(d,t)b(d,t) = 0_T \quad (18)$$
$$(a(d,t)\bar{r}(d,t) + b(d,t)\bar{s}(d,t) = 0_T), \quad (19)$$

where $\tilde{g}(d,t)$ ($\bar{g}(d,t)$) belonging to $\Re$ is the right (left) GCD, 0_T is the null T-periodic polynomial and

$$\tilde{l}(d,t) := \tilde{r}(d,t)a(d,t) = -\tilde{s}(d,t)b(d,t)$$
$$(\bar{l}(d,t) := a(d,t)\bar{r}(d,t) = -b(d,t)\bar{s}(d,t))$$

is the left (right) mcm.

The aim of the extended Euclidean algorithm is to solve the equations (16) and (18) by calculating the right (left) factorization

$$\bar{a}(d,t) = \tilde{s}(d,t) \quad (\tilde{a}(d,t) = \bar{s}(d,t)), \quad (20)$$
$$\bar{b}(d,t) = -\tilde{r}(d,t) \quad (\tilde{b}(d,t) = -\bar{r}(d,t)). \quad (21)$$

Thus the right and left factorizations are obtained by a series of multiplications and divisions and, in general, they are not equal.

Proposition 2.2 *If ($\tilde{a}(d,t)$,$\tilde{b}(d,t)$) and ($\bar{a}(d,t)$, $\bar{b}(d,t)$) are right and left factorizations of $a(d,t)$ and $b(d,t)$ belonging to $\Re$ respectively, then there are two pairs ($\tilde{x}(d,t)$,$\tilde{w}(d,t)$) and ($\bar{x}(d,t)$,$\bar{w}(d,t)$) such that*

$$\tilde{x}(d,t)\tilde{a}(d,t) + \tilde{w}(d,t)\tilde{b}(d,t) = 1_T, \quad (22)$$
$$\bar{a}(d,t)\bar{x}(d,t) + \bar{b}(d,t)\bar{w}(d,t) = 1_T. \quad (23)$$

3. CONTROL PROBLEM

Consider the LPDT SISO system Σ expressed by the polynomial form

$$y(t) = a(d,t)^{-1}b(d,t)u(t) + a(d,t)^{-1}c(d,t)\delta(t)\xi_0, \quad (24)$$

where $y(t), u(t) \in \mathbb{R}$, $t \in \mathbb{Z}$, $\delta(t)$ is the discrete unit impulse and $\xi_0 = \xi(0) \in \mathbb{R}^h$ is the initial state; the T-periodic polynomials have the forms

$$a(d,t) = a_0(t) - a_1(t)d - \ldots - a_n(t)d^n, \quad (25)$$
$$b(d,t) = b_1(t) + b_2(t)d + \ldots + b_m(t)d^m, \quad (26)$$

and the polynomial

$$c(d,t) = c_0(t) + c_1(t)d + \ldots + c_l(t)d^l \quad (27)$$

is time-invariant, i.e. $c_k(t) = c_k$, $k = 0, \ldots, l$ and for all $t \in \mathbb{Z}$.

The problem of finding a dynamic causal T-periodic controller $k(d,t) = \bar{s}(d,t)\bar{r}(d,t)^{-1}$ such that the closed-loop system is stable and satisfies input constraints $-u^- \leq u(t) \leq u^+$, with $t \neq 0$ and $u^-, u^+ \in \mathbb{R}$, and for all $\xi_0 \in P_N = \{\xi_0 \in \mathbb{R}^h : N\,\xi_0 \leq n$, with $N \in \mathbb{R}^{s \times h}$ and $n \in \mathbb{R}^s\}$ has been solved in (Colaneri *et al.*, 2000).

A stabilizing periodic controller is given by Youla-Kucera parameterization (Colaneri *et al.*, 2000), that in the right form is

$$\bar{r}(d,t) = \bar{x}(d,t) + \tilde{b}(d,t)q(d,t), \quad (28)$$
$$\bar{s}(d,t) = \bar{w}(d,t) - \tilde{a}(d,t)q(d,t). \quad (29)$$

The control input has the form

$$u(t) = -k(d,t)y(t) = \quad (30)$$
$$= [\tilde{a}(d,t)q(d,t) - \bar{w}(d,t)]\bar{c}(d,t)\delta(t)\xi_0, \quad (31)$$

where $q(d,t) = q_0(t) + q_1(t)d + \ldots + q_p(t)d^p$ is the T-periodic polynomial to calculate for satisfying the input constraints. The coefficients of polynomial $q(d,t)$ can be computed making use of an extension of the Farkas Lemma and of linear programming tools (Colaneri *et al.*, 2000).

4. INTRODUCTION TO PERPOL

For finding a solution to the above recalled control problem it is necessary to compute right or left periodic factorizations and to solve periodic diophantine equations. To simplify these computations a set of MATLAB functions has beeen developed and organized in the PERPOL toolbox. It is the extension of the Polynomial Toolbox (Henrion *et al.*, 1997), developed for time-invariant polynomials. For the sake of briefly, only the most significant functions are here introduced.

4.1 *Periodic polynomial representation*

The function $a = ppoly(T, dtype)$ computes the matrix representation of the T-periodic polynomial (1) according to the following form

$$\begin{bmatrix} a_0(0) & a_1(0) & \cdots & a_n(0) & n \\ a_0(1) & a_1(1) & \cdots & a_n(1) & 0 \\ \cdots & \cdots & \cdots & \cdots & \cdots \\ a_0(T-1) & a_1(T-1) & \cdots & a_n(T-1) & 0 \\ T & dtype & \cdots & 0 & NaN \end{bmatrix} \quad (32)$$

where *dtype* specifies the type of representation right or left one.

4.2 *Operations for periodic polynomials*

The function $s = ppadd(m1, m2, \ldots, m10)$ ($d = ppsub(m1, m2, \ldots, m10)$) adds (subtracts) a maximum of ten periodic polynomials.

The function $p = ppmul(a, b)$ computes multiplication between two periodic polynomials using rule (9).

The function $[q, r] = ppdiv(a, b, divtype, tol)$ computes the division between two periodic polynomials

$$a(d,t) = a_0(t) + a_1(t)d + \cdots + a_n(t)d^n, \quad (33)$$
$$b(d,t) = b_0(t) + b_1(t)d + \cdots + b_m(t)d^m, \quad (34)$$

by different rules for right and left case. The variable *divtype* specifies the rule. The variable *tol* specifies the considered tolerance in the numerical computations. The right division algorithm, for no zero divisor $b(d,t)$ and with time-invariant order, is composed by the following steps:
STEP 0 : for $t = 0, \ldots, T-1$,
if $n \geq m$

$$q_{n-m}(t)d^{n-m} = a_n(t)d^n[b_m(t)d^m]^{-1}$$
$$= a_n(t)b_m(t-n+m)^{-1}d^{n-m},$$
$$r_0(d,t) = a(d,t) - q_{n-m}(t)d^{n-m}b(d,t),$$

go to step 1;
else

$$\tilde{q}(d,t) = 0, \quad \tilde{r}(d,t) = a(d,t),$$

algorithm stops;
STEP i : for $t = 0, \ldots, T-1$,
if $n - i \geq m$ and $r_{i-1}(d,t) \neq 0$

$$\bar{q}_{n-m-i}(t)d^{n-m-i} = r_{i-1,n-i}(t)d^{n-i}[b_m(t)d^m]^{-1}$$
$$= r_{i-1,n-i}(t)b_m(t-n+m+i)^{-1}d^{n-m-i},$$
$$r_i(d,t) = r_{i-1}(d,t) - q_{n-m-i}(t)d^{n-m-i}b(d,t),$$

go to step i+1;
else

$$\tilde{q}(d,t) = q_{n-m}(t)d^{n-m} + q_{n-m-1}(t)d^{n-m-1} + \ldots$$
$$+ q_{n-m-i+1}(t)d^{n-m-i+1},$$
$$\tilde{r}(d,t) = r_{i-1}(d,t),$$

algorithm stops.
A symmetric algorithm is proposed for the left case.

4.3 *Algebraic periodic polynomial properties*

The function $z = pp0div(a)$ recognizes if the polynomial a is a zero divisor.

The function $[afac, bfac] = ppfac(a, b, factype, tol)$ computes the factorization between the polynomials a and b. The variable *factype* specifies the type of factorization. Euclidean algorithms, extended to the periodic case are used. The right extended Euclidean algorithm is composed by the following steps:
STEP 0 : for $t = 0, \ldots, T-1$,
if $b(d,t)$ is not a zero divisor and $deg\ b \geq m$, or if $b(d,t) = 0$

$$r_{-1}(d,t) = a(d,t), \quad r_0(d,t) = b(d,t),$$

go to step 1;
else algorithm failure;
STEP i $(i > 0)$: for $t = 0, \ldots, T-1$,
if $r_{i-1}(d,t)$ is not a zero divisor and its order is time-invariant

$$r_{i-2}(d,t) = q_i(d,t)r_{i-1}(d,t) + r_i(d,t),$$

that in matrix form is

$$\begin{bmatrix} 0 & 1 \\ 1 & -q_i(d,t) \end{bmatrix} \begin{bmatrix} r_{i-2}(d,t) \\ r_{i-1}(d,t) \end{bmatrix} = \begin{bmatrix} r_{i-1}(d,t) \\ r_i(d,t) \end{bmatrix},$$

go to step i+1;
else if $r_{i-1}(d,t) = 0$, go to step $n+1$ (final step of the algorithm, after $n = i-1$ steps of kind i);
else algorithm failure;
STEP n+1 : for $t = 0, \ldots, T-1$, execution of backward substitutions of matrix expressions graduates the following form

$$\prod_{i=1}^{n} \begin{bmatrix} 0 & 1 \\ 1 & -q_i(d,t) \end{bmatrix} \begin{bmatrix} a(d,t) \\ b(d,t) \end{bmatrix} = \begin{bmatrix} r_n(d,t) \\ 0 \end{bmatrix},$$

where $a(d,t) = r_{-1}(d,t)$ and $b(d,t) = r_0(d,t)$. The above expression can be represented with the following one

$$\begin{bmatrix} \tilde{p}(d,t) & \tilde{q}(d,t) \\ \tilde{r}(d,t) & \tilde{s}(d,t) \end{bmatrix} \begin{bmatrix} a(d,t) \\ b(d,t) \end{bmatrix} = \begin{bmatrix} \tilde{g}(d,t) \\ 0 \end{bmatrix},$$

which is equivalent to (16) and (18). This algorithm gives also the right GCD $\tilde{g}(d,t) = r_n(d,t)$ and the left mcm $\tilde{l}(d,t) = \tilde{r}(d,t)a(d,t) = -\tilde{s}(d,t)b(d,t)$. A symmetric algorithm is proposed for the left rule of the left case.

The function $[b, uni] = ppinv(a)$ verifies unimodularity of a periodic polynomial a and calculates its inverse b, if it exists, by the equation $a(d,t)b(d,t) = b(d,t)a(d,t) = 1_T$. The variable *uni* specifies if a is unimodular or not.

The function $[g, weakc] = ppweakc(a, b, coptype, tol)$ verifies weak coprimeness between a pair of periodic polynomials a and b. The variable *coptype* specifies the left or right coprimeness. The variable *tol* specifies the tolerance in the numerical computations. The variable g represents the GCD and *weakc* specifies if a and b are coprime or not.

The function $[x, w, strongc] = ppstrongc(a, b, coptype, tol)$ verifies strong coprimeness between a pair of periodic polynomials a and b and calculates a solution (x and w) for the left Bezout equation (23) or the right one (22). The variable *coptype* specifies the left or right coprimeness. The variable *tol* specifies the tolerance in the numerical computations. The variable *strongc* specifies if a and b are coprime or not.

4.4 *Periodic diophantine equations*

The function $[x, w] = ppunidio(a, b, c, eqtype, v, tol)$ let solve a linear unilateral periodic diophantine equation of the right (left) form

$$a(d,t)\bar{x}(d,t) + b(d,t)\bar{w}(d,t) = c(d,t) \quad (35)$$

$$(\tilde{x}(d,t)a(d,t) + \tilde{w}(d,t)b(d,t) = c(d,t)). \quad (36)$$

The variable *eqtype* specifies the right or left equation. The variable v represents an arbitrary periodic polynomial $v(d,t)$ useful to calculate a more general solution. The variable *tol* specifies the tolerance in the numerical computations.

The function $[x, w] = ppbezout(a, b, eqtype, tol)$ let solve a Bezout equation of the right (left) form

$$a(d,t)\bar{x}(d,t) + b(d,t)\bar{w}(d,t) = 1_T \quad (37)$$

$$(\tilde{x}(d,t)a(d,t) + \tilde{w}(d,t)b(d,t) = 1_T). \quad (38)$$

The variables *eqtype* and *tol* have the usual definitions.

4.5 *Constrained control problem solution*

With reference to the above recalled control problem, specific functions are developed for representing controlled system by polynomial equations (*pptf.m*, *ppss.m*, *ppss2tf.m*), constraints on initial state (*polyhedra.m*) and input control (*uconstr.m*). Other functions are developed for formulating and solving linear programming for periodic polynomials (*pplp.m*) and for realizing Youla-Kucera parameterization (*ppyoulak.m*). All these functions are organized in a main routine (*ppmain.m*) that let easily approach to this control design technique, by a sequence of command lines and menus.

A large set of not mentioned functions, adding to the previous ones, complete PERPOL (Ciferri, 2000). This toolbox let an easy manipulation of periodic polynomials and in some cases of periodic polynomial matrices too.

5. NUMERICAL EXAMPLE

Consider a process composed by a wedge, with a sliding-weights balancing mechanism, made of two identical weights sliding on a railway by a c.c. engine, that has to be able to keep its unstable equilibrium on its vertex (Hsu, 1992).

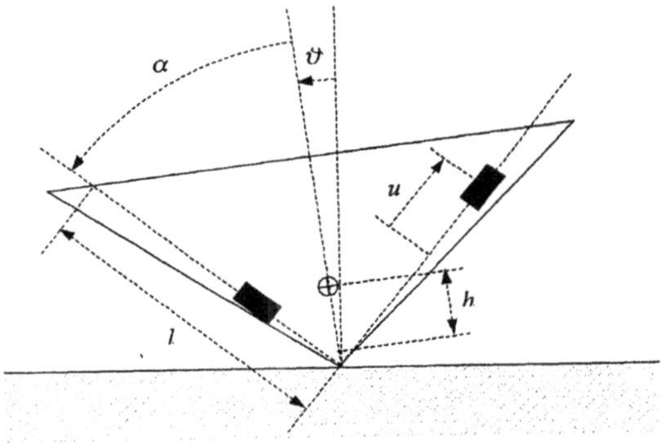

Fig. 1. Sliding-weights balancing mechanism.

The technological parameters of this system are: mass of the wedge $m_1 = 5.1\ kg$, mass of the translational weight $m_2 = 1\ kg$, inertia of the wedge about the pivot point $J_1 = 1\ kg\ m^2$, inertia of the balancing mechanism $J_2 = 0.024\ kg\ m^2$, railway length $l = 0.4\ m$, wedge maximum angular movement $\alpha = 45^0$, height of the center-of-mass of the wedge $h = 0.29\ m$, pulley radius $K_s = 0.031425\ m$, gravitational constant $g = 9.8\ m\ s^{-2}$. The input of the system is the weight displacement $u(t)$ from the wedge center line, while the output is the wedge angle $\theta(t)$, whose varying sets are $-0.2 \leq u(t) \leq 0.2\ m$ and $-\pi/4 \leq \theta(t) \leq \pi/4\ rad$ respectively. The translational weight mechanism is completed by an encoder, for imposing the control action $u(t)$ and whose dynamic is not significant, and an inclinometer, for the measure of $\theta(t)$. By a linearization about the unstable equilibrium point, $(u_0, \theta_0) = (0, 0)$, it is possible to obtain the following polynomial representation of the whole process (wedge + encoder + inclinometer):

$$v_i(t) = W(d)u(t) + \psi(d)\delta(t)\xi_0,$$

where $v_i(t)$ is the voltage output of the inclinometer proportional to the angle position $\theta(t)$, $\xi_0 = \xi(0) = [\theta(0)\ \dot{\theta}(0)]'$ is the initial state, $W(d) = a(d)^{-1}b(d)$ and $\psi(d) = a(d)^{-1}c(d)$ with

$$a(d) = 3.204 - 10.543d + 12.296d^2 - 5.978d^3 + d^4\,,$$

$$b(d) = 0.251d^2 + 0.197d^3 - 0.007d^4\,,$$

$$c(t) = [\,0.312d - 0.066d^2 - 0.266d^3 \;\; 0.021d^2 + 0.017d^3\,],$$

at a sample frequency of 15 Hz.

The admissible set for the initial state vector is given by the polyhedra

$$P_N = \{\xi_0 \in \mathbb{R}^h : N\xi_0 \leq \nu, N \in \mathbb{R}^{s \times h}, \nu \in \mathbb{R}^s\}$$

where

$$N = \begin{bmatrix} 1 & 0 & -1 & 0 \\ 0 & 1 & 0 & -1 \end{bmatrix}', \quad \nu = [\,\pi/36\ \ 10^{-9}\ \ -\pi/36\ \ -10^{-9}\,]'.$$

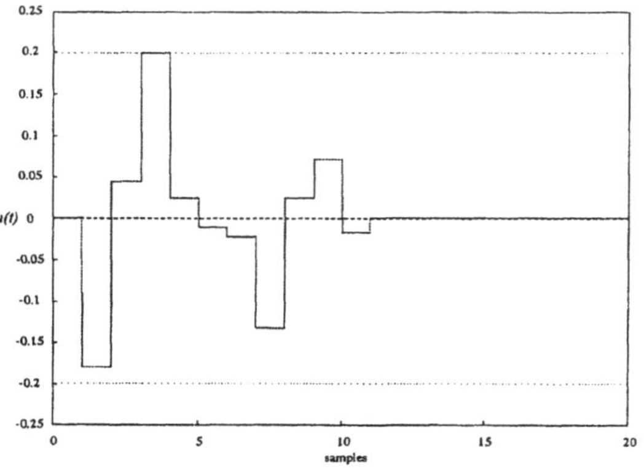

Fig. 2. Time-invariant controller: control input behaviour.

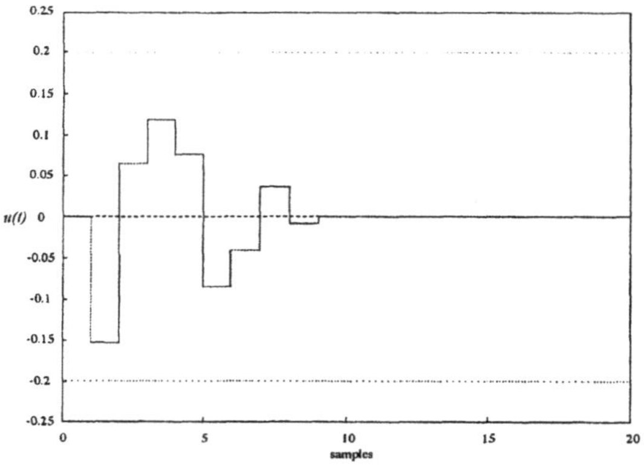

Fig. 3. Periodic controller ($T = 3$): control input behaviour.

Left and right factorizations for W(d) (given by *ppfac.m*) are identical

$$\tilde{a}(d) = \bar{a}(d) = 3.204 - 10.543d + 12.296d^2 - 5.978d^3 + d^4 ,$$
$$\tilde{b}(d) = \bar{b}(d) = 0.251d^2 + 0.197d^3 - 0.007d^4 ,$$
$$\tilde{c}(t) = [\, 0.312d - 0.066d^2 - 0.266d^3 \;\; 0.021d^2 + 0.017d^3 \,],$$

The solution of Bezout equations (given by *ppbezout.m*) are

$$\tilde{x}(d) = \bar{x}(d) = 1.312 + 1.027d + 0.620d^2 - 0.024d^3 ,$$
$$\tilde{w}(d) = \bar{w}(d) = 19.975 - 32.301d + 17.875d^2 - 3.185d^3 .$$

It is possible to design a time-invariant or a periodic controller for the process. The time-invariant controller is given by the function *pplp.m*, that computes the polynomial

$$q(d) = 4.174 + 3.732d + 2.391d^2 + 0.931d^3 .$$

For the bound initial state $\xi_0 = [-0.087\ 0]'$, $u(t)$ assumes the trend given in Fig. (2). The periodic controller is computed for different cases of increasing period T. The periodic polynomial $q(d, t)$, for $T = 2$, has the form

$$q(d,t) = 2.512d + 1.633d^2 , \; t = 2k,$$
$$= 3.942 + 0.323d + 1.261d^2 , \; t = 2k + 1,$$

while, for $T = 3$,

$$q(d,t) = 0 , \; t = 3k,$$
$$= 4.515 + 0.278d , \; t = 3k + 1,$$
$$= 0.388 + 4.212d , \; t = 3k + 2.$$

For $T \geq 3$ the degree of $q(d, t)$ does not decrease. For the bound initial state $\xi_0 = [-0.087\ 0]'$, $u(t)$ has the behaviour given in Fig. (3).

Note that all the controllers satisfy the input constraints; the number of samples necessary to set a dead-beat behaviour decreases by increasing period T. In particular, for $T = 3$, it is possible to satisfy more restrictive constraints. The order of the polynomial $q(d, t)$ and the order of the controller decrease by increasing the period T. All the controllers are stabilizing for the process.

6. CONCLUDING REMARKS

A toolbox for the design of periodic controllers subject to input constraints has been developed using a polynomial approach. An interface is also implemented for simplifying the use of the tool. The developed tool can be also used for the design of multirate control system which is a special case of the periodic systems.

The extension to the MIMO periodic systems with constraints on the state variables is under investigation.

7. REFERENCES

Henrion, D., S. Tarbouriech and V. Kucera (1999). Control of linear systems subject to input constraints: a polynomial approach. Part I - SISO plans. *Proc. of the 38th IEEE CDC*, Phoenix, USA, 2774–2779.

Colaneri, P., V. Kucera and S. Longhi (2000). Polynomial approach to the control of SISO periodic systems subject to input constraints *IFAC Conference on Control Systems Design*, Bratislava, Slovak Republic.

Kucera, V. (1979). *Discrete Linear Control: The Polynomial Equation Approach.* John Wiley and Sons. Chichester, England.

Kailath, T. (1980). *Linear Systems.* Prentice Hall, Inc.. Englewood Cliffs, New Jersey.

Blomberg, H. and R. Ylinen (1983). *Algebraic Theory for Multivariable Linear Systems.* Academic Press, Inc.. London, England.

El Mrabet, Y. and H. Bourles (1998). Periodic-polynomial interpretation for structural properties of linear periodic discrete-time systems *Systems & Control Letters*, Vol. 33. pp 241–251.

Hsu, P. (1992). Dynamics and Control Design Project Offers Taste of Real World *IEEE Control System Magazine*, Vol. 12. No. 3. pp 31–37.

Henrion, D., F. Kraffer, H. Kwakernaak, S. Pejchova, M. Sebek, R. C. W. Strijbos (1997). Polynomial Toolbox for MATLAB (c), Version 1.5 *The Mathworks Inc.*, http://www.math.utwente.nl/polbox.

Ciferri, R. (2000). Metologie CAD per la sintesi di controllori per sistemi lineari periodici a tempo discreto con vincoli sull'ingresso *Università degli Studi di Ancona, Facoltà di Ingegneria*, laurea.

Copyright © IFAC Periodic Control Systems,
Cernobbio-Como, Italy, 2001

PERIODIC OPTIMAL CONTROL OF MULTIRATE SAMPLED DATA SYSTEMS

J. Tornero, P. Albertos and J. Salt

Department of Systems Engineering and Control
Universidad Politécnica de Valencia, Valencia, Spain.
P.O.Box 22012 E-46071. Tel. 34-96-3879570 Fax 34-96-3879579
e-mail: {jtornero,pedro,julian}@isa.upv.es

Abstract: In this paper a detailed model of a MIMO multirate sampled data system is used for control design purposes. The process model is split into a time-invariant-single-rate part, involving the basic process dynamics, and two periodic blocks reflecting the external sample/hold strategy. The resulting periodic controller and state estimator design can be done using any control design technique suitable for periodic systems. In this paper, an LQG approach is followed. The optimal solution is found by solving two sequences of recurrent Riccati equations. The approach is validated by means of a simple but general and typical example. *Copyright © 2001 IFAC*

Keywords: Sampled data systems, linear quadratic regulator, optimal filtering, periodic systems, optimal multirate control.

1. INTRODUCTION

This contribution deals with the multirate control of linear time-invariant MIMO systems. Although the sampling schema can be totally random and/or asynchronous, it is generally accepted that there is a sampling/updating periodicity for each signal, and to simplify the notation, a frame period is also assumed.

This problem has been extensively treated in the last two decades and it is possible to find many contributions dealing with modeling and analysis, as well as controller design, for this kind of processes. One of the approaches to treat the modeling step is to assume an enlarged MIMO system, (Araki and Yamamoto, 1986), (Berg, et al., 1988), (Godbout, et al., 1990), (Albertos, et al., 1996). In these approaches, a discrete time-invariant state equation, for the frame period, T_0, with extended state, input and output vectors, is used. Similar to this one is the Vector Switch Decomposition (Kranc, 1957), later on implemented by (Thompson, 1986, 1988). Another viewpoint is taken in (Meyer and Burrus, 1975) and (Albertos, 1990). The state is not extended and block vectors are defined for the input and output signals. Finally, another popular technique is denoted as the lifting technique, developed and applied in (Bamieh, et al., 1991), (Voulgaris and Bamieh, 1992). In (Meyer, 1990) the equivalence among different techniques in multirate digital control is proved.

The approach in this work follows the original idea proposed by (Kalman and Bertram, 1959), based on a continuous time process model, time-events, and their interaction. This was the viewpoint initially used in (Tornero and Edmunds, 1983), (Tornero, 1985), later on also followed by (Longhi, 1994).

The application of the multirate schema to LQ control has been considered in many of the above referred contributions. Another notes of interest are (Glasson, 1983), usually cited as the first contribution in this field, and (Meyer, 1992), (Apostolakis and Jordan, 1991), and (AlRahmani and Franklin, 1992). A simplification for the cost function calculation in the LQ problem with multirate schemes, has been proposed (Tornero et al, 1999).

The paper is organized as follows. The multirate sampled Data (**SD**) model of the plant is described in the next section. The periodic optimal regulator design and the solution for the specific case we have considered are presented in section 3. The periodic state estimator is presented in section 4. An illustrative example is outlined in section 5. The paper ends with some conclusions.

2. MULTIRATE SD PLANT MODEL

Consider the linear time-invariant (**LTI**) continuous-time (**CT**) plant described by

$$\begin{aligned} \dot{x}(t) &= A_c \cdot x(t) + B_c \cdot u(t) \\ y(t) &= C_c \cdot x(t) \end{aligned} \tag{1}$$

where $x \in R^n$ is the state, $u \in R^m$ is the control input vector and $y \in R^p$ is the output vector. The single-rate SD system, at the base period T, is expressed as follows

$$\begin{aligned} x((k+1)T) &= A(T) \cdot x(kT) + B(T) \cdot u(kT) \\ y(kT) &= C \cdot x(kT) \end{aligned} \tag{2}$$

where $k \in Z^+$, $C = C_c$ and

$A(T) = \exp(A_c \cdot T)$, $B(T) = \int_0^T \exp(A_c \cdot \tau) \cdot B_c \cdot d\tau$.

In a compact form, let us denote for the SD plant

$$\Phi(T) = \begin{bmatrix} A(T) & B(T) \\ 0 & I \end{bmatrix} = \exp\begin{bmatrix} A_c \cdot T & B_c \\ 0 & 0 \end{bmatrix} \tag{3}$$

Now, let us introduce the multirate scenario where the plant inputs are updated according to a periodic hold policy and the output is accessed under a periodic sampling policy. The state is considered at any sampling instant of time, but the i-th component $u_i(t), i = 1,2,\ldots m$ of the input vector $u(t)$ can be only updated every $T.N_i$ time instants. Also, a sampling period $T.M_j$, is assumed for the output variable y_i, and $N_i, M_i \in Z^+$.

Let us denote by T, the basic period, the maximum common divisor of the updating/sampling periods. The arguments in (3) will be deleted in the following. Also assume that there exists a least common multiple (l.c.m.) T_0, the frame period, such as $T_0 = NT$, with N = l.c.m.(N_i, M_j). The whole system is T_0 –periodic.

Pursuing the ideas in (Tornero, 1985) and (Longhi, 1994), the multirate SD system can be expressed by two discrete time systems: one related to the plant and another one related to the input-hold mechanism,

$$\begin{aligned} x_{k+1} &= Ax_k + B(I - \Delta(k))v_k + B\Delta(k)u_k \\ v_{k+1} &= (I - \Delta(k))v_k + \Delta(k)u_k \end{aligned} \tag{4}$$

where $v \in R^m$ is a new state vector associated to the input-hold mechanism, and $\Delta(\bullet)$ is an N-periodic mxm matrix given by

$$\Delta(k) \equiv diag\{\delta_i(k), i = 1,2,\ldots,m\} \tag{5}$$

$$\delta_i(k) \equiv \begin{cases} 1, & k = j \cdot N_i, \\ 0 & k \neq j \cdot N_i, \end{cases} \quad j \in Z^+$$

Similarly, the measured output can be expressed by

$$m_k = \overline{\Delta}(k) \cdot y_k \tag{6}$$

where

$$\begin{aligned} \overline{\Delta}(k) &\equiv diag\{\overline{\delta}_i(k), i = 1,2,\ldots,p\} \\ \overline{\delta}_i(k) &\equiv \begin{cases} 1, & k = j \cdot M_i, \\ 0 & k \neq j \cdot M_i, \end{cases} \quad j \in Z^+ \end{aligned} \tag{7}$$

This multirate model (4) can be expressed in a more compact form, by

$$\overline{x}_{k+1} = A_M(k) \cdot \overline{x}_k + B_M(k) \cdot u_k \tag{8}$$

where $\overline{x}_k^{\,T} \equiv [x_k^{\,T} \;\; v_k^{\,T}] \in R^{n+m}$ is the enlarged state vector and the N-periodic matrices $A_M(k)$, $B_M(k)$ are given by

$$A_M(k) = \begin{bmatrix} A & B \cdot (I - \Delta(k)) \\ 0 & I - \Delta(k) \end{bmatrix} \quad B(k) = \begin{bmatrix} B \cdot \Delta(k) \\ \Delta(k) \end{bmatrix} \tag{9}$$

Eq. (8) can be written as follows

$$[\overline{x}]_k = [A_M(k) \;\; B_M(k)] \cdot \begin{bmatrix} \overline{x} \\ u \end{bmatrix}_k =$$

$$\Phi \cdot H \begin{bmatrix} \overline{x} \\ u \end{bmatrix}_k = \begin{bmatrix} A & B \\ 0 & I \end{bmatrix} \cdot \begin{bmatrix} I & 0 & 0 \\ 0 & I - \Delta(k) & \Delta(k) \end{bmatrix} \cdot \begin{bmatrix} x \\ v \\ u \end{bmatrix}_k \tag{10}$$

where Φ is given by (3), providing an easy way to compute it.

3. PERIODIC CONTROL

To design a control structure for this multirate system, using the separation principle, the control problem can be split into the controller design and the state estimator design. In both cases, the controller and the estimator will present a periodic feature.

To control the DT plant model, a discrete time quadratic optimization index, J, is defined by:

$$J=\frac{1}{2}\sum_{k=0}^{\infty}\left\{x_k{}^T Q x_k+u_k{}^T R u_k\right. \qquad (11)$$

This index can be also written in a more compact form as

$$J=\frac{1}{2}\sum_{k=0}^{\infty}\left[x_k{}^T \quad u_k{}^T\right]\cdot\begin{bmatrix}Q & 0\\ 0 & R\end{bmatrix}\cdot\begin{bmatrix}x\\ u\end{bmatrix}_k=$$
$$\frac{1}{2}\sum_{k=0}^{\infty}\left[x_k{}^T \quad u_k{}^T\right]\cdot\overline{Q}\cdot\begin{bmatrix}x\\ u\end{bmatrix}_k \qquad (12)$$

The state feedback optimal control

$$u_k=K.x_k \qquad (13)$$

is given by the solution of the well known discrete time algebraic Riccati equation.

A multirate DT quadratic optimization index J_M, equivalent to J, can be derived. Together with the multirate process model (8), it will determine an optimal control design problem. Similar to (12), this index should be written as

$$J_M=\frac{1}{2}\sum_{0}^{\infty}\left[x_k{}^T \quad v_k{}^T \quad u_k{}^T\right]\cdot\overline{Q}_M(k)\cdot\begin{bmatrix}x\\ v\\ u\end{bmatrix}_k \qquad (14)$$

Taking into account that the system input is either updated or held, according to the multirate updating schema, the original index can be rewritten as:

$$J=\frac{1}{2}\sum_{k=0}^{\infty}\{x_k{}^T\cdot Q\cdot x_k+u_k{}^T\Delta^T(k)\Delta(k)\cdot u_k+v_k{}^T[I-\Delta(k \qquad (15)$$

This gives the following matrix equivalence

$$\overline{Q}_M(k)=\begin{bmatrix}I & 0 & 0\\ 0 & I-\Delta(k) & \Delta(k)\end{bmatrix}^T\begin{bmatrix}Q & 0\\ 0 & R\end{bmatrix}\begin{bmatrix}I & 0 & 0\\ 0 & I-\Delta(k) & \Delta(k)\end{bmatrix}$$
$$\overline{Q}_M(k)=\begin{bmatrix}Q_M(k) & M_M(k)\\ M_M(k) & R_M(k)\end{bmatrix} \qquad (16)$$

Summarizing, the multirate SD LQR problem can be expressed as

$$J_M=\frac{1}{2}\sum_{k=0}^{\infty}\left\{\overline{x}_k^T Q_M(k)\overline{x}_k+2\overline{x}_k^T M_M(k)u_k+u_k^T R_M(k)u_k\right. \qquad (17)$$

subject to (8)

$$\overline{x}_{k+1}=A_M(k)\cdot\overline{x}_k+B_M(k)\cdot u_k$$

The index in (17) is a mixed one, involving the cross product of state and inputs. Nevertheless, as can be easily deduced from (16), if R is diagonal, the matrix $M_M(k)=0;\forall k$, leading to a basic form. If this is not the case, to reduce it to the basic form, a change in the input variable, taking

$$\overline{u}_k=u_k+R_M(k)^{-1}M_M(k)\cdot\overline{x}_k \qquad (18)$$

leads to the optimal control problem:

$$J_M==\frac{1}{2}\sum_{k=0}^{\infty}\left\{\overline{x}_k{}^T\cdot\overline{Q}_M(k)\cdot\overline{x}_k+\overline{u}_k{}^T\cdot R_M(k)\cdot\overline{u}_k\right. \qquad (19)$$

with the plant model

$$\overline{x}_{k+1}=\overline{A}_M(k)\cdot\overline{x}_k+B_M(k)\cdot\overline{u}_k \qquad (20a)$$
$$\overline{A}_M(k)=A_M(k)-B_M(k)R_M(k)^{-1}M_M(k)$$
$$\overline{Q}_M(k)=Q_M(k)-M_M(k)R_M(k)^{-1}M_M(k) \qquad (20b)$$

The general solution is the periodic control

$$\overline{u}_k=K(k)\cdot\overline{x}_k;$$
$$u_k=K(k)\overline{x}_k-R_M(k)^{-1}M_M(k)\cdot\overline{x}_k \qquad (21)$$

obtained as a result of the periodic Riccati equation

$$P(k)=\overline{A}_M(k)^T P(k+1)\overline{A}_M(k)+\overline{Q}_M(k)-$$
$$-\overline{A}_M(k)^T P(k+1)B_M(k)K(k) \qquad (22)$$

$$K(k)=\left[R_M(k)+B_M(k)^T P(k+1)B(k)\right]^{-1}\cdot$$
$$B_M(k)^T P(k+1)\overline{A}_M(k) \qquad (23)$$

4. PERIODIC STATE ESTIMATOR

Following a dual approach, a periodic state estimator can be derived. Assume an autonomous plant given by

$$x_{k+1}=A\cdot x_k+w_k$$
$$y_k=C(k)\cdot x_k+v_k \qquad (24)$$

with a gaussian initial state distribution with $\mathrm{cov}(x_0)=R_0$, where plant and measurement noises are white noises with

$$E\left(w_k{}^T w_k\right)=R_1$$
$$E\left(w_k{}^T v_k\right)=0 \qquad (25)$$
$$E\left(v_k{}^T v_k\right)=R_2$$

A constant matrix, C characterizes the output equation, followed by a periodic sampling device, as defined in (6-7), that is

$$C(k) = \overline{\Delta}(k) \cdot C \tag{26}$$

The periodic Kalman filter is giving by

$$\hat{x}_{k+1} = A \cdot \hat{x}_k + L(k)\left[y_k - C(k)\hat{x}_k\right] \tag{27}$$

where

$$\begin{aligned} L(k) &= AP(k)C(k)^T\left[R_2 + C(k)P(k)C(k)^T\right]^{-1} \\ P(k+1) &= AP(k)A^T + R_1 - L(k)C(k)P(k)A^T \end{aligned} \tag{28}$$

The sampling matrix, $\overline{\Delta}(k)$, determines the estimator N-periodicity.

5. ILLUSTRATIVE EXAMPLE

The application of this periodic control design to a multivariable system implies the generation of a periodic single rate system model at the basic period where either some inputs are not considered at certain instants of time or some outputs are not considered to estimate the state.

Consider the weakly coupled multivariable plant proposed in (Araki and Yamamoto, 1986) and (Godbout, *et al.*, 1990). The CT state-space model is given by:

$$A_c = \begin{bmatrix} -2.5 & 0 & 0 \\ 0 & -2 & 0 \\ 0 & 0 & 1 \end{bmatrix}; B_c = \begin{bmatrix} 2.5 & 0 \\ 10 & -1.2 \\ 5/6 & 1 \end{bmatrix}; C = \begin{bmatrix} -4 & 1 & 0 \\ -1/3 & 0 & 1 \end{bmatrix}$$

The inputs, u_1 and u_2, are updated every *0.1*sec and *0.15* sec, respectively, whereas the output sampling periods are 0.15 sec and 0.1 sec for y_1 and y_2, respectively. Thus, the basic sampling period is taken as *T=0.05*sec, being $T_0=0.3$ sec, the frame period, and *N*=6. The state is available at any *T*, being $N_1=3$, $N_2=2$, $M_1=2$ and $M_2=3$. In the cost index (11), the same weight has been assigned to both state and control variables.

In this case, although there are six basic periods per frame, T_0, there are only four different updating conditions, $\{0, 0.1, 0.15, 0.2\}$, repeated each period. The corresponding feedback matrices are computed by solving a set of six Riccati-like equations. In a compact form, the control law is:

$$u_k = K_i \overline{x}_k; i = k \,\mathrm{mod}(N)$$

where K_i is the state feedback matrix. In this case, the results are:

1) t=0

$$K_0 = \begin{bmatrix} 0.1343 & 0.6147 & 0.1130 & 0 & 0 \\ 0.0272 & -0.1029 & 0.4091 & 0 & 0 \end{bmatrix}$$

Both inputs are updated.

2) t=0.05 *sec* (and also t=0.25 *sec*)

$$K_1 = \begin{bmatrix} 0 & 0 & 0 & 0 & 0 \\ 0 & 0 & 0 & 0 & 0 \end{bmatrix} = K_5$$

3) t=0.1 *sec*

$$K_2 = \begin{bmatrix} 0.1325 & 0.6201 & 0.0875 & 0 & -0.0657 \\ 0 & 0 & 0 & 0 & 0 \end{bmatrix};$$

u_1 is updated, but u_2 is taken as an unchanged state variable, $v_2(lNT + 0.1) = u_2(lNT + 0.05)$.

4) t=0.15 *sec*

$$K_3 = \begin{bmatrix} 0 & 0 & 0 & 0 & 0 \\ 0.0155 & -0.1570 & 0.3983 & -0.1074 & 0 \end{bmatrix};$$

5) t=0.2 *sec*

$$K_4 = \begin{bmatrix} 0.1328 & 0.6195 & 0.0914 & 0 & -0.0544 \\ 0 & 0 & 0 & 0 & 0 \end{bmatrix};$$

As expected, at the sampling instants $\{lNT + 0.15\}$ the first raw of *K* is null, and at $\{lNT + 0.1, lNT + 0.2$ the second raw of *K* also vanishes.

The infinite time Riccati matrix is given by

$$P = \begin{bmatrix} 4.4268 & -0.4662 & -0.1625 & 0.2042 & 0.0211 \\ -0.4662 & 2.9427 & -0.2989 & 0.9467 & -0.1379 \\ -0.1625 & -0.2989 & 8.7367 & 0.1518 & 0.4155 \\ 0.2042 & 0.9467 & 0.1518 & 1.5315 & -0.0520 \\ 0.0211 & -0.1379 & 0.4155 & -0.0520 & 1.0300 \end{bmatrix}$$

Now, a periodic optimal filter is designed. The noise covariance matrices (25) have been taken as unity, $R_1 = R_2 = I$, as well as the initial state covariance matrix, R_0.

By (28), the filter gain matrices can be summarized (together with the previous control gains) in the Table 1.

The system response from the initial condition $x^T(0) = [0.5 \;\; -0.4 \;\; 0.3]$, assuming state feedback, is shown in Fig. 1. The multirate control input is plotted in Fig. 2.

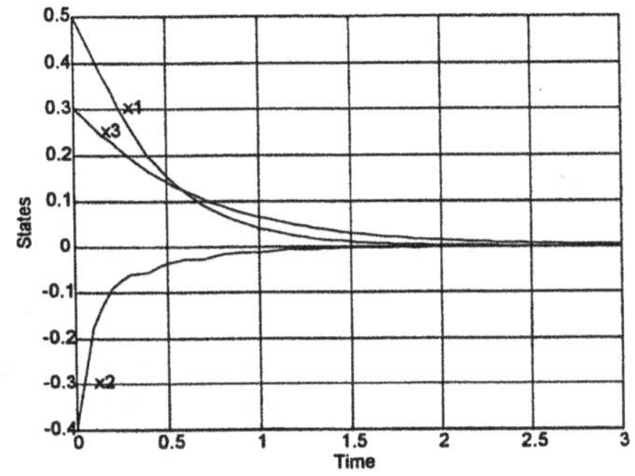

Figure 1. Optimal state trajectories.

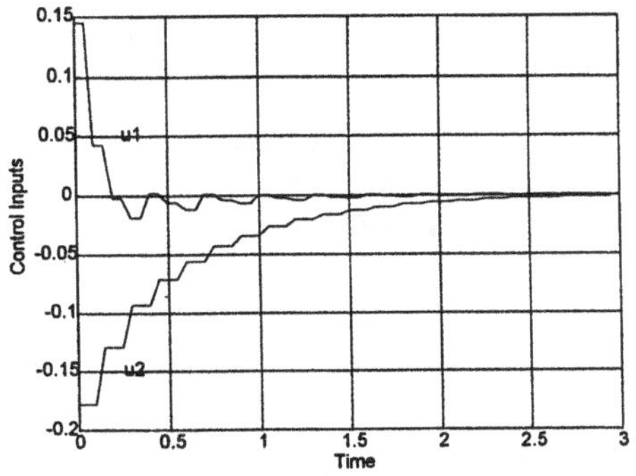

Figure 2. Optimal control actions

Table 1. Optimal gains

Time	Control Gain				
K(0)	0.1343	0.6147	0.1130	0	0
	0.0272	-0.1029	0.4091	0	0
K(1)	0	0	0	0	0
	0	0	0	0	0
K(2)	0.1325	0.6201	0.0875	0	-0.0657
	0	0	0	0	0
K(3)	0	0	0	0	0
	0.0155	-0.1570	0.3984	-0.1074	0
K(4)	0.1328	0.6195	0.0914	0	-0.0544
	0	0	0	0	0
K(5)	0	0	0	0	0
	0	0	0	0	0

Time	Filter Gain	
L(0)	-0.2002	-0.0125
	0.0614	-0.0360
	-0.0558	0.6781
L(1)	0	
L(2)	0	-0.1413
	0	0.0034
	0	0.6426
L(3)	-0.2006	0
	0.0601	0
	-0.0317	0
L(4)	0	-0.0860
	0	-0.0139
	0	0.6590
L(5)	0	

If an output feedback control is implemented, the effect of the observer in the state trajectories is shown in fig. 3.

6. CONCLUSIONS

In this paper, a compact model to represent discrete time systems with a multirate sampling/updating schema has been presented. This model enables an easy computation of the matrices involved in the multirate setting from the single rate system matrices and the periodic matrices including the features of the sampling/updating devices. In this way, it allows a great simplicity to change either the system parameters or the sampling events, in an independent way.

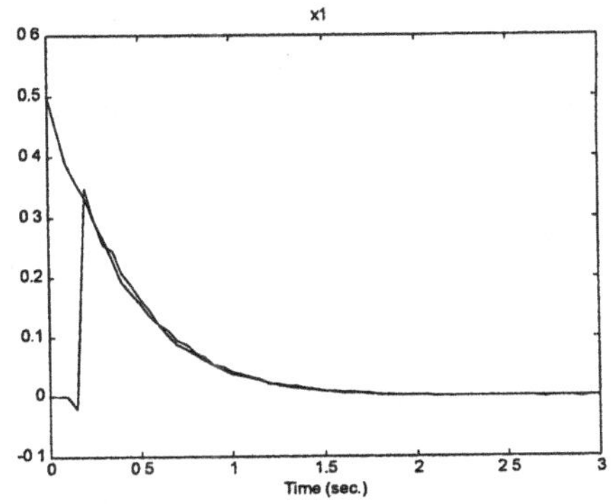

a)

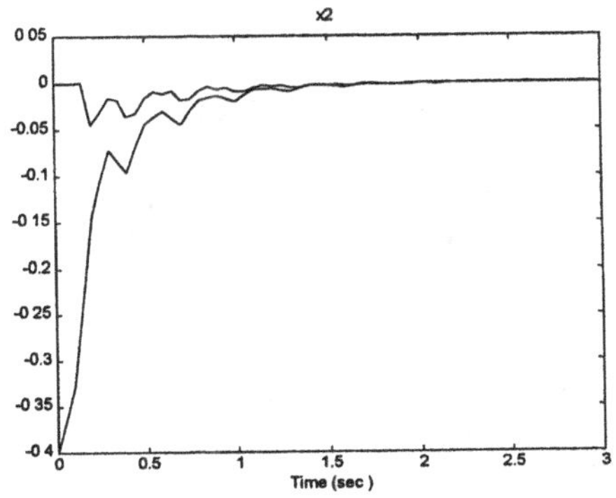

b)

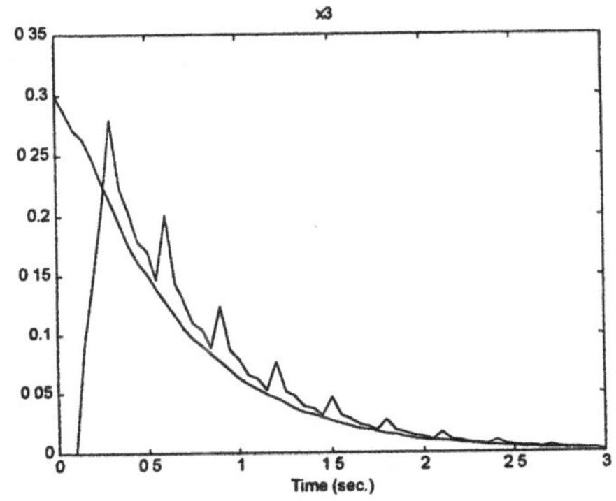

c)

Figure 3. State and estimated state trajectories for each state variable

This model has been also shown extremely useful to compute the LQR for SD systems with the multirate policy. The initial single rate performance index matrices are transformed into multirate index matrices, by means of the updating matrix. The result is a periodic state feedback control law that is computed by sequentially solving a set of Riccati-like equations.

In a dual way, the periodic Kalman filter to provide the state estimate has been developed. The result, again, is a periodic estimator. Altogether results in a periodic control structure with a clear interpretation. There are three blocks in the global system: the single rate fast model of the plant to be controlled, the periodic estimator involving the multirate sampling and the periodic state feedback controller involving the updating strategy. Changes in any of these components, i. e., the plant, the hold device or the sampling pattern are easily taken into account in the design.

An illustrative example points out the features of the proposed periodic subsystems.

ACKNOWLEDGMENTS

This work has been partially supported by CICYT (Comisión Interministerial de Ciencia y Tecnología) under grant CICYT TAP98-0252-C02-02.

REFERENCES

Albertos, P. (1990) Block Multirate Input-Output Model for sampled data Control Systems. *IEEE Trans. on Aut. Control.* Vol. **AC-35**, No 9, pp 1085-1088.

Albertos P., Salt J. ,Tornero J. (1996). Dual-rate adaptive control. *Automatica*, Vol. **32**, No 7, pp 1027-1030.

Al-Rahmani H. M., G. F. Franklin. (1992) Multirate Control: a new approach. *Automatica.* Vol. **28**, No.1, pp 35-44.

Apostolakis I.S., Jordan D. (1991). A time invariant approach to multirate optimal regulator design. *Int. J. of Control*, **53**, 1233-1254.

Araki M, Yamamoto K. (1986). Multivariable multirate sampled data systems: state-space description, transfer characteristics, and Nyquist criterion. *IEEE Trans. on Aut. Control*, **31**, 145-154.

Araki, M. (1993). Recent development in digital control theory. *Proc. 12th IFAC World Congr.*, Vol.**9**, pp. 951-960.

Bamieh B., Francis B., Tannenbaum A (1991). A Lifting Thecnique for Linear Periodic Systems with Aplications to Sampled – Data Control. *Systems and Control Letters*, vol **17**, pp 79-88.

Berg M.C., Amit N., Powell J.D. (1988). Multirate Digital Control System Design. *IEEE Trans. on Aut. Control*, Vol. **33**, No 12., pp 1139-1150.

Glasson D. P. (1983). Development and Applications of Multirate Digital Control.*Control System Magazine*, 2-8

Godbout L.F., Jordan D., Apostolakis I.S. (1990). Closed-loop model for general multirate digital control systems. *IEE Proceedings*, Part-D. **137**, 326-336.

Kalman R.E., Bertram J.E. (1959). A unified approach to the theory of sampling systems. *Journal Franklin Inst.*, vol. 267, pp 405-436, 1959.

Kranc G.M. (1957). Compensation of an error-sampled system by a multirate controller. *Trans. AIEE*, vol. **6**, M.6, part II, pp. 149-155, 1957.

Longhi, S. (1994). Structural Properties of Multirate Sampled Systems. *IEEE Trans. on Aut. Control*, Vol **AC-39**. No 3, pp 692-696.

Meyer R.A., Burrus C.S. (1975). A unified analysis of multirate and periodically time-invariant digital filters. *IEEE Trans. Circuits Syst.* **CAS-22**, pp.162-168.

Meyer D.G. (1990). A parameterization of stabilizing controllers for multirate sampled data systems. *IEEE Trans. on Aut. Control.* Vol. **AC-35**, No.2, pp 233-236.

Meyer D.G. (1992). Cost translation and a lifting approach to the multirate LQG problem. *IEEE Trans. on Aut. Control.* Vol. **AC-37**, No.9, pp 1411-1415.

Thompson P. M. (1986). Gain and phase margin of multirate sampled data feedback systems. *Int. J. Control*, **44**, 833-846.

Thompson P.M. (1988). Program CC Version 4: Reference Manual. *Hawthorne, Systems Technology, Inc.*, **II**, 566-585.

Tornero, J., Edmunds, J. (1983) Irregular but periodic Sampling. *Control System Centre Report n° 584/1983 University of Manchester (UMIST)*

Tornero, J. (1985) Non-conventional sampled data Systems Modelling. *Control System Centre Report n° 640/1985 University of Manchester (UMIST)*

Tornero, J., Salt, J. and Albertos, P. (1999). Lq Optimal Control For Multirate Sampled Data Systems. *IFAC World Congress*, Beijing.

Voulgaris P., Bamieh B (1992). Optimal H^∞ and H^2 Control of Hybrid Multirate Systems. *Systems and Control Letters*, vol. **20**, pp 249-261.

Copyright © IFAC Periodic Control Systems,
Cernobbio-Como, Italy, 2001

OPTIMALITY IN MULTICARRIER COMMUNICATION, MULTIPLE DESCRIPTION CODING AND THE SUBBAND CODING OF CYCLOSTATIONARY SIGNALS

Soura Dasgupta *Ashish Pandharipande**

ABSTRACT

This paper considers three different problems in signal processing/communications. The first involves Multiuser Discrete Multitone Transmission (DMT). The other two problems concerns variants of subband coding, specifically subband coding of cyclostationary signals and the multiple description coding. We show that underlying each is the same optimization problem that can be solved using the theory of majorization. *Copyright © 2001 IFAC*

1. INTRODUCTION

This paper considers three different problems in signal processing/communications. The first involves Multiuser Discrete Multitone Transmission (DMT), [1], also variously known as multicarrier modulation or Orthogonal frequency Division Multiplexed (OFDM) communication. The other two problems concerns variants of subband coding, specifically subband coding of cyclostationary signals and the multiple description coding. The paper demonstrates that despite the different antecedents of these three applications, underlying each is the same optimization problem. We show how this problem can be solved through the use of the theory of majorization, [8]. We begin by motivating the three problems in question.

1.1. Multiuser DMT:

DMT has been adopted as the signaling standard in Asymmetric Digital Subscriber Lines (ADSL), [12] and has been proposed as the modulation scheme of choice in the Mill Bahama and Magic Wand wireless ATM systems, [13]. Indeed in advocating DMT over CDMA, the following point is made in [14]: "A spreading factor of 85 (13 kb/s voice) or 128 (8kb/s voice) (for CDMA) is used with IS-95 to provide about 20 dB of processing gain. At much higher bit rates, CDMA systems must either reduce the processing gain or expand the bandwidth, but neither may be an attractive alternative."

We consider DMT in a multiuser environment. Thus the DMT system studied here supports multiple users, with varying quality of service (QoS) requirements, quantified by their respective bit rate and symbol error rate (SER) specifications. Specifically, consider the DMT system as in fig. 1 which depicts an M-subchannel filter bank model of a DMT system. We consider an overinterpolated ($N > M$) filter bank as the transceiver. We assume that the channel $C(z)$ is FIR of length κ (preequalization is assumed to have been done), and $v(n)$ is additive colored noise with known spectrum. Thus for example $v(n)$ could represent cochannel interference. Note, [10] provides models for cochannel interference in a variety of settings. To mitigate intersymbol interference (ISI), a form of redundancy is incorporated by choosing $N = M + \kappa$. The transmitting filter, $F_k(z) = \sum_{i=0}^{M-1} z^{-i} G_{ik}$, and $H_k(z) = \sum_{i=0}^{N-1} z^{i} S_{ki}$, act as modulating and demodulating transforms respectively. In a DFT based DMT implementation [1], the IDFT and DFT are used as the modulating and demodulating transforms respectively.

In this paper, as in [5] we will consider more general transformations leading to a *generalized DMT* system. To capture a multiuser environment, we assume that there are r-users each having been assigned L subchannels, i.e. $M = Lr$. Further the k-th user requires a bit rate of t_k, and an SER of no more than η_k. Our goal is to select F_i and H_i, and distribute the bit rates among the various sub-channels to achieve the above specifications with the *minimum possible transmitted power*, given the knowledge of $C(z)$ and the spectrum of $v(n)$. The problem addressed here thus directly generalizes that in [5], who also address the same power minmization issue, but assuming a single user subject to only one bit rate and SER constraint. The multiuser setting renders the optimization problem highly nontrivial in comparison to the single user case as will be shown.

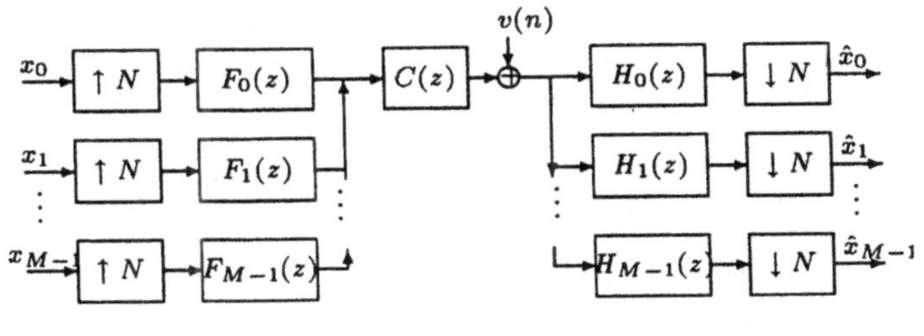

Figure 1. Filter bank based DMT model.

*Department of Electrical and Computer Engineering, The University of Iowa, Iowa City, IA-52242, USA, dasgupta@eng.uiowa.edu and pashish@icaen.uiowa.edu. Supported by ARO contract DAAD19-00-1-0534

1.2. Multiple Description Coding (MDC)

Multiple description coding is a variant of subband coding in which different parts of the signal to be coded have different bit budget requirements. Specifically in fig. 2 assume $M = Lr$ the Q_i are b_i bit quantizers that are subject to r separate average bit rate constraints. The goal in MDC is to have essentially r redundant coders as insurance against failure of one or more. Thus, channels $\{L(j-1)+1, \cdots, L(j-1)+L-1\}$ represent the j-th coder. Past work optimization related to MDC has focussed on the two coder case, with optimization directed at minimizing the average distortion subject to the failure of one coder, [15]-[16]. By contrast our goal is to optimize in the failure free case. Specifically, we select the LTI filters F_i, H_i, and allocate bits among the Q_i, subject to the bit rate constraints specified by the problem, to minimize the average output distortion variance. The optimization occurs under an orthonormality condition, specifically that the arrgangements to the left and right of the quantizers are *all pass*.

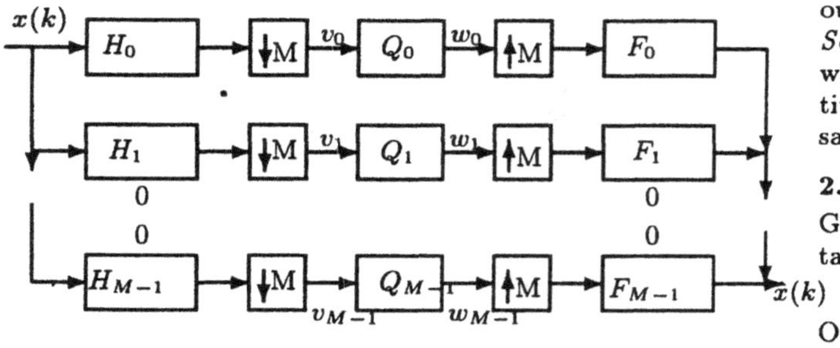

Figure 2. A Maximally Decimated Uniform Filter Bank

1.3. Subband Coding of Cyclostationary Signals

We consider optimum orthonormal subband coding of zero mean wide sense cyclostationary (WSCS) signals. A signal, $x(k)$ is WSCS with period N if for all k, l:

$$\mathcal{E}[x(k)x^*(l)] = \mathcal{E}[x(k+N)x^*(l+N)],$$

where $\mathcal{E}[\cdot]$ denotes the expectation operator.A wide variety of man made signals encountered in communication, telemetry, radar and sonar systems, as well as several generated by nature [6], are WSCS. Examples of manmade signals exhibiting cyclostationarity include signals found in amplitude, phase and frequency modulation systems, periodic keying of amplitude, phase and frequency in digital modulation systems, and periodic scanning in television, facsimile and some radar systems, [6]. Further, [7] demonstrates that WSCS models provide more accurate descriptions of speech signals than do traditional WSS models.

We assume that the filter bank is orthonormal. That is, for all square summable inputs $x(k)$, the combined energy of the M subband signals $v_i(k)$ equals the energy in $x(k)$, and in the absence of the quantizers the filter bank output $\hat{x}(k)$ matches $x(k)$ for all $x(k)$. It is easy to show that under these conditions the subband signals are themselves WSCS with period N. We adopt a *Periodically Dynamic Bit Allocation* (PDBA) scheme where we choose each $b_i(k)$ to be N-periodic, that is

$$b_i(k+N) = b_i(k). \qquad (1..1)$$

Our goal is to select $b_i(k)$ and the filters $H_i(k,z)$ and $F_i(k,z)$ so that the *average* variance of $\hat{x}(k) - x(k)$ is minimum, subject to orthonormality and the constraint that the average bit rate at each time instant is constant.

1.4. Outline

In Section 2. we expose the commonality of the underlying optimzation problems. Section 3. reviews the theory of majorization and explains its applicability to the solutions we seek. Section 4. describes the optima. Section 5. is the Conclusion.

2. FORMULATION

Underlying each of the three problems there are two essential tasks. *Optimum Bit Allocation (OBA)* that given a selection of the filters, distributes the bits among the various subchannels subject to the bit rate constarints. *Filter Selection* which involves selecting the filters in an optimal way. In this section we will focus on developing the objective functions and demonstrating that they reduce to the same form under OBA.

2.1. DMT

Generally, to preserve orthogonality $G = [G_{ij}]_{i,j=1}^{M}$ is unitary i.e.

$$GG^H = I. \qquad (2..2)$$

One can show, [5], under mild assumptions on $C(z)$, that given any G as in (2..2), $H_i(z)$ can be found to render the Perfect Reconstruction (PR), condition:

$$\hat{x}_i(n) = x_i(n), \quad \forall i \in \{0, \cdots, M-1\}.$$

Let the input power in the j-th subband of the k-th user be $\sigma^2_{x_{j,k}}$. Due to PR, this is also the output signal power $\sigma^2_{\hat{x}_{j,k}}$ in the j-th subband of the k-th user. Let the output noise power in this subband be $\sigma^2_{e_{j,k}}$, and $b_{j,k}$ be the number of bits allocated in this subchannel. Due to different QoS requirements, we may have different bit rate constraints for the users. The average number of bits for the k-th user is $b_k = \frac{1}{L}\sum_{j=0}^{L-1} b_{j,k}$. However we need to account for the reduction in bit rate due to the zero padding. The average bit budget for the k-th user is then $t_k = \frac{L}{N} b_k = \frac{1}{N}\sum_{j=0}^{L-1} b_{j,k}$.

With a high bit rate assumption made on the modulation system, we have, [5], for the k-th user

$$\sigma^2_{x_{j,k}} = c_k 2^{2b_{j,k}} \sigma^2_{e_{j,k}}$$

where the constant c_k depends on the SER η_k. We seek to minimize the average transmission power given by

$$f = \frac{1}{M}\sum_{k=1}^{r}\sum_{j=0}^{L-1} \sigma^2_{x_{j,k}} \qquad (2..3)$$

$$= \frac{1}{M}\sum_{k=1}^{r}\sum_{j=0}^{L-1} c_k 2^{2b_{j,k}} \sigma^2_{e_{j,k}} \qquad (2..4)$$

subject to the bit rate budgets

$$t_k = \frac{1}{N}\sum_{j=0}^{L-1} b_{j,k}, \quad k = 1, \dots, r, \tag{2..5}$$

and the PR requirement. Now apply the AM-GM inequality that states that the arithmetic mean is always greater than the geometric mean, with equality iff the numbers whose means they represent are identical. Thus,

$$f = \frac{1}{M}\sum_{k=1}^{r}\sum_{j=0}^{L-1} c_k 2^{2b_{j,k}}\sigma^2_{e_{j,k}} \tag{2..6}$$

$$\geq \frac{L}{M}\sum_{k=1}^{r} c_k (\prod_{j=0}^{L-1} 2^{2b_{j,k}}\sigma^2_{e_{j,k}})^{1/L} \tag{2..7}$$

$$= \frac{c}{r}\sum_{k=1}^{r} c_k (2^{2Nt_k}\prod_{j=0}^{L-1}\sigma^2_{e_{j,k}})^{1/L} \tag{2..8}$$

with equality holding iff for all i, k:

$$c_k(2^{2Nt_k}\prod_{j=0}^{L-1}\sigma^2_{e_{j,k}})^{1/L} = c_i(2^{2Nt_i}\prod_{j=0}^{L-1}\sigma^2_{e_{j,i}})^{1/L}. \tag{2..9}$$

This is the *optimum bit allocation strategy*. The optimal transceiver design is to find *unitary matrix* G so as to minimize

$$J = \sum_{k=1}^{r}(\alpha_k \prod_{j=0}^{L-1} a_{j,k})^{1/L} \tag{2..10}$$

where

$$\alpha_k = c_k 2^{2Nt_k} \quad a_{j,k} = \sigma^2_{e_{j,k}}. \tag{2..11}$$

One can show that the quantities $\sigma^2_{e_{j,k}}$ are the diagonal elements of R_e given by

$$R_e = G_0^T R_{\tilde{e}} G_0, \tag{2..12}$$

where $R_{\tilde{e}}$ is a known matrix obtained from the statistics of $v(n)$.

2.2. MDC

In this case the bit budget constraint is:

$$\sum_{i=0}^{L-1} b_{jL+i} = B_j \quad \forall j \in \{0, \cdots, r-1\}. \tag{2..13}$$

One must select the LTI filters F_i, H_i to be such that in the absence of quantizers $\hat{x}(k) = x(k)$, i.e. the filter bank is PR. In addition one imposes the requirement that $E(z)$, the $M \times M$, M-fold lifted equivalent of the arrangement to the left of the quantizers is *all pass*. Then the goal is to select all pass $E(z)$ and allocate bits among the subbands, subject to (2..13), and PR, so that the average quantizer induced mean-square distortion in $\hat{x}(k)$ is minimized. Under high bit rates the quantizer noise model is [11],

$$w_i(k) = v_i(k) + q_i(k)$$

where $q_i(k)$ is zero mean, white, independent from $v_i(k)$ and has variance

$$\sigma^2_{q_i} = c2^{-2b_i}\sigma^2_{v_i}. \tag{2..14}$$

The average output distortion is then given by:

$$\frac{1}{Lr}\sum_{k=0}^{r-1}\sum_{j=0}^{L-1}\sigma^2_{q_{jL+k}} \tag{2..15}$$

Because of (2..14) under optimum bit allocation one can show that the optimization problem reduces to finding an all pass $M \times M$, $E(z)$, so that (2..10) is minimized with $\alpha_k = 1$ and $a_{j,k}$ the variance of v_{jL+i}. Notice in this case one must find the all pass *operator* $E(z)$ and that the variance of v_{jL+i} are the diagonal elements of the matrix

$$\frac{1}{2\pi}\int_0^{2\pi} S_{\tilde{\mathbf{v}}}(\omega)d\omega \tag{2..16}$$

and that

$$S_{\tilde{\mathbf{v}}}(\omega) = E(e^{-j\omega})S_{\tilde{x}}(\omega)\left[E(e^{-j\omega})\right]^\dagger \tag{2..17}$$

where $S_{\tilde{x}}(\omega)$ is the known Power Spectral Density (PSD) matrix of the vector

$$[x(k), x(k-1), \cdots, x(k-Lr+1]^T.$$

Of course the all pass constraint reduces to

$$E(e^{-j\omega})\left[E(e^{-j\omega})\right]^\dagger = I. \tag{2..18}$$

2.3. WSCS

Suppose, now $x(k)$ in fig. 2 is has N-periodic second order statistics. Then the goal is to select N-periodic H_i and F_i, and bit allocation to minimize the distortion in $\hat{x}(k)$, subject to PR, and the condition that, $E(z)$ the $NM \times NM$, NM-fold lifted version of the arrangement to the left of the quantizers is *all pass*. In this case we select the b_i to be N-periodic and subject to the Periodically Dynamic Bit Allocation: with $b_i(k+N) = b_i(k)$ i.e.

$$b = \frac{1}{M}\sum_{i=0}^{M-1} b_i(k).$$

Clearly, the subband signals $v_i(k)$ are themselves WSCS with period N, that is

$$\sigma^2_{v_i}(k) = \sigma^2_{v_i}(k+N).$$

We will assume that the quantizers are modeled by additive zero mean noise sources, independent of the $v_i(k)$, with variances of the form

$$\sigma^2_{q_i}(k) = c2^{-2b_i(k)}\sigma^2_{v_i}(k). \tag{2..19}$$

Note that under (1..1), $\sigma^2_{q_i}(k)$ are N-periodic. Then the distortion $\hat{q}(k) = \hat{x}(k) - x(k)$ is *WSCS with period* MN, [17]. We propose to minimize the *average* variance of $\hat{q}(k)$,

$$\frac{1}{MN}\sum_{k=0}^{MN-1}\sigma^2_{\hat{q}}(k) = \frac{1}{MN}\sum_{k=0}^{N-1}\sum_{l=0}^{M-1}\sigma^2_{q_l}(k) \tag{2..20}$$

Under optimum bit allocation one can show that one must now find all pass $E(z)$ so that the following is minimized:

$$J_{SBC} = \sum_{j=0}^{N-1} \left(\prod_{i=0}^{M-1} \sigma_{v_i}^2(j) \right)^{1/M} . \qquad (2..21)$$

Note the similarity to (2..10). Again $\sigma_{v_i}^2(j)$ arte the diagonal elements of a matrix as in (2..16), with (2..17) and (2..18) in force. Now however, $S_{\tilde{x}}(\omega)$ and $S_{\tilde{V}}(\omega)$ are respectively, the PSD's of

$$\begin{aligned} \tilde{x}(k) &= [x_0(Nk), \ldots, x_0(NK - N + 1), \\ &\quad \ldots, x_{M-1}(Nk), \ldots, x_{M-1}(Nk - N + 1)], \end{aligned}$$

$$\begin{aligned} \tilde{\mathbf{V}}(k) &= [v_0(Nk), \ldots, v_0(NK - N + 1), \\ &\quad \ldots, v_{M-1}(Nk), \ldots, v_{M-1}(Nk - N + 1)]. \end{aligned}$$

3. MAJORIZATION

We define majorization and Schur concavity [8].

Definition 3..1 *Consider two sequences $x = \{x_i\}_{i=1}^n$ and $y = \{y_i\}_{i=1}^n$ with $x_i \geq x_{i+1}$ and $y_i \geq y_{i+1}$. Then we say that y majorizes x, denoted as $x \prec y$, if the following holds with equality at $k = n$*

$$\sum_{i=1}^{k} x_i \leq \sum_{i=1}^{k} y_i, \quad 1 \leq k \leq n.$$

Definition 3..2 *A real valued function $\phi(z) = \phi(z_1, \ldots, z_n)$ defined on a set $\mathcal{A} \subset R^n$ is said to be Schur concave on $\mathcal{A}$ if*

$$x \prec y \quad on\ \mathcal{A} \quad \Rightarrow \quad \phi(x) \geq \phi(y).$$

We will now state a theorem that results in a test for strict Schur concavity. We denote

$$\phi_{(k)}(z) = \frac{\partial \phi(z)}{\partial z_k} \quad and \quad \phi_{(i,j)}(z) = \frac{\partial^2 \phi(z)}{\partial z_i \partial z_j}.$$

Theorem 3..1 *Let $\phi(z)$ be a scalar real valued function defined and continuous on $\mathcal{D}$, and twice differentiable on the interior of $\mathcal{D}$. Then $\phi(z)$ is strictly Schur concave on $\mathcal{D}$ iff (i) ϕ is symmetric in its arguments, (ii) $\phi_{(k)}(z)$ is increasing in k, and (iii) $\phi_{(k)}(z) = \phi_{(k+1)}(z) \Rightarrow \phi_{(k,k)}(z) - \phi_{(k,k+1)}(z) - \phi_{(k+1,k)}(z) + \phi_{(k+1,k+1)}(z) < 0$.*

Theorem 3..2 *If H is an $n \times n$ hermitian matrix with diagonal elements $h_1, \ldots, h_n$ and eigenvalues $\lambda_1, \ldots, \lambda_n$, then $h \prec \lambda$ on R^n.*

Now turn to (2..10), the common objective function for all three problems. Suppose, given $\{a_{j,k}\}$ one were to seek the rearrangement of these to achieve the minimum value possible. Then it follows from [18] that the optimum arrangement must obey the following property:

$$a_{m,k_1} \geq a_{n,k_2} \Rightarrow \alpha_{k_1} \prod_{j \neq m}^{L-1} a_{j,k_1} \leq \alpha_{k_2} \prod_{j \neq n}^{L-1} a_{j,k_2} \qquad (3..22)$$

and

$$\alpha_m \geq \alpha_n \Rightarrow \prod_{j=0}^{L-1} a_{j,m} \leq \prod_{j=0}^{L-1} a_{j,n}. \qquad (3..23)$$

Call J under such an *optimum arrangement* J^*. Then one can show from Theorem 3..1 that:

Theorem 3..3 *The real valued scalar function J as defined in (2..10) under the optimality conditions (3..22-3..23) is strictly Schur concave.*

4. THE SOLUTION

All three problems have remarkably similar structure. Using the theory of majorization, and in particular Theorems 3..2 and 3..3, one can show the following.

- For DMT, the optimizing G is to within a permutation matrix the Karunen-Loeve Transform matrix of the autocorrelation matrix of the M-fold lifted version of $v_i(n)$.
- For Multiple Description Coding the solution is as follows. Suppose $S_{\tilde{V}}(\omega)$ is the Power Spectral Density (PSD) matrix of the vector of $v_i(n)$ in fig. 2. Suppose $\Lambda(\omega)$ is the diagonal matrix of the eigenvalues of $S_{\tilde{V}}(\omega)$, $\lambda_i(\omega)$, with $\lambda_i(\omega) \geq \lambda_{i+1}(\omega)$. Define $\Omega(\omega)$ to be the matrix that is unitary at all ω and in addition forces
$$\Omega(\omega) S_{\tilde{V}}(\omega) \Omega^H(\omega) = \Lambda(\omega).$$
Then for a constant permutation matrix, P, $E(e^{j\omega}) = P\Omega(\omega)$.
- The solution for the WSCS problem is trivially similar.

Here the frequency invariant permutation matrix is used to achieve the optimum arrangement exemplified in (3..22-3..23).

5. CONCLUSION

We have shown that three problems in signal processing and communications, with differing motivations and genesis have similar solutions. All three benefit from the powerful and elegant theory of Majorization. Given that both MDC and WSCS problems relate to subband coding similarity in their solutions is not a surprise. That they are also equivalent to the optimal DMT problem can be attributed to the fact that the optimum DMT can be interpreted as a dual problem to subband coding.

REFERENCES

[1] J.S. Chow, J.C. Tu, J.M. Cioffi, "A discrete multitone transceiver system for HDSL applications", *IEEE Journal on Selected Areas in Communications*, pp 895-908, Aug 1991.

[2] S. Dasgupta, C. Schwarz and B.D.O. Anderson, "Optimal subband coding of cyclostationary signals", *IEEE International Conference on Acoustics, Speech, and Signal Processing*, pp 1489-1492, Mar 1999.

[3] L.M.C. Hoo, J. Tellado, J.M. Cioffi, "Discrete dual QoS loading algorithms for multicarrier systems ", *IEEE International Conference on Communications*, pp 796-800, 1999.

[4] B.S. Krongold, K. Ramchandran, D.L. Jones, "Computationally efficient optimal power allocation algorithms for multicarrier communication systems", *IEEE Transactions on Communications*, pp 23-27, Jan 2000.

[5] Y. P. Lin, S.M. Phoong, "Perfect discrete multitone modulation with optimal transceivers", *IEEE Transactions on Signal Processing*, pp 1702 -1711, June 2000.

[6] W. A. Gardner "Exploitation of spectral redundancy in cyclostationary signals", *IEEE Signal Processing Magazine*, pp 14-37, April 1991.

[7] Y. Grenier, "Time dependent ARMA modelling of nonstationary signals", *IEEE Transactions on Accoustics, Speech, and Signal Processing*, pp 899-911, 1983.

[8] A. W. Marshall and I. Olkin, *Inequalities: Theory of Majorization and its applications*, Academic Press, 1979.

[9] A. Scaglione, S. Barbarossa, G.B. Giannakis, "Filterbank transceivers optimizing information rate in block transmissions over dispersive channels", *IEEE Transactions on Information Theory*, pp 1019-1032, Apr 1999.

[10] T. Starr, J.M. Cioffi, P. Silverman, *Understanding Digital Subscriber Line Technology*, Prentice Hall, 1999.

[11] P.P. Vaidyanathan, *Multirate Systems and Filter Banks*, Prentice Hall, 1992.

[12] Asymmetric Digital Subscriber Line (ADSL) Metallic Interface, ANSI T1E1.4/94-007R8, 1994.

[13] K. Pahlvan, A. Zahedi and P. Krishnamurthy, "Wideband local access: Wireless LAN and wireless ATM", *IEEE Communications Magazine*, pp 35-40, November 1997.

[14] L. J. Cimini, J. C. I. Chuang, and N. Sollenberger, "Advanced cellular internet service (ACIS)", *IEEE Communications Magazine*, pp 35-40, October 1998.

[15] X. Yang and K. Ramachandran, "Optimal subband filters for multidescription coding", *IEEE Transactions on Information Theory*, 2001.

[16] P.L. Dragotti, S.D. Servetto and M. Vetterli, "Analysis of optimal filter banks for multidescription quantization", *Proc. IEEE ICIP* , March 2000. 2000.

[17] V. P. Sathe and P. P. Vaidyanathan, "Effects of Multirate Systems on the Statistical Properties of Random Signals", *IEEE Transactions on Signal Processing* , pp 1766-1777, Aug 1995.

[18] G. M. Hardy, J. E. Littlewood and G. Polya, *Inequalities*, Cambridge University Press, 1934.

Copyright © IFAC Periodic Control Systems,
Cernobbio-Como, Italy, 2001

DESIGN FOR DIGITAL COMMUNICATION SYSTEMS VIA SAMPLED-DATA H^∞ CONTROL

M. Nagahara *[,1] Y. Yamamoto *[,2]

* *Department of Applied Analysis and Complex Dynamical Systems, Graduate School of Informatics, Kyoto University, Kyoto 606-8501, JAPAN*

Abstract: The design procedure for the equalization of digital communication channels is developed based on the sampled-data H^∞ control theory. The procedure provides transmitting/receiving filters so as to minimize the error between the original signal and the received signal with a time delay, and to reduce the noise added to the channel. While the system has an ideal sampler, a zero-order hold and a time delay, the design problem can be reduced to a finite-dimensional discrete-time problem using the FSFH (fast-sample and fast-hold) approximation. Numerical examples are presented to illustrate the effectiveness of the proposed method. *Copyright © 2001 IFAC*

Keywords: Sampled-data systems; Digital communications; Digital filters; H-infinity optimization; Time delay

1. INTRODUCTION

Nowadays the importance of digital communications is increasing owing to the rapid growth of the Internet, the cellular phones, and so on(Proakis, 1989). In digital communication, especially in pulse amplitude modulation (PAM) or in pulse code modulation (PCM), the analog signal which is to be transmitted is sampled and becomes a discrete-time signal. In the conventional way, the characteristic of the analog signal is not considered, and hence the whole system is regarded as a discrete-time system. For example, one usually assumes that the original analog signal is band-limited up to the Nyquist frequency. But in reality no signals are entirely band-limited.

This paper proposes a new design methodology based on the sampled-data control theory(Chen and Francis, 1995*b*) that takes account of inter-sample behaviors or frequency components beyond the Nyquist frequency in discrete-time. Recently, the sampled-data control theory is applied to some digital signal processing systems(Chen and Francis, 1995*a*; Khargonekar and Yamamoto, 1996; Yamamoto *et al.*, 1997; Nagahara and Yamamoto, 2000). We propose a design for the digital communication system via the sampled-data control. In (Erdogan *et al.*, 2000) a discrete-time H^∞ design for receiving filters or equalizers is introduced, but no design for transmitting filter is mentioned. However, it is difficult to attenuate both the signal reconstruction error and the additive noise by only an equalizing after the signal is received. Therefore we show the transmitting filter design which is executed in the same way as the receiving filter design. Moreover, we introduce the H^∞ method which takes account of a tradeoff between the signal reconstruction error and the energy of transmitted signal with an appropriate weighting function. Design examples are presented to illustrate the effectiveness of the proposed method.

---

[1] `nagahara@acs.i.kyoto-u.ac.jp`

[2] `yy@i.kyoto-u.ac.jp`; author to whom all correspondence should be addressed.

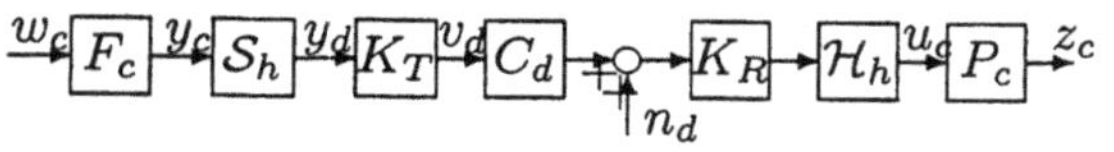

Fig. 1. Digital communication system.

2. DESIGN PROBLEM FORMULATION

The block diagram Figure 1 shows a digital communication system which is known as PAM or PCM. The incoming signal $w_c \in L^2[0,\infty)$ goes through an analog low-pass filter F_c and becomes y_c which is nearly (but not entirely) band limited. The filter F_c governs the frequency-domain characteristic of the analog signal y_c [3]. The signal y_c is then sampled by the sampler $\mathcal{S}_h$ to become a discrete-time signal (or PAM signal) y_d with sampling period h. Then the signal is shaped or enhanced by the transmitting digital filter K_T to the signal v_d to be transmitted to a communication channel.

The transmitted signal v_d is corrupted by the communication channel C_d and the additive noise n_d. In PCM communication, n_d is also considered as the noise generated by quantizing and coding error. The received signal goes through the receiving digital filter K_R which tries to attenuate the corruption and the noise, then becomes an analog signal u_c by the hold device $\mathcal{H}_h$ with sampling period h and smoothed by an analog low-pass filter P_c and finally we have the output signal z_c.

Our objective is to reconstruct the original analog signal y_c by the transmitting filter K_T and the receiving filter K_R against the corruption caused by the channel C_d and the additive noise n_d. Therefore consider the block diagram Figure 2 which is the signal reconstruction error system for the design. In the diagram the following points are taken into account:

- The time delay e^{-Ls} is introduced because we allow a certain amount of time delay for signal reconstruction.
- The transmitted signal v_d is estimated with a weighting function W_z because the energy or the amplitude of the transmitted signal v_d is usually limited.
- The noise has a frequency characteristic W_n.

Then our design problem is as follows:

Problem 1. Given stable analog filters $F(s)$ and $P(s)$, digital filters (weighting functions) $W_n(z)$ and $W_z(z)$ and a channel model $C_d(z)$, find digital filters $K_T(z)$ and $K_R(z)$ which minimizes

$$J^2 := \sup_{w_c \in L^2, n_d \in l^2} \frac{\|e_c\|_{L^2}^2 + \|z_d\|_{l^2}^2}{\|w_c\|_{L^2}^2 + \|n_d\|_{l^2}^2}. \quad (1)$$

[3] In the conventional design F_c is considered as an ideal filter which has a cut-off frequency up to the Nyquist frequency.

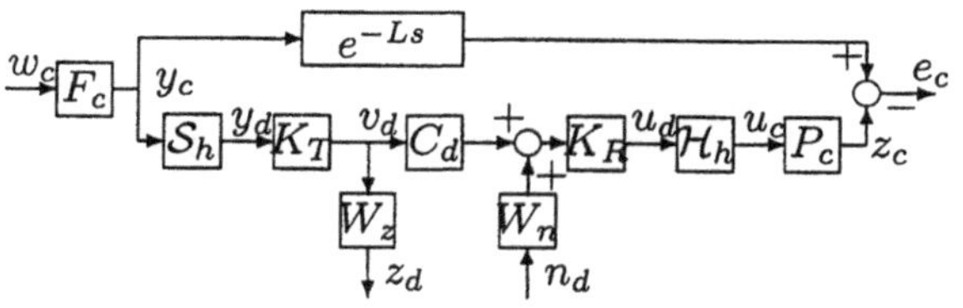

Fig. 2. Signal reconstruction error system.

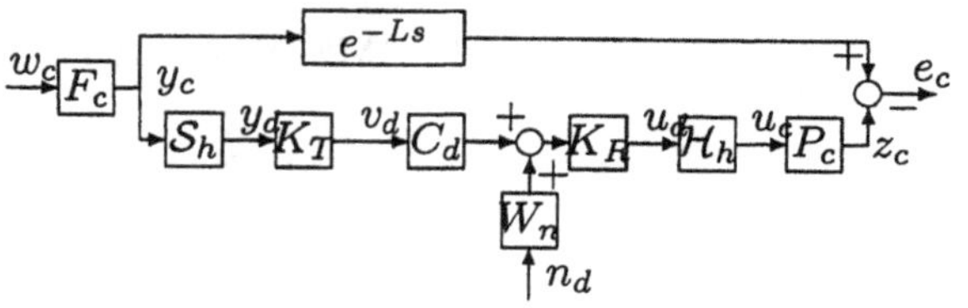

Fig. 3. Error System $\mathcal{T}_R$ for receiving filter design

3. DESIGN ALGORITHM

3.1 *Decomposing Design Problems*

Problem 1 is a simultaneous design problem of a transmitting filter and a receiving filter, and it is difficult to solve the problem directly. Therefore we introduce a decomposition of the design problem into two steps, that is the design for the receiving filter and that for the transmitting filter.

Obviously the transmitting filter K_T cannot attenuate the additive noise n_d, hence the receiving filter K_R has to play that role. Moreover K_R has to reconstruct the original signal from the corrupted signal (if K_R did not have to reconstruct, the optimal filter will be clearly $K_R = 0$) Therefore we first design the receiving filter K_R in order to reconstruct the original signal and to attenuate the noise by the block diagram Figure 2 with $W_z = 0$ and with $K_T = 1$. Then design the transmitting filter by the block diagram Figure 2 with $W_n = 0$ and with K_R which is obtained the previous design, that is we consider the channel as $K_R C_d$.

The design procedure is as follows:

Step 1(Design for receiving filter) Find a receiving filter K_R which minimizes

$$\|\mathcal{T}_R\|_\infty^2 := \sup_{w_c \in L^2, n_d \in l^2} \frac{\|e_c\|_{L^2}^2}{\|w_c\|_{L^2}^2 + \|n_d\|_{l^2}^2}, \quad (2)$$

in Figure 3 with fixed K_T (the initial filter is $K_T = 1$).

Step 2(Design for transmitting filter) Find a transmitting filter K_T which minimizes

$$\|\mathcal{T}_T\|_\infty^2 := \sup_{w_c \in L^2} \frac{\|e_c\|_{L^2}^2 + \|z_d\|_{l^2}^2}{\|w_c\|_{L^2}^2}, \quad (3)$$

in Figure 4 with K_R which is obtained in the previous step.

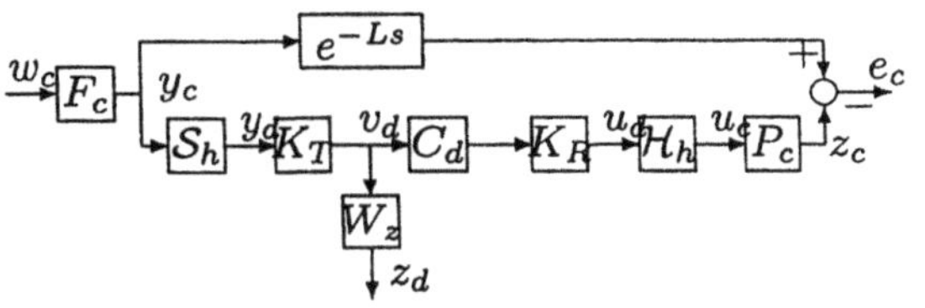

Fig. 4. Error System $\mathcal{T}_T$ for transmitting filter design

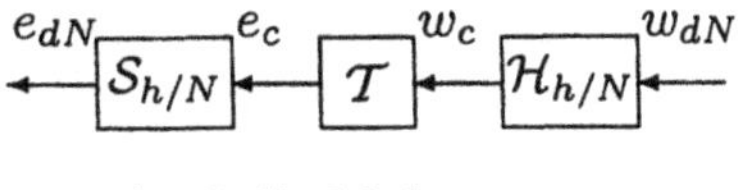

Fig. 5. fast sample/hold discretization

3.2 *Fast Sample/Hold Approximation*

The design problems (2) and (3) involve a continuous time delay component e^{-Ls}, and hence they are infinite-dimensional sampled-data problems. To avoid this difficulty, we employ the fast sample/hold approximation method (Keller and Anderson, 1992; Yamamoto *et al.*, 1999). By the method, our design problems (2) and (3) are approximated to finite-dimensional discrete-time problems assuming that the delay time L to be mh where m is a positive integer:

Theorem 2. Assume that $L = mh$, $m \in \mathbb{N}$. Then,

(1) for the error system $\mathcal{T}_R$ in Step 1, there exist finite-dimensional discrete-time systems $\{T_{R,N} : N = 1, 2, \ldots\}$ such that

$$\lim_{N\to\infty} \|T_{R,N}\|_\infty = \|\mathcal{T}_R\|_\infty.$$

(2) for the error system $\mathcal{T}_T$ in Step 2, there exist finite-dimensional discrete-time systems $\{T_{T,N} : N = 1, 2, \ldots\}$ such that

$$\lim_{N\to\infty} \|T_{T,N}\|_\infty = \|\mathcal{T}_T\|_\infty.$$

PROOF. By the fast sample/hold method, we approximate continuous-time inputs and outputs to discrete-time ones via the ideal sampler and the zero-order hold that operate in the period h/N (Figure 5). Then apply the discrete-time lifting(Yamamoto *et al.*, 1997) $\mathbf{L}_N$ to the discretized input/output signal e_{dN} and w_{dN}, we can get the lifted signals

$$\widetilde{e}_{dN} := \mathbf{L}_N(e_{dN}), \quad \widetilde{w}_{dN} := \mathbf{L}_N(w_{dN}).$$

Then we can approximate the continuous signal as

$$\|e_c\|_{L^2} \approx \sqrt{\frac{h}{N}}\|\widetilde{e}_{dN}\|_{l^2}, \quad \|w_c\|_{L^2} \approx \sqrt{\frac{h}{N}}\|\widetilde{w}_{dN}\|_{l^2}.$$

Moreover define

$$\|T_{R,N}\|_\infty^2 := \sup_{\widetilde{w}_{dN}, n_d \in l^2} \frac{\|\widetilde{e}_{dN}\|_{l^2}^2}{\|\widetilde{w}_{dN}\|_{l^2}^2 + \sqrt{\frac{N}{h}}\|n_d\|_{l^2}^2},$$

$$\|T_{T,N}\|_\infty := \sup_{\widetilde{w}_{dN} \in l^2} \frac{\|\widetilde{e}_{dN}\|_{l^2}^2 + \sqrt{\frac{N}{h}}\|z_d\|_{l^2}^2}{\|\widetilde{w}_{dN}\|_{l^2}},$$

where the systems $T_{R,N}$ and $T_{T,N}$ are approximated to finite-dimensional discrete-time systems, then we can show $\|T_{R,N}\|_\infty \to \|\mathcal{T}_R\|_\infty$, $\|T_{T,N}\|_\infty \to \|\mathcal{T}_T\|_\infty$ as $N \to \infty$ by using the method as shown in (Yamamoto *et al.*, 1999) under the assumption $L = mh$. □

Once the problems have been reduced to discrete-time problems, they can be solved by a control design toolbox such as those given by MATLAB. The resulting discrete-time approximant is given by the following:

Theorem 3. The approximated discrete-time systems $T_{R,N}$ and $T_{T,N}$ are given as follows:

$$T_{R,N} := \mathcal{F}_l(G_{R,N}, K_R),$$
$$T_{T,N} := \mathcal{F}_l(G_{T,N}, K_T),$$
$$G_{R,N} := \begin{bmatrix} [z^{-m}F_{dN}, 0], & -P_{dN} \\ [C_d K_T J F_{dN}, W_n], & 0 \end{bmatrix},$$
$$G_{T,N} := \begin{bmatrix} \begin{bmatrix} z^{-m}F_{dN} \\ 0 \end{bmatrix}, & \begin{bmatrix} -P_{dN}K_R C_d \\ W_z \end{bmatrix} \\ F_{dN}, & 0 \end{bmatrix},$$
$$J := [I, 0, \ldots, 0],$$

$$F_{dN}(z) := \left[\begin{array}{c|cccc} A_{Fd}^N & A_{Fd}^{N-1}B_{Fd}, & A_{Fd}^{N-2}B_{Fd}, & \ldots & B_{Fd} \\ \hline C_F & 0, & 0, & \ldots, & 0 \\ C_F A_{Fd} & C_F B_{Fd}, & 0, & \ldots, & 0 \\ \vdots & \vdots & \vdots & \ddots & \vdots \\ C_F A_{Fd}^{N-1} & C_F A_{Fd}^{N-2}B_{Fd} & C_{Fd}A_{Fd}^{N-3}B_{Fd}, & \ldots, & 0 \end{array}\right],$$

$$P_{dN}(z) := \left[\begin{array}{c|c} A_{Pd}^N & \sum_{k=1}^{N} A_{Pd}^{N-k}B_{Pd} \\ \hline C_P & D_P \\ C_P A_{Pd} & C_P B_{Pd} + D_P \\ \vdots & \vdots \\ C_P A_{Pd}^{N-1} & \sum_{k=2}^{N} C_P A_{Pd}^{N-k}B_{Pd} + D_P \end{array}\right],$$

$$A_{Fd} := e^{A_F \frac{h}{N}}, \quad B_{Fd} := \int_0^{\frac{h}{N}} e^{A_F t}B_F dt,$$

$$A_{Pd} := e^{A_P \frac{h}{N}}, \quad B_{Pd} := \int_0^{\frac{h}{N}} e^{A_P t}B_P dt,$$

$$F(s) =: \left[\begin{array}{c|c} A_F & B_F \\ \hline C_F & 0 \end{array}\right], \quad P(s) =: \left[\begin{array}{c|c} A_P & B_P \\ \hline C_P & D_P \end{array}\right].$$

where $\mathcal{F}_l(G, K)$ denotes the linear fractional transformation of plant G and filter K.

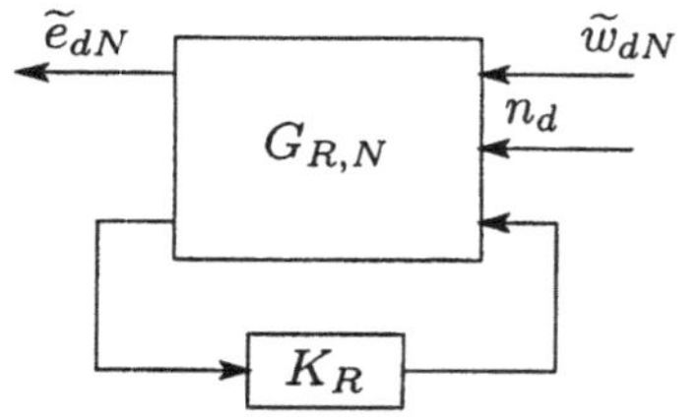

Fig. 6. Discrete-time H^∞ design problem for receiving filter K_R

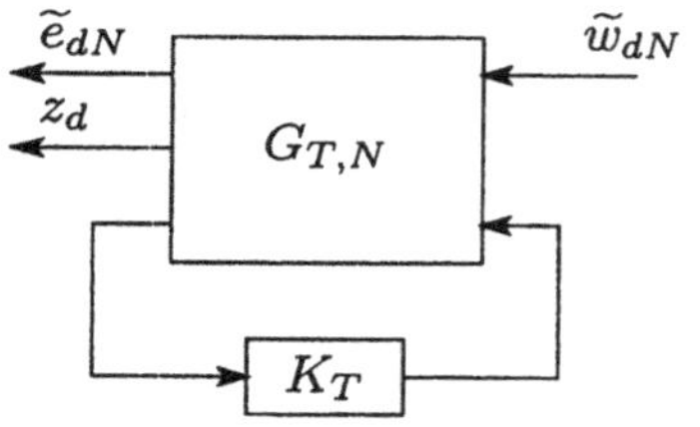

Fig. 7. Discrete-time H^∞ design problem for transmitting filter K_T

Then our design problems (2) and (3) are reduced to finite-dimensional discrete-time H^∞ problems, which are shown in Figure 6 and Figure 7.

4. DESIGN EXAMPLES

4.1 *Design for* $W_z = 0$

We present a design example for

$$F(s) := \frac{1}{10s+1}, \quad P(s) := 1, \quad W_n(z) := 1,$$
$$C_d(z) := 1 + 0.65z^{-1} - 0.52z^{-2} - 0.2975z^{-3},$$

with sampling period $h = 1$ and time delay $L = mh = 2$. An approximate design is executed here for $N = 8$. Here we design without considering the transmitting signal, that is $W_z(z) = 0$. For comparison, the discrete-time H^∞ design(Erdogan *et al.*, 2000) is also done.

Figure 8 shows the gain responses of the filters, and Figure 9 shows the frequency response of T_{ew} which is the system from the input w_c to the error e_c, and Figure 10 shows that of T_{zn} from the additive noise n_d to the output z_c. Compared with the discrete-time design, the sampled-data one shows better frequency response both in $\mathcal{T}_{ew}$ and in $\mathcal{T}_{zn}$. Moreover, we can say that only an equalizer is not able to attenuate the corruption caused by the channel and the additive noise, that is we need an appropriate transmitter for transmission.

To explain this fact, we show a simulation of these communication systems. The input signal y_c is the rectangular wave whose amplitude is 1, and the noise n_d is the discrete-time sinusoid $n_d[k] = \sin(2k)$. Figure 11 shows the output z_c

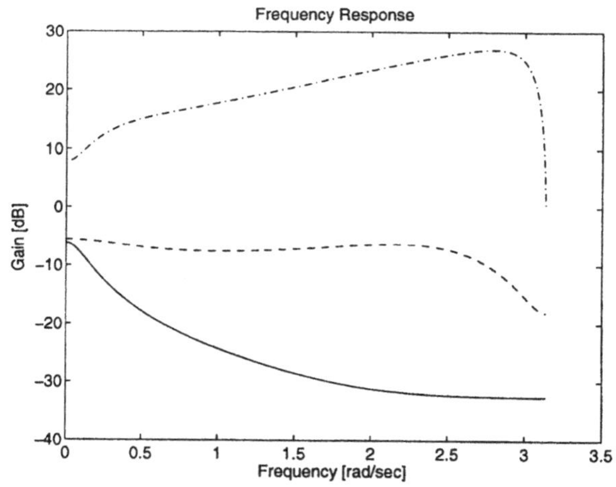

Fig. 8. Gain response of filters:sampled-data H^∞ design (transmitting filter: solid, receiving filter: dots) and discrete-time H^∞ design (dash).

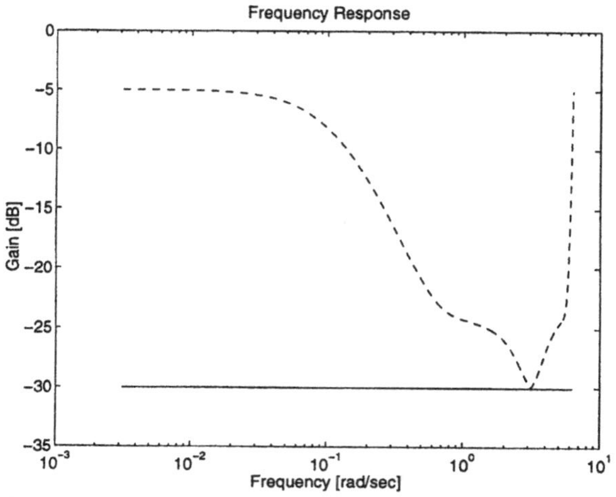

Fig. 9. Frequency response of $\mathcal{T}_{ew}$: sampled-data H^∞ design (solid) and discrete-time H^∞ design (dash).

with the receiving filter and the transmitting filter designed via sampled-data method, and Figure 12 shows that with the receiving filter designed in discrete-time (and without any transmitting filter). We see that the former shows much better reconstruction against the noise than the latter.

4.2 *Design for* $W_z(z) \neq 0$

Then we consider the design with the estimation of the transmitting signal v_d, that is $W_z(z) \neq 0$.

We observe from Figure 8 that the transmitting filter shows high gain around the Nyquist frequency (i.e. $\omega = \pi$), and hence we take

$$W_z(z) = r \cdot \frac{z-1}{z+0.5}$$

as the weighting function of the transmitting signal, whose gain characteristic is shown in Figure

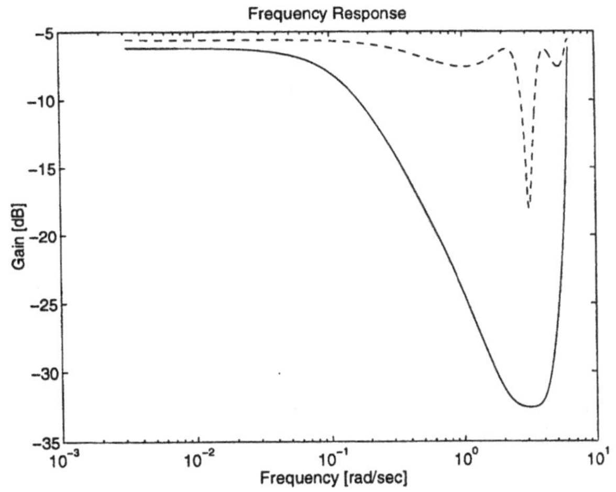

Fig. 10. Frequency response of $\mathcal{T}_{zn}$: sampled-data H^∞ design (solid) and discrete-time H^∞ design (dash).

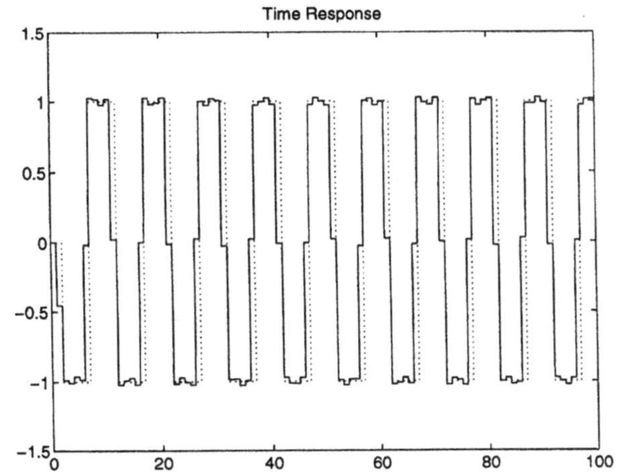

Fig. 11. Time response with sampled-data design.

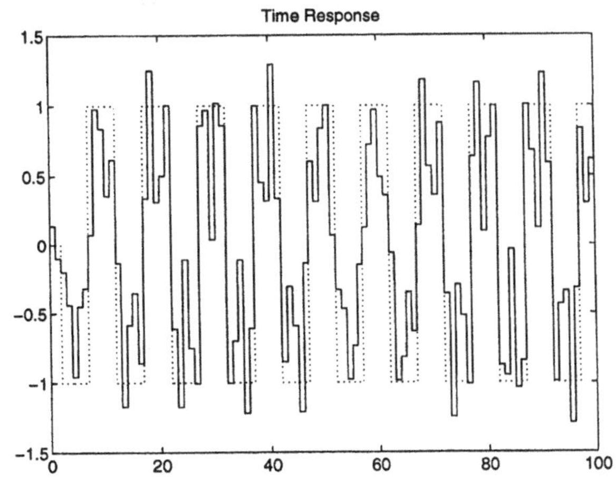

Fig. 12. Time response with discrete-time design.

13 where the parameter $r = 0.21$. The other design parameters are the same as the example above.

Figure 14 shows the H^∞ norm of $\mathcal{T}_{ew}$ and $\mathcal{T}_{vw}$ which is the system from w_c to v_d in Figure 2 which varies with $r \in [0, 5]$. We can take account

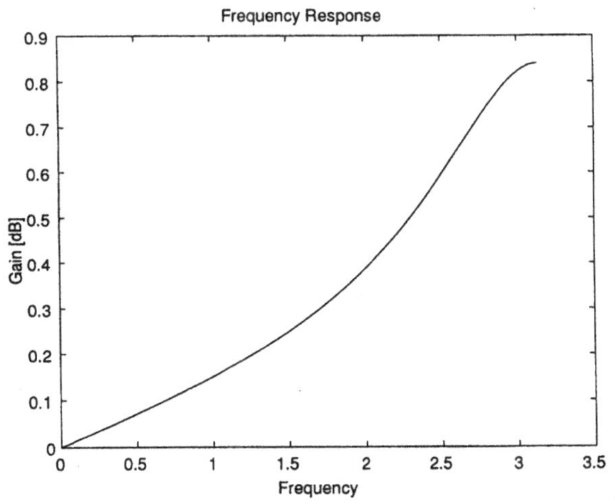

Fig. 13. Gain characteristic of the weighting filter $W_z(z)$.

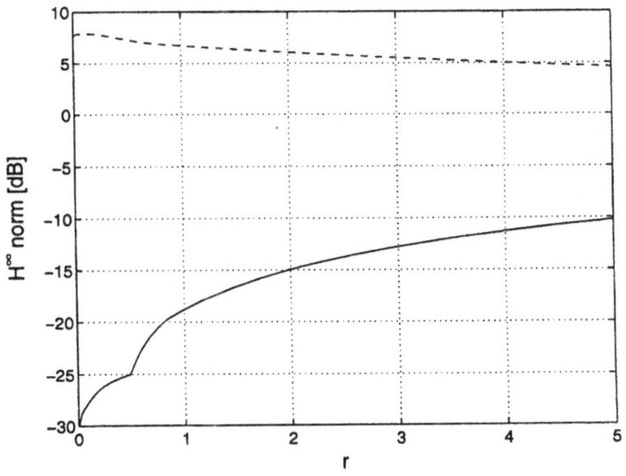

Fig. 14. Relation between r and $\|\mathcal{T}_{ew}\|_\infty$ (solid), $\|\mathcal{T}_{vw}\|_\infty$ (dash).

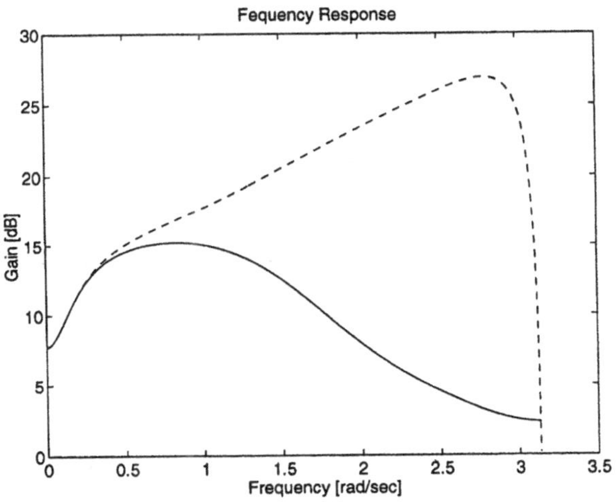

Fig. 15. Gain response of transmitting filters designed for $r = 0.21$ (solid) and $r = 0$ (dash)

of a trade-off between the error attenuation level and the amount of the transmitting signal with Figure 14. For example, we choose $r = 0.21$ in order to attenuate the error less than -26dB.

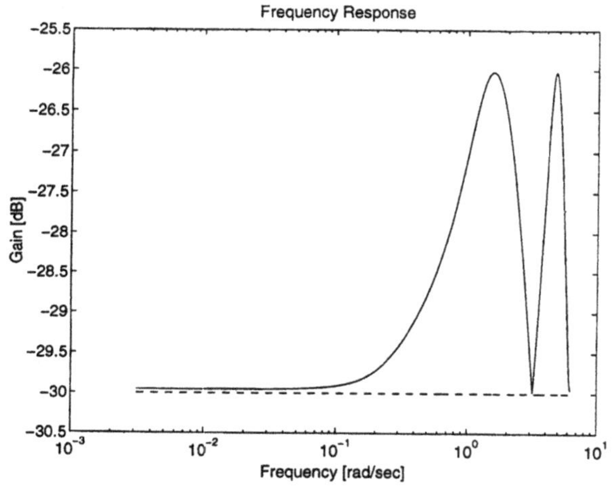

Fig. 16. Frequency response of $\mathcal{T}_{ew}$ designed for $r = 0.21$ (solid) and $r = 0$ (dash).

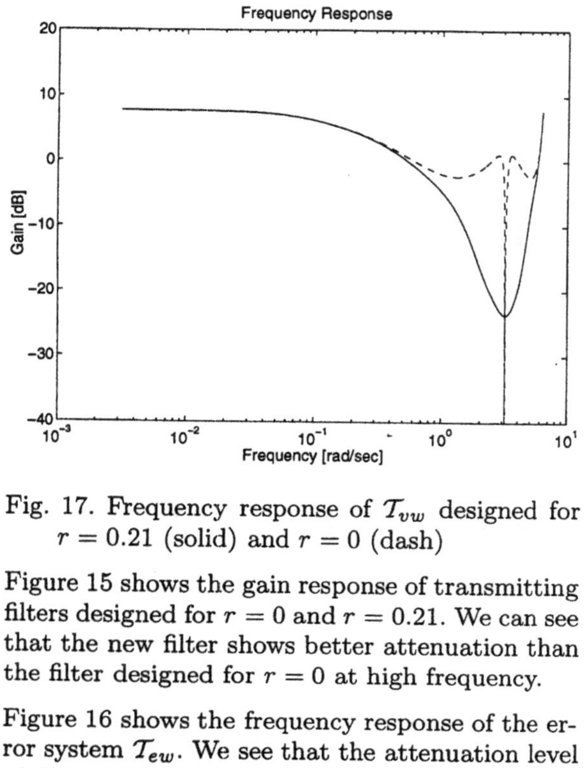

Fig. 17. Frequency response of $\mathcal{T}_{vw}$ designed for $r = 0.21$ (solid) and $r = 0$ (dash)

Figure 15 shows the gain response of transmitting filters designed for $r = 0$ and $r = 0.21$. We can see that the new filter shows better attenuation than the filter designed for $r = 0$ at high frequency.

Figure 16 shows the frequency response of the error system $\mathcal{T}_{ew}$. We see that the attenuation level of $\mathcal{T}_{ew}$ designed for $r = 0.21$ is less than −26dB. Figure 17 shows the frequency response of $\mathcal{T}_{vw}$. We can see that the amount of the transmitting signal is attenuated at high frequency.

5. CONCLUDING REMARKS

We have presented a new method of designing transmitting/receiving filter in digital communication. An advantage here is that an analog otimal performance can be obtained, and this can be advantageous in audio/speech signal transmission. Another advantage is that the trade-off between the attenuation of the reconstruction error and the energy of the transmitting signal is considered by the H^∞ design with an appropriate weighting function. By the fast sample/hold method, the design is reduced to a finite dimensional discrete time design, which can be easily implemented to CAD (e.g. MATLAB).

6. REFERENCES

Chen, T. and B. A. Francis (1995*a*). Design of multirate filter banks by $\mathcal{H}_\infty$ optimization. *IEEE Trans. on Signal Processing* **SP-43**, 2822–2830.

Chen, T. and B. A. Francis (1995*b*). *Optimal Sampled-Data Control Systems.* Springer.

Erdogan, A. T., B. Hassibi and T. Kailath (2000). On linear H^∞ equalization of communication cahnnels. *IEEE Trans. on Signal Processing* **SP-48**(11), 3227–3232.

Keller, J. P. and B. D. O. Anderson (1992). A new approach to the discretization of continuous-time controllers. *IEEE Trans. on Automatic Control* **AC-37**(2), 214–223.

Khargonekar, P. P. and Y. Yamamoto (1996). Delayed signal reconstruction using sampled-data control. *Proc. of 35th Conf. on Decision and Control* pp. 1259–1263.

Nagahara, M. and Y. Yamamoto (2000). A new design for sample-rate converters. *Proc. of 39th Conf. on Decision and Control* pp. 4296–4301.

Proakis, J. G. (1989). *Digital Communications.* McGraw Hill.

Yamamoto, Y., A. G. Madievski and B. D. O. Anderson (1999). Approximation of frequency response for sampled-data control systems. *Automatica* **35**, 729–734.

Yamamoto, Y., H. Fujioka and P. P. Khargonekar (1997). Signal reconstruction via sampled-data control with multirate filter banks. *Proc. of 36th Conf. on Decision and Control* pp. 3395–3400.

Copyright © IFAC Periodic Control Systems,
Cernobbio-Como, Italy, 2001

FUNCTION SPACE ANALYSIS OF MULTIRATE SAMPLED-DATA CONTROL SYSTEMS

Xiao Jian, Chen Tanglong

School of Electrical Eng. , Southwest Jiaotong University, Chengdu 610031, China

Abstract: Using two different lifting techniques, this paper first proposes a function space model of multirate sampled-data systems. This time-invariant discrete time model can describe the sampled-data system in continuous time sense, and is the complete description of the multirate sampled-data systems. Based on this model, the zeros and the stability properties of the multirate sampled-data systems are discussed, The relations between the zeros of the system and the asymptotic tracking are studied, which lead to the internal model principle of the multirate sampled-data systems. *Copyright © 2001 IFAC*

Keywords: multirate, sampled-data control, stability analysis, zero, tracking

1. INTRODUCTION

Due to rapidly developed computer science and technology, considerable attention has been focused on the study of sampled-data control systems. In the sampled-data control systems, it is usually assumed that all the signals in the system are sampled simultaneously in a single rate to simplify the analysis. However, when the plant under control is complex, it is neither reasonable nor implementable that every sampler in the various locations of the entire system must have the same sampling rate. In such cases, multirate sampling is unavoidable. Another reason for multirate sampled-data control system is that it is a period time-variant system, and can realize many control goals which single rate sampled-data control system fails to reach.

Researches on multirate sampled-data system began at late 50s. Kranc (1957) was among the first ones to study these systems. This topic has aroused a renewed interest in the recent years. Using lifting technique, the linear time-invariant state space model of multirate sampled-data system is obtained in (Araki and Yamamoto, 1986; Godbout et al., 1990). The applications of multirate sampled-data system in the various areas of control can be found in (Araki and Hagiwara,1986; Hagiwara and Araki,1988; Colaneri et al.,1990; Grasselli et al., 1995).

In a sampled-data control system, usually a continuous time plant is controlled by a digital controller. The current analysis and design methods only take the system behavior at the sampling points into consideration, and it is well known that the intersample behavior of the sampled-data control systems can be remarkably different from that predict from its sampling points (Patton et al., 1995).

Based on the work of Yamamoto (1994) and Bamieh and Pearson (1992), and by the combination of the lifting for the continuous time signals and the lifting for the multirate sampled signals, this paper presents the function space model of the multirate feedback control systems, then the stability properties and the zeros of the multirate sampling systems and its relations to the asymptotic tracking are outlined.

2. THE FNCTION SPACE MODEL OF THE MULTIRATE SAMPLED-DATA SYSTEM

Consider the multirate sampled-data control system

shown in Fig.1. Where P_c is the m-input p-output linear continuous plant; C_d is the multirate digital controller; H is the multirate zero-order hold; S is the multirate sampler; r_c, y and $e := r_c - y$ are the reference input, the output and the error of the system.

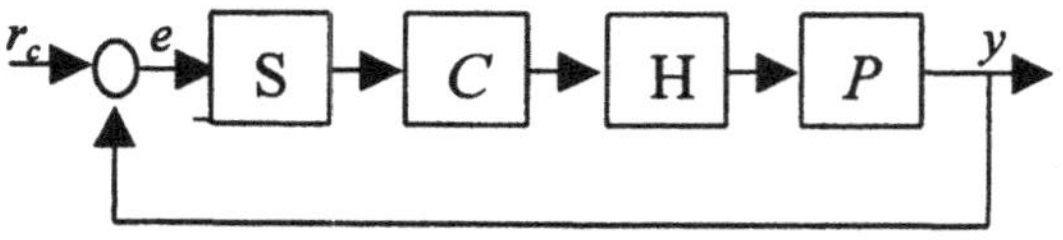

Fig. 1 Multirate sampled-data control system

Assume that the sampling periods for the input u and the error e are respectively defined as:

$T_{ui} = q_{ui}\,T$, $i = 1, 2, \ldots, m$; $T_{ei} = q_{ei}\,T$, $i = 1, 2, \ldots, p$.

Where all the q_{ui} s and q_{ei} s are positive integers and T is the base-rate sampling period. Thus, all sampling periods in the overall system are the integer multiples of the base-rate period and are all synchronized at the frame period $T_0 = qT$. Further define:

$r_i = q/q_{ui}$, $i=1, 2, \ldots, m$; $l_i = q/q_{ei}$, $i=1, 2, \ldots, p$.

Define W_T : $L_e^p[0,\infty) \to l_{L^p[0,h]}$ as:

$$W_T : \mapsto \{f_{kT}\}_{k=1}^{\infty}, \quad f_{kT}(\theta) = f((k-1)T + \theta) \qquad (1)$$

It is clear that $\{f_{kT}\}_{k=1}^{\infty}$ is a sequence, each element of which is a function of t in the interval (0, T]. Obviously, W_T is a bijective linear mapping and a linear isometry between $C[0,\infty)$ and $l_{L^p[0,h]}$.

Assume that the plant is represented by the linear time-invariant state space equation

$$P_c : \begin{cases} \dot{x}(t) = A_c x(t) + B_c u(t) \\ y(t) = C_c x(t) \end{cases} \qquad (2)$$

Let $x_{kT}(\theta)$, $u_{kT}(\theta)$, $e_{kT}(\theta)$ and $y_{kT}(\theta)$ be the elements of the sequences $\{x_{kT}\}_0^{\infty} \in l_{L^p[0,T]}$, $\{u_{kT}\}_0^{\infty} \in l_{L^p[0,h]}$, $\{e_{kT}\}_0^{\infty} \in l_{L^p[0,h]}$ and $\{y_{kT}\}_0^{\infty} \in l_{L^p[0,h]}$ obtained by applying the lifting W_T to $x(t)$, $u(t)$, $e(t)$ and $y(t)$, respectively. The state space equation of (2) then can be transform to the following equivalent state space realization

$$P_c \begin{cases} x_{(k+1)T}(\theta) = e^{A_c\theta}\delta_T x_{kT}(\theta) + \int_0^{\theta} e^{A_c(\theta-\tau)} B_c d\tau\, \delta_T u_{kT}(\theta) \\ y_{kT}(\theta) \;=\; C_c x_{kT}(\theta) \end{cases} \qquad (3)$$

Where δ_T is the sampling operator, i.e.

$$\delta_T f_{kt}(\theta) = f(kT)$$

Define the extended error and input vector as

$$\bar{E}(kT_0) = \begin{bmatrix} e(kT_0) \\ e(kT_0+T) \\ \vdots \\ e(kT_0+(q-1)T) \end{bmatrix}, \quad \bar{U}(kT_0) = \begin{bmatrix} u(kT_0) \\ u(kT_0+T) \\ \vdots \\ u(kT_0+(q-1)T) \end{bmatrix}$$

Using lifting technique, the following linear time-invariant state space description of the multirate digital controller can be formulated:

$$C_d \begin{cases} W((k+1)T_0) = A_e W(kT_0) + \bar{B}_e \bar{E}(kT_0) \\ \bar{U}(kT_0) = \bar{C}_e\, W(kT_0) + \bar{D}_e\, \bar{E}(kT_0) \end{cases} \qquad (4)$$

Similarly, by introducing extended vectors:

$$\bar{Y}_{kT_0}(\theta) = \begin{bmatrix} y_{kT_0}(\theta) \\ y_{kT_0+T}(\theta) \\ \cdots \\ y_{kT_0+(q-1)T}(\theta) \end{bmatrix}, \quad \bar{R}_{kT_0}(\theta) = \begin{bmatrix} r_{kT_0}(\theta) \\ r_{kT_0+T}(\theta) \\ \cdots \\ r_{kT_0+(q-1)T}(\theta) \end{bmatrix},$$

$$X_{kT_0}(\theta) = \begin{bmatrix} x_{(k-1)T_0+T}(\theta) \\ x_{(k-1)T_0+2T}(\theta) \\ \cdots \\ x_{kT_0}(\theta) \end{bmatrix}$$

and extended error vector $\bar{E}_{kT_0}(\theta) = \bar{R}_{kT_0}(\theta) - \bar{Y}_{kT_0}(\theta)$,

The function space model of the plant in the case of multirate sampling can be obtained:

$$P_c \begin{cases} X_{(k+1)T_0}(\theta) = F_e(\theta) X_{kT_0}(\theta) + \bar{G}_e(\theta)\bar{U}(kT_0) \\ \bar{Y}_{kT_0}(\theta) = \bar{M}_{e1}(\theta) X_{kT_0}(\theta) + \bar{M}_{e2}(\theta) X_{(k+1)T_0}(\theta) \end{cases} \qquad (5)$$

where

$$F_e(\theta) = \begin{bmatrix} 0 & \cdots & 0 & e^{A_c\theta}\delta_T \\ 0 & \cdots & 0 & e^{A_c(T+\theta)}\delta_T \\ \vdots & & \vdots & \vdots \\ 0 & \cdots & 0 & e^{A_c((q-1)T+\theta)}\delta_T \end{bmatrix}$$

$$\bar{G}_e(\theta) = \begin{bmatrix} \int_0^{\theta} e^{A_c(T-\tau)}B_c d\tau & 0 & \cdots & 0 \\ \int_0^{T} e^{A_c(T-\tau)}B_c d\tau & \int_0^{\theta} e^{A_c(T-\tau)}B_c d\tau & \cdots & 0 \\ \vdots & \vdots & & \vdots \\ \int_0^{T} e^{A_c(T-\tau)}B_c d\tau & \int_0^{T} e^{A_c(T-\tau)}B_c d\tau & \cdots & \int_0^{\theta} e^{A_c(T-\tau)}B_c d\tau \end{bmatrix}$$

$$\bar{M}_{e1} = \begin{bmatrix} 0 & \cdots & 0 & C_c \\ 0 & \cdots & 0 & 0 \\ \vdots & & \vdots & \vdots \\ 0 & \cdots & 0 & 0 \end{bmatrix}, \quad \bar{M}_{e2} = \begin{bmatrix} 0 & \cdots & 0 & 0 \\ C_c & \cdots & 0 & 0 \\ \vdots & & \vdots & \vdots \\ 0 & \cdots & C_c & 0 \end{bmatrix}$$

Equation (5) can be further simplified as

$$P_c \begin{cases} X_{(k+1)T_0}(\theta) = F_e(\theta) X_{kT_0}(\theta) + \bar{G}_e(\theta)\bar{U}(kT_0) \\ \bar{Y}_{kT_0}(\theta) \;=\; \bar{M}_e(\theta) X_{kT_0}(\theta) + \bar{N}_e(\theta)\bar{U}(kT_0) \end{cases} \qquad (6)$$

Where

$\overline{M}_e(\theta)=\overline{M}_{e1}+\overline{M}_{e2}F_e(\theta)$, $\overline{N}_e(\theta)=\overline{M}_{e2}\overline{G}_e(\theta)$

In (4) and (6), the extended vectors $\overline{Y}_{kT_0}(\theta)$, $\overline{E}(kT_0)$ and $\overline{U}(kT_0)$ contain some elements which do not corresponding to the sampling instants. The function space model (6) will be simplified by deleting these elements from corresponding extended vectors while keeping the corresponding equations unchanged. In this way, the equation (4) and (6) respectively can be simplified as

$$C_d:\begin{cases}W((k+1)T_0)=A_e W(kT_0)+B_e E(kT_0)\\ U(kT_0)=C_e W(kT_0)+D_e E(kT_0)\end{cases}\tag{7}$$

and

$$P_c\begin{cases}X_{(k+1)T_0}(\theta)=F_e(\theta)X_{kT_0}(\theta)+G_e(\theta)U(kT_0)\\ Y_{kT_0}(\theta)=M_e(\theta)X_{kT_0}(\theta)+N_e(\theta)U(kT_0)\end{cases}\tag{8}$$

Where $E(kT_0)$, $Y_{kT_0}(\theta)$ and $U(kT_0)$ are the reduced vectors of $\overline{E}(kT_0)$, $\overline{Y}_{kT_0}(\theta)$ and $\overline{U}(kT_0)$ respectively, and $G_e(\theta)$, $M_e(\theta)$ and $N_e(\theta)$ are respectively derived from $\overline{G}_e(\theta)$, $\overline{M}_e(\theta)$ and $\overline{N}_e(\theta)$. Combining the equation (7) and equation (8) together, one gets the closed-loop function space equation of the multirate sampled-data system:

$$\begin{cases}\begin{bmatrix}X_{(k+1)T_0}(\theta)\\ W((k+1)T_0)\end{bmatrix}=A_{cl}(\theta)\begin{bmatrix}X_{kT_0}(\theta)\\ W(kT_0)\end{bmatrix}+B_{cl}(\theta)R_{kT_0}(\theta)\\ \\ Y_{kT_0}(\theta)=C_{cl}(\theta)\begin{bmatrix}X_{kT_0}(\theta)\\ W(kT_0)\end{bmatrix}+D_{cl}(\theta)R_{kT_0}(\theta)\end{cases}\tag{9}$$

where

$$A_{cl}(\theta)=\begin{bmatrix}F_e(\theta)-G_e(\theta)HD_eM_e(T)\delta_T & G_e(\theta)HC_e\\ B_e(N_e(T)HD_e-I)M_e(T)\delta_T & A_e-B_eN_e(T)HC_e\end{bmatrix}$$

$$B_{cl}(\theta)=\begin{bmatrix}G_e(\theta)HD_e\delta_T\\ B_e(I-N_e(T)HD_e)\delta_T\end{bmatrix}$$

$$C_{cl}(\theta)=\begin{bmatrix}M_e(\theta)-N_e(\theta)HD_eM_e(T)\delta_T & N_e(\theta)HC_e\end{bmatrix}$$

$$D_{cl}(\theta)=N_e(\theta)HD_e \qquad H=(I+D_eN_e(T))^{-1}$$

3. STABILITY

It is beneficial to analysis the stability of the multirate sampled-data system from continuous time point of view. To this end, consider the unforced system of closed-loop system (9):

$$\begin{bmatrix}X_{(k+1)T_0}(\theta)\\ W((k+1)T_0)\end{bmatrix}=A_{cl}(\theta)\begin{bmatrix}X_{kT_0}(\theta)\\ W(kT_0)\end{bmatrix}\tag{10}$$

Lemma 3.1 The spectrum set of the closed loop multirate sampled-data system described by function space model (9) is the {0} and all the nonzero eigenvalues of the constant matrix

$$A_{cl}(T)=\begin{bmatrix}F_e(T)-G_e(T)HD_eM_e(T) & G_e(T)HC_e\\ B_e(N_e(T)HD_e-I)M_e(T) & A_e-B_eN_e(T)HC_e\end{bmatrix}$$

Proof: Clearly, $A_{cl}(\theta)$ is a bounded operator, since $A_{cl}(\theta)$ maps infinite dimensional space X into finite dimensional space $A_{cl}(\theta)X$, it is also a finite rank operator. Clearly, $A_{cl}(\theta)$ is a compact operator. According to the well-known Rieze-Szauder theory, The spectrum of $A_{cl}(\theta)$ is the union of {0} and the non-zero eigenvalues of $A_{cl}(\theta)$. According to the definition of the eigenvalue, the non-zero eigenvalue $\lambda\neq 0$ of $A_{cl}(\theta)$ can be obtained by the following formula:

$$[\lambda I-A_{cl}(\theta)]\begin{bmatrix}\widetilde{X}_{kT_0}(\theta)\\ \widetilde{W}(kT_0)\end{bmatrix}=0\tag{11}$$

where $\left[\widetilde{X}^T_{kT_0}(\theta)\ \ \widetilde{W}^T(kT_0)\right]^T$ is the eigenvector corresponding to λ. Consider the eigenvalue λ of the constant matrix $A_{cl}(T)$ and the associated eigenvector $\left[\widetilde{X}^T_{(k-1)T_0}(T)\ \ \widetilde{W}^T((k-1)T_0)\right]^T$, they satisfy

$$\lambda\begin{bmatrix}\widetilde{X}_{(k-1)T_0}(T)\\ \widetilde{W}((k-1)T_0)\end{bmatrix}-A_{cl}(T)\begin{bmatrix}\widetilde{X}_{(k-1)T_0}(T)\\ \widetilde{W}((k-1)T_0)\end{bmatrix}=0\tag{12}$$

which can be rewritten as

$$\lambda\begin{bmatrix}\widetilde{X}_{(k-1)T_0}(T)\\ \widetilde{W}((k-1)T_0)\end{bmatrix}=\begin{bmatrix}\widetilde{X}_{kT_0}(T)\\ \widetilde{W}(kT_0)\end{bmatrix}\tag{13}$$

where $\left[\widetilde{X}^T_{kT_0}(T)\ \ \widetilde{W}^T(kT_0)\right]^T$ is the vector representing the state vector of the system (10) at the instant kT_0, when the state vector at the instant $(k\text{-}1)T_0$ is $\left[\widetilde{X}^T_{(k-1)T_0}(T)\ \ \widetilde{W}^T((k-1)T_0)\right]^T$. Noticing the structure of $A_{cl}(\theta)$, the following equation holds:

$$A_{cl}(\theta)\begin{bmatrix}\widetilde{X}_{kT_0}(T)\\ \widetilde{W}(kT_0)\end{bmatrix}=A_{cl}(\theta)\begin{bmatrix}\widetilde{X}_{kT_0}(\theta)\\ \widetilde{W}(kT_0)\end{bmatrix}\tag{14}$$

Applying the operator $A_{cl}(\theta)$ to the both side of (13), and using the result of (14), we now have proved that the non-zero $\lambda\neq 0$ is also an eigenvalue of $A_{cl}(\theta)$. On the other hand, the eigenvalues of $A_{cl}(\theta)$ are also the eigenvalues of $A_{cl}(T)$. $A_{cl}(\theta)$ and

$A_{cl}(T)$ must have the same non-zero eigenvalue set.

Now study the internal stability of the closed-loop multirate sampling system described by the function space model (9). We say that the system (9) is internally stable if

$$\lim_{k\to\infty}\left\|A_{cl}^{k}(\theta)\right\|=0 \qquad (15)$$

Where the norm can be any induced one in the Banach space .

The following theorem is an adapted version of Yamamoto (1994), which gives the internal stability criterion of the closed-loop multirate sampling system (9). Its proof is omitted.

Theorem 3.1 The closed-loop multirate sampling system (9) is internally stable if and only if all the eigenvalues of the constant matrix $A_{cl}(T)$ are inside the unit circle in the complex plane.

4. ZEROS OF THE MULTIRATE SAMPLED-DATA SYSTEMS

One of the problems in the multirate sampled-data control systems, is that it may introduces ripples, as evidenced by Feuer and Goodwin (1995). The function model of (9) is a complete description of the multirate sampling-data control system, i.e. it describes the system in continuous time sense. Using this model, it can be more easily to study the reasons of the phenomenon of ripples and the methods to overcome it.

Based on the function space model (9), and making use of the definitions of the transfer function for finite-dimensional time-invariant systems, The transfer function of the system (9) can be defined:

$$\hat{W}_{er}(z)=-C_{cl}(zI-A_{cl})^{-1}B_{cl}\delta_T+(I-D_{cl}\delta_T) \qquad (16)$$

Theorem 4.1 Assume that the transfer function matrix of (16) is stable, and the reference input to the system (9) is $r_c(t)=e^{\mu t}v(\theta)$, $\mathrm{Re}\mu\geq 0$, then the response of the expended tracking error $E_{kT_0}(\theta)$ to this reference input asymptotically approaches

$$\lambda^{k-1}\hat{W}_{er}(\lambda)V_r(\theta)$$

Where $\lambda=e^{\mu T}$, $V_r(\theta)$ is a vector depending on θ, representing the direction in which the expanded reference vector $R_{kT_0}(\theta)$ enter the system:

$$V_r(\theta)=\begin{bmatrix} e^{\mu\theta}v_0 \\ e^{\mu(\theta+T)}v_0 \\ \vdots \\ e^{\mu(\theta+(q-1)T)}v_0 \end{bmatrix}=e^{\mu\theta}\begin{bmatrix} v_0 \\ e^{\mu T}v_0 \\ \vdots \\ e^{\mu(q-1)T}v_0 \end{bmatrix} \qquad (17)$$

Proof: For the reference $r_c(t)=e^{\mu t}v(\theta)$, it's z transform is:

$$Z\{R_{kT_0}(\theta)\}=\frac{1}{z-\lambda}e^{\mu\theta}\begin{bmatrix} v_0 \\ e^{\mu T}v_0 \\ \vdots \\ e^{\mu(q-1)T}v_0 \end{bmatrix}:=\frac{1}{z-\lambda}V_r(\theta) \qquad (18)$$

The response of the expanded error vector of the closed-loop system to this reference input is:

$$E_{kT_0}(\theta)=Z^{-1}\left\{\hat{W}_{er}(z)\frac{1}{z-\lambda}V_r(\theta)\right\}=Z^{-1}\left\{\hat{W}_{er}(\lambda)\frac{1}{z-\lambda}V_r(\theta)+W(z)\right\}$$

Where $W(z)$ is the response due to the nonzero initial conditions of the system. Since by assumption, the system (9) is asymptotically stable, $W(z)$ is analytic outside the unit disk, $E_{kT_0}(\theta)$ must asymptotically approaches $\lambda^{k-1}\hat{W}_{er}(\lambda)V_r(\theta)$.

Similar to the definition of zeros and zero directions in the finite-dimensional discete-time systems, we now give the definition of zeros and zero directions for the transfer matrix $\hat{W}_{er}(z)$:

Definition 4.1 If there exist a complex number λ and a nonzero vector $V(\theta)$, such that $\hat{W}_{er}(\lambda)V(\theta)=0$, then λ is called a transmission zero of $\hat{W}_{er}(z)$, and $V(\theta)$ the associate zero direction. If there exist a complex number λ and a nonzero vector $\left[X^T(\theta)\ \ W^T\ \ V^T(\theta)\right]^T$ satisfying the following equation (19):

$$\begin{bmatrix} \lambda I-F_e(\theta)+G_e(\theta)HD_eM_e(T)\delta_T & -G_e(\theta)HC_e & G_e(\theta)HD_e\delta_T \\ -B_e(N_e(T)HD_e-I)M_e(T)\delta_T & \lambda I-A_e+B_eN_e(T)HC_e & B_e(I-N_e(T)HD_e)\delta_T \\ M_e(\theta)-N_e(\theta)HD_eM_e(T)\delta_T & N_e(\theta)HC_e & I-N_e(\theta)HD_e \end{bmatrix}\begin{bmatrix} X(\theta) \\ w \\ V(\theta) \end{bmatrix}=0 \qquad (19)$$

then λ is called the invariant zero of $\hat{W}_{er}(z)$, and $V(\theta)$ the associate zero direction.

It is clear that, for the multirate sampled-data system (9) to realize the asymptotical tracking of the reference $r_c(t)=e^{\mu t}V(\theta)$, $\mathrm{Re}\mu\geq 0$, $\lambda=e^{\mu T}$ must be the transmission zero of the system, and the zero direction $V_r(\theta)$ must coincide with $V(\theta)$.

Comparing the definitions of the transmission zero and the invariant zero in the definition 4.1, obviously, when the closed-loop system (9) is stable and there is no unstable pole and zero cancellations between C_d and P_c, one can identify the concept of transmission zeros and the invariant zeros.

Theorem 4.2 Let $\hat{W}_{er}(z)$ be the close-loop transfer function matrix of the system of fig.1, and assume the asymptotic stability and no unstable pole and zero cancellations of the system, then the complex number $\lambda = e^{\mu T}$ outside of unit disk is a transmission zero of $\hat{W}_{er}(z)$, if and only if λ is a pole of the digital controller C_d, or μ is a pole of continuous plant P_c.

Proof: The last row of equation (19) gives

$$[I - N_e(\theta)HD_e]M_e(T)X(\theta) + N_e(\theta)HC_e w + [I - N_e(\theta)HD_e\delta_T]V(\theta) = 0 \quad (20)$$

Applying the operator $(I + N_e(\theta)D_e\delta_T)$ to the both side of equation (20) results

$$(I + H_e(\theta)D_e\delta_T)[M_e(\theta) - N_e(\theta)HD_e M_e(\theta)\delta_T]X(\theta) + (I + H_e(\theta)D_e\delta_T)N_e(\theta)HC_e w + (I + H_e(\theta)D_e\delta_T)[I - N_e(T)HD_e\delta_T]V(\theta) = 0 \quad (21)$$

which can be further simplified as

$$V(\theta) = -M_e(\theta)X(\theta) - N_e(\theta)C_e w \quad (22)$$

Substituting (22) and (20) into the first row and second row of equation (19), respectively, gives

$$\begin{cases} \lambda X(\theta) - F(\theta)X(T) - G_e(\theta)C_e w = 0 \\ w - A_e w = 0 \end{cases} \quad (23)$$

Equation (23) is a simultaneous equation of the vector $[X^T(\theta)\ \ w^T]^T$. Clearly, complex number λ outside of unit disk is a transmission zero of $\hat{W}_{er}(z)$, and $V(\theta) = -M_e(\theta)X(\theta) - N_e(\theta)C_e w$ is the corresponding zero direction, if and only if λ make the equation (23) having nonzero solution $[X^T(\theta)\ \ w^T]^T$. Now consider its form of solutions.

If the equation (23) admits a solution form $[X^T(\theta)\ \ w^T]^T$, where $w \neq 0$, then λ must be the eigenvalue of A_e and w is the corresponding eigenvector. On the other side, if the solution of the equation (23) has the form $[X^T(\theta)\ \ 0]^T$, λ must be the eigenvalue of $F_e(\theta)$ and $X(\theta)$ is the corresponding eigenvector. Inversely, if λ outside of unit disk is a eigenvalue of A_e, and let

$$X(\theta) = \frac{1}{\lambda}\{F_e(\theta)[\lambda I - F_e(T)]^{-1}G_e(T) + G_e(\theta)\}C_e w \quad (24)$$

Where w is the eigenvector of A_e corresponding to λ, then $[X^T(\theta)\ \ w^T]^T$ is a nonzero solution of (23). On the other hand, noticing that $F(\theta)X(T) = F(\theta)X(\theta)$, if λ outside of unit disk is a eigenvalue of $F_e(\theta)$, then $[X^T(\theta)\ \ 0]^T$, where $X(\theta)$ is the associate eigenvector, is a nonzero solution of (23).

5. ASYMPTOTIC TRACKING

Asymptotic tracking is an important topic in the control area. Using function space model, the ripple phenomena can be understood more deeply.

Consider the asymptotic tracking of the reference input $r_c(t) = e^{\mu t}v_0$, where $\mathrm{Re}\,\mu \geq 0$. By theorem 4.1 and theorem 4.2, the asymptotic tracking of this reference input is possible only if μ is the pole of the continuous plant P_c, or $\lambda = e^{\mu T}$ is the pole of A_e.

If $\lambda = e^{\mu T}$ is the pole of A_e, and w is the corresponding eigenvector, then $[X^T(\theta)\ \ w^T]^T$ is the nonzero solution of (23), where $X(\theta)$ is defined by (24). According to the theorem 4.1, λ is a transmission zero of $\hat{W}_{er}(z)$ and

$$V(\theta) = -M(\theta)X(\theta) - N_e(\theta)C w = -\{\tfrac{1}{\lambda}M(\theta)[F_e(\theta)(\lambda I - F_e(T)^{-1}G(T) + G(\theta)] + N_e(\theta)\}C_e w \quad (25)$$

is the corresponding zero direction. Since $\lambda = e^{\mu T}$ is not an eigenvalue of $F_e(\theta)$, μ is not an eigenvalue of A_c. Recalling the structure of the matrices $F_e(\theta)$, $G_e(\theta)$ and $N_e(\theta)$, obviously, they do not contain the mode $e^{\mu\theta}$. This will make every elements of $V(\theta)$ defined by (25) having no term of $e^{\mu\theta}$. In this case, although λ is a transmission zero of $\hat{W}_{er}(z)$, but the associate zero direction $V(\theta)$ can not coincide with the direction vector $V_r(\theta)$ defined by (17). By theorem 4.1, in this case, asymptotic tracking of the reference input is not possible.

Theorem 5.1 Assume that both of the digital controller C_d and the continuous plant P_c shown in fig.1 are stabilizable and detectable, and the closed-loop system (9) is stable and having no unstable pole-zero cancellations, then asymptotic ripple-free tracking of the reference signal $r_c\ (t)=\ e^{\mu t}v_0$, $\mathrm{Re}\,\mu \geq 0$, can be realized by incorporating μ as a pole of the continuous plant P_c.

Proof: For the reference signal $r_c\ (t)=\ e^{\mu t}v_0$, the corresponding z transformation of the extended reference input vector is given by (18). Since μ is an eigenvalue of A_c, $\lambda = e^{\mu T}$ is an eigenvalue of $F_e(\theta)$. Let x_0 be the eigenvector of A_c associated with λ , and

$$X_0 = \begin{bmatrix} 0 & \cdots & 0 & x_0^T \end{bmatrix}^T, \quad X(\theta) = \tfrac{1}{\lambda} F_e(\theta) X_0$$

Then $\begin{bmatrix} X^T(\theta) & 0 \end{bmatrix}^T$ is a nonzero solution of the equation (23). According to theorem 4.2, μ is a transmission zero of the closed-loop transfer function $\hat{W}_{er}(z)$, and the corresponding zero direction is given by

$$V(\theta) = -M_e(\theta)X(\theta) = -\frac{1}{\lambda} M_e(\theta)F_e(\theta)X_0$$

$$= \begin{bmatrix} 0 & \cdots & 0 & C_c e^{A_c((q-1)T+\theta)} \\ 0 & \cdots & 0 & C_c e^{A_c(qT+\theta)} \\ \vdots & & \vdots & \vdots \\ 0 & \cdots & 0 & C_c e^{A_c(2(q-1)T+\theta)} \end{bmatrix} \begin{bmatrix} 0 \\ \vdots \\ 0 \\ x_0 \end{bmatrix} = e^{\mu\theta} e^{\mu(q-1)T} \begin{bmatrix} C_c x_0 \\ e^{\mu T} C_c x_0 \\ \vdots \\ e^{\mu(q-1)T} C_c x_0 \end{bmatrix} \qquad (26)$$

Where the nonzeroness of $C_c x_0$ is guaranteed by the detectability of the continuous plant P_c. Comparing (26) and (17), if $v_0 = C_c x_0 e^{\mu(q-1)T}$ is selected, then $V(\theta)$ coincides with the vector $V_r(\theta)$ defined by (17). By theorem 4.1, the asymptotic ripple-free tracking of the reference signal $r_c\ (t)=\ e^{\mu t}v_0$, $\mathrm{Re}\,\mu \geq 0$, can be realized in this situation.

6 . CONCLUTION

Using two different lifting techniques, this paper proposes a function model for the multirate sampled-data control systems. Since the inherent nature of the proposed model, analysis and design based on this model are hopeful to overcome the performance deterioration caused by the intersample ripples in most multirate sampled-data control systems. As an example, the relations between the zeros of the system and the asymptotic tracking properties are studied in this paper.

ACKNOWLEDGEMENTS

This work was supported by the National Natural Science Foundation of China (No. 69774024)

REFERENCES

Araki M, and T.Hagiwara.(1986). Pole assignment by multi-rate sampled-data output feedback. *Int. J control,* **44**, 1661-1673

Araki M. and K.Yamamoto (1986). Multivariable multirate sampled-data systems: state-space description, transfer characteristics and Nyquist criterion. *IEEE Trans. Automat. Contr.,* **31**, 145-154

Bamieh B. and J.B.Pearson (1992) A general frame work for linear periodic systems with applications to H sampled-data control. *IEEE Trans. Automat. Contr.,* **37**, 418-435

Colaneri P., R.Scattolini and V.Schiavoni (1990). Stabilization of multirate sampled-data linear systems, *Automatica* , **26**, 377-380

Feuer A and G.C.Goodwin. Generalized sample hold functions-frequency domain analysis of robustness, sensitivity and intersample difficulties. *IEEE Trans. Automat. Contr.,* **39**, 1042-1047

Godbout L F, D.Jordan and I.S. Apostolakis (1990). Closed-loop model for general multirate digital control system. *IEE Proc. PtD* , **137**, 329-336

Grasselli O.M, L.Jetto and S.Longhi (1995). Ripple-free dead-beat tracking for multirate sampled-data systems. *Int. J. Control,* **65**, 1437-1455

Hagiwara T and M.Araki (1988) Design of a stable state feedback controller based on the multirate sampling of plant output. *IEEE Trans. Automat. Contr.,* , **33**, 812-819

Kranc G. M. (1957). Input-output analysis of multirate feedback system. *IRE Trans. Automat. Contr.,* **3**, 21-28

Patton R.J, G.P.Liu and P.Patel (1995). Sensitivity properties of multirate feedback control system based on eigen-structure assignment. *IEEE Trans. Automat. Contr.,* **40** , 337-342

Yamamoto Y. (1994). A function space approach to sampled data control systems and tracking problems. *IEEE Trans. Automat. Contr.,* **39**, 703-713

Copyright © IFAC Periodic Control Systems,
Cernobbio-Como, Italy, 2001

LEARNING CONTROL OF BATCH PROCESSES

Dennis Bonné[†] [*] and Sten Bay Jørgensen[‡] [*]

[*] *Department of Chemical Engineering*
Technical University of Denmark
DK-2800 Lyngby, Denmark
[†] *E-mail : db@kt.dtu.dk*
[‡] *E-mail : sbj@kt.dtu.dk*

Abstract: In this contribution it is established how the naturally occurring periodicity of batch and semi-batch processes may facilitate both multimodeling for control and self-improving control performance. It is shown how the highly nonlinear and time-varying dynamic behavior of batch processes may be approximated with a set of simple, linear models in a batch-to-batch modeling framework. A regularized identification scheme for identification of high dimensional models from noisy and sparse data is presented. Using the multimodels derived from data, it is shown how batch wise accumulated knowledge may be utilized to improve controller performance in an iterative learning control scheme formulated in a model predictive control setup. The modeling, identification, and control schemes are demonstrated through learning control of simulated fed-batch yeast fermentations. *Copyright ©2001 IFAC*

Keywords: Periodic systems, Iterative learning control, Modeling batch processes, Identification, Fermentation.

1. INTRODUCTION

Ever-increasing demands for flexible and specialized production methods has led to an augmented industrial need for tools which facilitates modeling and control design of batch processes. Traditional tools have failed in providing such a methodology since batch processes are often highly nonlinear, time-varying, and periodic.

However, because of their periodic nature, batch processes do offer system engineers additional facets which may facilitate modeling and control of batch processes. For the modeling part, data can be collected along an additional dimension, i.e., the batch index dimension, which offers possibilities for new modeling schemes. For the control part, information collected in previous batches can be utilized for improved control of present and future batches.

This paper first presents an extended batch modeling and identification scheme and subsequently how iterative learning control of batch processes may be set up in a model predictive framework. Finally, the paper illustrates the potential of the suggested modeling and control framework and conclusions are given.

2. MODELING BATCH PROCESSES

The implemented batch modeling approach is based on a framework presented by Chin *et al.* (1998). The complex and coherently time-varying behavior of batch processes can be approximated with a model set by modeling every sample time with a simple linear model. Chin *et al.* (1998) suggested using a set of Finite Impulse Response (FIR) models, here the model may be set up using either FIR models or AutoRegressive models with eXogenous inputs (ARX) (Gregersen, 2001).

Operation modes of batch processes will vary with time and hence, the number of measurements, which require trajectory tracking may vary with time. Often it will be necessary to include additional measurements for monitoring and estimation purposes, and to measure the quality of the end-product. Consequently, it seems reasonable to define the following variables:

- Control variable, $u(t) \in \mathbb{R}^{n_u(t)}$.
- Controlled output variable, $y(t) \in \mathbb{R}^{n_y(t)}$, which requires tracking control.
- Secondary output variable, $s(t) \in \mathbb{R}^{n_s(t)}$, used only for monitoring.
- Quality variable, $q \in \mathbb{R}^{n_q}$, only measured after a batch.

Let N be the batch length and define the controlled output sequence as:

$$\boldsymbol{y} \equiv \left[y^T(1) \quad y^T(2) \ldots y^T(N)\right]^T, \tag{1}$$

and the control sequence $\boldsymbol{u}$ (from 0 to $N-1$ though) and disturbance sequence $\boldsymbol{d}$ in a similar manner. Let the initial condition be defined as $y_{ini} \equiv y(0)$. Note, not all initial conditions are measurable and/or physically meaningful.

The output sequence is in general related to the input sequence, initial condition, and disturbance sequence by a nonlinear algebraic model:

$$\boldsymbol{y} = \mathcal{M}(\boldsymbol{u}, y_{ini}, \boldsymbol{d}) = \mathcal{N}(\boldsymbol{u}, \boldsymbol{d}) + \mathcal{N}_{ini}(y_{ini}). \tag{2}$$

Let $\bar{\boldsymbol{y}}$ be the specified and preferably optimal output reference trajectory and let k be the batch index. Then the output error trajectory $\boldsymbol{e}^y$ is defined as:

$$\begin{aligned} \tilde{\boldsymbol{e}}^y_{k+1} &= \tilde{\boldsymbol{e}}^y_k - \boldsymbol{G}^y \Delta \boldsymbol{u}_{k+1} \\ &\quad - \boldsymbol{G}^y_{ini} \Delta y_{ini,k+1} + \boldsymbol{w}^y_k \\ \boldsymbol{e}^y_k &\equiv \bar{\boldsymbol{y}} - \boldsymbol{y} = \tilde{\boldsymbol{e}}^y_k + \boldsymbol{v}^y_k, \end{aligned} \tag{3}$$

where $\Delta \boldsymbol{u}_{k+1} = \boldsymbol{u}_{k+1} - \boldsymbol{u}_k$ and $\Delta y_{ini,k+1} = y_{ini,k+1} - y_{ini,k}$. The matrices $\boldsymbol{G}^y$ and $\boldsymbol{G}^y_{ini}$ are linear system approximations and $\tilde{\boldsymbol{e}}^y_k$ is the part of $\boldsymbol{e}^y_k$ that reappears in $\boldsymbol{e}^y_{k+1}$. Both $\boldsymbol{w}^y_k$ and $\boldsymbol{v}^y_k$ are assumed zero-mean i.i.d. (independent and identically distributed) sequences with respect to k. To illustrate, $\boldsymbol{G}^y$ has the following periodic and causal structure:

$$\begin{aligned} \boldsymbol{G}^y &= \begin{bmatrix} g_{1,0} & 0 & \cdots & 0 \\ g_{2,0} & g_{2,1} & \cdots & 0 \\ \vdots & \vdots & \ddots & \vdots \\ g_{N,0} & g_{N,1} & \cdots & g_{N,N-1} \end{bmatrix} \\ &= \left[\boldsymbol{G}^y_0 \; \boldsymbol{G}^y_1 \cdots \boldsymbol{G}^y_{N-1} \right]. \end{aligned} \tag{4}$$

By applying the same modeling approach to the secondary and quality outputs the resulting combined model is:

$$\begin{aligned} \tilde{\boldsymbol{e}}_{k+1} &= \tilde{\boldsymbol{e}}_k - \boldsymbol{G} \Delta \boldsymbol{u}_{k+1} - \boldsymbol{G}_{ini} e^0_{k+1} + \boldsymbol{w}_k \\ \boldsymbol{e}_k &= \tilde{\boldsymbol{e}}_k + \boldsymbol{v}_k, \end{aligned} \tag{5}$$

where

$$\begin{aligned} \boldsymbol{e} &\equiv [e^{yT} \quad e^{sT} \quad e^{qT}]^T, \\ e^0 &\equiv [\Delta y^T_{ini} \quad \Delta s^T_{ini}]^T. \end{aligned} \tag{6}$$

To enable usage of a Kalman filter for estimation of the initial condition e^0, the initial condition is modelled with a simple batch wise random walk model:

$$\begin{aligned} e^0_{k+1} &= e^0_k + v^0_k \\ \mathfrak{e}^0_k &= e^0_k + \xi^0_k, \end{aligned} \tag{7}$$

where v^0_k is assumed a batch wise white disturbance term, $\mathfrak{e}^0_k$ is the measurement of e^0_k, and ξ^0_k is assumed batch wise white measurement noise.

3. IDENTIFICATION

One major drawback of the proposed modeling approach is the immense dimensionality of the resulting set of models. E.g., assuming causality a process with one output and one input sampled 100 times during a batch will be described by 5150 parameters using FIR models versus 198 model orders and assuming all model orders set equal to three then 594 parameters using ARX models. In practice, this immense dimensionality will render any identification problem ill-conditioned.

Fortunately, there is some structural information in the resulting set of models $\boldsymbol{G}$. Firstly, by causality the upper triangular is $\boldsymbol{0}$, secondly, the time variation can reasonably be assumed smooth ($\boldsymbol{L}_1$), i.e., the diagonal elements in $\boldsymbol{G}$ will vary in a smooth manner, and thirdly, the impulse responses can reasonably be assumed smooth ($\boldsymbol{L}_2$), i.e., the column elements in $\boldsymbol{G}$ will vary smoothly.

Regularization exploits such *a priory* knowledge, reducing the amount of data needed, by introducing constraints or penalties in order to constrain excessive degrees of freedom to values, which satisfy the above specified requirements. Unfortunately, regularization *will* result in biased estimates, i.e., there will be a tradeoff between variance and bias. The optimal tradeoff is found by selecting the regularization weights λ_i that minimize an average prediction error over the validation batches.

Given N_k data sets the unknown parameters x are determined by solving the regularized system in a Least Squares (LS) sense:

$$\begin{bmatrix} b \\ \boldsymbol{0} \\ \boldsymbol{0} \end{bmatrix} = \begin{bmatrix} \boldsymbol{A} \\ \lambda_1 \boldsymbol{L}_1 \\ \lambda_2 \boldsymbol{L}_2 \end{bmatrix} \boldsymbol{x}, \tag{8}$$

given

$$\boldsymbol{A} = \begin{bmatrix} \boldsymbol{A}_1 \\ \vdots \\ \boldsymbol{A}_{N_k} \end{bmatrix}, \quad \boldsymbol{b} = \begin{bmatrix} \boldsymbol{b}_1 \\ \vdots \\ \boldsymbol{b}_{N_k} \end{bmatrix}, \tag{9}$$

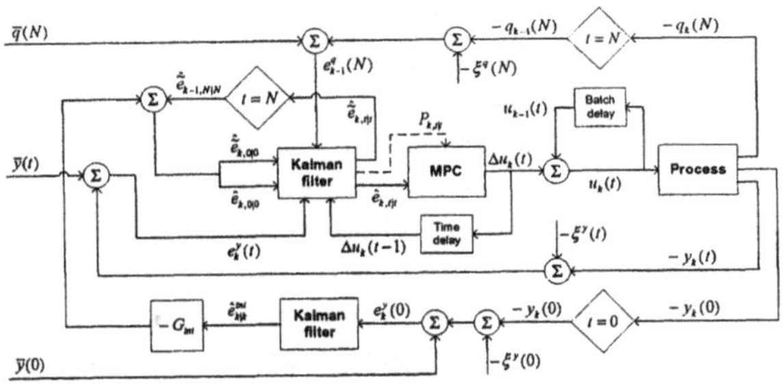

Fig. 1. Block diagram of an Iterative Learning batch Control (ILC) algorithm in a Model Predictive Control (MPC) framework with combined output trajectory tracking and end-product quality control.

where $\boldsymbol{b}$ is the system outputs and $\boldsymbol{A}$ is the system outputs, initial conditions, and inputs. In literature, several approaches to selection of scalar optimal regularization weights have been presented (Hansen, 1996). However, seemingly no work has been reported with vectorial regularization weights based on desired model properties.

Furthermore, usage of high quality data for the identification will decrease estimation variance. The quality of batch identification input data can be determined by statistical comparison to a *normal* reference model obtained through *multiway analysis*. For further discussion of quality or normal batch identification data the reader is referred to Gregersen and Jørgensen (1999), Bro (1998) and Nomikos (1996).

4. LEARNING CONTROL

Based on the above described batch transition model, Model Predictive Control (MPC) can be applied for trajectory tracking. Model predictive control has been comprehensively covered in literature (Morari and Lee, 1999; Muske and Rawlings, 1993; Clarke *et al.*, 1987) and most recently by Mayne *et al.* (2000) and Rao and Rawlings (2000). Chin *et al.* (1998) and Lee *et al.* (2000) presented an Iterative Learning batch Control (ILC) algorithm in a model predictive control framework combining trajectory tracking and end-product quality control. The block diagram of the control algorithm with the extensions introduced in the present paper, i.e., initial conditions and measurement noise, is shown in figure 1.

Having completed a batch run with control one can pose the question: What were the lessons learned? Lesson 1, it was learned how disturbances encountered in the batch run were rejected. This knowledge about the control profile can be used to improve the nominal control profile, giving better rejection of disturbances in the next batch run. Lesson 2, it was learned when and which

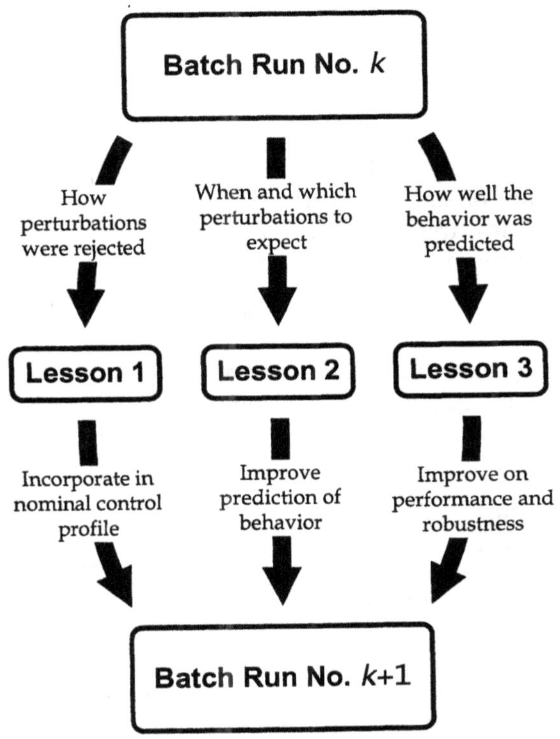

Fig. 2. Concept of Iterative Learning Control.

disturbances to expect in the next batch. This knowledge about the disturbance profile can be used to improve the prediction of the behavior in the next batch. Lesson 3, it was learned how well the behavior was predicted. This knowledge about the predictive capabilities can be used to improve both the performance and also the robustness of the controller for a batch as a whole, i.e., to obtain better but also safer disturbance rejection. These three lessons are depicted both in figure 1 and figure 2.

From equation (5) the following periodic, Linear Time-Varying (LTV) state-space model is obtained for control design:

$$\begin{aligned}\begin{bmatrix}\tilde{e}_{k,t}\\ e_{k,t}\end{bmatrix} &= \begin{bmatrix}\tilde{e}_{k,t-1}\\ e_{k,t-1}\end{bmatrix} - \begin{bmatrix}\boldsymbol{G}_{t-1}\\ \boldsymbol{G}_{t-1}\end{bmatrix}\Delta u_k(t-1),\\ \mathfrak{e}_k(t) &= \begin{bmatrix}\boldsymbol{0} & \boldsymbol{H}_t\end{bmatrix}\begin{bmatrix}\tilde{e}_{k,t}\\ e_{k,t}\end{bmatrix} + \xi(t),\end{aligned} \qquad (10)$$

for $t = 1, 2, \ldots, N$, where $\mathfrak{e}_k(t)$ is the measurement of the output error $e_k(t)$, and $\xi(t)$ is assumed zero-mean i.i.d. measurement noise. At the start of a batch the states in (10) satisfy:

$$\begin{aligned}\begin{bmatrix}\tilde{e}_{k,0}\\ e_{k,0}\end{bmatrix} &= \begin{bmatrix}\boldsymbol{I} & \boldsymbol{0}\\ \boldsymbol{I} & \boldsymbol{0}\end{bmatrix}\begin{bmatrix}\tilde{e}_{k-1,N}\\ e_{k-1,N}\end{bmatrix} - \begin{bmatrix}\boldsymbol{G}_{ini}\\ \boldsymbol{G}_{ini}\end{bmatrix} e_k^0\\ &+ \begin{bmatrix}w_{k-1}\\ w_{k-1}\end{bmatrix} + \begin{bmatrix}\boldsymbol{0}\\ v_k\end{bmatrix}.\end{aligned} \qquad (11)$$

Note that the structure of $\boldsymbol{H}_t$ and $e_k(t)$ depend on which measurements are available at sample t.

By introducing,

$$G_t^m = [G_t \quad \cdots \quad G_{t+m-1}],$$

$$\Delta u_{k,t}^m = \begin{bmatrix} \Delta u_k(t) \\ \vdots \\ \Delta u_k(t+m-1) \end{bmatrix}, \tag{12}$$

it may be realized from (10) that the optimal estimate of the error trajectory of the kth batch given measurements up to time t and m future control moves $\hat{e}_{k,t+m|t}$ is given by:

$$\hat{e}_{k,t+m|t} = \hat{e}_{k,t|t} - G_t^m \Delta u_{k,t}^m. \tag{13}$$

The optimal input movements are then determined by employing the standard quadratic objective function known from conventional MPC control.

$$\min_{\Delta u_{k,t}^m} \left[\frac{1}{2} \{ \hat{e}_{k,t+m|t}^T Q_t \hat{e}_{k,t+m|t} + \Delta u_{k,t}^{mT} R_t \Delta u_{k,t}^m \} \right], \tag{14}$$
$$s.t. \quad \hat{e}_{k,t+m|t} = \hat{e}_{k,t|t} - G_t^m \Delta u_{k,t}^m,$$

where Q_t and R_t are weighting matrices. The analytical solution to (14) is given by:

$$\Delta u_{k,t}^m = [G_t^{mT} Q_t G_t^m + R_t]^{-1} G_t^{mT} Q_t \hat{e}_{k,t|t}, \tag{15}$$

where only $\Delta u_k(t)$ from $\Delta u_{k,t}^m$ is implemented.

In practice input signals must be constrained to ensure non-provoking or system friendly and physically meaningful inputs, but also to dampen the effect of bad predictor estimates, which otherwise could result in very aggressive erroneous inputs. If possible the inputs should be constraint such that the process avoids regions of instability increasing the level of safety in the event that the controller should set out. Furthermore, the predicted outputs may need to be constrained to meet additional safety requirements and ensure operation in a physically meaningful operation region. Therefore, (14) should be extended to:

$$\min_{\Delta u_{k,t}^m, \epsilon_{k,t}} \left[\frac{1}{2} \{ \Delta u_{k,t}^{mT} \mathcal{Q}_t \Delta u_{k,t}^m + 2 \mathcal{R}_t^T \Delta u_{k,t}^m + \epsilon_{k,t}^T \mathcal{S}_t \epsilon_{k,t} \} \right], \tag{16}$$
$$s.t. \quad \mathcal{A}_t \Delta u_{k,t}^m \le \mathcal{B}_t(\epsilon_{k,t}),$$
$$\epsilon_{k,t} \ge \mathbf{0},$$

where $\mathcal{Q}_t = G_t^{mT} Q_t G_t^m + R_t$ and $\mathcal{R}_t^T = -\hat{e}_{k,t|t}^T Q_t G_t^m$. By proper configuration of $\mathcal{A}$ and $\mathcal{B}$ both the outputs and the inputs may be constrained. The *constraint softening* slack variable ϵ insures feasible solutions and $\mathcal{S}_t$ is the corresponding weighting matrix.

The presented control technique transfers the refined control error sequence as initial condition for the next batch. Furthermore, it computes the input sequence as a change from the previous batch based on the control error prediction. This use of information from previous batches provides the control algorithm with an offset elimination action along the batch index (Chin *et al.*, 1998). Thus, through this action the control algorithm is expected to eliminate offset errors both from disturbances that repeat themselves from batch to batch, and from model bias (Chin *et al.*, 1998), i.e., the disturbances represented by w. Lee *et al.* (1997) have proven convergence for a batch MPC controller based on state-space model (10). This proof can be extended to hold for this algorithm also.

5. STATE ESTIMATION

The optimal (i.e., minimum variance) estimates of the model states are obtained from a *Kalman filter*. Here the Kalman filter can be thought of as being made up of a set of Kalman filters or as an Extended Kalman Filter (EKF) without subsampling.

By defining R_w and R_v as the covariance matrices of w_k and v_k, respectively, the predictive part of the stabilized Kalman filter (Thornton and Bierman, 1978) is,

$$\begin{bmatrix} \hat{\hat{e}}_{k,t|t-1} \\ \hat{e}_{k,t|t-1} \end{bmatrix} = \begin{bmatrix} \hat{\hat{e}}_{k,t-1|t-1} \\ \hat{e}_{k,t-1|t-1} \end{bmatrix} - \begin{bmatrix} G_{t-1} \\ G_{t-1} \end{bmatrix} \Delta u_k(t-1), \tag{17}$$
$$\begin{bmatrix} \tilde{P}_{k,t|t-1} & \hat{P}_{k,t|t-1} \\ \hat{P}_{k,t|t-1} & P_{k,t|t-1} \end{bmatrix} = \begin{bmatrix} \tilde{P}_{k,t-1|t-1} & \hat{P}_{k,t-1|t-1} \\ \hat{P}_{k,t-1|t-1} & P_{k,t-1|t-1} \end{bmatrix},$$

and the updating part is,

$$\begin{bmatrix} \hat{\hat{e}}_{k,t|t} \\ \hat{e}_{k,t|t} \end{bmatrix} = \begin{bmatrix} \hat{\hat{e}}_{k,t|t-1} \\ \hat{e}_{k,t|t-1} \end{bmatrix} - \mathcal{K}_{k,t}[H_t \hat{e}_{k,t|t-1} - z_k(t)],$$
$$\mathcal{K}_{k,t} = \begin{bmatrix} \hat{P}_{k,t|t-1} \\ P_{k,t|t-1} \end{bmatrix} H_t^T,$$
$$\mathcal{P}_{k,t} = H_t P_{k,t|t-1} H_t^T + R_{n,t}, \tag{18}$$
$$\overline{\mathcal{K}}_{k,t} = 0.5 K_{k,t} \mathcal{P}_{k,t} - \mathcal{K}_{k,t},$$
$$\begin{bmatrix} \tilde{P}_{k,t|t} & \hat{P}_{k,t|t} \\ \hat{P}_{k,t|t} & P_{k,t|t} \end{bmatrix} = \begin{bmatrix} \tilde{P}_{k,t|t-1} & \hat{P}_{k,t|t-1} \\ \hat{P}_{k,t|t-1} & P_{k,t|t-1} \end{bmatrix} + K_{k,t} \overline{\mathcal{K}}_{k,t}^T + \overline{\mathcal{K}}_{k,t} K_{k,t}^T,$$
$$K_{k,t} = \begin{bmatrix} \hat{P}_{k,t|t-1} \\ P_{k,t|t-1} \end{bmatrix} H_t^T \times \left[H_t P_{k,t|t-1} H_t^T + R_{n,t} \right]^{-1},$$

with the initial conditions,

$$\begin{bmatrix} \hat{\tilde{e}}_{k,0|0} \\ \hat{e}_{k,0|0} \end{bmatrix} = \begin{bmatrix} \hat{\tilde{e}}_{k-1,N|N} \\ \hat{\tilde{e}}_{k-1,N|N} \end{bmatrix} - \begin{bmatrix} G_{ini} \\ G_{ini} \end{bmatrix} \hat{e}^0_{k,k}, \qquad (19)$$

$$\tilde{\boldsymbol{P}}_{k,0|0} = \tilde{\boldsymbol{P}}_{k-1,N|N} + \boldsymbol{G}_{ini} P_{e^0 k|k} \boldsymbol{G}_{ini}^T + \boldsymbol{R}_w,$$

$$\hat{\boldsymbol{P}}_{k,0|0} = \tilde{\boldsymbol{P}}_{k-1,N|N} + \boldsymbol{G}_{ini} P_{e^0 k|k} \boldsymbol{G}_{ini}^T + \boldsymbol{R}_w,$$

$$\boldsymbol{P}_{k,0|0} = \tilde{\boldsymbol{P}}_{k-1,N|N} + \boldsymbol{G}_{ini} P_{e^0 k|k} \boldsymbol{G}_{ini}^T + \boldsymbol{R}_w + \boldsymbol{R}_v.$$

In the above $\boldsymbol{P}$ is the prediction error covariance of the error trajectory, $\tilde{\boldsymbol{P}}$ is the prediction error covariance of the part of error trajectory that reappears, $\hat{\boldsymbol{P}}$ is the cross-covariance between $\boldsymbol{e}$ and $\tilde{\boldsymbol{e}}$, $\boldsymbol{\mathcal{P}}$ is the covariance of the innovations, $\boldsymbol{K}$ is the Kalman gain, $\boldsymbol{\mathcal{K}}$ is the normalized Kalman gain, and $R_{n,t}$ is the measurement noise covariance at time t. Note that the noise sequences are assumed both mutually independent and independent of the error sequences at time t.

The estimated initial condition $\hat{e}^0_{k,k}$ is obtained from:

$$\begin{aligned} \hat{e}^0_{k|k} &= (\boldsymbol{I} - \kappa_k)\, \hat{e}^0_{k-1|k-1} + \kappa_k z^0_k, \\ \kappa_k &= \left(P_{e^0 k-1|k-1} + R_{v^0}\right) \qquad (20) \\ &\quad \times \left(P_{e^0 k-1|k-1} + R_{v^0} + R_{n^0}\right)^{-1}, \\ P_{e^0 k|k} &= (\boldsymbol{I} - \kappa_k)\left(P_{e^0 k-1|k-1} + R_{v^0}\right), \end{aligned}$$

where $P_{e^0 k|k}$ is the prediction error covariance of the initial condition, κ_k is the Kalman gain, R_{v^0} is the covariance of v^0, and R_{n^0} is the measurement noise covariance.

For further discussion of numerically stable implementations of Kalman filters the reader is referred to Grewal and Andrews (1993) and Kailath *et al.* (2000).

6. APPLICATION

To validate the presented modeling and control framework a FIR model was identified on data from yeast (*Saccharomyces ceravisiae*) producing fed-batch fermentations simulated with a biochemically structured model (Lei *et al.*, 2001).

As controlled output variables the ethanol concentration (EtOH), Carbon dioxide Evolution Rate (CER), Oxygen Uptake Rate (OUR), and Respiratory Quotient (RQ) were chosen and the feed rate was chosen as control variable. In fact, RQ $\equiv$ CER/OUR, but here RQ was modelled as a measurement to see if its dynamics could be captured with the suggested modeling framework. White measurement noise was added to all inputs and outputs: EtOH $\pm 3\%$, OUR $\pm 2\%$, CER $\pm 2\%$, RQ $\pm 2\%$, feed rate $\pm 1\%$. The level of measurement noise was based on laboratory experiments and industrial data. It was assumed that there were no time delays.

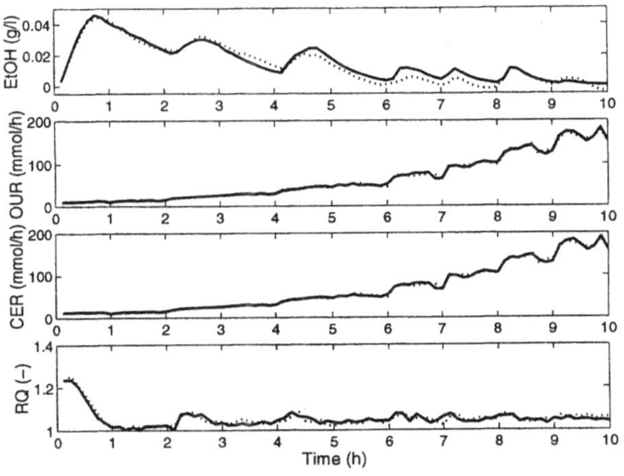

Fig. 3. Validation of noisy FIR model identification. True output, solid line; predicted output, dotted line.

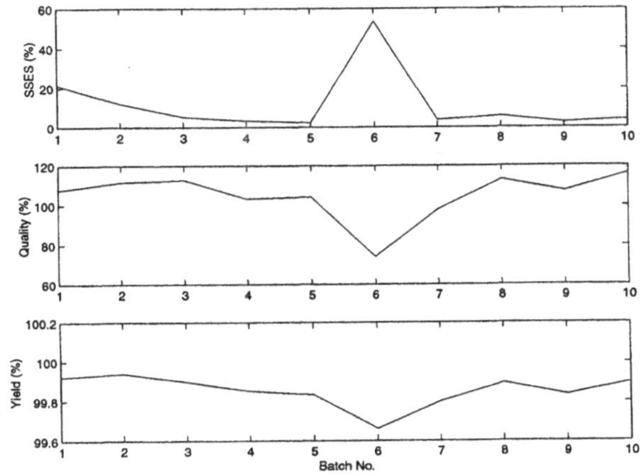

Fig. 4. Summed squared error sequence, quality, and yield evolution in noisy trajectory tracking of EtOH, OUR, CER, and RQ. No control corresponds to 100%.

The model was identified on data from 10 normal batch runs and cross-validated on 3 normal data sets. The identification was regularized to ensure smooth model parameter evolutions and smooth impulse responses. The regularization weights were optimized by trial and error, minimizing the mean prediction error from pure simulation model cross-validation. As demonstrated in figure 3, the identified model possesses reasonable predictive capabilities.

Based on the identified FIR model set the reference profiles of the four outputs were tracked for further model validation. In the control simulation scenario the initial conditions were randomly perturbed by $\pm 10\%$ in each batch run and the feed concentration was perturbed by a persistent 10% reduction in all batch runs.

In figure 4 it can be observed that the controller rejected approximately 80% of the Summed Squared Error Sequence (SSES) in the first batch run and that after having been trained on the first three batch runs more than 90% of SSES with exception of batch run no. 6. The sluggish performance in batch run no. 6 was due to a particularly intractable disturbance direction, i.e., the yeast had been starved before seeding. However, the controller still rejected approximately 40% of SSES and the performance in the subsequent batch run was not affected. In addition to SSES, also the end-product quality (i.e., high purity) and the yield are shown in figure 4, and it can be observed that the quality can be improved by output tracking and that the yield may endure a minute decrease as an expense of high tracking performance.

7. CONCLUSION

In this contribution it is established how the periodic nature of batch processes can be utilized for modeling and learning control of batch processes. For modeling purposes, the periodic nature of batch processes enables extreme multimodeling using one simple, linear model per sample instant. To obtain these multimodels of high dimensionality from sparse and noisy operation data, an identification scheme has been developed based on generalized ridge regression, in which the model parameters are constrained by desired model properties. Utilizing the resulting multimodels, an iterative learning batch control algorithm can be set up in a model predictive control framework, offering batch-to-batch improving feedback disturbance rejection. In this fashion, the framework provides reliable operation with respect to both process outputs and end-product quality. The potential of the presented tools have been demonstrated successfully applying iterative learning control to simulated yeast fermentations.

REFERENCES

Bro, Rasmus (1998). *Multi-way Analysis in the Food Industry.* Academish Proefschrift. Universiteit van Amsterdam.

Chin, In-Shik, Kwang S. Lee and Jay H. Lee (1998). A unified framework for control of batch processes. '98 AIChE Annual Meeting in Miami.

Clarke, D. W., C. Mohtadi and P. S. Tuffs (1987). Generalized predictive control — part i. the basic algorithm. *Automatica* **23**(2), 137–148.

Gregersen, Lars (2001). Monitoring and Fault Diagnosis of Fermentation Processes. PhD thesis. Technical University of Denmark.

Gregersen, Lars and Sten Bay Jørgensen (1999). Supervision of fed-batch fermentations. *Chemical Engineering Journal* **75**, 69–76.

Grewal, Mohinder S. and Angus P. Andrews (1993). *Kalman Filtering — Theory and Practice.* Prentice Hall. N. J. Englewood Cliffs.

Hansen, Per Christian (1996). *Rank-Deficient and Discrete Ill-Posed Problems.* Polyteknisk Forlag. Lyngby.

Kailath, Thomas, Ali H. Sayed and Babak Hassibi (2000). *Linear Estimation.* Prentice Hall. N. J. Englewood Cliffs.

Lee, Jay H., Kwang S. Lee and Won C. Kim (2000). Model-based iterative learning control with a quadratic criterion for time-varying linear systems. *Automatica* **36**(5), 641–657.

Lee, Kwang S., Jay H. Lee, In-Shik Chin and Hyuk J. Lee (1997). A model predictive control technique for batch processes and its application to temperature tracking control of an experimental batch reactor. '97 AIChE Annual Meeting in Los Angeles.

Lei, F., M. Rotbøll and S. B. Jørgensen (2001). A Biochemically Structured Model for *Saccharomyces cerrevisiae.* Accepted for Journal of Biotechnology.

Mayne, D. Q., J. B. Rawlings, C. V. Rao and P. O. M. Scokaert (2000). Constrained model predictive control: Stability and optimality. *Automatica* **36**, 789–814.

Morari, Manfred and Jay H. Lee (1999). Model predictive control: past, present and future. *Computers and Chemical Engineering* **23**, 667–682.

Muske, Kenneth R. and James B. Rawlings (1993). Model predictive control with linear models. *AIChE Journal* **39**(2), 262–287.

Nomikos, Paul (1996). Detection and diagnosis of abnormal batch operations based on multi-way principal component analysis. *ISA Transactions* **35**(3), 259–266.

Rao, Christopher V. and James B. Rawlings (2000). Linear programming and model predictive control. *Journal of Process Control* **10**, 283–289.

Thornton, C. L. and G. J. Bierman (1978). Filtering and error analysis via the UDU^T covariance factorization. *IEEE Transactions on Automatic Control* **23**(5), 901–907.

Copyright © IFAC Periodic Control Systems,
Cernobbio-Como, Italy, 2001

MIMO MULTI-PERIODIC REPETITIVE CONTROL SYSTEMS: A LYAPUNOV ANALYSIS

D.H.Owens, L.M.Li, S.P.Banks

Department of Automatic Control and Systems Engineering,
University of Sheffield, Mappin Street, Sheffield S1 3JD, UK
D.H.Owens@sheffield.ac.uk & L.Li@sheffield.ac.uk

Abstract: In this paper a new type of repetitive control problem where two or more periods exist in reference and disturbance signals is considered. A Lyapunov stability analysis under a positive real condition is provided and a new method of designing compensators in a negative feedback loop to create positive realness is outlined. *Copyright © 2001 IFAC.*

Keywords: MIMO, Repetitive Control, Lyapunov, Feedback Control.

1. *INTRODUCTION*

Signals associated with repetitive or rotating actions are often of periodic nature or are subjected to forms of periodic disturbances. In order to track or/and reject this kind of signals, an internal model that generates the corresponding periodic signals, should be included in the closed-loop (Francis and Wonham, 1975). This system is called a repetitive control system and has been widely used in control problems with robotic (Kaneko and Horowitz, 1997), motor (Kobayashi *et al.*, 1999), rolling process (Garimella and Srinivasan, 1996), rotating mechanisms (Fung, *et al.*, 2000), and much more (Tzou, *et al.*, 1999; Moon, 1998; Manayathara, *et al.*, 1996).

If the signal need to be tracked or rejected contains only one single frequency, a finite-dimensional model can serve as internal model (Bai and Wu, 1998). While more generally in practical situations a periodic signal contains many relevant frequencies—fundamental frequency and its harmonics within a certain bandwidth, and a (higher order) compensator is required but an infinite-dimensional internal model should be more appropriate. The one proposed is built on time-delay system. Perhaps the first infinite-dimensional repetitive controller was proposed by Inoue *et al.* (1981) and can be simplified in common form as illustrated in Figure 1(a), along with the poles of the controller. The infinitely many poles of the delay system on the imaginary axis are $ik\nu, k = 0, \pm 1, \ldots,$ where $\nu = 2\pi/\tau$ is called fundamental frequency, which make the system capable of generating signal containing frequency components of reference/disturbance. Hara *et al.* (1988) proposed a modified infinite-dimensional repetitive controller [see Figure 1(b)] by introducing a low-pass filter aiming at improving the system stability, at a cost of losing tracking accuracy at high frequencies.

In many cases the reference and/or disturbance signals may contain different fundamental frequencies. The ratio of these frequencies can be irrational. This leads to the idea of the so-called multi-periodic repetitive control, which has received very little attention (Weiss, 1997). In this paper the MIMO multi-periodic control problem is studied using Lyapunov techniques. It is proved that the stability of the multi-periodic repetitive control system is guaranteed if the plant satisfies a positive real condition. A new procedure to create positive realness by feedforward and feedback compensation is presented. A simulation example is also presented.

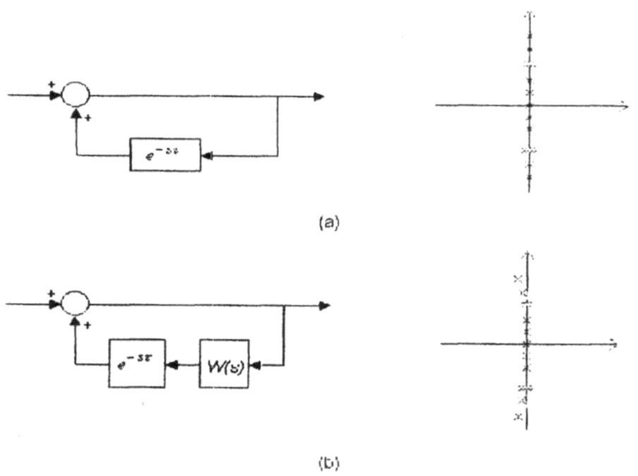

Fig. 1. Two common forms of infinite-dimensional repetitive controller and their poles location.

2. LYAPUNOV STABILITY ANALYSIS

The MIMO multi-periodic repetitive control system is shown in Figure 2.

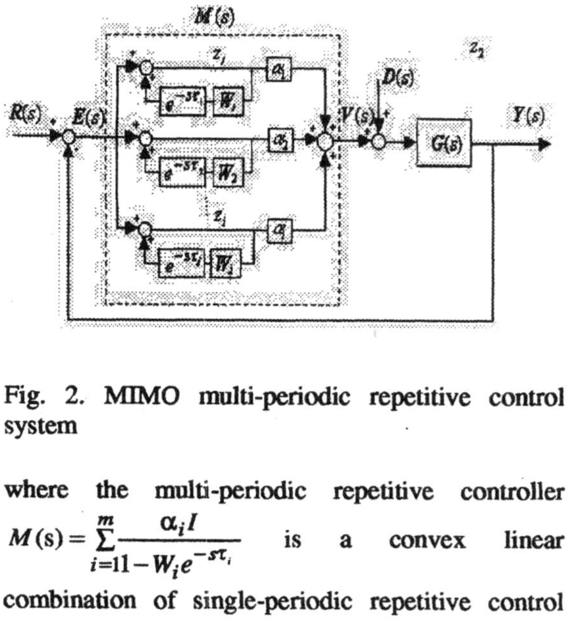

Fig. 2. MIMO multi-periodic repetitive control system

where the multi-periodic repetitive controller $M(s) = \sum_{i=1}^{m} \frac{\alpha_i I}{1 - W_i e^{-s\tau_i}}$ is a convex linear combination of single-periodic repetitive control elements. That is, $\sum_{i=1}^{m} \alpha_i = 1, \alpha_i > 0$ and $\tau_i, i = 1, \ldots, m$ (the periods of the components of the external signals reference r and disturbance d) are assumed known.

The plant Σ_G is finite-dimensional, linear time-invariant and its transfer function $G(s)$ is a $p \times p$ matrix. Initially the plant G is assumed to be positive real but later a procedure to create positive realness of a class of general plants is outlined.

Definition (Anderson and Vongpanitherd, 1973): The system Σ_G is said to be positive real(PR) if

1). All elements of its transfer function $G(s)$ are analytic in $\text{Re}[s] > 0$,

2). $G(s)$ is real for real positive s, and

3). $G(s) + G^*(s) \geq 0$ for $\text{Re}[s] > 0$.

where the superscript $*$ denotes complex conjugate transposition.

The following is the well-known positive real lemma.

Lemma (Anderson and Vongpanitherd, 1973): Assume

$$\dot{x} = Ax + Bu$$

$$y = Cx + Du, \quad x(0) = x_0$$

is a minimum realisation of $\Sigma_{G_{p \times q}}$. *Then* $G(s)$ *is positive real if and only if there exist matrices* $Q \in R^{n \times n}$, $L \in R^{n \times k}$, and $W \in R^{k \times p}$ *with* $Q^T = Q > 0$ *such that*

$$QA + A^T Q = -LL^T \quad (1)$$

$$QB = C^T - LW \quad (2)$$

$$D + D^T = H^T H \quad (3)$$

Here the superscript ^{T} denotes transposition. Note that LL^T is positive semidefinite.

It is obvious that if $G(s)$ is strictly proper, that is, $D = 0$, which means $W = 0$, then (1)~(3) become

$$QA + A^T Q = -LL^T \quad (4)$$

$$B = C^T \quad (5)$$

The main result of stability analysis can be stated below.

Theorem 1: Suppose that both reference r *and disturbance* d *are multi-periodic with components of period* $\tau_i, i = 1, \ldots, m$. Suppose *that* r *can be exactly tracked by some choice of input and that* Σ_G *is positive real and strict proper. Then the MIMO multi-periodic repetitive system in Figure 2 is globally asymptotically stable in the sense that the error signal* $e \in L_2^m(0, \infty) \cap L_\infty^m(0, \infty)$, *the state* $x(\cdot) \in L_\infty^n(0, \infty)$ *and* $Lx(\cdot) \in L_2^k(0, \infty)$.

Proof: Without loss of generality, take $r = 0$. The system Σ_G has the form

$$\dot{x} = Ax + B(v + d)$$

$$y = Cx \qquad x(0) = x_0$$

Let $d = \sum_{i=1}^{m} \alpha_i d_i$ where d_i has period τ_i. From Figure 2 it yields

$$v(t) = \sum_{i=1}^{m} \alpha_i z_i ; \qquad \sum_{i=1}^{m} \alpha_i = 1$$

$$z_i(t) = W_i z_i(t - \tau_i) + e(t) = W_i z_i(t - \tau_i) - y(t) \quad (6)$$

Noting that $v + d = \sum_{i=1}^{m} \alpha_i (z_i + d_i)$, it is easily verify that $\tilde{z}_i = z_i + d_i$ satisfies the same evolution equation as z_i. Hence, without loss of generality, it is possible to assume that $d_i = 0$, $1 \le i \le m$, (i.e. $d = 0$) for the rest of the proof.

Now introducing a positive definite Lyapunov function V of the form

$$V = x^T Q x + \sum_{i=1}^{m} \alpha_i \int_{t-\tau_i}^{t} \|z_i(\theta)\|^2 d\theta \quad (7)$$

By differentiating V along solutions and using the positive real lemma

$$\dot{V} = \dot{x}^T Q x + x^T Q \dot{x} + \sum_{i=1}^{m} \alpha_i [\|z_i(t)\|^2 - \|z_i(t - \tau_i)\|^2]$$

$$= -x^T L L^T x + 2 y^T v(t)$$

$$+ \sum_{i=1}^{m} \alpha_i [\|z_i(t)\|^2 - \|z_i(t - \tau_i)\|^2]$$

$$= -x^T L L^T x - \|y\|^2$$

$$+ \sum_{i=1}^{m} \alpha_i [\|y\|^2 + 2 y^T z_i(t) + \|z_i(t)\|^2 - \|z_i(t - \tau_i)\|^2]$$

(8)

Notice that from (6) the last term in (8) is $-\sum_{i=1}^{m} \alpha_i (1 - W_i^2) \|z_i(t - \tau_i)\|^2$ and notice that $0 \le W_i \le 1$, thus (8) becomes

$$\dot{V} = -x^T L L^T x - \|y\|^2$$

$$- \sum_{i=1}^{m} \alpha_i (1 - W_i^2) \|z_i(t - \tau_i)\|^2 < 0 \quad (9)$$

Integrating (9) and using (7) and the positivity of V yield

$$0 < V(0) = V + \int_0^t x^T L L^T x dt + \int_0^t \|y\|^2$$

$$+ \int_0^t \sum_{i=1}^{m} \alpha_i (1 - W_i) \|z_i(t - \tau_i)\|^2 < \infty$$

from which

$Lx(\cdot) \in L_2^k(0, \infty)$, $x(\cdot) \in L_\infty^n(0, \infty)$

and $y(\cdot) \in L_2^m(0, \infty)$ which proves the result. ■

Note: Extension to this result is possible to include low-pass filtering actions. There are omitted for brevity but will be included in the final version of the paper if space permits.

3. *CREATE POSITIVE REALNESS BY FEEDFORWARD AND FEEDBACK COMPENSATIONS*

It has been proved in section 2 that stability of closed-loop is guaranteed if the plant is positive real. This is, however, not necessarily true in most practical situations. Therefore the question arises: how to achieve positive realness by the use of compensators. In the literature this problem can be called 'positive real control' and efforts have been made by numerous authors. In this section a new and very simple algorithm to achieve positive realness for a class of strict proper plant by feedforward and output feedback compensation is introduced to indicate the potential of the approach.

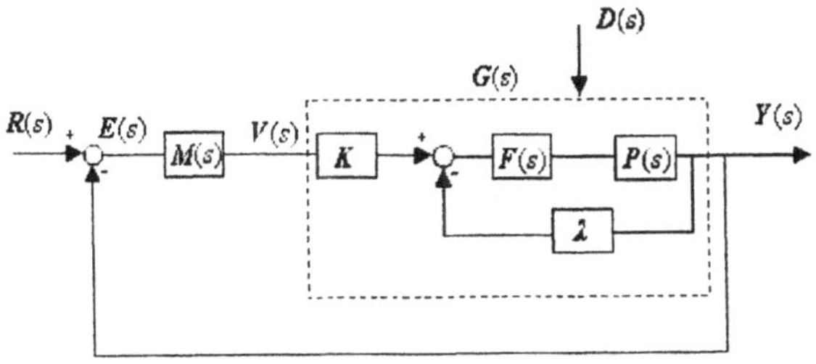

Fig. 3. PR compensation configuration

First the criterion for testing PR is reviewed.

Theorem 2 (Anderson and Vongpanitherd, 1973): Let $G(s)$ be a real rational matrix of functions of s. Then $G(s)$ is PR if and only if

1). No element of $G(s)$ has a pole in $\mathrm{Re}[s] > 0$,

2). $G(j\omega) + G^*(j\omega) \ge 0$ for all real ω, with $j\omega$ not a pole of any elements of $G(s)$.

Suppose that non-PR plant P is minimum phase and strictly proper with relative degree 1.

If define P as the state space model $P \sim (A, B, C, 0)$

then $\qquad P = C(sI - A)^{-1} B$

where we assume that that $(CB) = (CB)^T > 0$.

The inverse of P can be expressed in the form

$$L = P^{-1} = (CB)^{-1} s - \underbrace{(CB)^{-1}(CAB)}_{\text{Constant}} + \underbrace{H(s)}_{\text{Strict Proper and asymptotically stable}} \quad (10)$$

Then

$$\tilde{L} = L(j\omega) + L^*(j\omega)$$
$$= j\omega[(CB)^{-1} - ((CB)^{-1})^T] \qquad (11)$$
$$- 2(CB)^{-1}(CAB) + H(j\omega) + H^*(j\omega)$$

If we denote

$$F_0 = -\frac{1}{2}[-2(CB)^{-1}(CAB) + H(j\omega) + H^*(j\omega)] \qquad (12)$$

then from (11)

$$\tilde{L} + 2F_0 = j\omega[(CB)^{-1} - ((CB)^{-1})^T] = 0 \qquad (13)$$

Therefore choosing $\lambda = \lambda I$, with $\lambda I \geq F_0$, it follows that

$$\tilde{L} + 2\lambda I \geq 0, \; \omega > 0 \qquad (14)$$

which implies that G is PR where G is given by

$$G = (I + P\lambda)^{-1} PK \geq 0 \qquad (15)$$

This procedure can be graphically interpreted in Figure 4 for a SISO system. Equation (12) also gives a simple method of determining a least value of λ for P^{-1} in the case of SISO control.

Notice that if $(CB) \neq (CB)^T$ then it is always possible to choose $F = (CB)^{-1}$ to achieve a symmetry condition. In this case

$$L_{new} = (PF)^{-1} = sI - \underbrace{(CAB)}_{\text{Constant}} + \underbrace{(CB)H(s)}_{\text{Strict Proper}} \qquad (16)$$

which leads to the choice of λ_{new} so that

$$\tilde{L}_{new} + 2\lambda_{new} I \geq 0 \qquad (17)$$

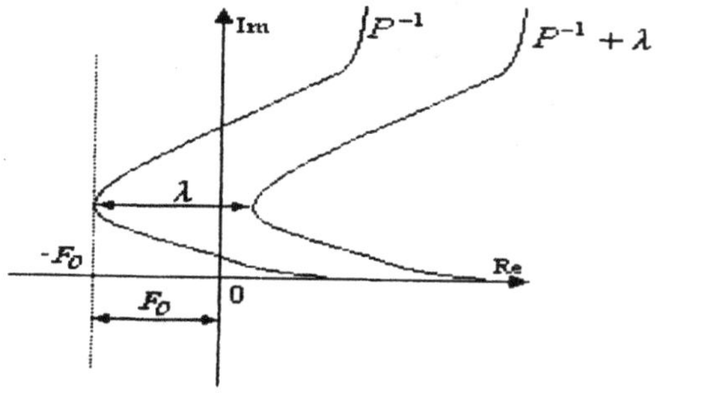

Fig. 4. PR interpretation in inverse polar plane

4. ISSUES ON THE CHOICES OF COMPENSATION FACTORS F, λ AND GAIN K

The design of a repetitive control system consists of two procedures, first classical stabilising controllers are designed independently from repetitive controller in order to obtain general performances, then the multi-periodic repetitive controller is added on to achieve high tracking/rejecting accuracy. Because there exists time-delays within the repetitive controller, the classical controllers play an important role in system transient performance before the repetitive controller cuts in and during the time-delay intervals. In this paper, apart from enforcing PR, which guarantees stability, controllers $F(s)$, λ and K can be also designed for robustness, poles assignment or other improvement, etc. Here just brief discussions are given on the choice of these controllers in terms of steady-state error. For simplicity, SISO form is used.

From Figure 3

$$G = \frac{KPF}{1 + \lambda PF} = \frac{KP'}{1 + \lambda P'} \qquad (18)$$

where $\lambda \geq -\inf_{\omega} \mathrm{Re}[P'^{-1}]$ is chosen in order to achieve a PR condition.

The steady state error for step reference can be determined as

$$E_{ss} = \lim_{s \to 0} \frac{1}{1+G} = \frac{1}{1+G(0)} \qquad (19)$$

By assuming $F(0) = 1$, (19) becomes

$$E_{ss} = \frac{1}{1 + \frac{KP(0)}{\lambda P(0)}} \qquad (21)$$

therefore the steady state error can be reduced either by increasing gain K or by reducing λ. The advantage of enforcing PR is that the value of K will never affect the stability of the system, but increasing K tends to reduce the relative stability margin, so K should be properly increased to achieve faster response and smaller steady state error. The feedforward compensator $F(s)$ could be designed in a manner to reduce the feedback compensating factor λ. This point can be better illustrated in a simulation example below.

5. SIMULATION EXAMPLES

For the sake of simplicity, a SISO system is examined to illustrate the control system performance. W_i is set to be constants close to one throughout the simulation. The plant under control is

$$P = \frac{4s^2 + 2s + 300}{s(s^2 + 4s + 40)}$$

The property of multi-periodic control system will be shown in two ways, namely, typical application to different fundamental frequencies, and an extension to the single fundamental frequency case.

5.1. Two Fundamental Frequencies Case

The reference is a mix of harmonics of two fundamental frequencies 0.5Hz and 1.7Hz, that is,

$$r = r_1 + r_2$$

where $r_1 = \sin\pi t + 2\sin 3\pi t + 0.7\sin 5\pi t$

$$r_2 = \sin 3.4\pi t + 3\sin 6.8\pi t$$

therefore the two fundamental periods are

$$\tau_1 = 1/0.5 = 2, \quad \tau_2 = 1/1.7$$

For simplicity, choose $\alpha_1 = 0.5$ and $\alpha_2 = 0.5$, which means equal weighting in the controller on two components of reference signal. The gain K is chosen as 100.

Two simulation experiments are conducted.

(1). Let $F \equiv 1$

In this case, $\hat{\lambda}$ can be determined as

$$\hat{\lambda} = -\inf_{\omega} \mathrm{Re}[(PF)^{-1}] \cong 20.27$$

so choose $\lambda = 22$ to satisfy the positive real condition. The simulation result of the convergent error $e(t)$ is shown in Figure 5.

(2). Choose $F = \dfrac{1+s}{1+0.1s}$, a phase-lead compensator. It is easy to verify that $F(0) = 1$.

In this case, $\hat{\lambda} = -\inf_{\omega} \mathrm{Re}[(PF)^{-1}] \cong 1.749$

which is significantly reduced compared with the above case. Choose $\lambda = 1.76$ in this case.

The simulation result is given in Figure 6(solid). It is clear that the accuracy in the second case has been almost improved by an order of magnitude compared to the first one during transient process, by simply designing feedforward compensator.

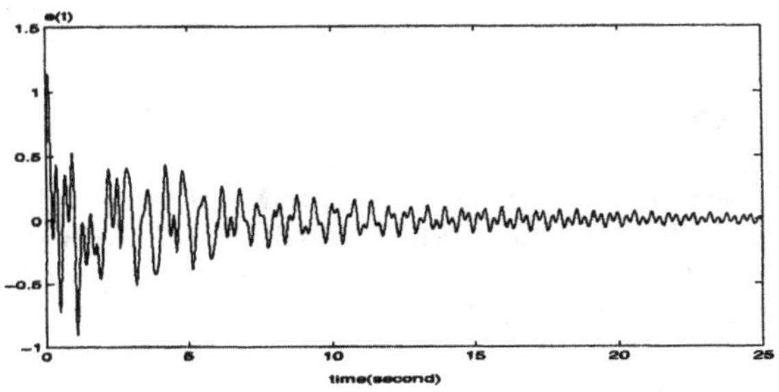

Fig. 5. The error when feedforward compensator $F \equiv 1$

Simulations show that this repetitive controller is also capable of effectively rejecting a periodic disturbance with the same frequency components.

Suppose in the followings that the disturbance is a square wave at a period of 3.4Hz and at peak value ± 10. A square wave is chosen to indicate that the scheme can cope with signals with an infinite (Fourier) frequency content. The tracking error is also demonstrated in Figure 6(dashed) and is almost identical to that of disturbance-free case (Figure 6, solid). The simulation result suggests that despite the significant existence of the disturbance signal, its influence is negligible.

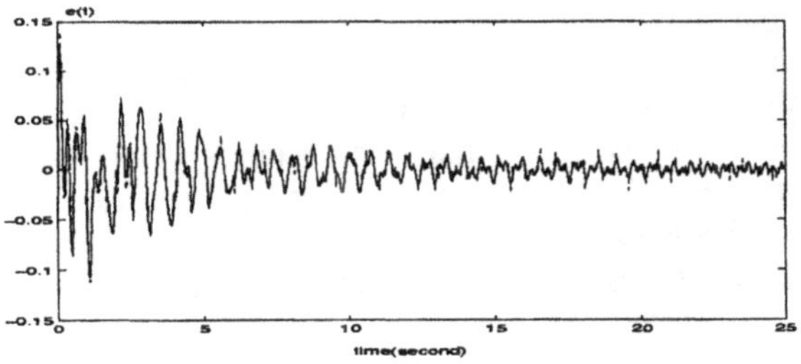

Fig. 6. The error when F is a phase-lead compensator—Solid for disturbance-free, and Dashed for square wave disturbance involved.

5.2. Single Fundamental Frequencies Case

It is interesting to notice that certain kinds of traditional single-periodic repetitive control systems can be extended to multi-periodic repetitive control with improved performance. The following example is shown based on above plants P, with controller parameters used in experiment (2).

Reference signal used in first simulation is taken as,

$$r = \sin\pi t + 2\sin 3\pi t + 0.7\sin 5\pi t$$

It can be considered either as a single period at fundamental period $\tau = 1/0.5$, or as a multi-periodic at three fundamental periods $\tau_1 = 1/0.5$, $\tau_2 = 1/1.5$, $\tau_3 = 1/2.5$. The simulation results for these two situations are shown in Figure 7.

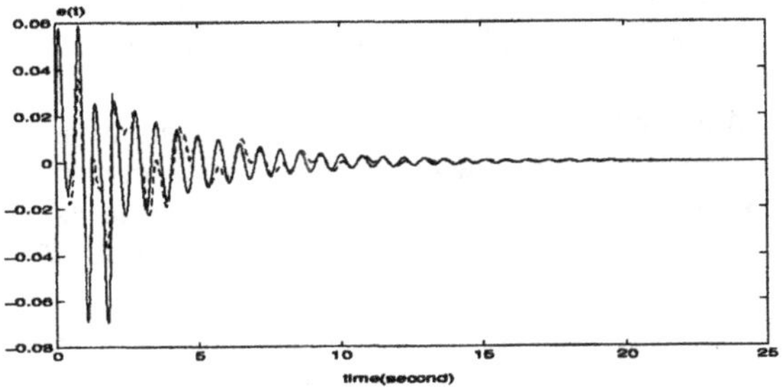

Fig. 7. Tracking errors for $r = \sin\pi t + 2\sin 3\pi t + 0.7\sin 5\pi t$ when using single-periodic repetitive control (_solid) and multi-periodic repetitive control (--dashed).

The Integral of square error are ISE1=0.3410(single-periodic) and ISE2=0.2437 (multi-periodic), which indicates that a better tracking performance is possible by reconfiguring the highest frequency content.

6. *CONCLUSIONS*

A new kind of MIMO multi-periodic repetitive control system is studied. These results are extensions of single periodic repetitive control problems. The stability is analysed in the sense of Lyapunov stability and it has been shown that asymptotic stability is guaranteed if the plant is positive real. While most real plants are not necessarily positive real, procedures can be employed to build positive realness by feedforward and feedback compensations. The simulation example shows that the new control system can be very effective.

Extensions to the work to relax the PR condition, prove exponential stability and include frequency dependent weighting $W_i(s)$ will be reported separately.

ACKNOWLEDGMENT

DHO, LL and SPB would like to thank the funding support from the EPSRC grant No. GR/M94106 and the EU TMR Nonlinear Control Network project. Thanks also to Dr. G Weiss of Imperial College for providing information in the early stage of the project.

REFERENCES

Anderson, B.D.O. and S. Vongpanitherd (1973). *Network analysis and synthesis: A modern system theory approach*, Prentice Hall, Englewood Cliffs, NJ.

Anderson, B.D.O., S. Dasgupta, P. Khargonekar, F.J. Kraus and M. Mansour (1990). Robust strict positive realness: characterization and construction, *IEEE Trans. on Circuits and Systems*, **37**, 869-875.

Bai, M.R. and T.Y. Wu (1998). Simulations of an internal model-based active noise control system for suppressing periodic disturbances, *Journal of Vibration and Acoustics-Trans. of the ASME*, **120**, 111-116.

Francis, B.A. and W.M. Wonham (1975). The internal model principal for linear multivariable regulators, *Appl. Math. Opt.*, **2**, 107-194.

Fung, R.F., J.S. Huang, C.G. Chien, and Y.C. Wang (2000). Design and application of a continuous repetitive controller for rotating mechanisms, *International journal of mechanical sciences*, **42**, 1805-1819.

Garimella, S.S. and K. Srinivasan (1996). Application of repetitive control to eccentricity compensation in rolling. *Journal of dynamic systems measurement and control-Transactions of the ASME*, **118**, 657-664.

Haddad, W.M. and D.S. Bernstein (1991). Robust stabilization with positive real uncertainty: Beyond the small gain theorem, *Systems & Control Letters*, **17**, 191-208.

Hara, S., Y. Yamamoto, T. Omata, and M. Nakano (1988). Repetitive control system: A new type servo system for periodic exogenous signals, *IEEE Trans. on Automatic Control*, **33**, 659-668.

Inoue, T., M. Nakano and S. Iwai (1981). High accuracy control of servomechanism for repeated contouring, *Proc. 10th Annual Symp. Incremental Motion Control Systems and Devices*, 258-292.

Inoue, T., M. Nakano, T. Kubo, S. Matsumoto and H. Baba (1981). High accuracy control of a proton synchrotron magnet power supply, *Proc. IFAC 8th World Congress*, 216-221.

Jia, Y., W. Gao and M. Cheng (1994). Robust strict positive realness design of interval systems, *Int. J. Syst. Sci.*, **25**, 1521-1534.

Kaneko, K. and R. Horowitz (1997). Repetitive and adaptive control of robot manipulators with velocity estimation, *IEEE Trans. on robotics and automation*, **13**, 204-217.

Kobayashi, Y., T. Kimura and S. Yanabe (1999). Robust speed control of ultrasonic motor based on H∞ control with repetitive compensator, *JSME int. journal Series C.*, **42**, 884-890.

Manayathara, T.J., T.C. Tsao, J. Bentsman and D. Ross (1996). Rejection of unknown periodic load disturbances in continuous steel casting process using learning repetitive control approach, *IEEE trans. on control systems technology*, **4**, 259-265.

Molander, P. and J.C. Willems (1980). Synthesis of state feedback control laws with a specified gain and phase margin, *IEEE Trans. on Automatic Control*, **25**, 928-931.

Moon, J.H., M.N. Lee and M.J. Chung (1998). Repetitive control for the track-following servo system of an Optical disk drive, *IEEE Trans on control systems technology*, **6**, 663-670.

Safonov, M.G., E.A. Jonckheere, M. Verma and D.J.N. Limebeer (1987). Synthesis of positive real multivariable feedback systems, *Int. J. Control*, **45**, 817-842.

Sun, W., P. Khargonekar and D. Shim (1994). Solution to the positive real control problem for linear time-invariant systems, *IEEE Trans. on Automatic Control*, **39**, 2034-2045.

Tzou, Y.Y., S.L. Jung and H.C. Yeh (1999). Adaptive control of PWM inverters for very low THD ac-voltage regulation with unknown loads, *IEEE Trans. on power electronics*, **14**, 973-981.

Weiss, G. (1997). Repetitive control systems: old and new ideas, In*: Systems and Control in the Twenty-first Century* (C. Byrnes, B.Datta, D. Gilliam and C. Martin, (Ed)), pp389-404, Birkhäuser-Verlag, Basel.

Copyright © IFAC Periodic Control Systems,
Cernobbio-Como, Italy, 2001

New Sampling Theorem Concerning the Ripple Phenomena

by Hisao KATOH and Yasuyuki FUNAHASHI

Dept.of Mechanical Engineering, Nagoya Institute of Technology,
Gokiso-cho, Showa-ku, Nagoya 466-8555 JAPAN
Fax: +81-52-735-5342 E-mail: kato@eine.mech.nitech.ac.jp

Abstract: In this paper, we propose a new approach to the ripple phenomena, by using open-loop analysis, Fourier analysis and new sampling theorem. Our result can explain explain the emergence of ripple phenomena and reduce them. *Copyright © 2001 IFAC*

1. Introduction

Throughout 1990's, the rigorous analysis and synthesis for sampled-data systems have been made, such as H_∞ optimization (Bamieh et.al, 1992), H_2 optimization (Chen et al., 1991), frequency response (Yamamoto et al. 1996; Araki et al., 1996). On the other hand, so-called ripple phenomena is one of the central theme of sampled-data control theory for many years. Do the results in 90's give much progress to the research of ripple? Unfortunately, it seems that we still do not reach the essential understanding for them. In this paper, we propose a new approach to the ripple phenomena by using open-loop analysis, Fourier analysis and new sampling theorem. Our result can explain the emergence of ripple phenomena and reduce them.

The outline of this paper is as follows. We restrict our attention to the steady-state ripple of sampled-data servosystems for periodical continuous-time reference signals. First, we will show that open-loop system analysis is sufficient for such analysis. Next, we will analyze the steady-state ripple mode by mode by using the Fourier series expansion. While doing it, we will encounter a new kind of sampling theorem. After that, we will show some reduction methods of such ripple.

In this paper, we denote Laplace (or Z) transforms by capital letters, and continuous-time(discrete-time) signals by the variables with(without) $\tilde{\cdot}$.

2. An example

In this section, we show a numerical example of ripple phenomenon given by Tetsuka et al.(1991). We consider the system in Fig.1. This system is called repetitive control system. S_T and H_T denote the sampler and zero-order hold with the sampling period T, respectively.

Assume that the continuous-time reference signal $\tilde{r}(t)$ is periodical with the period L (L-periodic), and the sampling period T satisfies

$$L = NT, \quad N: \text{natural number.}$$

Also, assume that $K(z)$ stabilizes the closed-loop system in Fig.1. Then, the discrete-time output y(k) converges to the discrete-time reference signal

$$r := S_T\tilde{r} ,$$

because of the existence of the discrete-time internal model $1/(1-z^{-1})$.

Let

$$\tilde{P}(s) = \frac{1}{(s+1)(3s+1)}$$

$$\tilde{r}(t) = \begin{cases} t, & t \in [0, L_0) \\ r_3t^3 + r_2t^2 + r_1t + r_0, & t \in [L_0, L) \end{cases}$$

$$L_0 = 0.96, \quad L = 1.$$

We determine the value of r_i (i=0,1,2,3) in order that $\tilde{r}$ is continuously differentiable. This reference signal is plotted in Fig.2. The one period of the steady-state responses of continuous-time input and output is shown in Figs. 5 and 6 in the cases with N=20 and N=40. Obviously, the ripple phenomena arise. Also, we can see that smaller sampling period does not necessarily mean smaller ripple.

3. Open-Loop Analysis

In this section, we establish the important principle for our analysis. To our knowledge, it has never been claimed explicitly. The principle is as follows: *If only the steady-state responses in Fig.1 are analyzed, then it is sufficient to consider the open-loop system in Fig.3.*

In fact, this principle holds quite generally. For example, consider the continuous-time system in Fig. 4. Assume this system is stable. Since

$$\tilde{Y}(s)=\frac{\tilde{P}(s)\tilde{K}(s)}{1+\tilde{P}(s)\tilde{K}(s)}\tilde{R}(s) \tag{1}$$

and

$$\tilde{U}(s)=\frac{\tilde{K}(s)}{1+\tilde{P}(s)\tilde{K}(s)}\tilde{R}(s), \tag{2}$$

both $\tilde{Y}$ and $\tilde{U}$ have the modes of $\tilde{R}$ and the closed-loop system. So, putting

$$\tilde{Y}(s)=\tilde{Y}_{ss}(s)+\tilde{Y}_{tr}(s) \tag{3}$$

$$\tilde{U}(s)=\tilde{U}_{ss}(s)+\tilde{U}_{tr}(s) \tag{4}$$

$\tilde{Y}_{ss},\tilde{U}_{ss}$: (antistable) modes of $\tilde{R}$

$\tilde{Y}_{tr},\tilde{U}_{tr}$: (stable) modes of the closed-loop system

("ss" and "tr" mean steady state and transient, respectively), we have

$$\tilde{Y}_{ss}+\tilde{Y}_{tr}=\tilde{P}(\tilde{U}_{ss}+\tilde{U}_{tr}). \tag{5}$$

The modes of $\tilde{P}$ do not appear on the left-hand side. The reason of it is that $\tilde{U}$ has the zeros identical to the poles of $\tilde{P}$ (See (2)).

For sampled-data systems, same fact holds (proof is omitted).

Theorem 1: For the system in Fig.1,

$$\tilde{U}(p)=0,\quad \forall p:\tilde{P}(p)=\infty \tag{6}$$

holds. ■

Furthermore, putting

$$\tilde{Y}_{ss}(s)=\sum_k\frac{\tilde{y}_k}{s-s_k} \tag{7}$$

and

$$\tilde{U}_{ss}(s)=\sum_k\frac{\tilde{u}_k}{s-s_k} \tag{8}$$

we obtain

$$\tilde{y}_k=\tilde{P}(s_k)\tilde{u}_k,\quad \forall k. \tag{9}$$

Note that this equation has nothing to do with controller. This means that the open-loop analysis is sufficient for the analysis of steady-state ripple.

In the following section, we will see that (9) is a quite convenient tool for the analysis of ripple phenomena.

4. Ripple Coefficients

Assume that the system in Fig.1 is stable. Then, the continuous-time steady-state error

$$\tilde{e}_{ss}:=\tilde{r}-\tilde{y}_{ss} \tag{10}$$

is L-periodic. So, the Fourier series expansion for it is written in the form

$$\tilde{e}_{ss}(t)=\sum_{m=-\infty}^{\infty}\beta(m)e^{j\frac{2\pi m}{L}t}. \tag{11}$$

We consider that the information about the appearance of ripple must be included in the coefficients $\beta(m), m=0,\pm1,\pm2,\cdots$. Therefore, in this paper, we call the coefficients $\beta(m), m=0,\pm1,\pm2,\cdots$ ripple coefficients. Analyzing $\beta(m)$ is equivalent to analyze the ripple phenomena mode by mode.

In the followings, we calculate $\beta(m)$ by the analysis of the open-loop system given in Fig.3.

Step 1 Calculate the mode decomposition of $y_{SS}(=r)$. Since

$$Y_{ss}(z)=R(z)=\frac{R'(z)}{1-z^{-N}} \tag{12}$$

$$R'(z):=r(0)+\cdots+r(N-1)z^{-(N-1)}$$

$$r(k):=\tilde{r}(kT),$$

Y_{ss} has N modes on the unit circle in the complex plane. Therefore

$$R(z)=\sum_{k=0}^{N-1}\frac{r_k}{1-\Omega^k z^{-1}} \tag{13}$$

$$r_k=\frac{1}{N}R'(\Omega^k),\qquad \Omega:=e^{j\frac{2\pi}{N}}.$$

Step 2 Calculate the mode decomposition of u_{ss}. By the relationship like (4), we obtain

$$U_{ss}(z)=\sum_k\frac{1}{N}\frac{R'(\Omega^k)}{P(\Omega^k)}\frac{1}{1-\Omega^k z^{-1}}. \tag{14}$$

Step 3 Calculate the mode decomposition (Fourier series) of $\tilde{u}_{SS}$. The result is

$$\tilde{u}_{ss}(t) = \sum_{m=-\infty}^{\infty} \alpha(m) \frac{R'(\Omega^m)}{P_T(\Omega^m)} e^{j\frac{2\pi m}{L}t} \tag{15}$$

$$\alpha(m) := \begin{cases} \frac{1}{N}, & if\ m = 0 \\ \frac{1-\Omega^{-m}}{j2\pi m}, & if\ m \in \Lambda \\ 0, & otherwise \end{cases}$$

$\Lambda := \bigcup \Lambda_k$

$\Lambda_k := \{m \in kZ : m = k \pmod m\}$

$kZ := \{kl : l = 0, \pm 1, \pm 2, \cdots\}$

k : nonzero integers.

Step 4 Calculate the mode decomposition (Fourier series) of $\tilde{y}_{SS}$. By using the relationship like (9) again, we obtain

$$\tilde{y}_{ss}(t) = \sum_{m=-\infty}^{\infty} \alpha(m) \tilde{P}\left(j\frac{2\pi m}{L}\right) \frac{R'(\Omega^m)}{P(\Omega^m)} e^{j\frac{2\pi m}{L}t} . \tag{16}$$

This is the end of our analysis of the system in Fig.3.

Next, for the comparison with (16), we calculate the mode decomposition (Fourier series) of $\tilde{r}$. The result is

$$\tilde{r}(t) = \sum_{m=-\infty}^{\infty} \frac{1}{L} \tilde{R}\left(j\frac{2\pi m}{L}\right) e^{j\frac{2\pi m}{L}t} \tag{17}$$

$$\tilde{R}'(s) := \int_0^L \tilde{r}(t) e^{-st} dt .$$

By (10),(11),(16) and (17), we have the following result.

Theorem 2: The ripple coefficients are given by

$$\beta(m) = \frac{1}{L} \tilde{R}'\left(j\frac{2\pi m}{L}\right) - \alpha(m) R'(\Omega^m) \frac{\tilde{P}\left(j\frac{2\pi m}{L}\right)}{P(\Omega^m)} \tag{18}$$

■

5. New Sampling Theorem

The equation (16) given above is interesting itself. It describe how $y(t)$ is interpolated between $y(iT), i \in Z$. In the celebrated paper by Shannon (1948), the equation is derived that determines $y(t)$ under the situation that $y(iT), i \in Z$ are all known. That result is called sampling theorem. In the same meaning, we consider that (16) should be called another type of sampling theorem. We restate it as follows.

Theorem 3: Assume that $\tilde{y}$ is L-repetitive function in Fig.3. Then,

$$\tilde{y}(t) = \sum_{m=-\infty}^{\infty} \sum_{n=0}^{N-1} \tilde{y}(nT) \Omega^{-mn} \alpha(m) \frac{\tilde{P}(j2\pi m/L)}{P(\Omega^m)} e^{j\frac{2\pi m}{L}t}$$

$$= \sum_{m=-\infty}^{\infty} c_m e^{j\frac{2\pi m}{L}t} \tag{19}$$

holds, where

$$\begin{bmatrix} c_0 \\ c_1 \\ c_2 \\ \vdots \end{bmatrix} = \Sigma \Lambda \begin{bmatrix} \tilde{y}(0) \\ \tilde{y}(T) \\ \vdots \\ \tilde{y}((N-1)T) \end{bmatrix} \tag{20}$$

$$\Sigma := diag\{\alpha(m) \tilde{P}(j2\pi m/L)/P(\Omega^m) : m = 0,1,2,\cdots\}$$

$$\Lambda := \begin{bmatrix} 1 & 1 & 1 & \cdots & 1 \\ 1 & \Omega^{-1} & \Omega^{-2} & \cdots & \Omega^{-(N-1)} \\ 1 & \Omega^{-2} & \Omega^{-4} & \cdots & \Omega^{-2(N-1)} \\ \vdots & \vdots & \vdots & & \vdots \end{bmatrix}$$

■

6. Analysis and Reduction of Ripple

In this section, we give an analysis of the ripple coefficients for the example in Section 2, and later propose some reduction methods of ripple.

First, we calculate the ripple coefficients in the case with N=20. The result is given in Table 1. Note that the value with m=10 is very large. Obviously, it corresponds to the ripple.

Next, we consider the method to reduce it. For this purpose, we give up

$$y_{ss} = r .$$

By manipulating $y_{ss}(k), k = 0,1,\cdots,N-1$, we aim to reduce $\beta(m), m = 0,\pm 1,\pm 2,\cdots$. Many approaches can be considered. Here, we propose two methods.

Method A: Reduce

$$\beta(m), \quad |m| \le \frac{N}{2} .$$

If N is odd, we can achieve

$$\beta(m) = 0, \quad \forall |m| \le \frac{N}{2} . \tag{21}$$

However, if N is even, we cannot generally achieve (21) (Proof is omitted). The result with N=20 is shown in Fig.7.

Method B: We do not like to move y_{ss} far away from r. So, maintain

$$y_{ss}(k) = r(k), \quad k = 1, 2, \cdots, N-2$$

and manipulate only $y_{ss}(0)$ and $y_{ss}(N-1)$ in order to achieve

$$\beta(N/2) = 0$$

(We consider N=20). The result is given in Fig.8.

We can see that, by both methods, the ripple phenomena are dramatically reduced.

REFERENCES

Bamieh,B and Pearson,J.B. (1992). A General Framework for Linear Periodic Systems with Applications to H_∞ Sampled-Data Control, *IEEE Trans. Automat. Contr.*, vol.37, pp.418-435.

Chen,T. and Francis, B.A. (1991). H_2-Optimal Sampled-Data Control, *IEEE Trans. Automat. Contr.*, vol. 36, pp.387-397.

Yamamoto,Y. and Khargonekar,P.P. (1996). Frequency Response of Sampled-Data Systems, *IEEE Trans. Automat. Contr.*, vol.41, pp.166-176.

Araki,M. Ito,Y. and Hagiwara,T. (1996). Frequency Response of Sampled-Data Systems, *Automatica*, vol. 32, pp, 483-497.

Tetsuka,M. Hara,S. and Kondo,R. (1991). Ripple Phenomena in Digital Repetitive Control, *Trans. Soc. Instr. Contr. Engin.*, vol.27, pp.915-921 (in Japanese).

Shannon,C.E. (1948). A mathematical theory of communication, *Bell System Technical Journal*, vol.27, pp.379-423.

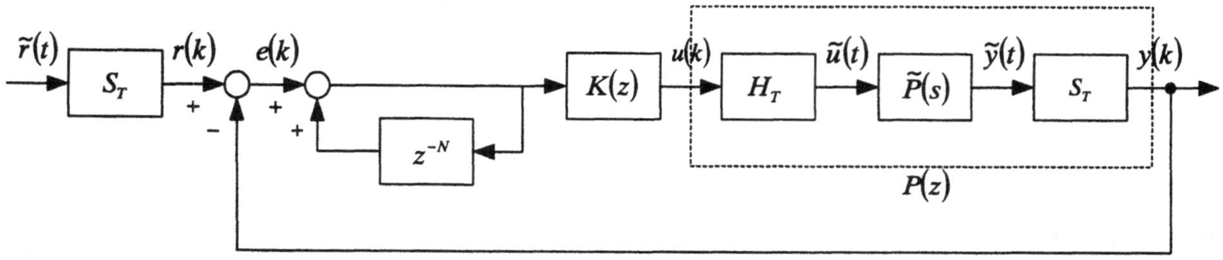

Fig. 1 Sampled-data repetitive control system

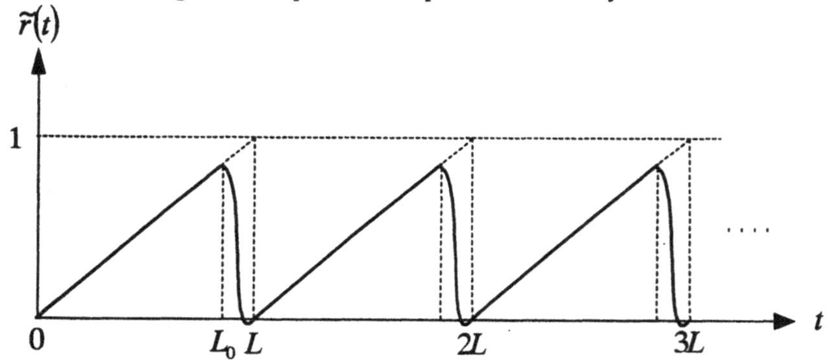

Fig.2 Reference signal

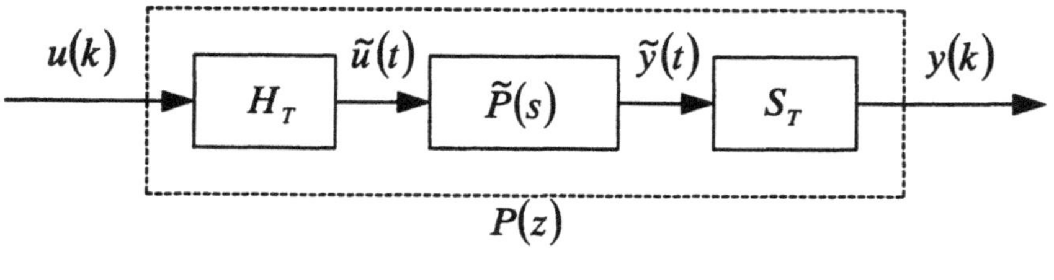

Fig.3 Open-loop system

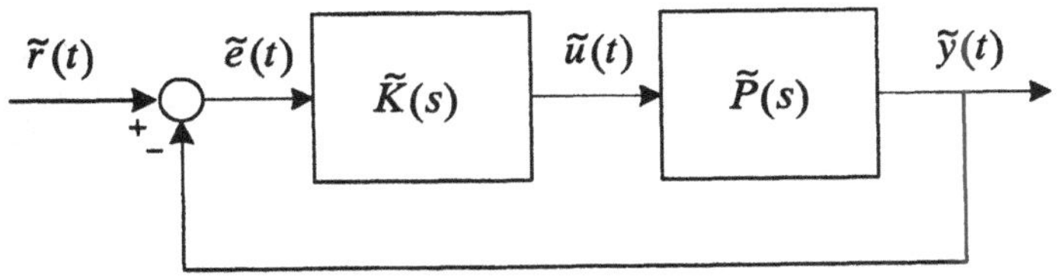

Fig. 4 Continuous-time system

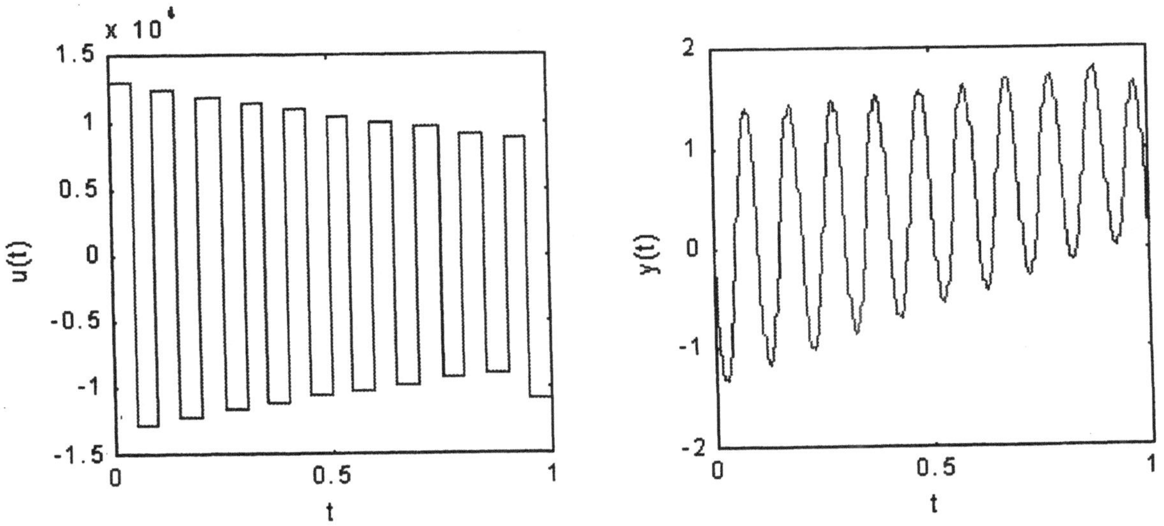

Fig. 5 Responses (N=20)

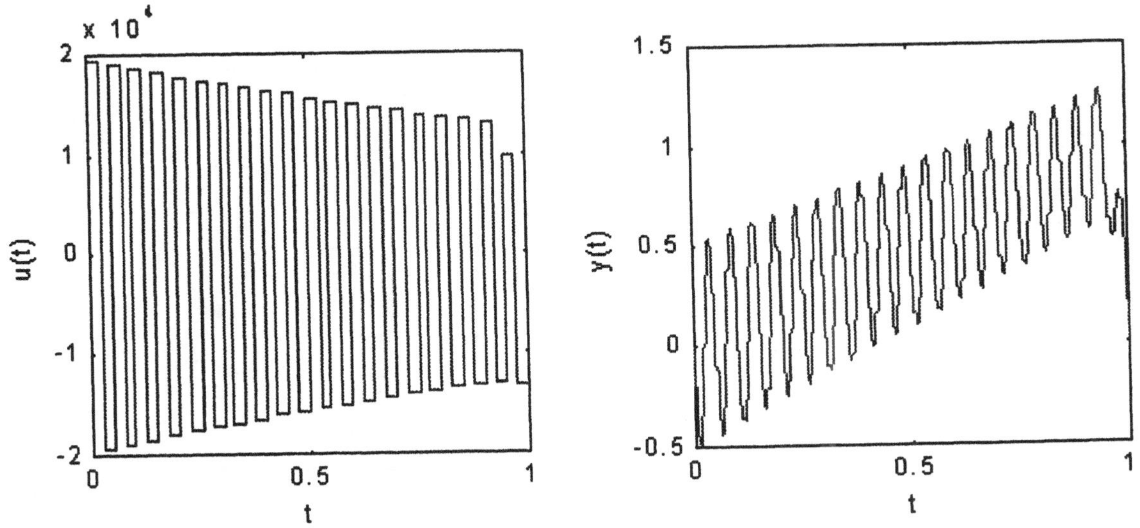

Fig. 6 Responses (N=40)

m	$\beta(m)$	m	$\beta(m)$
0	0.0050	7	0.0131
1	0.0051	8	0.0198
2	0.0054	9	0.0401
3	0.0059	10	0.5763
4	0.0066	11	0.0119
5	0.0078	12	0.0105
6	0.0098	13	0.0093

Table 1

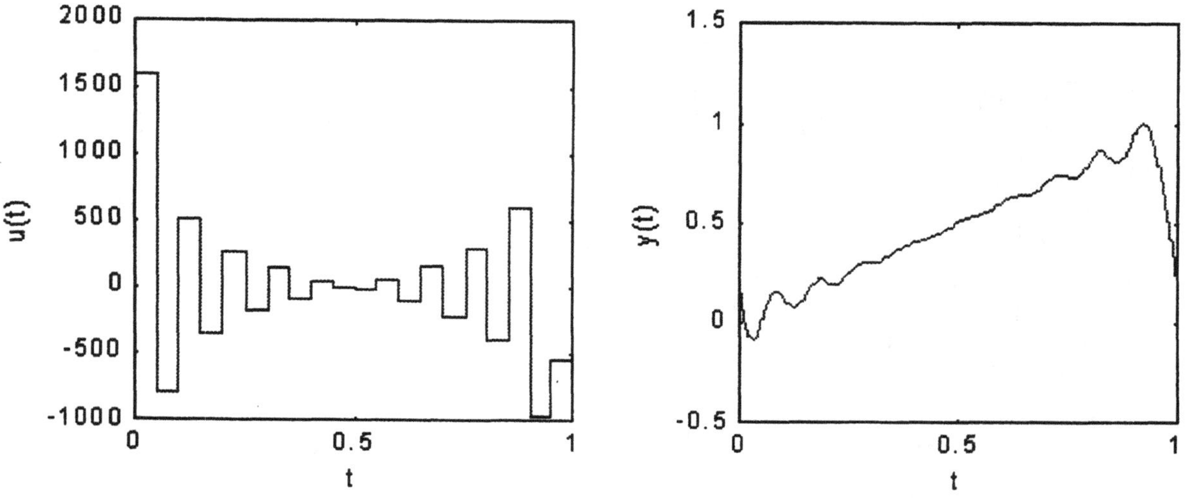

Fig.7 Responses by Method A

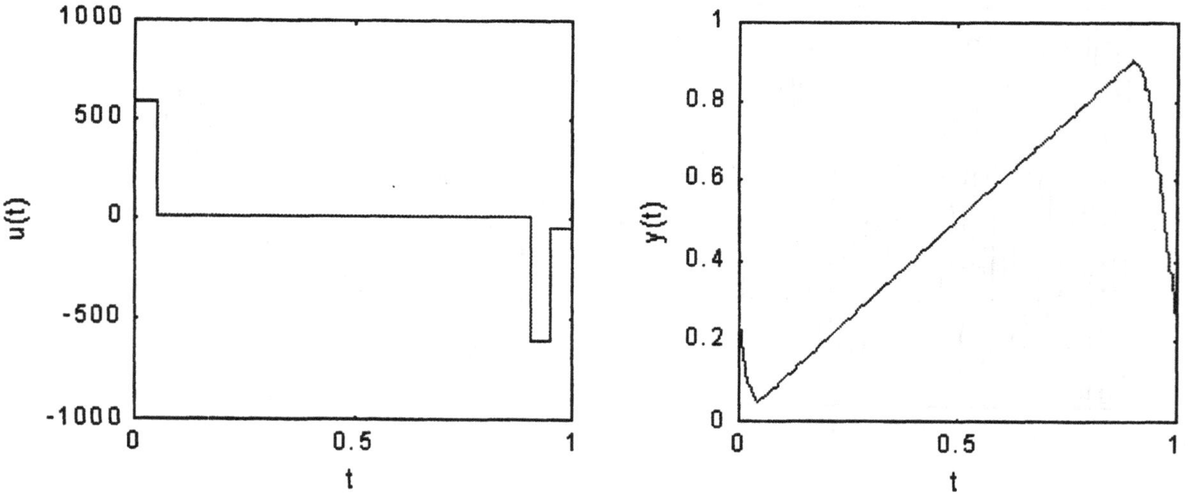

Fig. 8 Responses by Method B

Copyright © IFAC Periodic Control Systems,
Cernobbio-Como, Italy, 2001

COMPENSATION OF OSCILLATIONS USING FEEDBACK CONTROL WITH ADAPTIVE TUNING

Ryszard Gessing

Politechnika Śląska Instytut Automatyki,
ul. Akademicka 16, 44-101 Gliwice, Poland,
fax: +4832 372127, email: gessing@ia.gliwice.edu.pl

Abstract: In the paper the compensation of the oscillations caused by the disturbance in considered. The disturbance is close to the sinusoidal signal with varying frequency. The modified LQ control with internal model describing an integral–resonant corrector is applied. The varying frequency is identified and the parameters of the corrector are re-tuned when it is needed. Owing to this for any constant and sinusoidal disturbance the error of the system in steady state is close to zero. *Copyright © 2001 IFAC*

Keywords: Resonant corrector, integral model, LQ control, adaptive tuning.

1. INTRODUCTION

Many engineering structures are influenced by periodical excitations causing un-demanded vibrations. One way to decrease these vibrations is to implemented active approach basing on control technique.

In the paper an original idea of compensation of oscillations caused by the disturbance z is presented. The disturbance may have varying frequency ω, but it is close to a sinusoidal signal.

During the design the original plant transfer function (TF) is modified by adding to the plant an integral-resonant corrector in accordance with the internal model approach. Then an augmented, over-dimensional state space model is formulated and the LQ technique is appropriately used giving the dynamic regulator with output feedback. It is shown that by means of some appropriate choice of the weighting matrix Q in the performance index it is possible to obtain a partially assumed pole placement of the Closed loop (CL) system (Gessing, 2000).

During the implementation of the control the integral-resonant corrector is included to the regulator.

The applied approach makes it possible to re-tune the parameters of the regulator-corrector to the varying frequency of the disturbance, when this frequency is is identified.

The advantage of the proposed approach is that in steady state caused by a sinusoidal and constant disturbance the error in the CL system is close to zero.

The contribution of the paper is in applying the modified LQ approach with integral-resonant corrector, in the manner making it possible to re-tune the regulator-corrector parameters, when the frequency of the disturbance is changing.

2. APPLYING THE INTERNAL MODEL

Consider the closed loop CL system shown in Fig. 1. Assume that the plant is described by a strictly proper, rational transfer function (TF) $G(s)$, with p-th order polynomial in denominator

$$G(s) = \frac{\bar{b}_1 s^{p-1} + \bar{b}_2 s^{p-1} + \ldots + \bar{b}_p}{s^p + \bar{a}_1 s^{p-1} + \ldots + \bar{a}_p} \qquad (1)$$

where some coefficients $\bar{b}_i$ my be equal to zero.

Assume that the set point $w(t)$ and disturbance $z(t)$ fulfill the following differential equation

$$\tilde{u}^{(q)} + d_1\tilde{u}^{(q-1)} + ... + d_q\tilde{u} = 0 \qquad (2)$$

This means that the equation (2) is fulfilled if we substitute $\tilde{u} = w(t)$ or $\tilde{u} = z(t)$. The fact that the signals w and z are different results from different initial conditions of the equation (2) used for generating both the signals.

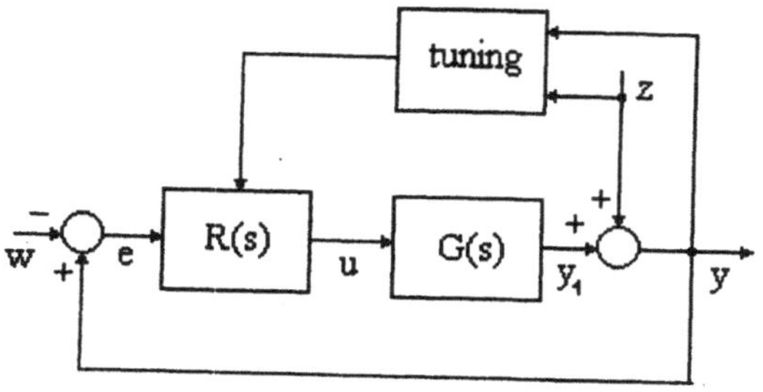

Fig. 1. Closed loop system with adaptive tuning of the corrector.

If for instance $w_1 = \bar{w} = const$ and $z_1 = Asin(\omega t + \varphi)$ then the equation (2) takes the form

$$\tilde{u}^{(3)} + \omega^2\tilde{u}^{(1)} = 0 \qquad (3)$$

The signals w_1 and z_1 with any $\bar{w}$, A and φ are the solutions of the equation (3) with appropriate initial conditions.

Consider the corrector described by the transfer function (TF)

$$\frac{U(s)}{V(s)} = C(s) = \frac{1}{s^q + d_1 s^{q-1} + ... + d_q} \qquad (4)$$

which in denominator has the polynomial corresponding to the equation (2); $V(s) = \mathcal{L}[v(t)]$ is the input of the corrector. The corrector having in denominator the polynomial determined by the equation (3) may be called the integral–resonant corrector.

The augmented plant described by the TF

$$\frac{E(s)}{V(s)} = C(s)G(s) = \frac{b_{n-l}s^l + b_{n-l+1}s^{l-1} + ... + b_n}{s^n + a_1 s^{n-1} + ... + a_n} \qquad (5)$$

is of $n = (p+q)$ order and has all the modes of the plant $G(s)$ and of the corrector (4). Thus, the error $e = y_1 + z - w$ appearing in the closed loop (CL) system shown in Fig.1 can be obtained directly as the output of the augmented plant (5) for some appropriate initial conditions. This fact justifies the used notation $E(s) = \mathcal{L}[e(t)]$ for the augmented plant output.

3. LQ PROBLEM FORMULATION

Consider the system described by TF (5). Determine the state variables in the form (Gessing, 2000).

$$\begin{array}{lll} x_1 = e^{(n-m-1)} & x_2 = e^{(n-m-2)} ... & x_{n-m} = e \\ x_{n-m+1} = e^{(-1)} & x_{n-m+2} = e^{(-2)} ... & x_n = e^{(-m)} \\ x_{n+1} = v^{(-1)} & x_{n+2} = v^{(-2)} ... & x_{n+m} = v^{(-m)} \end{array} \qquad (6)$$

where $l \le m \le n-1$ and $e^{(-i)}$, $v^{(-i)}$ determines the multiple integral with multiplicity i.

Differentiating both sides of (6), using notation (6) and the resulting from (5) equation

$$\begin{array}{r} e^{(n-m)} + ... + a_{n-m}e + \\ + a_{n-m+1}e^{(-1)} + ... + a_n e^{(-m)} = \\ = b_{n-l}v^{(l-m)} + \\ + b_{n-l+1}v^{(l-m-1)} + ... + b_n v^{(-m)} \end{array} \qquad (7)$$

we obtain the overdimensional state space model in the form

$$\dot{x} = Ax + Bu, \qquad e = Cx \qquad (8)$$

where $x = [x_1, x_2, ..., x_{n+m}]$ is $(n+m)$ dimensional state. State equations (6) for particular components x_i take the form

$$\begin{array}{rl} \dot{x}_1(t) = & -a_1x_1 - a_2x_2 - ... - a_nx_n + \\ & + b_{n-m+1}x_{n+1} + ... + b_nx_{n+m} + b_{n-m}v \\ \dot{x}_i = & x_{i-1}, \text{ for } 2 \le i \le n+m, i \ne n+1 \\ \dot{x}_{n+1} = & v \end{array} \qquad (9)$$

where $b_{n-j} = 0$ for $l < j \le m$, which results from (5). The first equation of (9) results from accounting (6) in (7); the remaining equations of (9) result directly from determination (6). The matrices A, B result from (9); the elements of the row vector C are determined by

$$C_{n-m} = 1 \text{ and } C_i = 0 \text{ for } i \ne n-m \qquad (10)$$

It may be noted that the model (8) is controllable but non observable.

Assume the quadratic performance index

$$J = \int_0^{t_f} (x^TQx + Rv^2)dt, \qquad t_f = \infty \qquad (11)$$

with the matrix $Q = f^T f$, where f is a $(n+m)$ row vector and R is a small positive number. Let (A, B) is controllable and (A, f) is observable. Then the solution of LQR problem (8), (11) in the form of feedback law is (Dorato *et al.*, 1995)

$$v = -kx = -k_1x_1 - k_2x_2 - ... - k_{n+m}x_{n+m} \quad (12)$$

The regulator TF results from substituting (6) into (12) and has the form

$$R(s) = \frac{V(s)}{E(s)} = -\frac{k_1s^{n-1} + k_2s^{n-2} + ... + k_n}{s^m + k_{n+1}s^{m-1} + ... + k_{n+m}} \quad (13)$$

The regulator–corrector TF is described by

$$R_c = R(s) \cdot C(s) \quad (14)$$

It is easy to note that if $m + q = n - 1$ then the regulator $R_c(s)$ has the same order of the numerator and denominator. This means that this case occurs when $m = n - q - 1 = p - 1$.

It is known (Dorato *et al.*, 1995) that the CL system composed of the plant (8) and the LQ regulator (12), described by a state space model is stable. One can note that the CL system composed of the plant $G(s)$ and the regulator $R_c(s)$ described by the TF model is equivalent to the previous one and is also stable.

In the paper a special role will play the integral resonant correctors. The appearance of this kind of the corrector in the regulator $R_c(s)$ means that in steady state for any constant and sinusoidal signals z and w the error e is equal to zero.

4. POLE PLACEMENT CHOICE

By means of the appropriate choice of the matrix Q in the performance index (22) it is possible to establish an appropriate placement of $(n-1)$ poles of CL system. Really, if we choose the matrix Q so that

$$Q = f^Tf, \qquad f = [f_1, f_2, ..., f_n, 0, ..., 0] \quad (15)$$

(where $dimf = n + m$ and f_i, i=1, 2, ..., n are appropriately chosen) then the performance index (11) takes the form

$$J = \int_0^{t_f} (\varepsilon^2 + ru^2)dt, \qquad t_f \to \infty \quad (16)$$

where

$$\varepsilon = fx, \qquad \varepsilon^2 = x^TQx \quad (17)$$

Substituting (6) and (15) into (17) and then differentiating m-times both sides of (17) we obtain

$$f_1y^{(n-1)} + f_2y^{(n-2)} + ... + f_ny = \varepsilon^{(m)} \quad (18)$$

Minimization of (16) especially for $r \to 0$ means that $\varepsilon(t)$ and also $\varepsilon^{(m)}(t)$ tends to zero when $t \to \infty$. This means that when r tends to zero then $(n-1)$ roots of the characteristic equation of the CL system determined by the steady state LQR solution of the problem (8), (11) tend to the roots of the equation

$$f_1s^{n-1} + f_2s^{(n-2)} + ... + f_n = 0 \quad (19)$$

Thus, by means of an appropriate choice of the coefficients f_i, $i = 1, 2, ..., n$, we can place $(n-1)$ poles of the CL system, freely. Owing to this it is possible to improve the transients appearing in the CL system.

5. ADAPTIVE TUNING OF THE CORRECTOR

The proposed system should compensate the oscillations caused by the disturbance z also then, when the frequency of the disturbance z is varying. To improve the compensation, when the frequency has been changed the corrector parameters should be appropriately suited to the new frequency of the disturbance. When these parameters are well suited to the frequency then the oscillations of the output y disappear. When some small oscillations appear in the output, it means that the frequency of the disturbance has been changed and there arises the need of re-tuning the corrector parameters. To do this the frequency is measured and the corrector parameters are re-tuned. The tuning starts only then, when there appear oscillations of the output. After each change of the corrector parameters the regulator parameters are also suited to the new situation, using the LQ approach described above.

6. EXAMPLE

Consider the system shown in Fig. 1 in which

$$G(s) = \frac{40}{s^2 + 3s + 2} \quad (20)$$

and

$$z = a\sin(\omega t + \psi) \quad (21)$$

where A, ω and ψ are unknown. Assume the integral–resonant corrector (4) in the form

$$C(s) = \frac{1}{s^3 + \omega^2 s} \quad (22)$$

which means that the CL system in steady state will have zero error for any constant and sinusoidal disturbance with the frequency ω.

The augmented plant (5) is described by the TF

$$C(s)G(s) = \frac{b_5}{s^5 + a_1 s^4 + a_2 s^3 + a_3 s^2 + a_4 s + a_5} \quad (23)$$

where the denominator of (23) results from multiplication of the denominators of (1) and (22).

Note that the frequency ω is unknown, therefore the coefficients $a_i = i = 1, 2, ..., 5$ are unknown too. In the simulations the frequency ω will be identified.

Assuming the state components x_i in the form (6) with $m = 1$ we obtain the state space model (8) of 6–th order for which

$$A = \begin{bmatrix} -a_1 & -a_2 & -a_3 & -a_4 & -a_5 & b_5 \\ 1 & 0 & 0 & 0 & 0 & 0 \\ 0 & 1 & 0 & 0 & 0 & 0 \\ 0 & 0 & 1 & 0 & 0 & 0 \\ 0 & 0 & 0 & 1 & 0 & 0 \\ 0 & 0 & 0 & 0 & 0 & 0 \end{bmatrix}$$

$$B = [0\ 0\ 0\ 0\ 0\ 1]' \quad (24)$$
$$C = [0\ 0\ 0\ 1\ 0\ 0]$$

The mattrix Q and the number R were assumed as

$$G = f'f, \quad f = [0\ 0\ 1\ 1\ 0.5\ 0] \quad R = 0.001 \quad (25)$$

This assures that the pair (A, f) is observable. Since the pair (A, B) is controllable then the CL system composed of the plant (8) and the LQR (12) is stable and has two poles close to $-0.5 \pm j0.5$.

The coefficients k_i, $i = 1, 2, ..., 6$ of the control law (12) are calculated using the *lqr* MATLAB command. The regulator (13) takes the form

$$R(s) = -\frac{k_1 s^4 + k_2 s^3 + k_3 s^2 + k_4 s + k_5}{s + k_6} \quad (26)$$

Finally the regulator–corrector $R_c(s)$ results from multiplying the right–hand sides of (26) and (22).

The SIMULINK block diagram of the system is shown in Fig. 2. The main part of it is the block *Adaptation* based on the MATLAB function *compos* shown in Fig. 3. The function determines the frequency ω and calculates the TF $R_c(s)$ using the above formulas. The period T_z of the disturbance z results from sampling of the disturbance with the sampling period $h = 0.005$, calculating several successive crossings of zero by z and averaging. The frequency is calculated from $\omega = 2\pi/T_z$ The function *compos* is activated when the averaged absolute value of the error overcome a dead-zone 0.1. The latter averaging is performed by means of the first order lag filter $1/(s+1)$ in the *Adaptation* block.

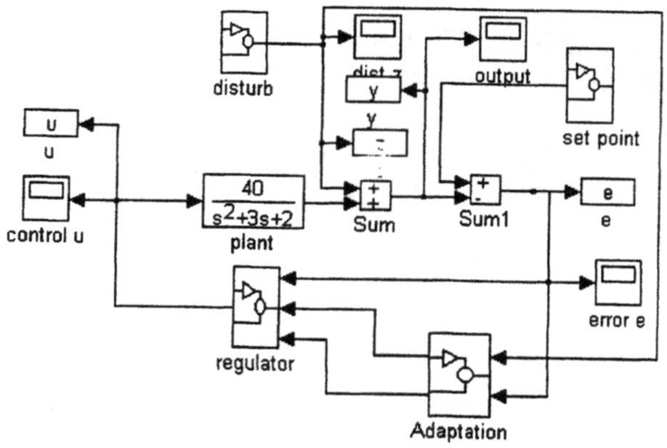

Fig. 2. SIMULINK block diagram of the closed loop system.

```
function r=compos(u)
z=u(1);z1=u(2);t=u(3);i=u(4);lrf(1)=u(5);
lrf(2)=u(6);lrf(3)=u(7);lrf(4)=u(8);
lrf(5)=u(9);mrf(1)=u(10);mrf(2)=u(11);
mrf(3)=u(12);mrf(4)=u(13);t1=u(14);

if t==0
 rf=[2.2463,14.6708,42.6887,35.8343,15.8114];
 mrf1=conv([1,13.4053],[1,0,2^2,0]);
 mrf=mrf1(2:5);
end
if z~=0
 if i<5
  if sign(z)>sign(z1)
   i=i+1;
   if i==1
    t1=t;
   end
   if i==5
    t5=t;
   end
  end
 end
 if i==5
  T=(t5-t1)/4; omega=2*pi/T;
  mf=[1,0,omega^2,0]; lf=1;
  mo=[1,3,2];
  mof=conv(mo,mf);
  lo=40;b5=lo;
  lof=lo;
  A=[-mof(2:6),b5;eye(4),zeros(4,2);zeros(1,6)];
  B=[zeros(1,5),1]';
  C=[0,0,0,1,0,0];
  f=[0,0,1,1,0.5,0];Q=f'*f;R=.001;
  k=lqr(A,B,Q,R);
  lr=k(1:5);mr=[1,k(6)];
  lrf=lr;
  mrf1=conv(mr,mf);i=0;
  mrf=mrf1(2:5);
 end
end
r=[i,lrf,mrf,t1];
```

Fig. 3. MATLAB function used in the *Adaptation* block.

Some results of simulations are presented in Fig. 4 and Fig. 5. The figures show the responses of the closed loop system to the set point variation described by the function $w(t) = 5\mathbf{1}(t)$, where $\mathbf{1}(t)$ is the unit step function. The disturbance takes the form of the sinusoidal function $z(t) = 5sin(\omega t)$, which in the time interval (0,32) has the frequency $\omega = 10$ and at the instant $t = 32$ it occurs a rapid change of the frequency to the value $\omega = 5$. At $t = 0$ the system starts with the regulator-corrector tuned to the freely chosen value $\omega = 2$. Since the latter value is different from the real frequency of the signal z, in the initial time interval (0,5) it appears identification of the real value of ω as well as adaptation of the regulator-corrector to this value. In the interval (24,32) it appears the steady state in which the maximal value of the sinusoidal error is equal to 0.06, which results from Fig. 5. The amplitude of the disturbance is decreased then $(9/0.06) \approx 83$ times. After the rapid change of the frequency in the interval (32,39) the new frequency is identified and the corrector is re-tuned. Then in the in the interval (53,60) it appears a new steady state with the error amplitude equal to 0,005. Thus the amplitude of the disturbance is decreased then (5/0.005)=1000 times.

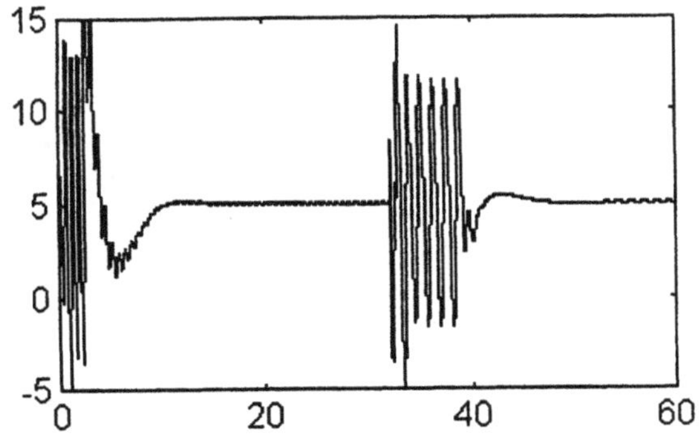

Fig. 4. The output y obtained from simulations.

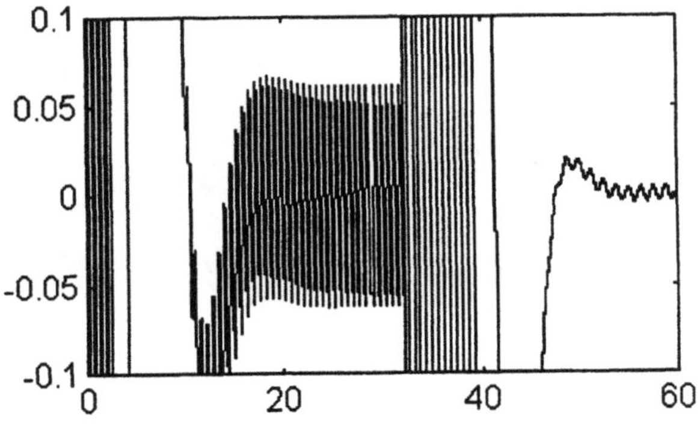

Fig. 5. The error e obtained from simulations.

From other performed simulations it results that the error amplitude for higher frequency is more sensitive to identification accuracy of ω. Then this amplitude may be decreased by means of more accurate identification of ω.

From the performed simulations it results that for the considered example it is possible to obtain the compensation of z with approximately 1% of accuracy for the frequences ω lying in the interval $0 \leq \omega \leq 15$.

7. CONCLUSIONS

The proposed method based on the modified LQ problem statement and integral–resonant corrector makes it possible to design a CL system which for any constant and sinusoidal disturbances has close to zero error in steady state. By means of the appropriate choice of the weighting matrix Q it is possible to obtain the partially, assumed pole placement of the CL system. This improves its transients.

The system may work if the frequency is varying; the frequency is identified and the parameters of the corrector are suited to the new value.

The advantage of the proposed solution is the small values of the CL system error in steady state.

Some disadvantage is the high order of the regulator–corrector transfer function. This results from including to the regulator the integral–resonant corrector.

ACKNOWLEDGMENT

The paper was partially supported by the INCO COPERNICUS CP977022 DYCOMANS programme.

8. REFERENCES

Dorato, P., C. Abdallah and V. Cerone (1995). *Linear- quadratic Control, An Introduction.* Prentice Hall, New Jersey.

Gessing, R. (2000). Continuous-Time Linear-Quadratic Regulator with Output Feedback. Proceedings of the American Control Conference ACC2000, Chicago, Illinois, pp. 877-881

Copyright © IFAC Periodic Control Systems,
Cernobbio-Como, Italy, 2001

SUBOPTIMAL PERIODICAL VS OPTIMAL BANG-BANG CONTROL FOR A CERTAIN CLASS OF THE INFINITE DIMENSIONAL SYSTEMS

J. Smieja, A. Swierniak

Department of AutomaticControl, Silesian University of Technology
Akademicka 16, 44-100 Gliwice, Poland
Tel.: (+48 32) 372750, fax: (+48 32) 371165
E-mail: jsmieja@ia.polsl.gliwice.pl

Abstract: The paper is concerned with optimal control of dynamical systems, described by state equation with infinite dimensional, tridiagonal system matrix. First, a method of model analysis is briefly outlined. Afterwards, optimization problem is stated and necessary conditions for optimal control are shown. Due to their particular form and control constraints, the solution to the problem is a bang-bang control. Following that conclusion, two gradient methods are presented – one for finding optimal switching times and another for finding suboptimal, periodical solution. Exemplary numerical results are presented and some remarks on applicability of both the model and the solutions are made. *Copyright © 2001 IFAC*

Keywords: optimal control, infinite-dimensional systems

1.INTRODUCTION

Despite long history of research and rich literature devoted to optimization problems, relatively few works concerning optimal control for infinite dimensional systems can be found. Moreover, they mostly present approaches suitable for PDE models and optimization solutions are often limited to LQ problems. Although that class of problems is arguably the most important in practical applications and undoubtedly covers the broadest area of research, there exist different challenges for which , as presented in the following sections, new approaches are required, tailored to particular models and goals to be achieved.

Models based on infinite number of state equations may be applied to a variety of systems. They may describe e.g. RC ladders (or RLC ladders, which require only slight modification of presented methods, as far as modeling is concerned) that are approximation of long transmission lines (Mitkowski, 1997; Zadeh, 1963), different models arising in biomathematical field (Kimmel and Axelrod, 1990; Polanski *et al.*, 1997; Swierniak *et al.* 1998, 2001) or some queuing systems (Kleinrock, 1976). Especially the two latter fields seem to be promising regarding application of obtained results. Usually, analysis of such models is limited to their finite-dimensional approximation. However, in that case, some dynamical properties may be neglected. Moreover, as shown in our previous papers (Polanski *et al.*, 1997; Swierniak *et al.*, 1999), work on infinite dimensional models may lead to compact results, convenient in further analysis, which would be impossible or very difficult to obtain in finite dimensional approximation.

As mentioned above, most of works on optimal control consider some kind of LQ problems, since they are the most common application of the theory. Relatively little research addresses other than quadratic performance indices, despite the fact that they also are applicable in some areas, and create some theoretical challenges as well. This work is devoted to a solution to an optimal control problem, in which the performance index is defined in l^1 space with respect the state variable and L^1 with respect to control variable. The main application of obtained results is in the field of biomedical modeling and queuing systems, from which the form of the performance index in the analyzed problem stems. Nevertheless, taking into account the broad class of systems described by presented model, it should easily be adapted into other areas.

2. THE CLASS OF INVESTIGATED MODELS

The system is described by following state equation

$$\dot{x} = \mathbf{A}x + \mathbf{B}(x)u, \tag{1}$$

where x – infinite dimensional state vector ($x^T=[x_0\ x_1\ x_2 \ldots x_i \ldots]^T$), $\mathbf{A}$ – tridiagonal system matrix

$$\mathbf{A} = \begin{bmatrix} c_0 & c_2 & 0 & 0 & 0 & 0 & \cdots \\ c_1 & a_2 & a_3 & 0 & 0 & 0 & \cdots \\ 0 & a_1 & a_2 & a_3 & 0 & 0 & \cdots \\ 0 & 0 & a_1 & a_2 & a_3 & 0 & \cdots \\ \vdots & \vdots & \ddots & \ddots & \ddots & \ddots & \ddots \end{bmatrix}, \tag{2}$$

$\mathbf{B}(x)$ – infinite dimensional vector

$$\mathbf{B}(x) = \begin{bmatrix} b_1 x_0 \\ 0 \\ 0 \\ \vdots \end{bmatrix}, \tag{3}$$

$u(t)$ – bounded scalar control variable

$$0 \le u(t) \le u_{max}. \tag{4}$$

It is also assumed that $a_1 a_2 a_3 \neq 0$. This assumption is made only to avoid discussing different analytical cases and is usually satisfied in known model applications.

Taking into account a broad class of systems that could be described by that model, the performance index can take various forms, depending on its application. This work is concerned with searching optimal control that minimizes the following performance index:

$$\min \leftarrow J = \sum_{i \ge 1} x_i(t_f) + r\int_0^{t_f} u(t)\mathrm{d}t, \tag{5}$$

where r is a positive weighing factor. The form of the performance index stems from biomedical modeling and its meaning is discussed further on. However, it could be used in other applications involving systems in which state values are nonnegative, e.g. in modeling of queuing systems.

3. THE METHOD OF MODEL ANALYSIS

For dealing with the model introduced in the previous section, a special methodology has been developed. It consists in analyzing an autonomous model first, then incorporating control into it and finally transforming the system description into an integro-differential equation.

Let us consider the following model without control:

$$\begin{cases} \dot{x}_1(t) = a_2 x_1(t) + a_3 x_2(t) \\ \dot{x}_2(t) = a_1 x_1(t) + a_2 x_2(t) + a_3 x_3(t) \\ \ldots \\ \dot{x}_i(t) = a_1 x_{i-1}(t) + a_2 x_i(t) + a_3 x_{i+1}(t) \\ \ldots \end{cases} \tag{6}$$

It can be proved (Smieja, 2000), that for initial condition $x_i(0) = \delta_{ik}$ (Kronecker delta), i.e. $x_k(0) = 1$, $x_i(0) = 0$ for $i \neq k$, following relations hold true:

$$X_1^k(s) = \frac{1}{a_3}\left(\frac{s - a_2 - \sqrt{(s-a_2)^2 - 4a_1 a_3}}{2a_1}\right)^k \tag{7}$$

$$X_\Sigma^k(s) = \frac{1}{s+\theta}\left[1 - \left(\frac{s - a_2 - \sqrt{(s-a_2)^2 - 4a_1 a_3}}{2a_1}\right)^k\right] \tag{8}$$

where $\theta = -(a_1+a_2+a_3)$, $X_1^k(s)$, $X_\Sigma^k(s)$ - Laplace transforms of $x_1(t)$ and $\sum_{i \ge 1} x_i(t)$, respectively (superscript k is introduced to underscore the number of state variable with non-zero initial condition). Relations (7) and (8) lead to the following formulae:

$$x_1^k(t) = \frac{k}{a_3}\left(\sqrt{\frac{a_3}{a_1}}\right)^k \frac{I_k\left(2\sqrt{a_1 a_3}\, t\right)}{t}\exp(a_2 t) \tag{9}$$

$$x_{\Sigma}^{k}(t)=\sum_{i\geq 1}x_i(t)=\exp[(a_1+a_2+a_3)t]\cdot\left[1-k\left(\sqrt{\frac{a_3}{a_1}}\right)^k\int_0^t\frac{I_k\left(2\sqrt{a_1a_3}\,\tau\right)}{\tau}\exp[-(a_1+a_3)\tau]d\tau\right] \tag{10}$$

where $I_k(t)$ – modified Bessel function of the k-th order.

It is important to notice that the assumption about initial condition does not introduce any additional constraints to applicability of the model. The model (6) is linear, hence, in case of finite number of non-zero initial conditions the superposition principle can be applied.

The relation (10) will be used in the solution of optimization problem later in this paper. The relation (7), in turn, can be used to determine the following transfer function in the model (1):

$$\frac{X_1(s)}{X_0(s)}=\frac{c_1}{a_3}\left(\frac{s-a_2-\sqrt{(s-a_2)^2-4a_1a_3}}{2a_1}\right), \tag{11}$$

which will be subsequently utilized to transform system description into one integro-differential equation.

Let us assume that the initial conditions are equal to zero, except for finite number of state variables, i.e. $K=\{k: x_k(0)\neq 0\}$, $\overline{\overline{K}}<\aleph_0$. Taking into account (9) and (11), it is possible to obtain the following description of the model (1):

$$\dot{x}_0(t)=(c_0+b_1u(t))x_0(t)+ +c_1c_2\int_0^t\phi_1(t-\tau)x_0(\tau)d\tau+c_2\sum_{k\in K}x_k(0)\phi_k(t) \tag{12}$$

where $\phi_k(t)=x_1^k(t)$ is defined by (9).

Moreover (Smieja, 2000),

$$x_{\Sigma}(t)=\sum_{i\geq 1}x_i(t)= =\sum_{k\in K}x_{\Sigma}^k(t)+c_2\int_0^t x_{\Sigma}^1(t-\tau)x_0(\tau)d\tau \tag{13}$$

where $x_{\Sigma}^k(t)$ is defined by (10).

4. OPTIMAL CONTROL PROBLEM

As indicated earlier, this paper is concerned with minimization of the performance index (5). Its choice is justifiable in many applications, in which state variables are nonnegative, e.g. in biomedical modeling or some queuing systems. In the latter case, the aim is to minimize the probability of existence of queue and simultaneously try not to interfere in normal process service. In biomedical applications, the goal is to minimize number of drug resistant cancer cells with simultaneous minimization of negative accumulated effect of the drug represented by the integral component.

It has been shown (Smieja *et al.*, 1999) that necessary conditions for optimal control for model (12) and performance index (5) are as follows:

$$u^{opt}(t)=\arg\min_u\left[\left(r-b_1p(t)x_0(t)\right)u(t)\right], \tag{14}$$

$$\dot{p}(t)=-\left[rc_1x_{\Sigma}^1(t_f-t)+p(t)(b_1u(t)+c_0)+ +c_1c_2\int_t^{t_f}p(\tau)\phi_1(t-\tau)d\tau\right], \tag{15}$$

$$p(t_f)=1, \tag{16}$$

$p(t)$ – adjoint variable.

Taking into account the constraint (4), it can be easily noticed that, in order to satisfy (14), the optimal control must be bang-bang one, determined by following relation

$$u(t)=\begin{cases}0 & for \quad r-2p(t)\lambda x_0(t)>0\\ u_{\max} & for \quad r-2p(t)\lambda x_0(t)<0\end{cases} \tag{17}$$

Therefore, the problem is reduced to finding optimal number of switches and optimal switching times.

Let M be number of switches, τ_j $(j=0,1,\ldots,M)$ – switching times, with $\tau_0=0$, $\tau_M\leq t_f$. Taking into account that even number of switching time corresponds to the rising edge of control and odd number to the trailing edge, it is sufficient to limit considerations to the case of odd number of switches (in other case $\tau_M=t_f$). Bang-bang control can be then presented in the following form:

$$u(t)=\sum_{j=0}^{\frac{M-1}{2}}\left[\mathbf{1}(t-\tau_{2j})-\mathbf{1}(t-\tau_{2j+1})\right] \tag{18}$$

It is crucial to notice that (Smieja *et al.*, 1999)

$$\Delta J=J(u+\delta u)-J(u)\approx \approx 2\sum_{j=1}^{M}(-1)^{j+1}\left(r-b_1p(t)x_0(t)\right)\Big|_{\tau=\tau_j^M}\delta\tau_j^M \tag{19}$$

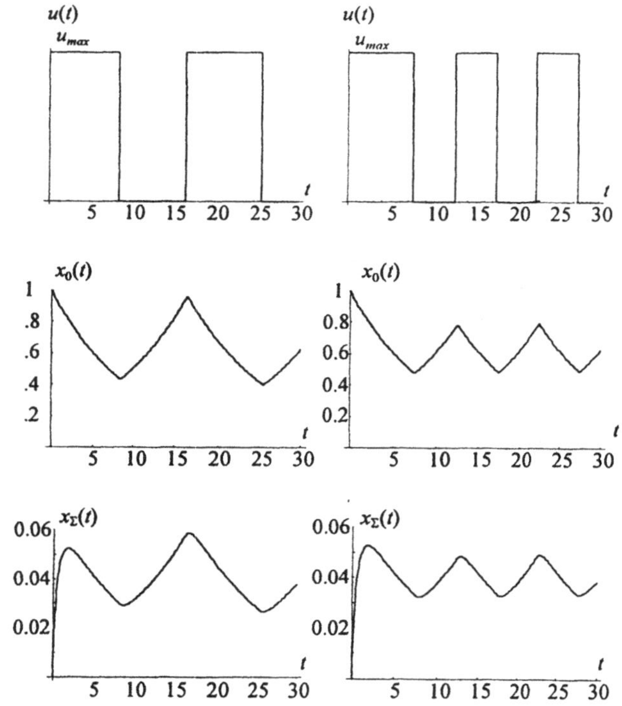

Fig. 1. Optimal control $u(t)$ and respective trajectories for $M = 3$ and $M = 3$ switches

5. NUMERICAL RESULTS

In order to apply effectively necessary conditions (14)-(16), a special gradient method has been developed (Smieja *et al.*, 1999):

1. Assume number of switches M and initial switching times τ_i^M, $i = 1,2, \ldots, M$.
2. Solve the equation describing the system (12) for bang-bang control with switching times τ_i^M.
3. Compute $p_2(t)$ from the adjoint equation (15) integrating it backward in time.
4. Assume $\delta\tau_j^M = (-1)^j k_j \left.\dfrac{\partial \tilde{H}}{\partial u}\right|_{\tau=\tau_j^M}$ where k_j is a positive coefficient
5. Compute new switching times $\tau_i^M + \delta\tau_i^M$.
6. Repeat steps 2-5 until some stop criterion is met, e.g. $\sum_{i=1}^{M} (\delta\tau_i^M)^2 < \varepsilon$, ε-small given number.

However, as discussed in (Smieja et al., 2000), this method allows only finding optimal switching times for arbitrarily chosen number of switches. In general, in the case of infinite dimensional model and bang-bang control, finding the optimal number of switches is almost impossible. Of course, it is possible to analyze the effect of this number on the performance index, nevertheless the conclusions regarding the global minimum are a very delicate matter.

Exemplary results of this method are presented in the Fig. 1. Two different numbers of switches have been considered: $M = 3$ and $M = 5$, while the weighing coefficient has been constant $r = 0.1$. The initial condition was $x_0(0) = 1$, $x_i(0) = 0$ for $i > 0$. In subsequent rows the optimal control, $x_0(t)$ and $x_\Sigma(t)$ are shown. The numerical values of model parameters result from modeling of drug resistance evolution in cancer cells (Smieja *et al.*, 2000) ($a_1 = 0.2$, $a_2 = -2.1$, $c_2 = a_3 = 2$, $c_0 = c_1 = 0.01$)

The values of the performance index for both solutions are similar. Similarly, computational results for greater number of switches suggest that the difference in performance index is negligible for at least some of them. However, comparison of time responses makes clear that more regular switching periods are preferable, since they lead to smaller variation in values of both $x_0(t)$ and $x_\Sigma(t)$. That, in turn, suggests that some periodical control could be preferable in certain cases over the optimal one, if it gives comparable value of the performance index.

Let T, γ denote parameters of periodical control as illustrated in the Fig. 2. Periodical control is then determined by the following relation

$$u(t) = \sum_{j=0}^{n} \left[\mathbf{1}(t - jT) - \mathbf{1}(t - (jT + \gamma)) \right], \tag{20}$$

$$(n-1) \le t_f \le nT \tag{21}$$

The variation of control, caused by change of the parameters δT, $\delta\gamma$ can be described as follows

$$\begin{aligned} \delta u(t) = (n-1)\sum_{j=1}^{n-1} \Big\{ & \delta^*\left[t-(j-1)T-\gamma\right]\delta\gamma \\ & - \delta^*\left[t - jT\right]\delta T \Big\} \\ + \delta^*\left[t - nT - \gamma\right] & \delta\gamma \max\left\{0, t_f - nT - \gamma\right\} \\ - \delta^*\left[t - nT - \gamma\right] & \delta T \max\left\{0, t_f - nT\right\} \end{aligned} \tag{22}$$

where δ^* denotes Dirrac delta. Subsequently,

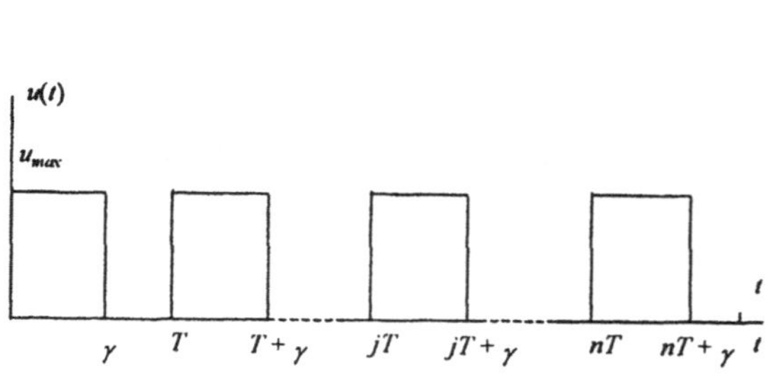

Fig.2. Periodical control

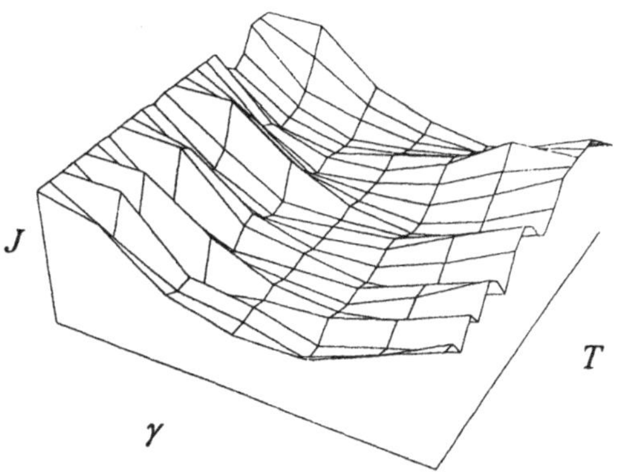

Fig. 3. Value of the performance index as a function of the parameters of periodic control

$$\Delta J \approx (n-1)\delta\gamma\left[\sum_{j=0}^{n-1}\left(r - b_1 p(jT+\gamma)x_0(jT+\gamma)\right) + \right.$$
$$\left. + \max\left\{0, t_f - nT - \gamma\right\}\right] + \tag{23}$$
$$-(n-1)\delta T\left[\sum_{j=0}^{n-1}\left(r - b_1 p(jT)x_0(jT)\right) + \right.$$
$$\left. + \max\left\{0, t_f - nT - \gamma\right\}\right]$$

Following the same line of reasoning that lead to development of the gradient method described above, it is possible to determine formulae describing δT and $\delta\gamma$ that would be later used in searching for optimal values of the parameters of periodical control.

However, in case of periodical bang-bang control the performance index is a function of only two independent variables, i.e. period and pulse duration. Hence it is possible to examine its shape numerically (see Fig. 3) and find not only the optimal one but also discuss other possible solutions. Exemplary suboptimal solution is presented in the Fig. 4.

6. CONCLUSIONS

In this paper we are concerned with a particular infinite dimensional bilinear model of some dynamical systems and its optimal control. It has been shown that decomposition of the model into two subsystems, one with explicit influence of control variable and the other being described by tridiagonal infinite dimensional system matrix, enables analytical analysis of some of their dynamical properties. Furthermore, this approach allows transformation of the system description into the integro-differential form, which is crucial to efficiently address an optimal control problem with the performance index defined in l^1 space of summable sequences. It has been shown that, most likely, the optimal protocol should have a bang-bang form (the problems of singular solutions are beyond the scope of this paper). Hence, it is possible to apply a particular gradient method to finding optimal switching times. However, it is impossible to define the optimal number of switches in the infinite dimensional system. Nevertheless, this problem can be solved introducing an upper limit for number of switches, which is reasonable in most of applications. Numerical analysis leads to conclusion that, at least in some cases, suboptimal periodical control may be more preferable than the optimal solution due to qualitatively better trajectories of the solution. Although it is possible to develop an appropriate method also in case of periodic control, it is much more convenient to proceed with numerical analysis of all possible solutions.

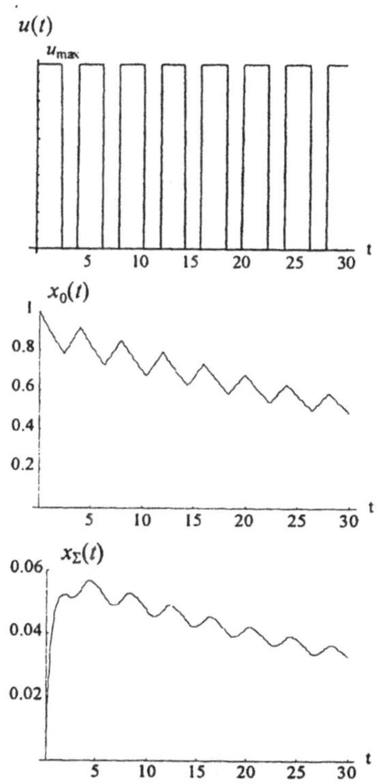

Fig. 4. Exemplary periodic control and respective trajectories

ACKNOWLEDGEMENT

The research has been supported by the KBN grant 8T11E01319

REFERENCES

Kimmel M. and D.E. Axelrod (1990). Mathematical models of gene amplification with applications to cellular drug resistance and tumorigenicity. *Genetics*, vol. 125, pp. 633-644.

Kleinrock L.(1976). *Queuing Systems. Vol. 1: Theory*. Wiley, New York.

Mitkowski W. (1997). Analysis of ladder and ring RC-network, *Bull. Pol. Acad. Techn. Sci*, vol. 45, pp. 445-450.

Polanski A., M. Kimmel and A. Swierniak (1997). Qualitative analysis of the infinite-dimensional model of evolution of drug resistance, *Advances in Mathematical Population Dynamics - Molecules, Cells and Man*, World Scientific, pp. 595-612.

Smieja J. (2000). Dynamical properties and control of infinite dimensional models with tridiagonal system matrix. Ph.D. dissertation Silesian University of Technology (in Polish).

Smieja J., Z. Duda and A. Swierniak (1999). Optimal Control for the Model of Drug Resistance Resulting from Gene Amplification, *Preprints of 14th World Congress of IFAC*, vol. L, pp. 71-75.

Smieja J., A.Swierniak and Z.Duda (2000). Gradient method for finding optimal scheduling in infinite dimensional models of chemotherapy. *Journal of Theoretical Medicine*, vol. 3, pp. 25-36.

Swierniak A., M. Kimmel and A. Polanski (1998). Infinite dimensional model of evolution of drug resistance of cancer cells. *Journal of Mathematical Systems, Estimation and Control*, vol. 8(1), pp. 1–17.

Swierniak A., A. Polanski, M. Kimmel, A. Bobrowski and J. Smieja (1999). Qualitative analysis of controlled drug resistance model - inverse Laplace and semigroup approach, *Control and Cybernetics*, vol. 28, pp. 61-74.

Swierniak A., J. Smieja, J. Rzeszowska-Wolny and M. Kimmel (2001). Random Branching Walk Models arising in molecular biology – control theoretic approach. *Proc. of IASTED MIC Conference*, vol. II, pp. 584-589.

Zadeh L.A. and C.A. Desoer (1963). *Linear System Theory. The State Space Approach*. Mc. Graw-Hill, New York.

Copyright © IFAC Periodic Control Systems,
Cernobbio-Como, Italy, 2001

BALANCING FOR DISCRETE PERIODIC NONLINEAR SYSTEMS

Erik I. Verriest

School of Electrical and Computer Engineering, Georgia Institute of Technology, Atlanta, GA 30332-0250, USA.

Abstract: This paper extends earlier work on balanced realizations for smooth nonlinear systems The main idea behind the global balanced structure, if it exists, is that the linearized system along a nominal solution is balanced in the usual linear time-varying sense. Hence the global nonlinear balanced system is defined as the one that closes the commutative diagram between linearization and balancing. It was shown earlier that such a *global* balanced realization may not exist in dimensions larger than one (or, if one relaxes the notion of balancedness to *uncorrelatedness*, in dimensions larger than two.) In this paper, it is shown that by focusing on the orbit points, balancedness or uncorrelatedness can be obtained via interpolation. Extensions for chain recurrent systems are given. *Copyright ©2001 IFAC*

Keywords: Periodic orbit, nonlinear reachability, nonlinear observability, balanced realization, Mayer-Lie system.

1. INTRODUCTION

Balanced realizations for periodic linear systems were defined by Varga (1997) and Verriest and Helmke (1998). Recently, different groups of researchers have extended the notion of a balanced realization to the general nonlinear realm. One of the first contributions was the Ph. D. Dissertation of J. Scherpen (1994) extended in (Gray and Mesko, 1997; 1999; Gray and Scherpen, 1998) Newman and Krishnaprasad (1998) offer a stochastic approach.

In earlier work, we used the idea of balancing a continuous time system locally near the nominal trajectories, and tried to derive a globally defined diffeomorphism matching the local balancing Jacobians. This forms a so-called Mayer-Lie system, for which it is known that only under certain conditions a solution can exist. Consequently, global balancing may not be defined. It was shown in (Verriest and Gray, 2000) that the notion of balancing may be relaxed while still retaining the relevant information for e.g., model reduction, thus allowing a generic balancing for second order systems, but generically the integrability problem presents an obstruction to higher order ones. The discrete analog is presented in (Verriest and Gray, 2001). The idea of using a periodic orbit, rather than a fixed equilibrium point has also been introduced in (Verriest and Gray, 1999), and along general trajectories in (Mavrikis and Vintner, 1997). This paper proposes an idea to circumvent this integrability problem. If a local balancing transformation is only specified at a finite set of separated points, then an *interpolation* is defined, for which the resulting Mayer-Lie system is integrable, and a global balanced realization can be defined. This is exactly possible if one is interested in obtaining reduced order models along periodic orbits for discrete time systems, and in the case of a pseudo-orbit (a chain recurrent set) of the nominal system.

The class of *input affine* systems

$$\dot{x} = f(x) + g(x)u \quad (1)$$

$$y = h(x) \quad (2)$$

has the property that invertible state space transformations of the form $\xi = \hat{\xi}(x)$ leave the above system in the same class. For discrete affine systems, this is no longer true as the following example illustrates: Let

$$x_{k+1} = x_k^3 + u_k.$$

The homeomorphism $\xi = x^3$ transforms the affine system to

$$\xi_{k+1} = (x_k^3 + u_k)^3 = (\xi_k + u_k)^3$$

which is no longer affine. Constructing a transformation $\hat{\xi}(x, u)$ also does not help, as this mixes u_{k+1} and u_k in the dynamical equation. Therefore in the sequel we shall work with general smooth discrete time nonlinear systems

$$x_{k+1} = f(x_k, u_k), \quad (3)$$
$$y = h(x_k, u_k). \quad (4)$$

Let $\hat{\xi}$ be a diffeomorphism (we will need differentiation). With $\xi = \hat{\xi}(x)$ and its inverse $x = \hat{x}(\xi)$, the system (4) transforms to

$$\xi_{k+1} = \hat{\xi}(f(\hat{x}(\xi_k)), u_k)) = \overline{f}(\xi_k, u_k) \quad (5)$$
$$y_k = h(\hat{x}(\xi)) = \overline{h}(\xi_k). \quad (6)$$

If one considers the notion of balancedness in linear systems as suggested by Moore (1981), then it is immediate that the main thrust in its definition are the notion of energy used to reach a state *from the zero state*, and the energy available from its output, when the system settles back to the equilibrium state $x = 0$ from some initial state x_0, with zero input. In this sense, balancing could only be defined for stable linear systems.
The rationale of our balancing philosophy is that the balancing should not make reference to just one solution of the system (the equilibrium solution), but should be defined for all nominal solutions (iterated maps). Without loss of generality, one may assume that the nominal solution, the sequence $\{\overline{x}_k\}$, is governed by the map

$$\overline{x}_{k+1} = f(\overline{x}_k, 0) \quad (7)$$

This f incorporates already the effect of any nominal input. Let $\overline{x}_0$ be the *nominal state* we are initially interested in. The perturbation equations follow from the controlled map

$$x_{k+1} = f(\overline{x}_k + \tilde{x}_k, u_k) \quad (8)$$

where $\tilde{x_k}$ and u_k are assumed *small*. Thus:

$$\tilde{x}_{k+1} = d\overline{f}\big|_k \tilde{x_k} + \frac{\partial f}{\partial u}\bigg|_k u_k \quad (9)$$

where $[d\overline{f}_k]_i \stackrel{\text{def}}{=} \left[\frac{\partial f_i}{\partial x_1}, \cdots, \frac{\partial f_i}{\partial x_n}\right]$ evaluated at $\overline{x}(k)$. Likewise the output perturbation equation is

$$\begin{aligned} y_k - \overline{y}_k &= h(\overline{x}_k + \tilde{x}_k, u_k) - h(\overline{x}_k, 0) \\ &= d\overline{h}_k \tilde{x}_k + \left[\frac{\partial h}{\partial u}\right]_k u_k. \end{aligned}$$

Some remarks regarding consistency are in order: First linearization near a nominal solution will yield a time-varying linear system. In order to balance such a system, it is necessary to use the extension of balanced realizations to time-variant systems as described by Verriest and Kailath (1983). Second, for the linearized equations to remain valid it is required that the actual (perturbed) state remains in the neighborhood of the nominal state. This prompts: i) the consideration of a small number of iterations only for the computation of the gramians of the perturbed system, and ii) the assumptions of small perturbation inputs, u_k.
The use of small-time gramians is certainly favorable from a computational point of view. Of course, there is a lower limit, n, the order of the system, for the gramians to be nonsingular.
The reason why one might be interested in balancing stems from the success enjoyed in the arena of linear system theory. Indeed balanced realizations have been used in problems of in problems of identification, parameterization, model reduction, and robust design See the references in (Verriest and Gray, 2000). It is only natural to extend these ideas further to the nonlinear realm. The periodic case may provide the least complexity for doing this.
The paper is organized as follows: in Section 2, local realization properties, i.e., reachability and observability of discrete time nonlinear systems are reviewed. More details may be found in (Verriest and Gray 2001). In section 3, the balanced realization for a class of nonlinear systems is given. Section 4 discusses a specific interpolation for periodic orbits of planary systems.

2. REALIZATION PROPERTIES

The rationale of our balancing method is first explained. Next the reachability and observability properties are discussed in a the discrete setting.

2.1 *Flow Balancing*

The rationale behind our approach to balancing in the continuous time case was the idea that balancing and linearization should commute as in the following diagram

$$\begin{array}{ccc} (f,g,h) & \stackrel{\text{linearization}}{\longrightarrow} & (A_P,b_P,c_P) \\ \text{global balancing} \downarrow & & \text{local balancing} \downarrow \\ (\hat{f},\hat{g},\hat{h}) & \stackrel{\text{linearization}}{\longrightarrow} & (\hat{A}_P,\hat{b}_P,\hat{c}_P) \end{array}$$

It needs to be emphasized that in general, linearization near a nominal trajectory will yield a time-varying linear system. In order to balance such a system, it is necessary to use the proper extension of balanced realizations to time-variant systems as described in (Verriest and kailath, 1983). Finally, for the linearized equations to remain valid it is required that the actually perturbed state remains in the neighborhood of the nominal state. This prompts the consideration of small time-intervals for the computation of the gramians of the perturbed system. Essentially, the *small-time* gramians need to be used. These ideas carry also over to the discrete system case.

2.2 *Local Reachability and Observability*

Let us start here from the *perturbation*-model for (4), assuming a periodic nominal orbit, $\overline{x}_0, \overline{x}_1, \ldots \overline{x}_{N-1}$, and small excursions. Introduce the notation

$$df|_k = \left.\frac{\partial f}{\partial x}\right|_k \tag{10}$$

$$g|_k = \left.\frac{\partial f}{\partial u}\right|_k \tag{11}$$

for the gradients and the iterated symbol (analogous to the "Ad"-operator in continuous time), defined as $\left(\mathrm{it}_f^k g\right)_\ell \stackrel{\text{def}}{=} df|_\ell \cdots df|_{\ell-k+1}\, g|_{\ell-k}$. This satisfies the recursion $\left(\mathrm{it}_f^k g\right)_\ell = \left(\mathrm{it}_f^{k-1}(df g)\right)_\ell$ with $\left(\mathrm{it}_f^0 g\right)_\ell = g|_\ell$. The local ℓ-step reachability map for *small inputs* is given by ("h.o.t." are higher order terms)

$$x_\ell = R_{\mathrm{loc}}^{(\ell)}(\ell)\mathcal{U}_\ell + f^{o\ell} + \mathrm{it}_f^\ell \tilde{x}_0 + h.o.t. \tag{12}$$

Here, the matrix

$$R_{\mathrm{loc}}^{(\ell)}(\ell) = \left[g|_\ell,\ \mathrm{it}_f^1 g|_{\ell-1}\ \cdots\ \mathrm{it}_f^{\ell-1} g|_1\ \mathrm{it}_f^\ell g|_0\right], \tag{13}$$

is the *local (ℓ-step) reachability matrix* (reaching x_ℓ), and $\mathcal{U}_\ell = [u_{\ell-1}, \ldots, u_0]'$. Hence if $R_{\mathrm{loc}}^{(\ell)}(\ell)$ is nonsingular, the sequence $\mathcal{U}_\ell = [R_{\mathrm{loc}}^{(\ell)(\ell)}]^{-1}\tilde{x}_f$ will steer the event $(\overline{x}_0, 0)$ to a neighborhood of $f^{o\ell}(\overline{x}_0) + \tilde{x}_f$. More precisely, $\tilde{x}_f$ will be the deviation, up to first order from the nominal state $\overline{x}_\ell$ in ℓ steps.

Remarks:

(1) Our notation already reflects the fact that the nonlinear perturbation system is represented as a time-varying linear system. Indeed the "$\cdot|_k$" may be interpreted as "at step k" or "at $f^{ok}(\overline{x}_0)$".

(2) It is obvious that for a system of order n, not all perturbations of $\overline{x}_k$ will be reachable unless the number of steps, $k > n$, and a reachability condition holds: $\operatorname{rank} R_{\mathrm{loc}}^{(k)}(\cdot) = n$.

(3) If more than n steps are taken, generically many different input perturbations will bring the state sufficiently close to a desired perturbation state $\tilde{x}_f$. This freedom may be exploited, for instance for taking the minimum norm input.

Likewise, the output perturbation (deviation from the nominal output), in the absence of input perturbations, but with nonzero $\tilde{x}_0$, is found as

$$\tilde{y}_k = d\overline{h}|_k \left(\mathrm{it}_f^k\right)_{k-1} \tilde{x}_0 \tag{14}$$

and thus, with $\mathcal{Y}_\ell = [y_{\ell-1}, \ldots, y_0]$,

$$\mathcal{Y}_\ell = O_{\mathrm{loc}}^{(\ell)}(0)\,[\tilde{x}_0] \tag{15}$$

Here

$$O_{\mathrm{loc}}^{(\ell)}(0) = \begin{bmatrix} dh|_0 \\ dh|_1 df|_0 \\ \vdots \\ dh|_{\ell-1} df|_{\ell-2} \cdots df|_0 \end{bmatrix} \tag{16}$$

is the *local ℓ -step observability matrix* (observing x_0).

It is shown in (Verriest and Gray,2000) that the *local reachability Gramian*

$$\mathcal{R}_{\mathrm{loc}}(n) = R_{\mathrm{loc}}^{(n)}(n) R_{\mathrm{loc}}^{(n)'}(n) \tag{17}$$

and the *local observability Gramian* matrix

$$\mathrm{O}_{\mathrm{loc}}(0) = \left[O_{\mathrm{loc}}^{(n)'}(0) O_{\mathrm{loc}}^{(n)}(0)\right]. \tag{18}$$

play a fundamental role: If input and output energies are respectively measured by

$$\mathcal{E}_{\mathcal{U}} = \sum_{k=0}^{\ell-1} u_k^2, \tag{19}$$

and

$$\mathcal{E}_{\mathcal{Y}} = \sum_{k=0}^{n-1} \tilde{y}_k^2, \tag{20}$$

then the *minimum energy* transfer from the $(x_0, 0)$ to $(f^{on}(x_0) + \tilde{x}_n)$ is given by the quadratic form

$$\mathcal{E}_{\mathcal{U}} = \tilde{x}_n' \left[\mathcal{R}_{\mathrm{loc}}(n)\right]^{-1} \tilde{x}_n$$

Likewise, the energy available for observation of an initial state deviation $\tilde{x}_0$ is

$$\mathcal{E}_{\mathcal{Y}} = \tilde{x}_0' \mathcal{O}_{\text{loc}}(n)\tilde{x}_0$$

Finally, for reason of *time symmetry* reachability and observability problems should both be related to the *same* space-time event. Hence for the event $(x_0, 0)$, the reachability problem must be *advanced* by n steps, in order to be able to speak about the reachability in the neighborhood of x_0 at step 0. Consequently, one needs to start from the nominal space-time event $(x_{-n}, -n)$, chosen in such a way that

$$f^{\circ n}(\overline{x}_{-n}) = \overline{x}_0 \tag{21}$$

Of course, one is now implying that $\overline{x}_0$ must have come from somewhere, using the given (nominal) dynamics. What this means is that the *inverse map* must be well defined. For instance the map $f : x \to x^2$ does not have an inverse in all of $\mathbb{R}$: the state $x = -1$ cannot be obtained from a previous transition. Furthermore, the state $x = 1$ has two predecessors: -1 and $+1$. This fact does not pose too much of a problem: one could always choose a representant in the inverse image $f^{-1}\{\overline{x}\}$ and so define a unique inverse. This is a standard problem in determining the *maximal injective restriction* of a map. In view of the applications, one proper choice would be to select the element $\overline{x}^*_{-n}$ in the set $f^{-n}\{x\}$ for which the requisite energy for the transition $\overline{x}^*_{-n} \to \overline{x}_0$ is minimal among all $x \in f^{-n}\{x\}$. This entails considering $\mathcal{R}_n(0)$ instead of $\mathcal{R}_n(n)$.

3. LOCAL AND GLOBAL BALANCING

3.1 *Why Balancing?*

In the above section, a pair of gramians was attached with each state in the state space. Their significance was that $\mathcal{O}_{\text{loc}}$ and $\mathcal{R}_{\text{loc}}$ are perturbation weight matrices characterizing the ease with which the perturbation was either obtained, or will be observed. This procedure of *balancing* is well known, and must here be applied in a time variant setting. Locally, a state transformation T is derived such that the gramians after transformation are equal and diagonal (this gramian was referred to as the *canonical gramian.*)

3.2 *State transformation*

Definition : The discrete time varying system (A_k, B_k, C_k) of order n is called *minimum time balanced* (MTB) if its local (minimum time) gramians satisfy $\mathcal{R}_g(k) = \mathcal{O}_g(k) = \Lambda_n(k)$ where $\Lambda_\delta(\cdot)$ is a diagonal matrix with nonnegative valued functions on its diagonal.

If the diagonal elements $\lambda_i(k)$ are all distinct at t, then a *canonical* gramian may be defined as the gramian Λ_n (actually a sequence) for which the values on the diagonal are ordered $\forall k$: i.e., $\lambda_1(k) > \lambda_2(k) > \cdots > \lambda_n(k)$.

Theorem 1: *The canonical gramian is an invariant (function) for the system.*
Proof: Indeed, a time variant state transformation, T_k, has the following effect on the minimum time gramians

$$\mathcal{R}_g(k) \to T_k \mathcal{R}_g(k) T_k'$$
$$\mathcal{O}_g(k) \to T_k^{-T} \mathcal{O}_g(k) T_k^{-1}.$$

Hence the product $\mathcal{R}_g(k)\mathcal{O}_g(k)$ transforms by similarity, and therefore has invariant eigenvalues $\lambda_i^2(k)$, $i = 1, \ldots n$. □

MT-balancing is performed by simultaneous diagonalization of the reachability and observability gramians. For a realization in the balanced coordinates, in any arbitrary direction the same quantitative measure for its reachability and observability will be found.

In this discrete time setting, the solution to this problem is relatively simple: Let T be the balancing transformation. Then $TR_{\text{loc}}^{(n)} = \Lambda V'$ and $O_{\text{loc}}^{(n)} T^{-1} = U\Lambda$ for some orthogonal matrices U and V. But upon defining the *local Hankel matrix* $\mathcal{H}(t)$ by

$$\mathcal{H}_{\text{loc}} = O_{\text{loc}}^{(n)} R_{\text{loc}}^{(n)} \tag{22}$$

we get

$$\mathcal{H}_{\text{loc}} = O_{\text{loc}}^{(n)} T^{-1} T R_{\text{loc}}^{(n)} = U\Lambda^2 V'.$$

It follows that one only has to obtain a singular value decomposition of the local Hankel matrix (23): and deduce $T_{\text{bal}} = [\Lambda V'] R_{\text{loc}}^{(n)}$ or $T_{\text{bal}}^{-1} \left(O^{(n)}\right)_{\text{loc}}^{-1} [U\Lambda]$.

Remarks:

(1) The local Hankel matrix $\mathcal{H}_{\text{loc}}(x)$ at x is obviously the time varying Hankel matrix (linear sense) for the perturbation system. It therefore inherits the input-output properties this represents.
(2) We defined MT balancing since it is easily obtained (computationally the simplest.) It should be clear that balancing may be de-

fined for an arbitrary number of steps. However, because the map may exhibit sensitive dependence on initial conditions, it may not be a good idea to stray too far.

3.3 *Global Balancing*

The problem is now to extend these local (at $\bar{x}$) balancing transformations $T(\bar{x})$ to a transformation on at least some open subset of the state space. To this effect, the equation

$$\frac{\partial \xi}{\partial x} = T(x) \tag{23}$$

needs to be solved. This is a set of n partial differential equations of first order in n variables. It is a special case of a Mayer-Lie system (Frankel, 1997). It is known that such a system of equations is not generically solvable. The necessary and sufficient conditions for solvability are

$$\frac{\partial T_{ij}(x)}{\partial x_k} - \frac{\partial T_{ik}(x)}{\partial x_j} = 0. \tag{24}$$

for all $i, j, k = 1, \ldots, n$. Hence, for scalar systems, there is no obstruction to global MT-balancing. For second order systems, $n = 2$, generically the conditions (25) will not hold. However, one can always determine integrating factors, $s_i(\bar{x})$, $i = 1, 2$ for which the Jacobian matrix $ST(\bar{x})$ is integrable, where $S = \mathrm{diag}\,[s_1, s_2]$. The effect of the additional non-uniform scaling transformation on the reachability and observability gramian in the (local) balanced form is

$$\mathcal{R}_{gs}(\bar{x}) = S(\bar{x})\mathcal{R}_g(\bar{x})S(\bar{x}) = S^2(\bar{x})\Lambda(\bar{x})$$
$$\mathcal{O}_{gs}(\bar{x}) = S^{-1}(\bar{x})\mathcal{O}_g(\bar{x})S^{-1}(\bar{x}) = S^{-2}(\bar{x})\Lambda(\bar{x})$$

Hence, the uncorrelatedness (diagonality) is retained, as well as the fact that the product of the two gramians correctly specifies the canonical gramian. This ensures that the information about the relative importance of each coordinate direction is retained as in the original notion of a balanced realization. Since this is precisely the requisite information for model reduction, the scaled balanced realization is still useful, and we shall define

Definition: A realization for which the reachability and observability gramians are both diagonal is called an *uncorrelated* realization.

Thus balanced realizations are special cases of uncorrelated realizations.
For $n \geq 3$ it is not always possible to find a set of integrating factors. We summarize the above into:

Theorem 2: i) *A first order minimal system can be balanced.*
ii) *A second order minimal system can be brought to uncorrelated form.*
iii) *A higher order system can be uncorrelated if and only if integrating factors exist for which ST is integrable.*

If one realizes that perhaps too much was asked for, it may be possible to relax the conditions. This is the case of balancing in the neighborhood of a periodic orbit or a pseudo periodic orbit (chain recurrent set) as discussed in the next section.

4. PERIODIC ORBITS

Assume that the nominal system (7) exhibits a periodic orbit $x_0, x_1, \ldots, x_{N-1}$ where N is assumed to exceed the system order. In this case one should only be concerned about the behavior of the system at the N discrete points $x_0, x_1, \ldots, x_{N-1}$. In other words, we shall try to find a suitable diffeomorphism which matches the locally defined balancing transformations at these points. This constitutes an *interpolation problem*: Find a diffeomorphism ξ such that

$$\left.\frac{\partial \xi}{\partial x}\right|_{\bar{x}_i} = T(\bar{x}_i) = T^{(i)} \quad ; \quad i = 0, \ldots, N-1.$$

This leaves a lot of freedom in the choice of the diffeomorphism. This nonuniqueness is not a problem, as the behavior away from the sample points is never needed anyway. On the other hand it simplifies the problem drastically: one can start with a parametrized set of diffeomorphisms, and specify the parameters that match the finite number of Jacobians.

We study the planar systems $n = 2$. The original coordinates are denoted as x and y, the new coordinates are ξ and η. The case N odd will only be discussed here, the case for even N is similar, but presents some additional freedom. Let $\xi(x, y)$ and $\eta(x, y)$ be homogeneous polynomials in x and y of degree $2N - 1$.

$$\xi(x, y) = c_1^{(1)}x^{2N-1} + c_2^{(1)}x^{2N-2}y + \cdots + c_{2N}^{(1)}y^{2N-1}$$
$$\eta(x, y) = c_1^{(2)}x^{2N-1} + c_2^{(2)}x^{2N-2}y + \cdots + c_{2N}^{(2)}y^{2N-1}$$

then

$$\frac{\partial(\xi, \eta)}{\partial(x, y)}^T = \mathcal{Z}(x, y)\, C$$

where $\mathcal{Z}(x, y)$ is

$$\begin{bmatrix} (2N-1)x^{2N-2} & \cdots & 0 \\ 0 & \cdots & (2N-1)y^{2N-2} \end{bmatrix}$$

and

$$C = \begin{bmatrix} c_1^{(1)} & c_1^{(2)} \\ \vdots & \vdots \\ c_{2N}^{(1)} & c_{2N}^{(2)} \end{bmatrix}$$

The two times $2N$ coefficients are determined by matching the Jacobian at the N points. Thus

$$\begin{bmatrix} T'(x_0, y_0) \\ \vdots \\ T'(x_{N-1}, y_{N-1}) \end{bmatrix} = \begin{bmatrix} \mathcal{Z}(x_1, y_1) \\ \vdots \\ \mathcal{Z}(x_{N-1}, y_{N-1}) \end{bmatrix} C.$$

or simply $\mathcal{T}^T = \mathbf{Z}(x_0, y_0, \ldots x_{N-1}, y_{N-1})\, C$. If $\mathbf{Z}(x_0, y_0, \ldots x_{N-1}, y_{N-1})$ is invertible, then $C = \mathbf{Z}(x_0, y_0, \ldots x_{N-1}, y_{N-1})^{-1} \mathcal{T}^T$ and we obtain the candidate diffeomorphism

$$\begin{bmatrix} \xi(x,y) \\ \eta(x,y) \end{bmatrix} = \mathcal{T}\mathbf{Z}^{-T} \begin{bmatrix} x^{2N-1} \\ \vdots \\ y^{2N-1} \end{bmatrix}.$$

Since $\mathbf{Z}$ is homogeneous of degree $2N-2$, the Jacobian determinant will have degree $4(N-1)$. Consequently, there are at most $4(N-1)$ lines through the origin where full rankness will fail. These lines define wedges in the original state space coordinates. Hence, if all states $\overline{x}_0, \ldots, \overline{x}_{N-1}$ fall inside a wedge, a *globally* defined balanced realization can be defined. It remains to check the invertibility of the matrix $\mathbf{Z}(x_0, y_0, \ldots, x_{N-1}, y_{N-1})$.

Lemma *If no two states are collinear with the origin, then the matrix* $\mathbf{Z}(x_0, y_0, \ldots, x_{N-1}, y_{N-1})$ *is invertible.*
Proof: omitted for space reasons.

5. CONCLUDING REMARKS

Balancing of discrete time nonlinear systems has been developed. The key ideas are the use of mobile frames, entrained with the nominal map, and the commutation of balancing and linearization. Interpolation may be a viable method to generate a globally defined balanced realization if only a finite number of points are specified where the realization needs to be balanced, thus bypassing the local integrability problem. This is the case when the discrete nominal system behavior exhibits a periodic or quasi-periodic orbit (i.e., there is a chain recurrent point).

Acknowledgement: The support of the NSF-CNRS collaborative grant INT-9818312 is gratefully acknowledged.

6. REFERENCES

Frankel, T. (1997). *The Geometry of Physics: An Introduction*, Cambridge University Press.

Gray, W. S. and J. Mesko. (1997). General Input Balancing and Model Reduction for Linear and Nonlinear Systems. *Proceedings of the 1997 ECC* Brussels, Belgium.

Gray, W.S. and J. Mesko. (1999). Observability Functions for Linear and Nonlinear Systems. *Systems and Control Letters*, vol. 38, no. 2, pp. 99-113.

Gray, W.S. and J.M.A. Scherpen. (1998). Hankel Operators and Gramians for Nonlinear Systems. *Proceedings of the 1998 Conference on Decision and Control*, Tampa, FL.

Gray, W.S. and E.I. Verriest. (1999). Balancing Non-Linear Systems Near Attracting Invariant Manifolds. *Proceedings of the 1999 ECC*, Karlsruhe, Germany.

Mavrikis, P. and R.B. Vintner. (1997). Trajectory Specific Model Reduction. *Proceedings of the 1997 Conference on Decision and Control*, San Diego, CA.

Moore, B.C. (1981). Principal Component Analysis in Linear Systems: Controllability, Observability, and Model Reduction. *IEEE Transactions on Automatic Control*, Vol. AC-26, No. 1, pp. 17-32.

Newman, A.J, and P.S. Krishnaprasad. (1998). Computation for Nonlinear Balancing. *Proc. 37-th IEEE Conference on Decision and Control,*, Tampa, FL, pp.4103-4104.

Scherpen, J.M.A. (1994). *Balancing for Nonlinear Systems,* Ph.D. Dissertation, University of Twente.

Varga, A. (1997). Solution of Positive Periodic Discrete Lyapunov Equations with Applications to the Balancing of Periodic Systems. *Proceedings of the 1997 ECC*, Brussels, Belgium.

Verriest, E.I. (1980). *On Balanced Realizations for Time Variant Linear Systems*, Ph.D. Dissertation, Department of Electrical Engineering, Stanford University.

Verriest, E.I. (2001). Discrete Nonlinear Balanced Realizations. *Proc. NOLCOS 2001*, St Petersburg, Russia.

Verriest, E.I and W.S. Gray. (2000). Flow Balancing Nonlinear Systems. *Proceedings of the MTNS-2000*, Perpignan, France.

Verriest, E.I., and U. Helmke. (1998). Periodic Balanced Realizations. *Proceedings of the IFAC Workshop on System Structure and Control*. Nantes, France.

Verriest, E.I. and T. Kailath. (1983). On Generalized Balanced Realizations. *IEEE Transactions on Automatic Control*, Vol. **28**, pp. 833-844.

AUTHOR INDEX

www.ingramcontent.com/pod-product-compliance
Lightning Source LLC
Chambersburg PA
CBHW082305210326
41598CB00028B/4447
* 9 7 8 0 0 8 0 4 3 6 8 2 1 *